Proteins and Nucleic Acids in Plant Systematics

Edited by
U. Jensen and D. E. Fairbrothers

With 148 Figures

Springer-Verlag
Berlin Heidelberg New York Tokyo 1983

Prof. Dr. U. Jensen
Lehrstuhl für Pflanzenökologie
und Systematik
Universität Bayreuth
8580 Bayreuth
FRG

Prof. Dr. D. E. Fairbrothers
Department of Biological Sciences (Botany)
Rutgers University
Piscataway, N.J. 08854
USA

ISBN-13:978-3-642-69289-5 e-ISBN-13:978-3-642-69287-1
DOI: 10.1007/978-3-642-69287-1

Softcover reprint of the hardcover 1st edition 1983

2131/3130-543210

Preface

The continued rapid expansion of molecular biology, genetics, and macromolecular biochemistry has provided significant data for analyses, interpretations, and incorporation into plant phylogenetic research and plant classification. These disciplines have produced techniques and methods which enable the evolutionary biologists to obtain new and provocative information, especially about those substances which are coupled with the genetic material. In July, 1982 these important biochemical substances extracted from living plant organs, tissues or cells were the subject of an International Symposium held at the University of Bayreuth, Federal Republic of Germany, entitled *Proteins and Nucleic Acids in Plant Systematics*. At this Symposium German scientists communicated with leading scientists from eleven other countries. The Deutsche Forschungsgemeinschaft generously supported this symposium and thus enabled the exchange of data, ideas, and new scientific proposals. This book contains 26 contributions delivered at the Symposium, which review the present status of Plant Macromolecular Systematics. The two editors acknowledge the effort of the Springer Verlag and their indispensable help with the preparation of this publication.

Bayreuth, FRG and
New Brunswick, NJ, USA

U. JENSEN and
D.E. FAIRBROTHERS

November 1983

Particular Proteins Contributing to Phylogeny and Taxonomy

Serological Protein Properties Contributing to Phylogeny and Taxonomy

Contents

Nucleic Acids

Proteins: Structural Properties

Contributors

You will find the addresses at the beginning of the respective contribution

Nucleic Acids

Quantitative and Qualitative Differentiation of Nuclear DNA in Relation to Plant Systematics and Evolution

F. EHRENDORFER[1]

Abstract. The present state of contributions from biochemistry, karyology, and eukaryote cytogenetics to the phylogenetic differentiation of nuclear DNA, chromatin, and nucleotype in plants is surveyed. Major methodologies include cytophotometry, CsCl-ultracentrifugation, HAP-chromatography, DNA denaturation and reassociation kinetics, DNA/DNA and DNA/RNA hybridization, restriction endonuclease digestion, DNA cloning etc.; their applicability and results are briefly considered.

Amounts of nuclear DNA usually are specific to species or even larger taxa, but extremes may diverge by as much as 1:10 and 1:50 within certain genera and families, and 1:500 within the angiosperms as a whole. Both DNA increase and decrease occur, and these quantitative changes partly influence the (ultra)structure of interphase nuclei and chromatin. Giemsa, fluorochromes and other stains demonstrate eu- and heterochromatin differentiation as chromosome banding. Up to 99% and even more of nuclear DNA in eukaryotic plants is not transcribed, is more or less redundant and often highly repetitive (particularly in the heterochromatin). Changes in non-repetitive, repetitive, and total DNA are correlated but follow different ratios in different species groups, genera or families of plants and in animals. Examples from Ranunculaceae, Rutaceae, Fabaceae, Asteraceae-Anthemideae and -Microseridinae, Liliaceae, Poaceae etc. illustrate the systematic relevance of these nucleotype parameters and their correlations with cell cycles, life forms, karyotypic stability etc.

Differences in DNA base composition are important systematic indicators in lower plants only. More detailed comparisons of DNA from different species as revealed by bulk or nonrepetitive DNA/DNA hybridization are available for a few plant groups (e.g. *Chlorella, Osmunda*, cereal grasses, and *Atriplex*), but still difficult to interpret. Hybridizing cRNA from particular satellite or cloned repetitive DNA sequences in situ onto denatured chromosomal DNA (and partly also onto endonuclease digested DNA tracks) gives remarkably precise informations on their location within the karyotype and on their homologies within and between species of *Scilla, Secale, Triticum,* and *Aegilops.* These data complement information from reassociation experiments on characteristic interspersion patterns of unique and medium to highly repeated sequences for several angiosperms. Even the physical mapping of several cloned repeats and subrepeats recently has become possible in rye. The genomes of cereal grasses (oats, barley, rye, and wheat) thus are shown to exhibit a hierarchical structure reflecting their evolutionary differentiation. The repeated DNA families are simple or interspersed in a more and more complex way, and they are either species-specific or common to smaller or larger groups of common descent. All this suggests cycles of amplifications, interspersions, and replacements of relatively "movable" DNA sequences as decisive mechanisms of speciation and evolutionary divergence.

1 Institut für Botanik und Botanischer Garten der Universität Wien, Rennweg 14, A-1030 Vienna, Austria

Proteins and Nucleic Acids in Plant Systematics
(ed. by U. Jensen and D.E. Fairbrothers)

1 Introduction

Nuclear DNA is the main component of the hereditary system in all living organisms. Quantitative and qualitative differentiations of nuclear DNA bring about changes of genophores, chromosomes, and cell nuclei, and underly speciation and organismic evolution. These phenomena are of paramount importance to biology. It is not surprising, therefore, that we have witnessed during the last decade an explosive development of DNA biochemistry, karyology, and eukaryote cytogenetics. This trend leads to the gradual merging of these research fields, as is obvious from all recent surveys (e.g. Busch 1974–1979, Nagl 1976, v. Sengbusch 1979, Nagl et al. 1979, Bradbury et al. 1981, Knippers 1982, Parthier and Boulter 1982). As an illustration one can point to the growing importance and consideration of the "nucleotype" (Bennett 1972, 1973), a new term complementing the well known concepts of "karyotype" and "genotype". These developments have also been largely responsible for the current revolution in our understanding of evolutionary mechanisms (Stebbins 1971, Ayala 1976, K. Jones 1978, Flavell 1981, Barigozzi 1982, Dover 1982b, Dover and Flavell 1982), culminating in anti-Darwinian slogans like "selfish DNA" (Doolittle and Sapienza 1980, Orgel and Crick 1980) or "molecular drive" (Dover 1982a).

So far only very little of the new methods and results of nucleotype and DNA research has been applied to systematic botany and the phylogenetic-evolutionary interpretation of plant diversity. It is the purpose of this review to stimulate such studies and to extend earlier proposals (Nagl and Capesius 1976, Capesius and Nagl 1978) in several directions. In view of the breadth and continuing expansion of these fields I can only try to give a very general survey and illustrate it with a limited number of examples. Furthermore, the limited space does not allow for methodological details. In this respect the reader is referred to Nagl (1976), Birnie and Rickwood (1978), Adams et al. (1981), Parthier and Boulter (1982), Rickwood and Hames (1982), Knippers (1982), and other text- and handbooks.

Quantitative and structural aspects of DNA: amounts per genome, the appearance of interphase nuclei and chromatin, and other parameters of the nucleotype are discussed in the following paragraph. The central problem of repetitive and unique DNA sequences and their orderly interspersion follows in paragraph three. Finally, approaches towards a more detailed qualitative characterization and comparison of nuclear DNA are presented in paragraph four. This forms a link with the following contributions on DNA and RNA nucleotide sequencing, an aspect which is not dealt with here.

2 Amounts of Nuclear DNA, Chromatin Structure, and Nucleotype

2.1 Nuclear DNA Values

The cytophotometry (microdensitometry) of DNA-specific Feulgen or fluorochrome stained nuclei has widely opened the field of quantitative determinations and comparisons of nuclear DNA values. (For technical details and recent improvements using scanners and on-line computers see e.g. Nagl 1976, Geber und Hasibeder 1980). Results are presented in DNA pg or nucleotide pairs (= NTP) of DNA per haploid prereplication G_1 (1C), diploid G_1 (2C) or postreplication G_2 (4C) nuclei. A survey of nuclear DNA-values (2C) of comparable meristematic cells of diploids demonstrates either a relative stability within species, as in various conifers (Teoh and Rees 1976), in *Secale cereale* (15.8–16.6 pg: Bennett et al. 1977), and other *Triticinae* (Furuta et al. 1975, 1977, 1978) or some racial variation, e.g. in *Helianthus annuus* (Nagl and Capesius 1976) and species of *Microseris* (Price et al. 1980, 1981a,b), or an often remarkable variation within more comprehensive taxonomic groups, e.g. within various species of *Vicia* (4.0–28.8 pg), *Anemone* (11.9–45.4 pg; Fig. 1) and *Allium* (15.2–45.4 pg), within various genera of Ranunculaceae (1.1–50.2 pg) and Liliaceae s.lat. (9.3 to 254.8 pg), and extreme variability among the angiosperms as a whole (0.5–254.8 pg;

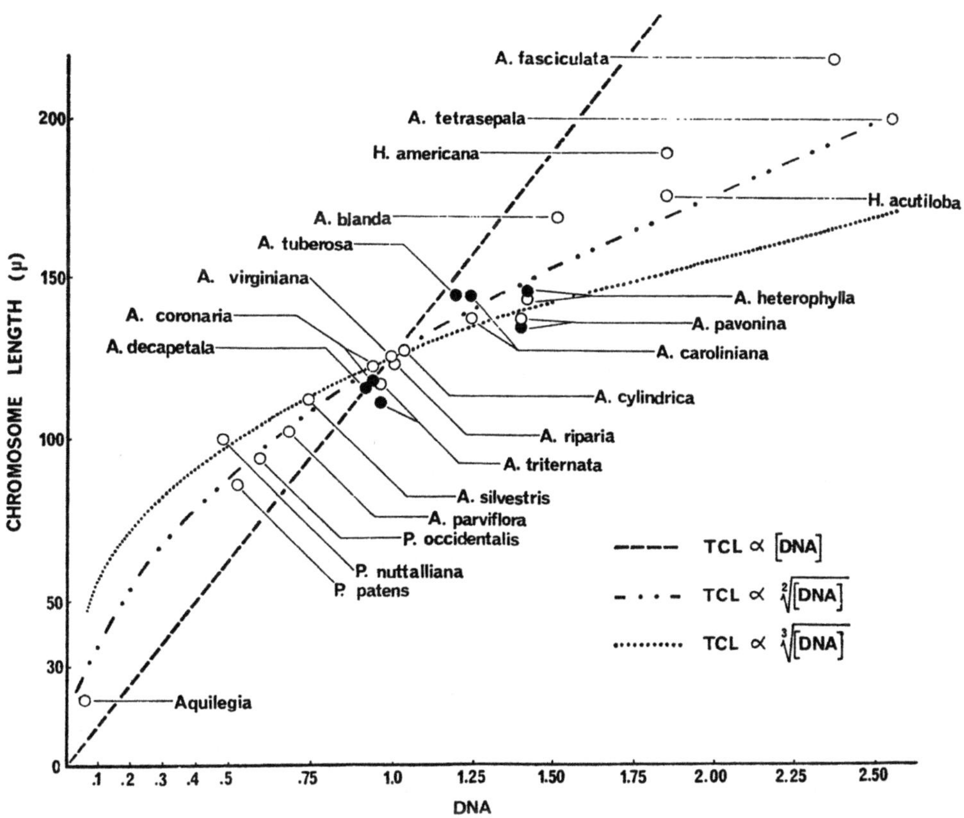

Fig. 1. Correlation between relative amounts of DNA per nucleus and total chromosome complement length for *Aquilegia* and the genera *Pulsatilla* (P.), *Anemone* (A.), and *Hepatica* (H.). Absolute 2C DNA values in pg range from *Aquilegia* (1.1) to *Pulsatilla* (8.7–10.4), *Anemone* sect. *Eriocephalus* (11.9–25.5), *Hepatica* (33.1–33.2), and *Anemone* sect. *Homalocarpus* (42.3–45.4). The three theoretical curves are obtained when total chromosome length **TCL** is correlated to DNA content in a linear, square root or cube root fashion (Rothfels et al. 1966, absolute DNA values from Bennett and Smith 1976)

Bennett and Smith 1976, Bennett et al. 1982). Mean DNA values and, particularly, minimum DNA values of mosses, algae, fungi, and bacteria range progressively at lower levels and suggest a certain positive correlation between organismic complexity and DNA quantity (Price 1976). But this correlation is obscured by the great variation among members of comparable organization level (e.g. among angiosperms), a fact which has led to the concept of the so-called "C-value paradox". An explanation for this discrepancy is that only a small fraction of nuclear DNA appears as absolutely indispensible during eukaryote evolution, while much of it is highly repetitive or else "redundant" (see Sect. 3), and often has been in- or decreased during phylogeny. Such events of quantitative DNA change are of great interest to the systematist and evolutionist, and will concern us throughout this contribution.

2.2 Levels of Chromatin Organization

Another important aspect of eukaryote diversification concerns the structural appearance of chromatin in interphase nuclei and during the mitotic cycle. Both light (= LM) and electron microscopical (= TEM) analyses have greatly contributed to this field since the classical studies by Heitz and Geitler. LM data available up to the early sixties have been summarized by Tschermak-Woess (1963), including a discussion of eu- and heterochromatin differentiation and the description of various nuclear types. More recent attempts to correlate ultrastructures apparent from TEM studies with these LM observations, and to differentiate between active or condensed euchromatin and facultative or constitutive heterochromatin, are surveyed by Nagl (1982). Apart from certain changes which may occur during cell cycles and cell differentiation, the species-specific LM and TEM chromatin structure of interphase nuclei has been well documented for plants in contrast to animals which apparently exhibit more plasticity in this respect. But so far, relevant information has been used little for systematic purposes and evolutionary studies. Thus Cruciferae mostly have prochromosomic nuclei, and Commelinaceae have chromomeric nuclei, often provided with chromocentres (Tschermak-Woess 1963). Different nuclear types have been found relevant for the taxonomy of Saxifragaceae (Hamel 1953) and Rutaceae (Guerra 1980, Ehrendorfer 1982, cf. p. 11) or for separating Flacourtiaceae and Tiliaceae (Morawetz 1981). Among angiosperms correlations have been suggested between chromatin condensation and DNA content (Nagl and Fusenig 1979, Bachmann 1979, Nagl and Bachmann 1980) and also between euchromatin condensation, percentage of mean repetitive DNA and systematic affinities (Nagl 1982). For lower plants the peculiar eukaryotic (or "mesokaryotic") nucleus of Dinophyceae ("dinokaryon") can be mentioned (for recent reviews see Rizzo and Burghardt 1980, Spector and Triemer 1981, Raikov 1982). Its DNA-fibers are not linked to histones in the normal form of nucleosomes. These fibers are somewhat reminiscent of bacterial genophores but are organized into chromosomes.

A breakthrough in the karyosystematics of plants, particularly angiosperms, has been due to the developments of methods for chromosome banding, based on the differential staining of constitutive heterochromatin with Giemsa and fluorochrome stains. These methods have improved the chromatin analysis of interphase nuclei; in mitotic and meiotic chromosomes they produce characteristic heterochromatic bands and allow the elaboration of banded karyotypes. (For these and other banding techniques and their application to plant material the reader is referred e.g. to Greilhuber 1975, 1982, 1983, Comings 1978, Schweizer 1980). Even in taxonomic groups with great karyotype variation the general species-specificity of Giemsa C-banding and fluorochrome-banding patterns has been repeatedly documented, e.g. in *Secale cereale* (Singh and Röbbelen 1975) or in *Scilla bifolia* and *S. vindobonensis* (Greilhuber and Speta 1977). Nevertheless, structural heterozygosity and polymorphism with regard to heterochromatic bands may occur, particularly in long-lived plants. Well-known examples are the polymorphism of cold-sensitive chromosome segments in populations of *Trillium* (Haga and Kurabayashi 1954, Kurabayashi 1963, Fukuda and Grant 1980) or of Giemsa C-bands in *Leopoldia* (Bentzer and Landström 1975) and some *Scilla* species (Greilhuber and Speta 1978). How such polymorphisms may be linked with aspects of speciation is discussed by Kaina and Rieger (1979). The application of banding techniques to problems of chromosome evolution and systematics can be illustrated by recent studies on the diploid taxa of *Anacyclus* (Asteraceae-Anthemideae; Schweizer and Ehrendorfer 1976, Ehrendorfer et al. 1977, Humphries 1979, 1981), and *Tulipa* (Blakey and Vosa 1981, 1982), the dysploid orchid genus *Cephalanthera* (Schwarzacher et al. 1980, Schwarzacher and Schweizer 1982) or on the polyploid origins of wheat (Gill and Kimber 1974) and Asiatic barley cytotypes (Linde-Laursen et al. 1980). Other examples, to be discussed later, concern the Ranunculaceae, Asteraceae-Anthemideae, *Scilla*, and Poaceae-Triticeae.

2.3 Nucleotype

Nuclear DNA content and nuclear volume together with chromatin and chromosomal organization are constituents of the so-called nucleotype (Bennett 1972, 1973) and influence the phenotype of an organism as physical parameters. The functional and evolutionary implications of these nucleotypic parameters should become evident from the following examples. One of the first angiosperm groups where a wide differentiation of chromosome size was used as a phylogenetic marker to improve a hitherto rather artificial systematic arrangement was the Ranunculaceae. Gregory (1941) clearly demonstrated that a separation of *Aquilegia* (with follicles) and *Thalictrum* (with nutlets) into different tribes was contradicted by their similarities in chromosome size and base number. His concept of Coptideae, Thalictreae (including *Thalictrum* and *Aquilegia*), Helleboreae and Anemoneae was later confirmed by phytochemical and serodiagnostic findings (Jensen 1968). Within the Anemoneae Rothfels et al. (1966) have verified a close correlation between chromosome size and DNA values, which apparently is of general validity. From their data (Fig. 1) it is obvious that even smaller nucleotypic differences are of phylogenetic significance, e.g. the relatively low DNA values of *Pulsatilla*, the central block of species belonging to *Anemone* sect. *Eriocephalus* (*A. parviflora* to *A. heterophylla*) and sect. *Anemonthea* (*A. blanda*), the peak occupied by sect. *Homalocarpus* (with *A. tetrasepala* and *A. fasciculata*) and the block of somewhat lower values belonging to the two species of *Hepatica*. The relative increase of absolute DNA amounts (2C) in *A. pavonina* (24.9 pg) in relation to the closely related *Anemone coronaria* (16.9 pg) may be due, at least partly, to an increase in constitutive heterochromatin and chromosome band material from 8% to

11% (Marks and Schweizer 1974). There is also an increase in repetitive DNA concerning these two species (53% → 62%). However, amounts of heterochromatin and repetitive DNA are not always correlated, as in more distantly related taxa of *Anemone* (Cullis and Schweizer 1974).

In the *Scilla bifolia* alliance (including the former genus *Chionodoxa*, corresponding to the *S. luciliae* subgroup), Greilhuber and collaborators (Greilhuber 1979, Greilhuber et al. 1981) have demonstrated that the stepwise morphological and embryological differentiation and expansion of taxa from the Balkan peninsula (e.g. with the primordial *S. messeniaca* or *S. kladnii*) to SE. Europe (*S. taurica* etc.), C. and W. Europe (*S. bifolia*), and the Aegaeis (*S. luciliae, S. tmoli* etc.) is clearly linked with nucleotypic changes on the 2x-level (Fig. 2): First a reduction and then a stabilization in genome size and DNA content has occurred, but side line taxa are characterized by the addition of constitutive heterochromatin and massive chromosome bands (e.g. in the *S. vindobonensis* subgroup and in *S. tmoli*); additionally some polyploids (4x, 6x) have become established. In species with numerous chromosome bands there is a striking similarity of band patterns throughout the karyotype ("equilocal heterochromatin position"), particularly among arms for which spatial interphase proximity can be postulated (Greilhuber and Loidl 1983).

In several angiosperm genera, e.g. *Lathyrus* (Rees and Hazarika 1969), *Crepis* (Jones and Brown 1976), and *Microseris* (Price and Bachmann 1975, Bachmann et al. 1979) reduction of DNA amounts and genome size (with reduced chromosome size and/or number) is clearly linked with the well known phylogenetic trend from relatively primitive, often outbreeding perennials to specialized, often inbreeding annuals. Size of cells and organs usually is reduced in annuals, and their development is speeded up in comparison with related perennials. This is understandable, since both cell size and the duration of nuclear and cell divisions are positively correlated with amounts of nuclear DNA (Evans et al. 1972, Nagl 1976). For *Lathyrus* (Narayan and Rees 1977, Narayan 1982) and for *Microseris* with related genera (Asteraceae-Microseridinae; Bachmann and Price 1977) there is also evidence that changes in total DNA are mainly due to changes in its repetitive fraction and in heterochromatin (see Sect. 3). That such changes do not directly affect the core of coding genes has been shown for seven diploid *Crepis* species where approximately 19 genes for electrophoretically detectable enzymes have been conserved notwithstanding a seven-fold variation in nuclear DNA (Roose and Gottlieb 1978). Our examples concerning the origin of annual life forms linked to DNA reduction thus support the concept of selective regulation and adaptive importance of the nucleotype and its redundant and repetitive DNA component (Bennett 1972, 1973, Nagl 1976, Cavalier-Smith 1978). Differences in DNA amounts among races of *Microseris douglasii* and *M. bigelowii* (up to 25%; Price et al. 1980, 1981a,b) illustrate the dynamics of such changes.

Nevertheless, one has to be cautious with too simple conclusions about the probably quite complex functional aspects of nucleotypic changes (Price 1976). Within the Asteraceae-Microseridinae the correlation between the DNA amounts of perennial and outbreeding versus annual and inbreeding species in *Pyrrhopappus* is the reverse to what has been outlined for *Microseris* (and other genera). *Pyrrhopappus* consists of annual species which have nearly twice the DNA content of perennial Microseridinae and up to 5.5 times that of annuals in *Microseris* and *Agoseris*, yet they display

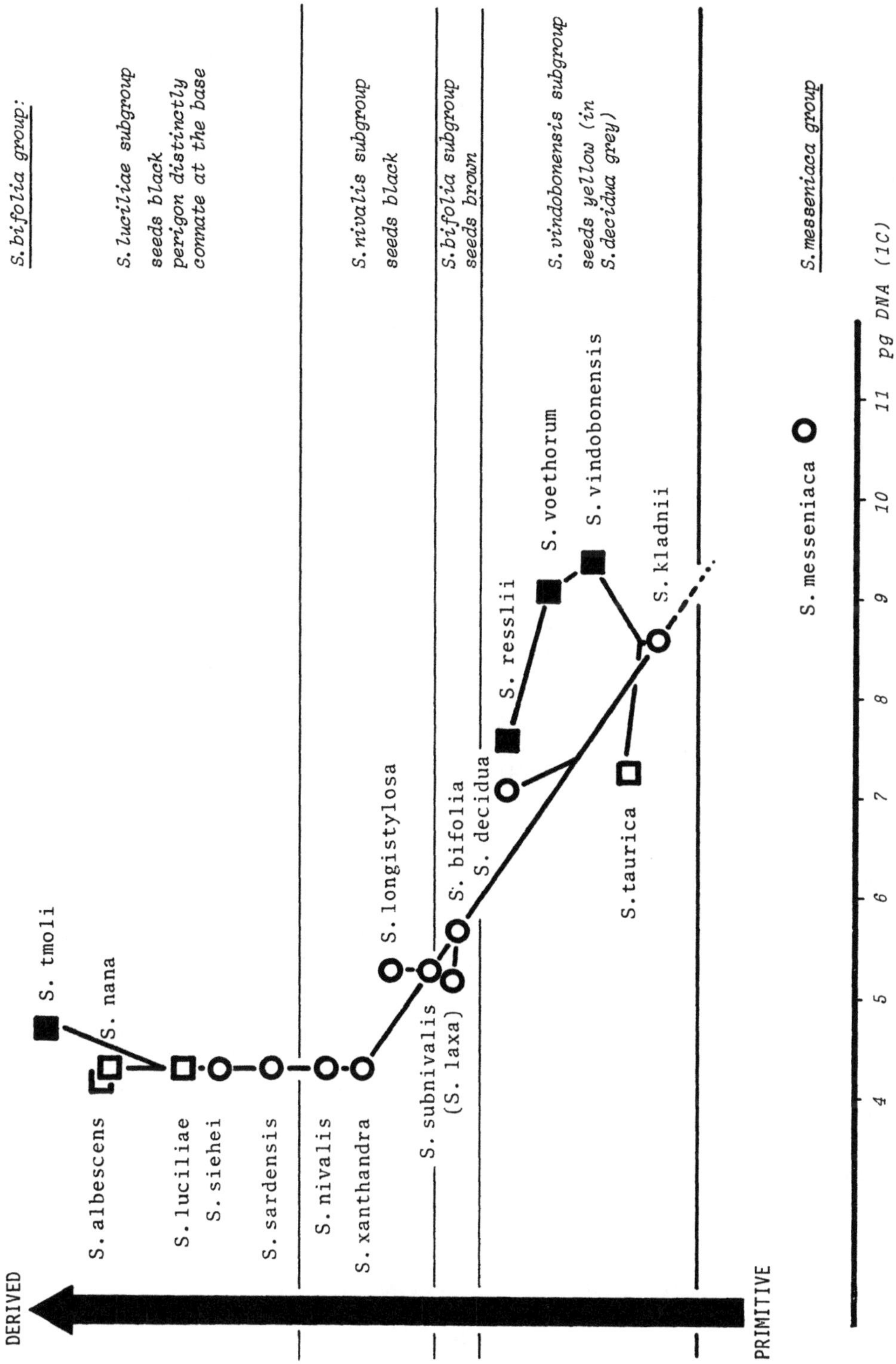

Fig. 2. Diagram of evolutionary relationships among diploid members of the *Scilla bifolia* group (n=9) based on DNA amounts, heterochromatin amounts, morphology, and embrylology. *Empty symbols:* species with little heterochromatin; ***black symbols:*** heterochromatin-rich species; ***circles:*** species with monosporic embryo sacs; ***squares:*** species with tetrasporic embryo sacs. (Greilhuber et al. 1981; embryological data according to E. Sroma)

the most rapid growth rate and shortest generation time within the subtribe. Their proportionately longer mitotic cycle time is compensated by large cell volume and the potential to differentiate more mass in a given period of time (Price and Bachmann 1976). A similar situation is found in the grass genus *Lolium*, where an inspection of gross morphology and Giemsa C-band karyotypes reveals three natural groups (Rees and Jones 1972, Hutchinson et al. 1979, Thomas 1981). The first with the perennial *L. perenne* (2C = 8.3 pg), the closely related short-lived perennial *L. multiflorum* (2C = 8.6 pg), and the annual *L. rigidum* (2C = 8.7 pg) exhibits very similar DNA values; the other two include more distantly related annuals with markedly higher amounts, i.e. *L. remotum* and *L. temulentum* (2C = 12.1–12.5 pg) and *L. loliaceum* (2C = 11.0 pg). Here, as well as in the related genus *Festuca* and in many other instances, the overall increase in DNA is almost evenly distributed over all chromosomes of the karyotype (Seal and Rees 1982).

The Asteraceae-Anthemideae offer several examples for quite divergent nucleotypic strategies in closely related groups of perennials and annuals. Within genera checked (*Artemisia, Anacyclus, Anthemis, Chamaemelum,* and *Leucanthemum*), annuals always exhibit faster growth rate, higher mitotic activity, shorter mitotic cycle time, and larger cell size than related perennials, irrespective of their larger or smaller nuclear DNA amounts (Nagl and Ehrendorfer 1974, Nagl 1974). Thus, genic regulation seems to override nucleotypic effects in some groups. Among different lines leading from perennials to biennials and annuals, DNA reduction has occurred in two groups of *Artemisia*, but both DNA increase and reduction are observed within closely related annual members of *Anacyclus* sect. *Anacyclus* and *Anthemis* subgen. *Cota* (Table 1).

Table 1. DNA-values (2C in pg) from selected genera of Asteraceae-Anthemideae with closely related perennial (♄,♃), biennial (⊖), and annual (⊙) species. Numbers in brackets refer to literature sources: (1) Nagl and Ehrendorfer (1974), (2) Humphries (1981), (3) Geber and Hasibeder (1980), (4) Geber (1979)

ANACYCLUS							
pyrethrum	(♃)	13.21	(2)	*radiatus*	(⊙)	16.92	(1)
		11.66	(2)			16.04	(2)
"depressus"	(♃)	13.61	(2)	*"coronatus"*	(⊙)	14.21	(2)
		12.85	(2)			14.03	(2)
		12.42	(1)				
				clavatus	(⊙)	12.71	(2)
						11.55	(2)
						10.48	(1)
				homogamos	(⊙)	9.58	(2)
ARTEMISIA							
absinthium	(♃)	7.28	(1)	*judaica*	(♄)	11.73	(3)
sieversiana	(⊖)	5.11	(3)	*atrata*	(♃)	6.37	(4)
capillaris		3.65	(4)	*annua*	(⊙)	4.05	(1)
						3.85	(3)
ANTHEMIS							
tinctoria	(♃)	7.46	(1)	*cota* (= *altissima*)	(⊙)	15.78	(1)
				austriaca	(⊙)	9.63	(1)

Proportions of repetitive DNA are not clearly correlated to total amounts of DNA in these groups (Fuhrmann and Nagl 1979). Heterochromatin decreases with DNA in *Artemisia annua* but it increases at different rates in annuals of *Anacyclus* and *Anthemis*, again without obvious correlations to total DNA, e.g. *Anacyclus radiatus* and *A. "coronatus"* with 8.3%–9.8% and 15.0%, respectively, or *Anthemis cota* and *A. austriaca* with 4.4% and 23.0% heterochromatic band material relative to total karyotype length (Schweizer and Ehrendorfer 1976, 1983). Different types of heterochromatin clearly have different functions and different evolutionary stabilities in Anthemideae as in other plant groups: while centromeric and also nucleolar heterochromatin are almost omnipresent, terminal and particularly interstitial heterochromatin tend towards increasing variability. The latter components may actually contribute to overall karyotype instability because of the well established higher rates of breakage and fusion as well as unspecific exchanges associated with heterochromatin (see Sect. 4). In contrast to the very stable karyotypes of perennial groups of Anthemideae with a strong tendency towards hybridization (e.g. *Achillea*: Tohidast-Akrad 1982), annuals have much more differentiated and diversified chromosome structures and banding patterns. This is apparently responsible for their much more effective crossing barriers (Mitsuoka and Ehrendorfer 1972), but is possibly also connected with their increased chiasma frequency (Uitz 1970). From other observations we know that chiasmata normally do not form within heterochromatic regions, and that these regions obviously have regulating influences on the location and frequency of chiasma formation (G.H. Jones 1978, John and Miklos 1979, Loidl 1979, 1981, John 1981, Ambros 1983).

The predominantly tropical and woody Rutaceae offer possibilities to study the correlations between nucleotypic parameters (amounts of nuclear DNA, ploidy level, structure of interphase nuclei, and euchromatin versus heterochromatin and/or condensed chromatin) and aspects of evolutionary diversification on a family level (Guerra 1980, Ehrendorfer 1982). Among 17 representatives from seven major tribes, nuclear DNA amounts range from 0.39 to 17.42 pg (2C), and chromosome numbers from n=7–68 with diploid base numbers from x=7–12, and polyploidy from 2x to 16x. To make Rutaceae DNA values comparable, they are presented in Table 2 and Fig. 3 for single average 1C-chromosomes and in 10^{-2} pg, together with other relevant information. From these data it becomes clear that natural groups of genera can be characterized by their nucleotype features, as Zanthoxyleae s.str. (taxa 1–3), Ruteae s.str. (9–12), Aurantieae (16–17), etc. Amounts of chromosomal and nuclear DNA are positively correlated with amounts of condensed chromatin and/or heterochromatin. With regard to karyotype and chromosome base number stability, it appears that both excessively high and low DNA amounts have a buffering effect, while medium values of DNA and heterochromatin seem to enhance dysploid differentiation, particularly in groups extending into new habitats (as the Boronieae and Diosmeae in the sclerophyll vegetation of Australia and S. Africa). While the majority of (sub)tropical Rutaceae exhibit a medium range of DNA values, extratropical genera tend to have either very low or very high amounts. This suggests divergent and progressive evolutionary lines, and contradicts the postulate of Levin and Funderburg (1979) that there is a general trend towards increasing DNA values from tropical to temperate angiosperms.

Table 2. DNA values (per average 1C-chromosome in 0.01 pg) and ploidy level for 17 species of Rutaceae representing seven major tribes: Zanthoxyleae (1–4), Boronieae (5), Diosmeae (6, 7), Cusparieae (8), Ruteae s.str. (9–12), Dictamneae (13), Toddalieae s.lat. (14. 15), and Aurantieae (16, 17). *Arrows* indicate minimum and maximum values (Guerra and Filho 1980, Ehrendorfer 1982)

	Rutaceae	Ploidy	DNA:1 c/nx (10^{-2} pg)
1	*Fagara zanthoxyloides*	8x	9.10
2	*Zanthoxylum alatum*	12x	16.43
3	– *piperitum*	8x	10.26
4	*Melicope ternata*	4x	5.17
5	*Correa virens*	4x	8.27
6	*Coleonema album*	4x	5.65
7	– *pulchrum*	4x	6.01
8	*Erytrochiton brasiliensis*	12x	8.57
9	*Boenninghausenia albiflora*	2x	2.55
10	*Ruta chalepensis*	4x	1.63
11	– *graveolens*	8x	1.94
12	– *montana*	4x	0.98 ←
13	*Dictamnus albus*	4x	19.10
14	*Ptelea baldwinii*	6x	4.94
15	*Skimmia japonica*	4x	21.43 ←
16	*Citrus sinensis*	2x	6.86
17	*Murraya paniculata*	2x	5.56

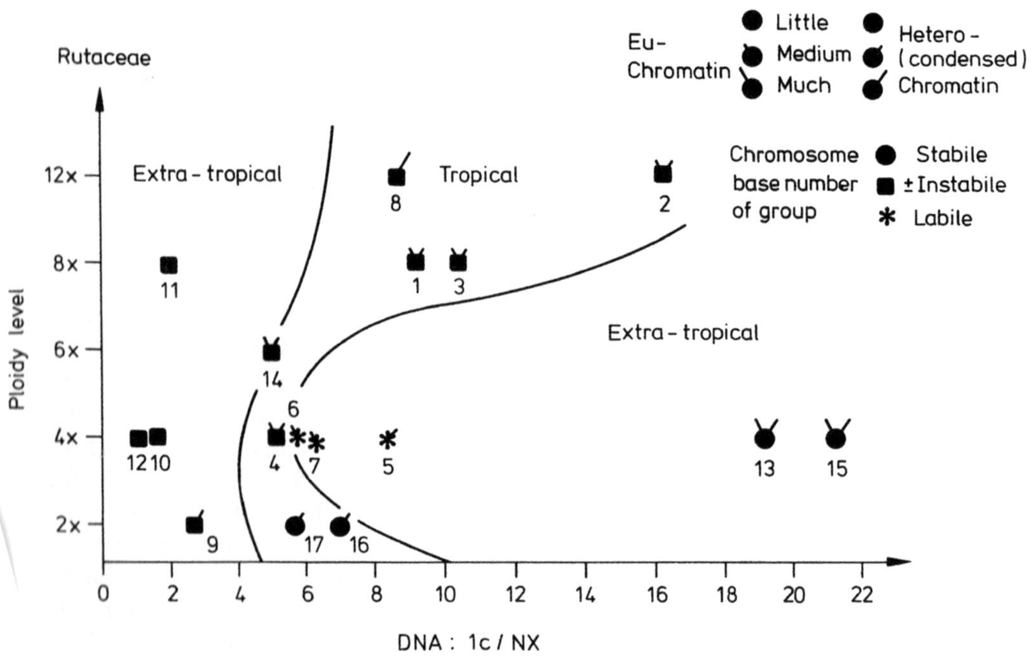

Fig. 3. Nucleotypic parameters, systematics and evolution in the Rutaceae. Co-ordinates: DNA in 10^{-2} pg per chromosome and ploidy level. *Symbols* indicate relative amounts of chromatin types n interphase nuclei, and stability of chromosome number in the groups involved; *lines* separate opical and extra-tropical groups. (Ehrendorfer 1982, species numbers and data are from Table 2 d from Guerra 1980)

3 Unique Versus Medium and Highly Repetitive Nuclear DNA, Proportions and Interspersion Patterns

One of the most important steps in our understanding the cytogenetic mechanisms of evolution during the last two decades has been the demonstration that in eukaryotes (in contrast to prokaryotes) usually only a small fraction of nuclear DNA makes up structural and regulatory genes actively involved in transcription and translation, while up to 99% and even more is "silent", more or less redundant and often highly repetitive (Britten and Kohne 1968). This has triggered an avalanche of research on the nature, chromosomal arrangement and importance of repeated DNA (see Flavell 1982b, and references cited).

3.1 Reassociation Kinetics

Decisive data on repeated DNA sequences come from renaturation experiments documenting the reassociation kinetics of nuclear DNA. Basically, the speed at which complementary strands of denatured DNA reassociate depends on their concentration and complexity. In suspensions of single stranded DNA sheared to fragments of comparable length, unique (i.e. single- or few-copy) sequences reassociate slowly, medium to highly repetitive correspondingly faster, and palindromes (inverted repeats within one strand) immediately. The extent of reassociation usually is determined from the hypochromicity in a spectrophotometer or from the preferential binding of double versus single stranded DNA in hydroxyapatite (HAP) columns measured by spectrophotometry and/or in conjunction with radio-labelled probes, S_1 nuclease digestion, electron microscopy etc. Results are usually presented in two-dimensional graphs with c_0t (start-concentration × time) on the abscissa, and the proportion of single versus double stranded DNA on the ordinate (Fig. 4). The resulting reassociation kinetics can be interpreted in an additive way from several partial c_0t curves which correspond to kinetic subfractions such as unique DNA, slow and fast repeats.

It is generally acknowledged today that, in eukaryotes, single copy and medium to highly repeated DNA sequences are arranged in highly ordered and species-specific interspersion patterns (Dover and Flavell 1982, Flavell 1982b). Proof for this notion mainly has come from HAP-experiments: Long or shorter and radioactively labelled "tracer" DNA are reassociated with an excess of unlabelled short fragments of "driver" DNA at stringent conditions, allowing only for pairing of repetitive sequences. Curves obtained from varying the length of fragments result in "plateaus" indicating average lengths of unique sequences attached to repetitive sequences (Fig. 13, upper curve). Actual observation and measurements of such single and double stranded DNA complexes in TEM preparations are used to verify and extend such biochemical results.

3.2 Unique and Repetitive DNA

With the following examples we will try to illustrate the special and general evolutionary and systematic relevance of comparative reassociation studies, first with regard to

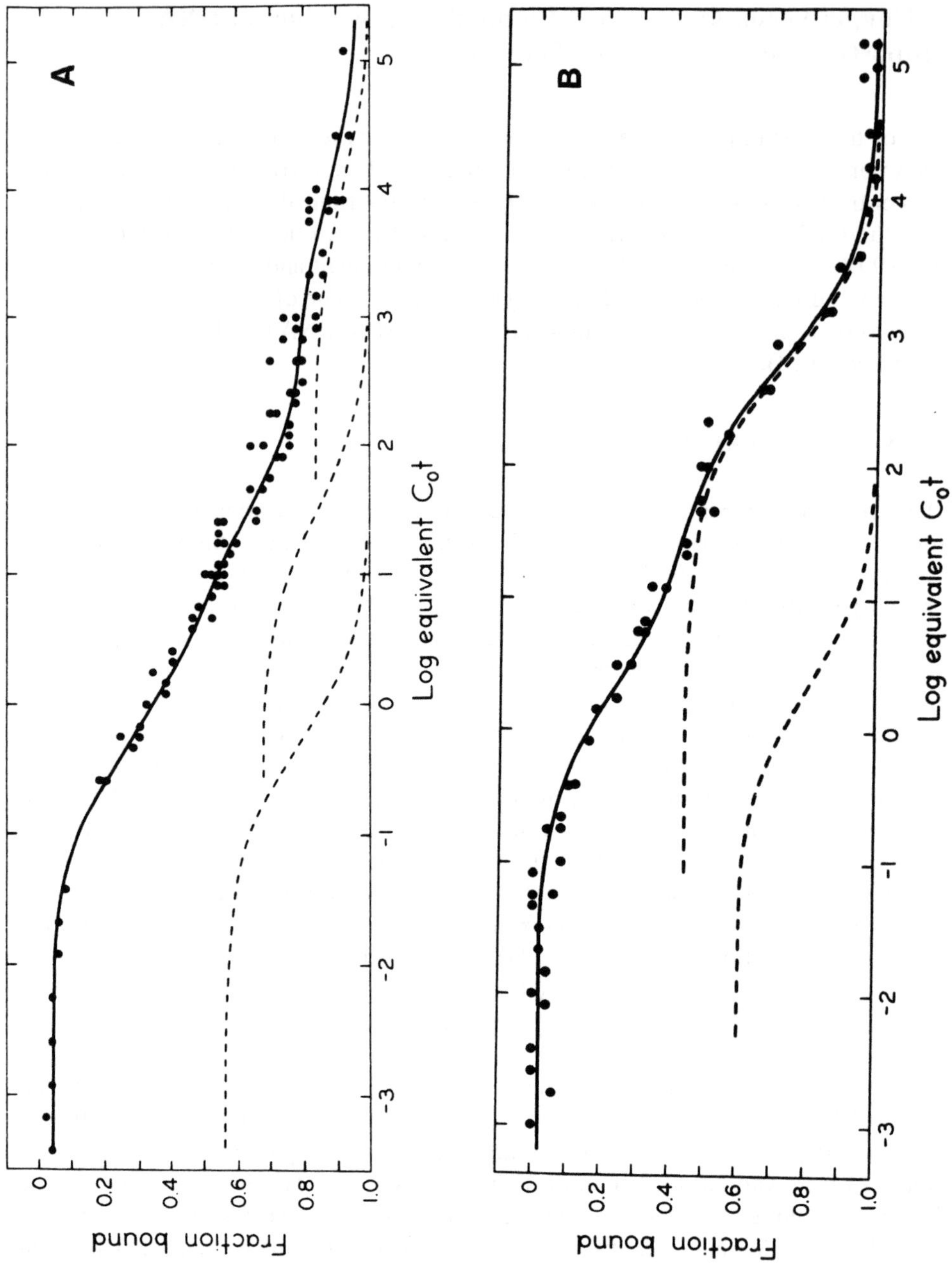

Fig. 4. Reassociation kinetics of total leaf DNA from two legumes (Fabaceae): **A** pea (*Pisum sativum*, Vicieae), **B** mung bean (*Vigna radiata*, Phaseoleae). Fragments with an average length of 300 base pairs (=bp) were reassociated to various equivalent c_0 t values and fractionated on HAP. *Solid lines* are the simplest least-squares fit to the data using three **A** or two **B** theoretical second order components *(dashed lines)* for repetitive (fast and medium: *left*) and non-repetitive, i.e. single (or few) copy sequences (slow: *right*) (Murray et al. 1978, 1979)

unique and repetitive DNA. Among Fabaceae, members of the tribe Vicieae tend to have on the average 10 times more DNA than Phaseoleae; this also applies to such typical representatives as pea (*Pisum sativum*, 1 C = 4.6–4.8 pg) and mung bean (*Vigna radiata* = *Phaseolus aureus*, 1 C = 0.48–0.53 pg). As in many other angiosperms (see also Sects. 2 and 4), this change in total DNA is brought about by changes not only in repetitive, but also in non-repetitive DNA. Pea has 79% repetitive (46% fast and 33% medium) and 17% single copy DNA (Fig. 4A); mung bean, 32% repetitive and 65% single copy DNA (Fig. 4B); the remaining percents are "very fast" palindromes (Thompson and Murray 1980).

In the genus *Oryza*, Iyengar et al. (1979) have studied 8 taxa with a 1.6 fold range of DNA (0.6–2.1 pg/1 C). This is due to parallel changes in repetitive and non-repetitive DNA at a fairly constant ratio of 2.2: 1. Similar constraints upon the composition of supplementary DNA have been reported by Hutchinson et al. (1980b) for other taxa (with increase ratios of repetitive to non-repetitive DNA in brackets): *Allium* (1.08), *Anemone* (1.09), several Asteraceae (2.02), *Lathyrus* (3.91), *Lolium* (5.51), and other Poaceae (5.22) (Fig. 5). This signals orderly and group-specific processes underlying both loss or gain of nuclear DNA and makes correlations between DNA components and chromatin structures more intelligible.

Wenzel and Hemleben (1979, 1982) have applied ratios of unique to repeated nuclear DNA (U/R) to a larger number of angiosperms. Generally, their data also suggest relative group-specificity, a certain interdependence of unique, medium and highly repetitive components of DNA, and hyperbolic function between U/R and total genome size. The latter function is also documented by Flavell (1982b) together with the trend for an increasing frequency of single copies shorter than 2,000 base pairs (= bp) in larger genomes. Since numbers of active genes in angiosperms generally are estimated to be in the order of 30,000 (i.e. ca. 3×10^7 bp) or less, it is obvious that this corresponds to only a small fraction of what is classified as "single copy" DNA by methods of reassociation kinetics. Thus, the majority of this non-repeated DNA must be regarded as diverged derivatives of repeated DNA.

Many lower eukaryotic plants, particularly fungi, not only have small genome sizes, but also low proportions of repetitive DNA which are often less than 15% (Nagl 1976, John and Miklos 1979). Among angiosperms, woody Magnoliidae (Schann and Nagl 1979) and other groups regarded as relatively "primitive" also seem to have relatively low amounts of total and repetitive DNA (taxa with < 2 pg), in contrast to "advanced" taxa (with > 2 pg). Nevertheless, a reversal of this trend also often has occurred in angiosperms, e.g. in the progressive annual groups discussed in Sect. 2.

While genomes of higher plants and animals up to a 1 C-size of about 10^9 bp exhibit a similar correlation between single copy and total DNA, functions clearly diverge at larger genomes for the two groups of organisms (Thompson 1978, Fig. 6). If an increase above this level is due to the addition of repeated DNA, the only explanation for this phenomenon apparently is the faster reversion of such sequences to secondarily single (or few-) copy DNA in the relatively smaller genomes of animals as compared to the often much larger genomes in plants.

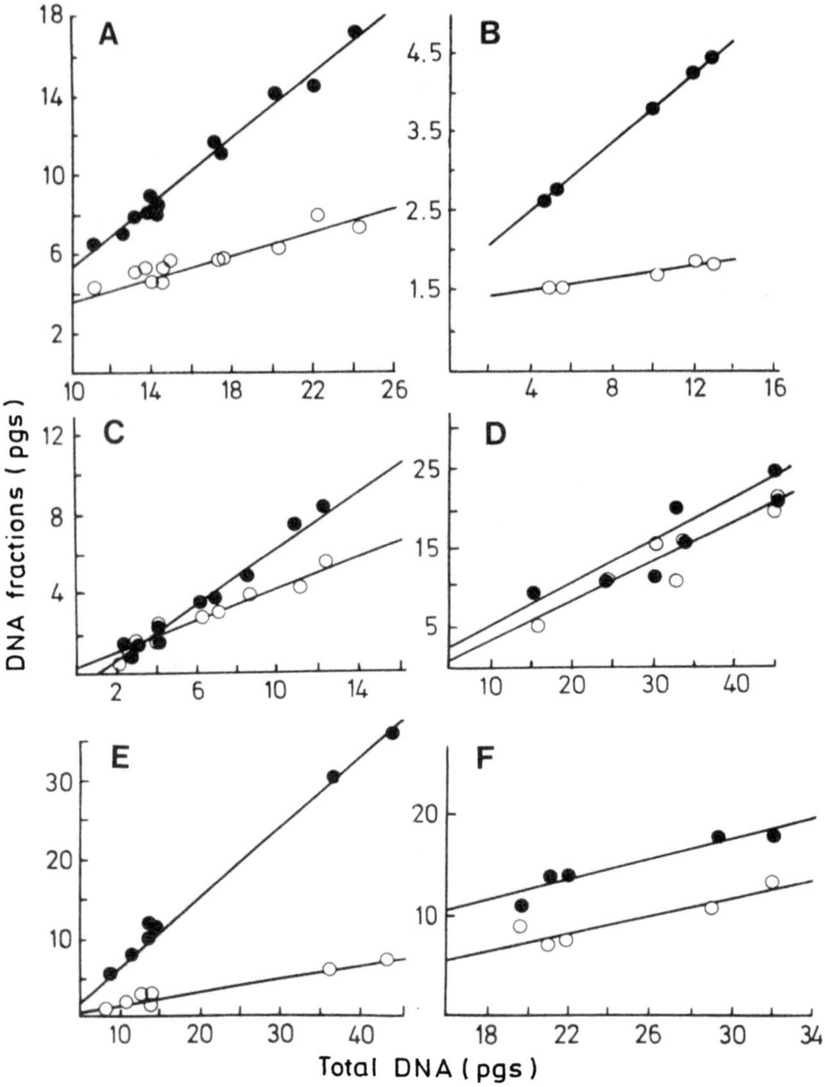

Fig. 5. Repetitive (•) and non-repetitive (○) nuclear DNA change with different but constant ratios compared with total DNA in various groups of angiosperms: **A** *Lathyrus;* **B** *Lolium;* **C** Asteraceae-Cichorieae; **D** *Allium;* **E** Poaceae; **F** *Anemone.* (Hutchinson et al. 1980b, data from different authors)

3.3 Interspersion Patterns

Interspersion patterns so far have been mostly studied and evaluated for animals (see e.g. Schmidtke and Epplen 1980). For plants relevant information is only available for some angiosperms so far. Among Fabaceae a comparison of *Pisum* (Vicieae) with *Glycine, Vigna,* and *Phaseolus* (Phaseoleae) is instructive. *Pisum sativum* has a typical short-period interspersion pattern: the majority of unique (evidently to a large extent

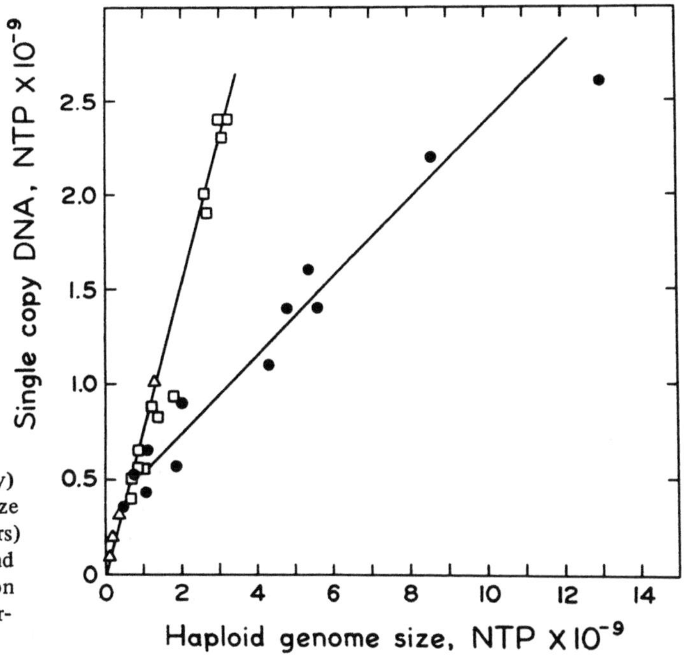

Fig. 6. Different ratios of non-repetitive (single copy) DNA and total genome size (1 C: NTP=nucleotide pairs) in eukaryotic plants (•) and animals (△, □). (Thompson 1978, Thompson and Murray 1980)

not transcriptive) and the repetitive DNA sequences are only 300–400 bp long, the latter include 46% fast repeats with hundreds of "families" each with an average of 10,000 copies (Murray et al. 1977, 1978, Thompson and Murray 1980). In contrast, *Vigna radiata,* strongly tends towards a long-period interspersion: the majority of unique and repetitive DNA sequences are longer than 1,200 bp, nearly half of the single copies are not interspersed for 6,700 bp and more, whereas the rest is separated by short to long repeats; the latter include only 5% of fast DNA with high copy number "families" (Murray et al. 1977, 1979, Preisler and Thompson 1978, Thompson and Murray 1980). Soybean, *Glycine max,* is more similar to *Vigna radiata,* judging from reassociation and TEM studies (Goldberg 1978, Pellegrini and Goldberg 1979), but has more DNA (1 C = 1.1 pg) and only 23% single copy DNA, with an average length of 1,150 bp. Interspersed with short to middle repeats (500–2,500 bp), this makes up about half of the genome. Much of the remaining DNA consists of long blocks (> 1,500 pg) of medium and some (3%) highly repetitive sequences. *Phaseolus vulgaris* can be included here, but has even more DNA (1 C = 1.8 pg), and possibly less long-period sequences (Seshadri and Ranjekar 1980).

The data discussed for legumes, but also some new informations on grasses (Wimpee and Rawson 1979, Hake and Walbot 1980) are in line with some generalizations concerning interspersion patterns of nuclear DNA in angiosperms: positive and linear correlations between number of interspersion segments and the ratio U/R, and between complexity of repetitive DNA and total DNA (Wenzel and Hemleben 1982), as well as between palindromes and total DNA (Flavell 1982b). Furthermore, these data tend to support the following two generalizations: (1) Long-period interspersion seems to be correlated with small genomes, short-period interspersion with larger genomes (Flavell 1982b) and (2) there is less "turning over" of DNA sequences in smaller than in larger genomes (Thompson and Murray 1980).

4 Base Composition, Hybridization, and Cloning of Nuclear DNA

4.1 Methodological Aspects

Further progress in nuclear DNA research was due to a general shift from quantitative to more qualitative analytical methods. An early approach of this type was the development of buoyant density determinations by ultracentrifugation in a CsCl gradient (Birnie and Rickwood 1978). Results give information on base composition, as GC-rich DNA is heavier and sediments faster than AT-rich DNA. This also allows the separation of satellite DNAs if AT/GC-ratios differ between satellite and main bulk DNA. "Cryptic" DNA satellites often are better isolated in a gradient of Cs_2SO_4 supplemented with Ag^+ or Hg^{2+}. An example of the application of these techniques to the different nuclear DNAs of cucumber and radish is offered by Ranjekar et al. (1978). Surveys of nuclear satellite DNAs among plants, particularly angiosperms (e.g. Ingle et al. 1973, 1975, Nagl 1976) demonstrate a certain group-specificity and systematic relevance, e.g. with a characteristic ϱ 1,705–1,708 satellite in Cucurbitaceae or the nearly total lack of any satellite DNA in Ranunculaceae and in the majority of monocotyledons.

Because DNA of different base composition also denatures at different temperatures, melting curves obtained from spectrophotometer measurements of hyperchromicity or from HAP columns indicate "thermic" satellites and the nuclear DNA components differing in this respect. Both types of satellite DNA, whether isolated by their different density or melting temperature, usually prove to be quite short and highly repeated sequences (10^4–10^6 copies), which represent up to 30% and more of the total nuclear DNA. On the other hand, highly repetitive sequences often do not show up as DNA satellites, particularly in plants with much nuclear DNA. In situ hybridization experiments have demonstrated in many plants (and animals) that there is a clear correspondence between this highly repetitive DNA component and constitutive heterochromatin (as described throughout Sect. 2) (see Peacock et al. 1978, John and Miklos 1979, John 1981).

With a number of fluorochrome stains it is possible to differentiate between AT- and GC-rich constitutive heterochromatin in LM-preparations of interphase nuclei and chromosome band material (Schweizer 1980, Fig. 10). Quinacrine and diamidinophenylindol (DAPI), e.g., are AT-specific, chromomycin is GC-specific. In most plant fluorochrome banded karyotypes studied with this method, several heterochromatin types can be separated (Fig. 10). Between taxa there is a considerable variation, but NOR-adjacent heterochromatin apparently is always GC-rich. First applications to systematic problems are promising, e.g. in *Allium* (Vosa 1976a,b: diploids, hybrids, polyploids; Loidl 1981: 30%–70% quinacrine-positive bands in the *A. paniculatum* group but not in other species studied), *Ornithogalum* (Ambros 1983), *Scilla* (Fig. 12), and *Cephalanthera*.

Most important and precise methods for DNA analysis and comparisons are based on the potential of denatured single stranded DNA sequences to recombine with other and more or less homologous DNA (or RNA) strands. This is not only fundamental for reassociation kinetics (Sect. 3) but also for DNA/DNA and DNA/RNA hybridization techniques (Faires and Boswell 1981, Adams et al. 1981, Knippers 1982). These techniques include the radioactive labelling of one of the hybridizing strands which allows e.g. (1) measurement of the amount of hybrid duplexes formed and of single stranded DNA eluted (e.g. from HAP columns), (2) location of particular hybrid duplexes in situ on chromosomes or on gel electrophoresis tracks from restriction enzyme digested DNA.

Hybridiziation experiments using DNA from different species rest on the basic premise that relative DNA homologies are proportional to evolutionary divergence. Early studies using bulk DNA have suffered from several shortcomings: the contamination of nuclear with plastid and mitochondrial DNA, the formation of intricate assemblies when too long DNA sequences are reannealed, the excessive quantities and erratic changes of repetitive DNA (particularly in plants; Fig. 6), the overestimates of homologies due to the reassociation either of palindromes or long tandem arrays within one strand or of short and highly repetitive homologous sequences between different strands with attached long single stranded "tails". More recent studies (e.g. Flavell et al. 1977, Rimpau et al. 1978, 1980, Stein et al. 1979, Belford and Thompson 1981a, b, Belford et al. 1981, Kashevarov and Antonov 1982) consider these aspects; they often use DNA from isolated nuclei only, DNA strands of determined length (usually quite short, i.e. 300–400 bp) and an excess of unlabelled "driver" DNA, and consider repetitive and non-repetitive (single or few copy) DNA separately. Two separate parameters of DNA homology can be compared at different conditions: the rate of hybrid duplex formation (Figs. 12, 13) and their relative thermal stability (Fig. 8). This double checking adds to the reliability of results. Some controversial results (e.g. partly closer affinities between *Spinacia* and *Atriplex* than within *Atriplex*: Belford and Thompson 1981a,b) demonstrate the necessity for further methodological improvements.

For in situ hybridization experiments tritium labelled complementary RNA (^{3}H cRNA) is produced from a defined DNA probe. This is hybridized onto the denatured DNA in chromosome preparations from the same or from different species. Then the preparations are covered by a radiation sensitive film emulsion: Precipitation of silver grains indicates location and amounts of homologous, i.e. hybridizing, DNA and RNA (Nagl 1976, Bedbrook et al. 1980, Jones and Flavell 1982a,b, Deumling and Greilhuber 1982). The in situ hybridization technique allows the mitotic and meiotic chromosomes of one parent in a hybrid to be marked (e.g. Bedbrook et al. 1980, Hutchinson et al. 1980a, Appels 1982). It also helps to study the distribution of highly repetitive DNA sequences among the chromosomes of one species, and to compare the occurrence of such sequences and their placement in the karyotypes of several species by interspecific DNA combinations (Figs. 10 and 11, Bedbrook et al. 1980).

Most recent contributions to nuclear DNA research in plants have come from the application of restriction endonuclease digestion to DNA extracts and the utilization of fragments produced for cloning, in situ hybridization, gene mapping, and nucleotide sequencing (Brutlag 1980, Bedbrook et al. 1980, Flavell et al. 1980, 1981, Jones and

Flavell 1982a,b, Flavell 1982a,b, Appels 1982, Kössel et al. this volume). Different endonucleases (e.g. Alu I, Bam HI, Eco RI, Hae III, Mbo I and II, Taq I) cut DNA at different recognition sites and then produce different and defined fragments. These can be separated by gel electrophoresis. When stained they produce DNA "finger prints" with specific bands corresponding to fragments from repeated sequences, superimposed on a smear of single or few copy fragments. With blotting procedures a transfer of such DNA tracks on nitrocellulose and "probing" by hybridization with radioactive marker DNA or RNA from the same or different species is possible. Furthermore, endonucleae-produced fragments can be ligated into a vector, e.g. *E. coli* plasmid DNA, inserted by transformation into the bacteria, cloned, screened, and extracted for further analysis and physical mapping. We will illustrate these most advanced techniques by results from recent cereal genome analyses (cf. Figs. 15-17).

4.2 Elucidation of Plant Relationships and Evolution

First, let us turn to several other examples illustrating the application of qualitative nuclear DNA analyses to systematic and evolutionary problems and consider their potential for more general conclusions. In prokaryotes base composition varies widely, in *Bacteria* from about 23-73 GC Mol%, but there are particular ranges for natural genera, e.g. for *Clostridium* 23%-43%, *Escherichia* 50%-51%, *Streptomyces* 69%-73%. AT/GC ratios thus represent a prominent molecular characteristic in bacterial systematics (cf. Buchanan and Gibbons, 1974). Among lower eukaryotic plants base composition of total cell DNA has been studied in the green algal genus *Chlorella* (Kessler 1982). Base composition varies excessively from 43%-79% GC. Together with many other data this suggests that the genus is heterogeneous and probably should be split up into natural groups with more homogeneous GC-values. This is supported by the results of bulk DNA hybridization experiments (Fig. 7). In higher plants and within angiosperms base composition becomes much more uniform and thus has little promise for systematics. The reasons for this trend towards relative fixation of AT/GC-ratios of nuclear DNA in seed plants (as in mammals) is still uncertain (Träger 1975).

In vascular plants relationships among species of the ancient and relic fern genus *Osmunda* were studied with the help of DNA hybridization experiments using fragments from total nuclear DNA (Stein and Thompson 1977, Stein et al. 1979). Comparisons of homologous and heterologous duplex DNA by thermal elution from HAP columns suggest that *O. cinnamomea* (subgen. *Osmundastrum*), *O. claytoniana* and *O. regalis* (both placed into subgen. *Osmunda*) are - contrary to some anatomical evidence - about evenly distant from each other. Their genome sizes and reassociation kinetics are very similar, and also their morphology and anatomy has not changed much since the late Cretaceous and early Tertiary, i.e. during the last 60-70 million years (Miller 1971).

Among angiosperms the polymorphic genus *Lathyrus* with nearly 300 perennial and annual species has evolved extensively since the Tertiary. Separate hybridization experiments with non-repetitive and repetitive DNA between *L. hirsutus* and several other taxa, and thermal stability analyses of the resulting heterologous duplex DNA (Narayan and Rees 1977) demonstrate parallel divergence, both in non-repetitive and

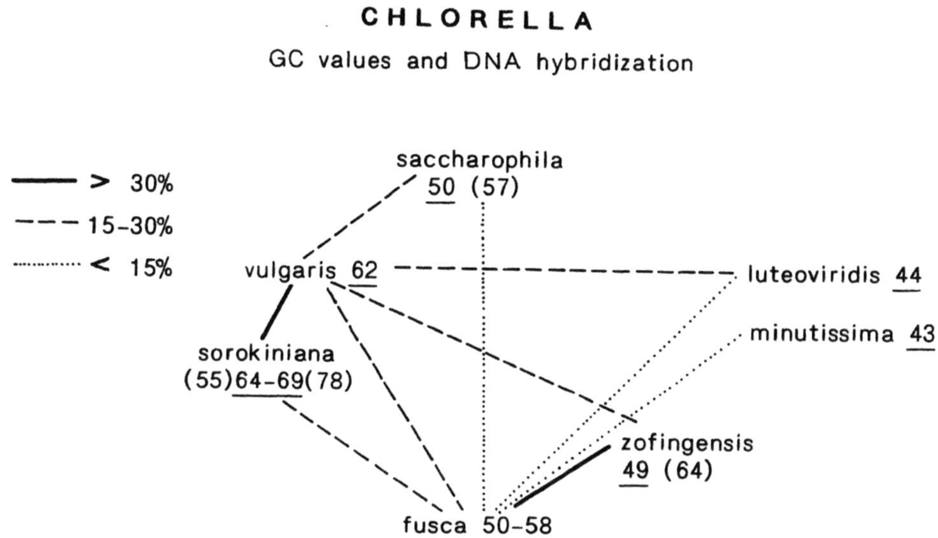

Fig. 7. Molecular systematics in the genus *Chlorella* from the green algae (Chlorococcales): GC values (in % of total nucleotides; typical values *underlined*, exceptional values *in brackets*) and results of interspecific DNA hybridization (heterologous duplexes formed in % of autologous duplexes within *C. vulgaris* and *C. fusca*). Compare the species groups of *C. vulgaris* and *C. fusca; C. saccharophila,* and particularly *C. luteoviridis* and *C. minutissima* are more isolated. (Original draft, data from Kessler 1982)

repetitive DNA, and correspond well with the accepted taxonomic system. Changes in total nuclear DNA have been dramatic (1 C = 3.4–14.6 pg) and may have occurred step-wise (Narayan 1982). Whereas eu- and heterochromatin change at about the same rate (Rees and Narayan 1977), changes of repetitive and non-repetitive DNA follow a steady ratio of ≈ 3.9 (Hutchinson et al. 1980b: Fig. 5 A). Strongly increased satellite sequences in *L. tingitanus* are shown by in situ hybridization to be distributed over all chromosomes (Narayan 1982).

The occurrence of species with the C_3 and other species with the C_4 photosynthetic pathway within the halophilous genus *Atriplex* (Chenopodiaceae) has thrown doubts on its taxonomic classification by Hall and Clements (1923). This has prompted interspecific reassociation and melting experiments with single copy DNA alone (Belford and Thompson 1981a,b, Fig. 8). This fraction accounts for about one third of the total DNA (1 C = 0.36–0.75 pg) and exhibits a remarkable interspecific divergence of homologies. This result suggests that a large proportion of even the single copy DNA is of secondary origin and therefore rather easily interchangeable. From the complete set of (not always unequivocal) data (Fig. 9), it can be deduced that among the species studied *A. hortensis* and *A. triangularis* represent strongly divergent C_3 members of sect. *Atriplex. A. phyllostegia* seems to have originated from a C_3 ancestor near the divergence of sect. *Atriplex* and sect. *Obione*, and remained isolated since. Among similar taxa, and possibly in connection with the world-wide extension of dry habitats during the Oligocene/Miocene, the early switch to C_4 photosynthesis must have occurred. One branch, morphologically still similar to sect. *Atriplex*, is seen in the re-

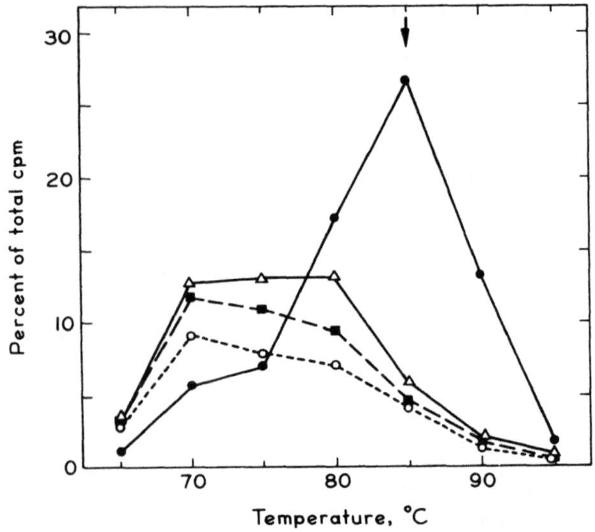

Fig. 8. Thermal stability of intraspecies and interspecies tracer-driver DNA-duplexes in *Atriplex* as seen from elution profiles (in % of total HAP-bound duplexes). Single copy ^{125}I-tracer DNA with an average length of 350–400 bp from *A. hortensis* was first combined with a 3,000-fold excess of unlabelled sheared total DNA as driver from various species at low temperature to minimize self-reassociation. The hybrid duplexes formed from *A. hortensis* tracer with driver DNA from *A. hortensis* (●), *A. triangularis* (△), *A. sabulosa* (■), and *A. serenana* (○) were then eluted with 0.12 M Na-phosphate buffer at increasing temperatures. (Belford and Thompson 1976)

lated *A. sabulosa* and *A. rosea*. Other typical members of sect. *Obione* include the somewhat isolated *A. truncata* and the phylogenetically more advanced close pair of *A. fruticulosa* and *A. serenana*. Crossing experiments tend to support this phylogenetic interpretation.

Interesting data are also available for several species of the *Scilla siberica* alliance, geophytes from the Near East, all with 2n=12 karyotype (Deumling 1981, Deumling and Greilhuber 1982, Greilhuber 1982, 1983, Fig. 10). Total DNA (1 C) ranges from *S. mischtschenkoana* (21.6 pg) to *S. amoena* (23.7 pg), *S. ingridae* (23.9 pg), and *S. siberica* (32.3 pg). Staining with Giemsa and various fluorochromes reveals marked differences of *S. mischtschenkoana* from the other species; *S. ingridae* and *S. amoena* share many similarities; *S. siberica* is somewhat more remote. The satellite DNA of the three latter species is highly GC-rich. In *S. siberica* it represents about 19% of the total nuclear DNA and contains i.a. a dominant and tandemly repeated inverted sequence of 33 bp which was successfully sequenced. Satellite DNA of all the species was transcribed into ^{3}H-cRNA and used for autologous and reciprocal interspecific in situ hybridization (Fig. 11). The NOR-regions react identically in all species (Fig. 11 C) but the majority of heterochromatic bands in terminal and intercalary regions correspond in reciprocal combinations only with the homologous satellite DNA of *S. ingridae, S. amoena*, and *S. siberica* (Fig. 11 A). In contrast, the satellite DNA of *S. mischtschenkoana* is different, and only some of its sequences reappear (probably independently amplified) in the paracentromeric heterochromatin of

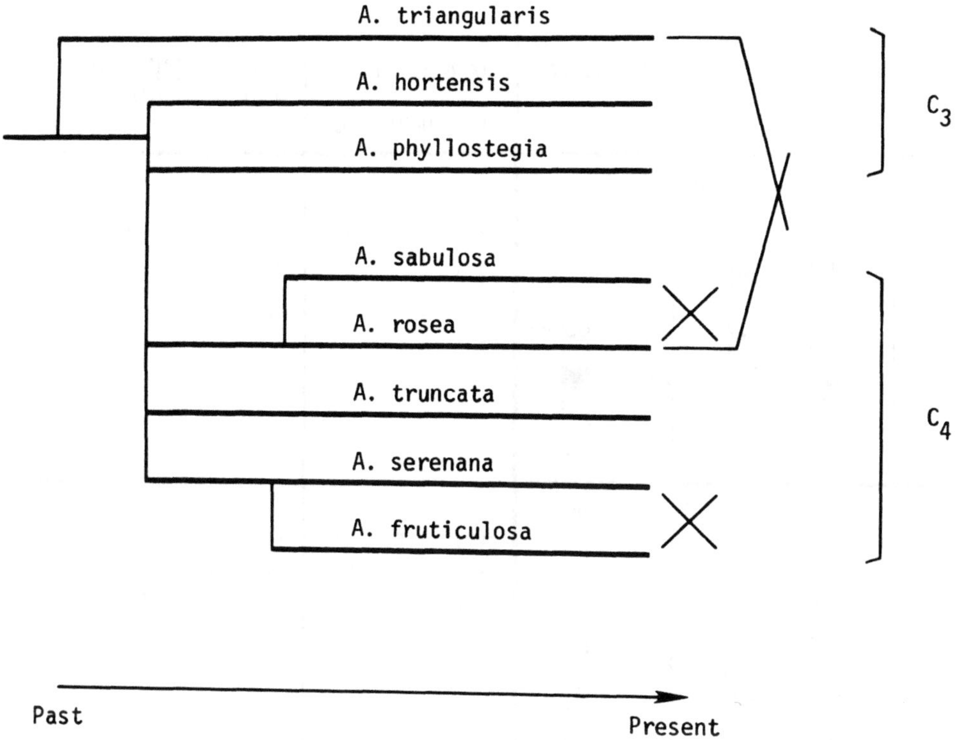

Fig. 9. Phylogenetic scheme of selected *Atriplex* species with C_3 or C_4 photosynthetic pathways, based on nucleotide sequence divergence as measured from interspecific hybridization of non-repetitive DNA. DNA divergence and branching of "phylogenetic tree" are shown in correct proportions. *Crossed lines* (above "Present") indicate successful sexual hybridization. (Belford et al. 1981)

S. siberica and of that species only (Fig. 11 B). Thus, a stepwise differentiation process of highly repetitive satellite DNA becomes apparent in the *S. siberica* alliance: first an extensive divergence separating *S. mischtschenkoana* and *S. amoena* plus *S. ingridae*, and secondly, a slight change towards *S. siberica*.

A comparison of four representatives of tropical epiphytic Orchidaceae suggests that their evolution has been linked to changes in chromosome numbers, total nuclear DNA and the addition or loss of various satellite DNA fractions apparent as 3%–7% heterochromatin (Nagl and Capesius 1977, Capesius and Nagl 1978). There is a trend from conservative taxa with diffuse chromatin structure and low values of DNA with uniform derivative melting curves (e.g. *Phalenopsis*) to progressive ones with chromocentric nuclei and higher values of DNA showing several thermic peaks corresponding to satellites (e.g. *Cymbidium*). Among species of the temperate terrestrial orchid genus *Cephalanthera* different types of heterochromatic bands have been identified by sequential fluorochrome staining (Schwarzacher et al. 1980, Schwarzacher and Schweizer 1982). Comparable processes of heterochromatin insertion or deletion, and changes in chromosome numbers by Robertsonian events must have occurred in this genus as well.

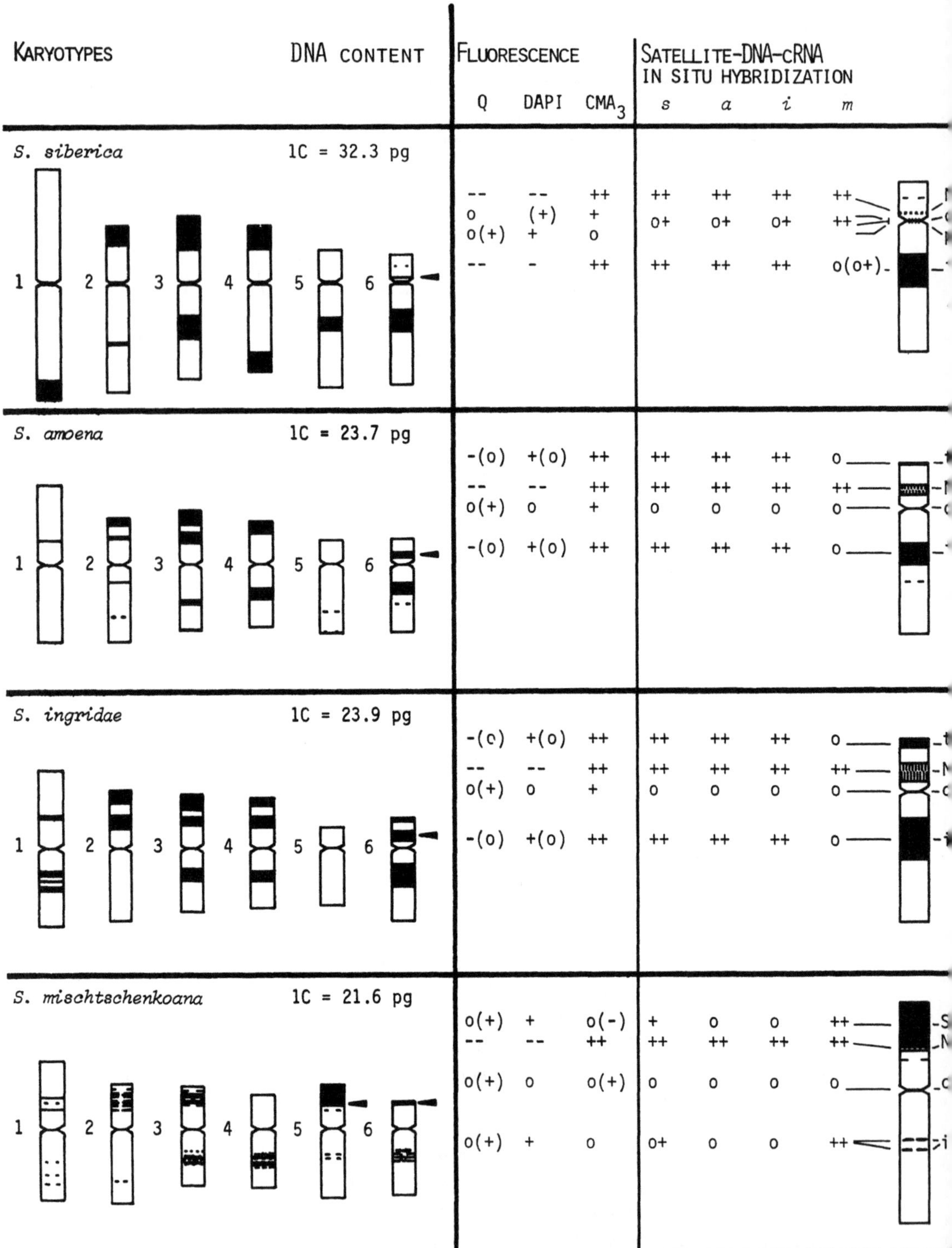

Fig. 10. (Legend see p. 25)

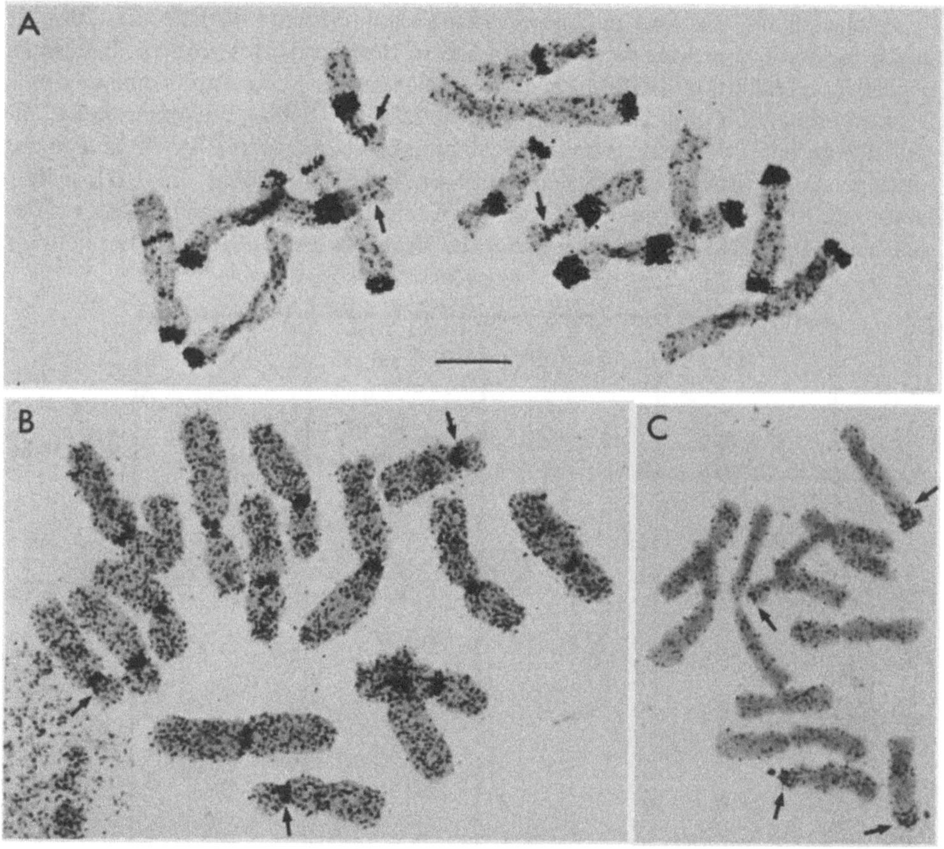

Fig. 11. Interspecific in situ hybridization between representatives of the *Scilla siberica* alliance with satellite DNA-cRNA: **A** from *S. amoena* onto chromosomes of *S. siberica* (heavy labelling of heterochromatic intercalary, terminal, and NOR bands); **B** from *S. mischtschenkoana* onto chromosomes of *S. siberica* (labelling of paracentromeric and NOR bands only); **C** from *S. siberica* onto chromosomes of *S. mischtschenkoana* (labelling of NOR bands only); *arrows* mark NORs; *bar* represents 10 μm. (Modified from Deumling and Greilhuber 1982)

Fig. 10. Total nuclear DNA (1 C) and heterochromatin differentiation among four representatives of the *Scilla siberica* alliance. Giemsa C-banded karyotypes (*left;* NOR-zones marked by *arrows*); heterochromatic bands of NOR-chromosomes stained with the fluorochromes quinacrin (*Q*), DAPI and chromomycin A_3 (*CMA_3*) (*middle*), and marked by in situ hybridization with labelled satellite DNA-cRNA from the same and the other species (*right*): (*s*) *S. siberica,* (*a*) *S. amoena,* (*i*) *S. ingridae,* (*m*) *S. mischtschenkoana;* reaction ++ = strongly positive, + = positive, o = neutral, – = negative, – – = strongly negative, () = rare expressions; centromeric region = *c*; nucleolus organizing region = NOR; heterochromatin terminal = *t*, intercalary = *i* or paracentromeric = *p*; heterochromatic satellite = *Sat*. (Greilhuber 1982)

Among plants, the new and more sophisticated methods of molecular DNA research for obvious reasons were first applied to the cereal grasses. In the late sixties Bendich and McCarthy (1970a,b) tested results from base composition analyses, bulk DNA hybridization and thermal stability profiles of hybrid duplexes against the generally accepted views of relationships among genera of cereals (Fig. 14) and among wheat relatives (with the genome formulas AA, AABB, AABBDD, and DD), and obtained positive results. Later and more sophisticated studies have basically verified their data (Fig. 12) but have shown that the greater or lesser DNA affinities between

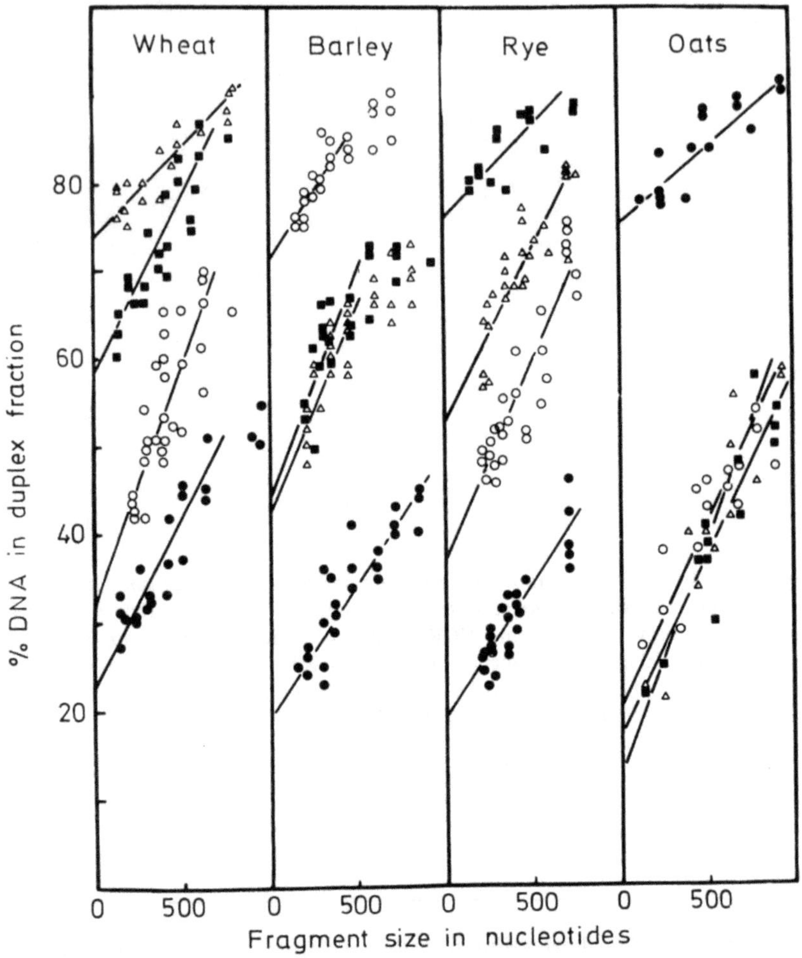

Fig. 12. Intraspecies and interspecies DNA hybridization in cereal grasses. Unlabelled DNA sheared to an average length of 300–400 NTP from *wheat* (△), *rye* (■), *barley* (○), and *oats* (●) was mixed as "driver" at a ratio of 8,000: 1 with labelled DNA of various lengths from the same four species as "tracer" and renatured to a c_0t of approximately 120 on HAP; duplex formation is between repetitive sequences only; percentages reflect relative homology or divergence but also depend on fragment size: longer fragments produce DNA duplexes with longer single-stranded "tails". (Flavell et al. 1977)

these cereals are not really due to simple copy sequences but to the predominant and variously interspersed repeats which amount to more than 75% of their genomes (Flavell et al. 1977). These interspersion patterns were mainly revealed by DNA renaturation experiments using different fragment lengths of "tracer" with short "driver" DNA (pp. 19f.) from the same or from different species (homologous and heterologous pairing: Fig. 13). Generally, there are 25%–40% short non-repeats interspersed with short repeats, 55%–70% interspersed or tandem repeats, and 2%–9% long non-repeats (Flavell et al. 1981). Figure 14 illustrates details about the seven classes of interspersed repeats detected (Rimpau et al. 1978, 1980, Flavell et al. 1979, 1981): I is common to all four species, II characterizes related Triticeae, III the close

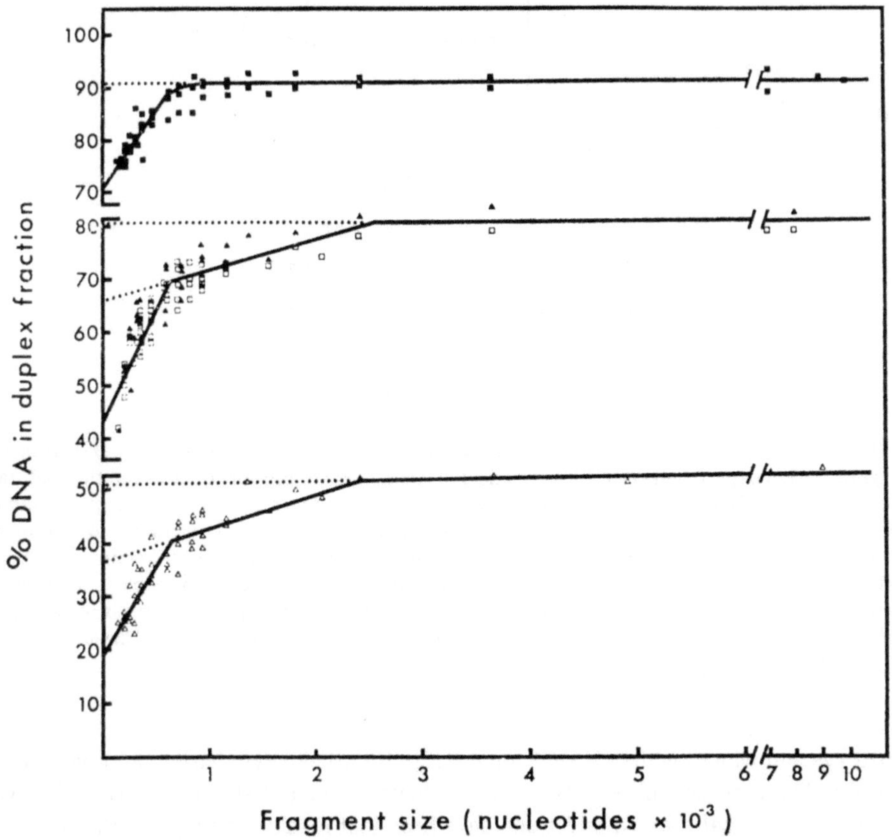

Fig. 13. DNA interspersion patterns in cereal grasses as revealed by reassociation experiments. Labelled barley DNA of various average fragment size was mixed as "tracer" at a ratio of 1:8,000 with unlabelled "driver" DNA fragments of 300–400 NTP from barley (■), wheat (▲), rye (□), and oats (△), renatured to c_0t 120 and then fractionated on HAP. Correlations between fragment size and slopes, shoulders and plateaus of curves suggests average lengths of interspersed sequences, extrapolation towards ordinate indicates proportions of various interspersion patterns within the barley genome, e.g. from upper curve: 9% non-repeated DNA of ca. 700 NTP interspersed with repeats I common to all four species, or from middle curve: 7% non-repeated DNA interspersed with 23% repeats II common to barley, wheat and rye. (Rimpau et al. 1980)

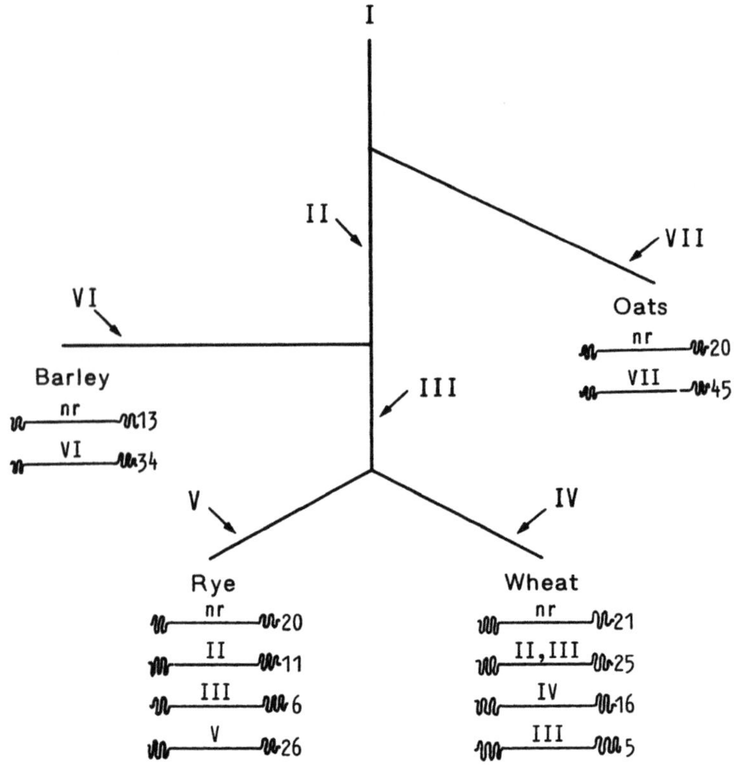

Fig. 14. Phylogeny of cereal grasses as suggested from morphological affinities and from DNA hybridization experiments (see **Figs. 12** and **13**). Distances are proportional to the thermal stability of hybrid DNA duplexes formed between homologous repeated sequences. *Roman numbers* refer to seven classes of interspersed repeats which have been generated during the evolution from a common ancestor: *I* repeated sequences common to all four species; *II* common to barley, rye, and wheat; *III* common to rye and wheat only; *IV–VII* specific for each species; *nr* nonrepeated sequences; the figure attached to each interspersion pattern corresponds to its approximate proportion in the respective genome. (Combined from data presented by Flavell et al. 1977, 1981, Rimpau et al. 1978, 1980)

pair rye and wheat, while IV–VII are specific for wheat, rye, barley, and oats, respectively. These classes of repeats obviously have originated stepwise during the evolution of the cereal grasses, and none have been lost totally, as they are all represented in the four species with different frequency and in somewhat different interspersion. Homologous repeats have diverged more between than within species.

Even more precise informations on the genome evolution of cereal grasses became available with the restriction enzyme digestion of their DNAs, the cloning of particular repeats, and hybridization experiments on nitrocellulose filters, on tracks (Fig. 15) or in situ on denatured chromosome DNA (Bedbrook et al. 1980). The relative simple repeat family with 120 bp from *Secale cereale* (Fig. 16), for instance was shown to be present also in the wheat group, thus being a member of the ancient repeat classes II or III (Flavell 1982a, Jones and Flavell 1982b). A similar situation has been

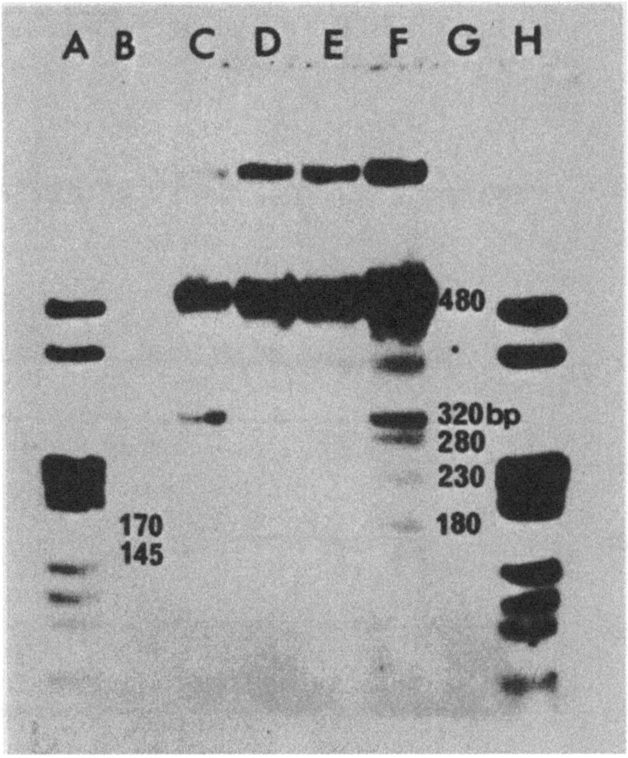

Fig. 15. Intraspecies and interspecies hybridization of endonuclease digested and cloned DNA in *Secale*. Nuclear DNA was digested with Taq I, separated by gel electrophoresis, blotted on nitrocellulose paper and then hybridized with cloned and labelled *S. cereale* DNA from plasmid p SC 210; tracks **B** *S. silvestre;* **C** *S. montanum;* **D** *S. vavilovii;* **E** *S. iranicum;* **F** *S. cereale;* **A, H** cloned *S. cereale* DNA from plasmid p BR 322 and digested with Hae III (only partially homologous to p SC 210). Compare interspecific DNA homologies in the 480 and 320 bp repeat families. (Jones and Flavell 1982b)

found with another, very short and highly repeated 5–10 bp repeat which occurs in wheat and barley (Peacock et al. 1978). In contrast, the complex 480 bp and 610 bp repeats belong to class V and are found in *Secale* only. *S. silvestre,* obviously a very early offshoot within this genus, has the 120 bp but not yet these 480 bp and 610 bp repeats, which are found i.a. in the primitive perennial *S. montanum* as 1.2%–4.3% ($2–7 \times 10^5$ copies) and 0.5% ($\approx 6.5 \times 10^4$ copies) of the total genome. In *S. cereale* these two repeats have been strongly increased, accounting for 6.1% (10^6 copies) and 2.7% (3.5×10^5 copies) of its genome. Of particular interest is the 630 bp repeat because its stepwise origin, incorporating portions of the 120 bp and a 356 bp repeat (Fig. 16), and subsequent distribution within the genus can be followed in some detail: From *S. montanum* (0.16%) to the advanced annuals *S. vavilovii* and *S. iranicum* (0.01%–0.04%) and to *S. cereale* (0.06%). From these data by Jones and Flavell (1982a,b) the relatively high DNA value and heterochromatin proportion (12%) of *S. cereale* (2C = 15.8–16.6 pg) is thus seen as the result of accumulation of several repeat families and their placement as additional heterochromatin in interstitial, but predominantly terminal position on *all* chromosomes of its karyotype (Fig. 17). This process evidently has gone on relatively independent from the classical translocations which characterize the gross karyotype evolution of *Secale* species (Singh and Röbbelen 1977, Appels 1982).

The studies on nuclear DNA differentiation of cereal grasses, both on lower (within *Secale*, but also in the *Triticum-Aegilops* group) as well as on higher taxonomic levels

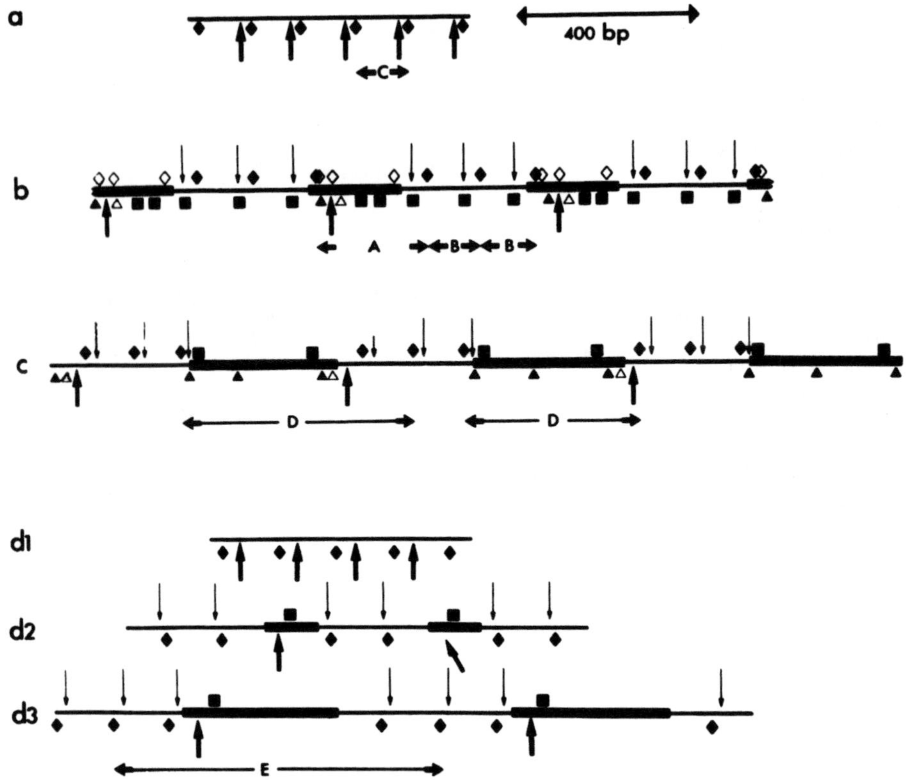

Fig. 16. Physical maps for the major repeated sequences families of *Secale cereale* telomeric heterochromatin: **a** 120 bp family; **b** 480 bp family; **c** 610 bp family; **d** three families that have partial homology and are stepwise interspersed: d_1 120 bp; d_2 356 bp; d_3 630 bp. Restriction enzyme recognition sites are shown by symbols: ◆ = Hae III, ◇ = Mbo I, ▲ = Mbo II, △ = Taq I, ■ = Alu I. *Heavy arrows* pointing upwards indicate the length of the repeating unit; *light arrows* pointing downwards the length of subrepeats within a complex repeating unit, *heavy horzion-tal lines* mark other uncorrelated subrepeats. The letters *A–E* denote different clone classes from Hae III fragmentation. (Bedbrook et al. 1980)

(within and outside of *Triticeae*) reveal a strategy which apparently applies to all higher plants: Stepwise amplification of non-coding DNA sequences, interspersion with single-copy DNA or with other repeats, new amplification cycles leading to complex repeats, and partial deletion or further diffusion of repeats over the whole karyotype as relatively "movable elements" by unspecific contacts or other mechanisms still insufficiently understood. It is likely that these processes are also responsible for the gradual evolutionary divergence of karyotypes and their decline of meiotic pairing capacity (Dvořák and McGuire 1981, Flavell 1982a). This phenomenon, in former speculations often designated as "cryptic structural differentiation" (Stebbins 1971), obviously is a most important aspect of speciation.

Acknowledgements. The author is greatly indepted to Prof. D. Schweizer, Doc. J. Greilhuber, and Prof. E. Woess for many valuable suggestions, critical remarks and literature references. Relevant

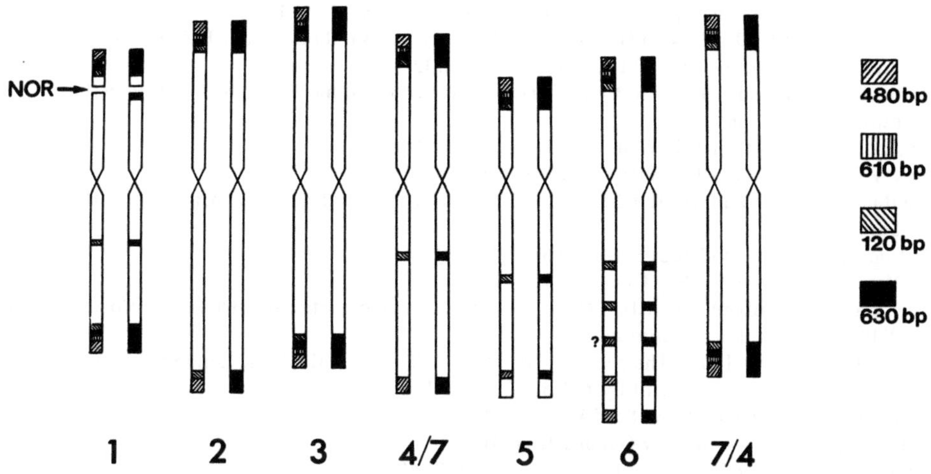

Fig. 17. The seven chromosome pairs of rye, *Secale cereale*, each with a schematic representation of Giemsa C-bands (*right*) and the probable position of repeated sequence families with 480, 610, 120, and 630 bp at the telomeric and some intercalary bands, according to the results of in situ hybridization (*left*) (Jones and Flavell 1982a)

research was supported by the Fonds zur Förderung der wissenschaftlichen Forschung in Österreich, in particular by grants P 4052 and P 4799.

References

Adams RLP, Burdon RH, Cambell AM, Leader DP, Smellie RMS (1981) The biochemistry of the nucleic acids. Chapman and Hall, London
Ambros P (1983) Ph thesis, Univ Vienna
Appels R (1982) Int Rev Cytol 80:93–132
Ayala FJ (1976) Molecular evolution. Sinauer, Sunderland, Mass
Bachmann K (1979) Theor Appl Genet 57:107–111
Bachmann K, Price J (1977) Chromosoma 61:267–275
Bachmann K, Chambers KL, Price HJ (1979) Plant Syst Evol Suppl 2:41–66
Barigozzi C (ed) (1982) Mechanisms of speciation. Symp Acad Nac Lincei, 1980. Liss, New York
Bedbrook JR, Jones J et al (1980) Cell 19:545–560
Bendich AJ, McCarthy BJ (1970a) Genetics 65:545–565
Bendich AJ, McCarthy BJ (1970b) Genetics 65:567–573
Belford HS, Thompson WF (1976) Carnegie Inst Washington Yearb 75:362–367
Belford HS, Thompson WF (1981a) Heredity 46:91–108
Belford HS, Thompson WF (1981b) Heredity 46:109–122
Belford HS, Thompson WF, Stein DB (1981) In: Young DA, Seigler DS (eds) Phytochemistry and angiosperm phylogeny. Praeger, New York, pp 1–18
Bennett MD (1972) Proc R Soc London Ser B 181:109–135
Bennett MD (1973) Brookhaven Symp Biol 25:344–366
Bennett MD (1982) In: Dover GA, Flavell RB (eds) Genome evolution. Academic Press, London New York, pp 239–261
Bennett MD, Smith JB (1976) Philos Trans R Soc London Ser B 274:227–274

Bennett MD, Gustafson JP, Smith JB (1977) Chromosome 61:149–176
Bennett MD, Smith JB, Heslop-Harrison JS (1982) Proc R Soc London Ser B 216:179–199
Bentzer B, Landström T (1975) Hereditas 80:219–232
Birnie GD, Rickwood D (1978) Centrifugation separations in molecular biology and cell biology. Butterworths, London
Blakey DH, Vosa CG (1981) Plant Syst Evol 139:47–55
Blakey DH, Vosa CG (1982) Plant Syst Evol 139:163–178
Bradbury EM, McLean N, Matthews H (1981) DNA, chromatin, and chromosomes. Blackwell, London
Britten RJ, Kohne DE (1968) Science 161:529–540
Brutlag DL (1980) Annu Rev Genet 14:121–144
Buchanan RE, Gibbons NE (eds) (1974) Bergey's manual of determinative bacteriology, 8th edn Williams and Wilkins, Baltimore
Busch H (ed) (1974–1979) The cell nucleus, vols I–VII. Academic Press, London New York
Capesius I, Nagl W (1978) Plant Syst Evol 129:143–166
Cavalier-Smith T (1978) J Cell Sci 34:247–278
Comings DE (1978) Annu Rev Genet 12:25–46
Cullis CA, Schweizer D (1974) Chromosoma 44:417–421
Deumling B (1981) Proc Natl Acad Sci USA 78:338–342
Deumling B, Greilhuber J (1982) Chromosoma 84:535–555
Doolittle WF, Sapienza C (1980) Nature (London) 284:601–603
Dover G (1982a) Nature (London) 299:111–117
Dover G (1982b) In: Barigozzi (ed) Mechanisms of speciation. Symp Acad Nac Lincei, Roma 1980, pp 435–459
Dover GA, Flavell RB (eds) (1982) Genome evolution. Academic Press, London New York
Dvořák J, McGuire PE (1981) Genetics 97:391–414
Ehrendorfer F (1982) In: Barigozzi (ed) Mechanisms of speciation. Symp Acad Nac Lincei, Roma 1980, pp 479–509
Ehrendorfer F, Schweizer D, Greger H, Humphries CJ (1977) Taxon 26:387–394
Evans GM, Rees H, Snell CL, Sun S (1972) Chromosomes Today 3:24–37
Faires RA, Boswell GGJ (1981) Radioisotope laboratory techniques. Butterworths, London
Flavell RB (1981) Chromosomes Today 7:42–54
Flavell RB (1982a) In: Dover GA, Flavell RB (eds) Genome evolution. Academic Press, London New York, pp 301–323
Flavell RB (1982b) In: Parthier B, Boulter D (eds) Nucleic acids and proteins in plants, II. Encycl Plant Physiol, vol 14B. Springer, Berlin Heidelberg New York, pp 46–74
Flavell RB, Rimpau J, Smith DB (1977) Chromosoma 63:205–222
Flavell RB, O'Dell M, Smith D (1979) Heredity 42:309–322
Flavell RB, Bedbrook J, Jones J, O'Dell M, Gerlach WL, Dyer TA, Thompson RD (1980) In: Davies DR, Hopwood DA (eds) The plant genome. 4th John Innes Symp, Norwich 1979, John Innes Charity, pp 15–30
Flavell RB, O'Dell M, Hutchinson J (1981) Cold Spring Harbor Symp Quant Biol 45:501–508
Fuhrmann B, Nagl W (1979) Plant Syst Evol Suppl 2:235–245
Fukuda I, Grant WF (1980) Can J Genet Cytol 22:81–91
Furuta Y, Nishikawa K, Makino T (1975) Jpn J Genet 50:257–263
Furuta Y, Nishikawa K, Kimizuka T (1977) Jpn J Genet 52:107–115
Furuta Y, Nishikawa K, Haji T (1978) Jpn J Genet 53:361–366
Geber G (1979) M Sc Dipl Work, Univ Vienna
Geber G, Hasibeder G (1980) Microsc Acta Suppl 4:31–35
Gill BS, Kimber G (1974) Proc Natl Acad Sci USA 71:4086–4090
Goldberg RB (1978) Biochem Genet 16:45–68
Gregory WC (1941) Transact Am Philos Soc 31:443–521
Greilhuber J (1975) Plant Syst Evol 124:139–156
Greilhuber J (1979) Plant Syst Evol Suppl 2:263–280
Greilhuber J (1982) Stapfia 10:11–51
Greilhuber J (1983) In: Heywood VH (ed) Current topics in plant taxonomy. Academic Press, London New York, in press

Greilhuber J, Speta F (1977) Plant Syst Evol 127:171–190
Greilhuber J, Speta F (1978) Plant Syst Evol 129:63–109
Greilhuber J, Deumling B, Speta F (1981) Ber Dtsch Bot Ges 94:249–266
Greilhuber J, Loidl J (1983) In: Brandham P, Bennett MD (eds) Proc 2nd Kew Chromosome Conf. Allen and Unwin, London, in press
Guerra Filho M (1980) PhD thesis, Univ Vienna
Haga T, Kurabayashi M (1954) Mem Fac Sci Kyushu Univ Ser E 1:159–185
Hake S, Walbot V (1980) Chromosoma 79:251–270
Hall HM, Clements FE (1923) In: The North American species of Artemisia, Chrysothamnus, and Atriplex. Carnegie Inst Washington Publ 326. Judd and Detweiler, Washington DC, pp 235–247
Hamel JL (1953) Rev Cytol Biol Veg 14:115–313
Humphries CJ (1979) Bull Br Mus (Nat Hist) Bot 7:83–142
Humphries CJ (1981) Nord J Bot 1:83–96
Hutchinson J, Rees H, Seal AG (1979) Heredity 43:411–421
Hutchinson J, Chapman V, Miller TN (1980a) Heredity 45:245–254
Hutchinson J, Narayan RKJ, Rees H (1980b) Chromosoma 78:135–145
Ingle J, Pearson GG, Sinclair J (1973) Nature (London) New Biol 242:193–197
Ingle J, Timmis JN, Sinclair J (1975) Plant Physiol 55:496–501
Iyengar GAS, Gaddipati JP, Sen SK (1979) Theor Appl Genet 54:219–224
Jensen U (1968) Bot Jahrb 88:204–268
John B (1981) Chromosomes Today 7:128–137
John B, Miklos GLG (1979) Int Rev Cytol 58:1–114
Jones GH (1978) Chromosoma 66:45–57
Jones JDG, Flavell RB (1982a) Chromosoma 86:595–612
Jones JDG, Flavell RB (1982b) Chromosoma 86:613–641
Jones K (1978) Adv Bot Res 6:119–194
Jones RN, Brown LM (1976) Heredity 36:91–104
Kaina B, Rieger R (1979) Biol Zentralbl 98:661–697
Kashevarov GP, Antonov AS (1982) Bot Zh (Leningrad) 67:537–543
Kessler R (1982) Prog Phycol Res 1:111–135
Knippers R (1982) Molekulare Genetik. Georg Thieme, Stuttgart
Kössel H, Edwards K, Fritzsche E, Koch W, Schwarz Zs (1983) In: Jensen U, Fairbrothers DE (eds) Proteins and nucleic acids in plant systematics. Springer, Berlin Heidelberg New York
Kurabayashi M (1963) Evolution 17:296–306
Levin DA, Funderburg SW (1979) Am Nat 114:784–795
Linde-Laursen I, v Bothmer R, Jacobsen N (1980) Hereditas 93:235–254
Loidl J (1979) Chromosoma 73:45–51
Loidl J (1981) PhD thesis, Univ Vienna
Marks GE, Schweizer D (1974) Chromosoma 44:405–416
Miller CN (1971) Contrib Mus Paleontol Univ Mich 23:105–169
Mitsuoka S, Ehrendorfer F (1972) Oesterr Bot Z 120:155–200
Morawetz W (1981) Plant Syst Evol 139:57–76
Murray MG, Preisler RS, Thompson WF (1977) Carnegie Inst Washington Yearb 76:240–246
Murray MG, Cuellar RE, Thompson WF (1978) Biochemistry 17:5781–5790
Murray MG, Palmer JD, Cuellar RE, Thompson WF (1979) Biochemistry 18:5259–5266
Nagl W (1974) Dev Biol 39:342–346
Nagl W (1976) Zellkern und Zellzyklen. Ulmer, Stuttgart
Nagl W (1982) In: Parthier B, Boulter D (eds) Nucleic acids in plants II. Encycl Plant Physiol, vol 14B. Springer, Berlin Heidelberg New York, pp 1–45
Nagl W, Ehrendorfer F (1974) Plant Syst Evol 123:35–54
Nagl W, Capesius I (1976) Plant Syst Evol 126:221–237
Nagl W, Capesius I (1977) Chromosomes Today 6:141–150
Nagl W, Hemleben V, Ehrendorfer F (eds) (1979) Genome and chromatin: Organization, evolution, function. Symp Kaiserslautern. Plant Syst Evol Suppl 2
Nagl W, Fusenig HP (1979) Plant Syst Evol Suppl 2:221–233
Nagl W, Bachmann K (1980) Theor Appl Genet 57:107–111

Narayan RKJ (1982) Evolution 36:877–891
Narayan RKJ, Rees H (1977) Chromosoma 63:101–107
Orgel LE, Crick FHC (1980) Nature (London) 284:604–607
Parthier B, Boulter D (1982) Nucleic acids and proteins in plants II. Structure, biochemistry, and physiology of nucleic acids. Encycl Plant Physiol, vol 14B. Springer, Berlin Heidelberg New York
Peacock WJ, Lohe AR, Gerlach WL, Dunsmuir P, Dennis ES, Appels R (1978) Cold Spring Harbor Symp Quant Biol 42:1121–1136
Pellegrini M, Goldberg RB (1979) Chromosoma 75:309–326
Preisler RF, Thompson WF (1978) Carnegie Inst Washington Yearb 77:323–330
Price HJ (1976) Bot Rev 42:27–52
Price HJ, Bachmann K (1975) Am J Bot 62:262–267
Price HJ, Bachmann K (1976) Plant Syst Evol 126:323–330
Price HJ, Bachmann K, Chambers KL, Riggs J (1980) Bot Gaz 141:156–159
Price HJ, Chambers KL, Bachmann K (1981a) Bot Gaz 142:156–159
Price HJ, Chambers KL, Bachmann K (1981b) Bot Gaz 142:415–426
Raikov IB (1982) The protozoan nucleus. Cell Biol Monogr 9. Springer, Berlin Heidelberg New York
Ranjekar PK, Pallotta D, LaFontaine JG (1978) Can J Biochem 56:808–815
Rees H, Hazarika MH (1969) Chromosomes Today 2:158–165
Rees H, Jones RN (1972) Int Rev Cytol 32:53–92
Rees H, Narayan RKJ (1977) Chromosomes Today 6:131–139
Rickwood D, Hames BD (1982) Gel electrophoresis of nucleic acids: A practical approach. IRL Press Ltd, Oxford, Wash
Rimpau J, Smith DB, Flavell RB (1978) J Mol Biol 123:327–350
Rimpau J, Smith DB, Flavell RB (1980) Heredity 44:131–149
Rizzo PJ, Burghardt RC (1980) Chromosoma 76:91–99
Roose ML, Gottlieb LD (1978) Heredity 40:159–163
Rothfels K, Sexsmith E, Heimburger M, Krause MO (1966) Chromosoma 20:54–74
Schaan ME, Nagl W (1979) Plant Syst Evol Suppl 2:67–71
Schmidtke J, Epplen JT (1980) Hum Genet 55:1–18
Schwarzacher T, Ambros P, Schweizer D (1980) Plant Syst Evol 134:293–297
Schwarzacher T, Schweizer D (1982) Plant Syst Evol 141:91–113
Schweizer D (1980) In: Davies DR, Hopwood DA (eds) The plant genome. Proc 4th John Innes Symp, Norwich 1979. John Innes Charity, pp 61–72
Schweizer D, Ehrendorfer F (1976) Plant Syst Evol 126:107–148
Schweizer D, Ehrendorfer F (1983) Biol Zentralbl (in press)
Seal AG, Rees H (1982) Heredity 49:179–190
Sengbusch P v (1979) Molekular- und Zellbiologie. Springer, Berlin Heidelberg New York
Seshadri M, Ranjekar PK (1980) Biochim Biophys Acta 610:211–220
Singh RJ, Röbbelen G (1975) Z Pflanzenzücht 75:270–285
Singh RJ, Röbbelen G (1977) Chromosoma 59:217–225
Spector DL, Triemer RE (1981) Biosystems 14:289–298
Stebbins GL (1971) Chromosomal evolution in higher plants. Arnold, London
Stein DB, Thompson WF (1977) Carnegie Inst Washington Yearb 76:252–255
Stein DB, Thompson WF, Belford HS (1979) J Mol Evol 13:215–232
Teoh SB, Rees H (1976) Heredity 36:123–137
Thomas HM (1981) Heredity 46:263–267
Thompson WF (1978) Carnegie Inst Washington Yearb 77:310–316
Thompson WF, Murray MG (1980) In: Davies DR, Hopwood DA (eds) The plant genome. 4th John Innes Symp, Norwich 1979. John Innes Charity, pp 31–45
Tohidast-Akrad M (1982) PhD thesis, Univ Vienna
Träger L (1975) Einführung in die Molekularbiologie. Fischer, Stuttgart
Tschermak-Woess R (1963) Strukturtypen der Ruhekerne von Pflanzen und Tieren. Protoplasmatologia 5/1. Springer, Wien

Uitz H (1970) PhD thesis, Univ Graz
Vosa CG (1976a) Heredity 36:383–393
Vosa CG (1976b) Chromosoma 57:119–133
Wenzel W, Hemleben V (1979) Plant Syst Evol Suppl 2:29–40
Wenzel W, Hemleben V (1982) Plant Syst Evol 139:209–227
Wimpee CF, Rawson JRY (1979) Biochim Biophys Acta 562:192–206

Phylogenetic Significance of Nucleotide Sequence Analysis

H. KÖSSEL, K. EDWARDS, E. FRITZSCHE, W. KOCH, and Zs. SCHWARZ[1]

Abstract. An evaluation for the use of various parameters of DNA and RNA structures as criteria for phylogenetic relationships is presented. Hybridization data and restriction endonuclease patterns generally allow only limited conclusions. However, T_1-oligonucleotide mapping, and more recently, sequence analysis of entire large ribosomal RNA and DNA species have been used to quantify the relationship of chloroplast ribosomes with prokaryotic ribosomes. S_{AB}-values deduced from oligonucleotide mapping appear more suitable for closely related species, whereas comparison of complete sequences reveal homologies also between more distantly related species or within less conserved regions of DNA and RNA molecules.

A comparison of rRNA operons from chloroplasts of various algae and higher plants is presented on the following four levels:

a) *number* and *arrangement* of operon copies
b) relative *positions* of individual components (i.e. rRNA genes, flanking, and spacer tRNA genes, small rRNA genes) within the rRNA operons
c) *nucleotide sequences* of the individual components
d) *secondary structures* of the individual components.

It is concluded that useful criteria for the evaluation of phylogenetic relationships can be obtained on any of the four levels; however, sequence comparison is the most suitable method for quantification.

1 Introduction

Amino acid sequences of several proteins have been used as a criterion for establishing phylogenetic relationships (Schwartz and Dayhoff 1978). For plant systematics the amino acid sequences of cytochrome c have been introduced by Boulter and his collegues (Boulter 1974). This is well documented in a recent review by Jensen (1981), in which amino acid sequence data from other proteins are also included. Due to a lag in the development of RNA and DNA sequencing techniques comparison of nucleotide sequences has become feasible only recently. Thus phylogenetic trees based on 5S rRNA sequences have been published by Schwartz and Dayhoff (1978) and by Osawa and Hori (1979) and a phylogenetic tree based on tRNA sequences was presented by Cedergren and his group in 1981 (for review see also Sprinzl, this volume). The plant kingdom is represented in both of these trees by only a few species, but this gap is likely to be diminished in the near future.

1 Institut für Biologie III, Universität Freiburg, Schaenzlestr. 1, D-7800 Freiburg, FRG

Proteins and Nucleic Acids in Plant Systematics
ed. by U. Jensen and D.E. Fairbrothers

Comparison of DNA sequences has already proven to be a powerful phylogenetic criterion in vertebrate systems as, for instance, is evident from DNA sequences of globin genes, which have been used for deduction of a phylogenetic tree of the globin gene family (Efstratiadis et al. 1980). Only three years ago the first DNA sequences of plant genes became known (Schwarz and Kössel 1979, 1980, McIntosh et al. 1980). The publication since then of many more plant DNA sequences, and the likelihood of still more in the forthcoming years, indicates that they will play an increasingly important role in the establishment of phylogenetic relationships within the plant kingdom.

In the first part of this article we will present a general survey of the methods available for RNA and DNA sequence comparison. In the second part we will discuss chloroplast rRNA operons and their DNA sequences in order to demonstrate their usefulness in establishing phylogenetic relationships on the various structural levels of rDNA and rRNA.

2 Experimental Methods underlying DNA and RNA Sequence Comparisons

Most of the parameters determined by earlier methods of DNA characterization such as base compositions (determined directly or by buoyant density or melting profiles), nearest neighbour frequencies, DNA content per cell and molecular hybridization, are of very limited taxonomic value (for review see Ehrendorfer, this volume). This limitation is not unexpected as these parameters are only very grossly related to the genetic information contained in the respective DNAs. Also, from the precision of these methods and from the variation ranges of the parameters obtained, only rough estimates of phylogenetic relationship could be expected. Hybridization data, as for instance that obtained from renaturation of DNAs from different species, are difficult to evaluate quantitatively. On the one hand a certain number of mismatches – depending on the stringency of hybridization conditions – can never be excluded from the hybrid DNAs and on the other hand relatively short interspersed stretches of homologous sequences – 50 or 100 basepairs, especially when they are rich in G/C – may be sufficient to provoke strong hybridization, even if the remaining portions of the two DNAs to be compared show very little homology. A powerful improvement to the hybridization technique, which is based on the hybridization of specific RNA or DNA probes with DNA fragmented by restriction endonucleases, was introduced by Southern (1975). The taxonomic value of this variant will be discussed below in connection with restriction analysis (Sect. 2.2).

2.1 T_1-Oligonucleotide Mapping of 16S rRNA

This method was introduced by Woese and his co-workers in order to establish a quantitative method for the taxonomy of micro-organisms (Fox et al. 1980). ^{32}P labelled small ribosomal subunit RNAs (usually 16S rRNAs) are isolated and digested with T_1-ribonuclease. The resulting mixtures of oligonucleotides, which all have 3′ terminal G residues, are separated in a two dimensional system and all oligonucleotides of chain length six and greater, are then sequenced and combined in oligonucleotide catalogues characteristic for each individual species (see e.g. Zablen et al. 1975). For the comparison of two rRNA species an association coefficient is defined as $S_{AB} = \frac{2\,N_{AB}}{N_A + N_B}$ in which N_{AB} is the number of oligonucleotides common to both the species and N_A and N_B are the numbers of all oligonucleotides (chain lengths ≥ 6) which are found in species A and B, respectively. This method is particularly useful for the taxonomy of micro-organisms as they lack the abundance of morphological criteria generally used for the taxonomy of higher organisms. Application of this method together with other biochemically defined taxonomic criteria has led to detailed dendrograms of micro-organisms which indicate two major bacterial kingdoms, the eubacteria and the archaebacteria (Fox et al. 1980). A dendrogram which correlates S_{AB}-values from several eubacterial 16S rRNAs and from chloroplast 16S rRNAs of two algae (*Euglena gracilis* and *Porphyridium*) and of the plant *Lemna minor* is shown in Fig. 1.

While the ^{32}P labelling of 16S rRNA is easily achieved with micro-organisms grown in vitro, labelling of animal or plant rRNA is not possible in the quantity necessary for this method. This is the main reason why oligonucleotide mapping has so far not been applied for the taxonomy of higher organisms including plants (with the exception of *Lemna minor,* see Fig. 1). It is , however, possible to obtain T_1-oligonucleotide catalogues from nonradioactive rRNAs, if the ^{32}P label is introduced into the oligo-

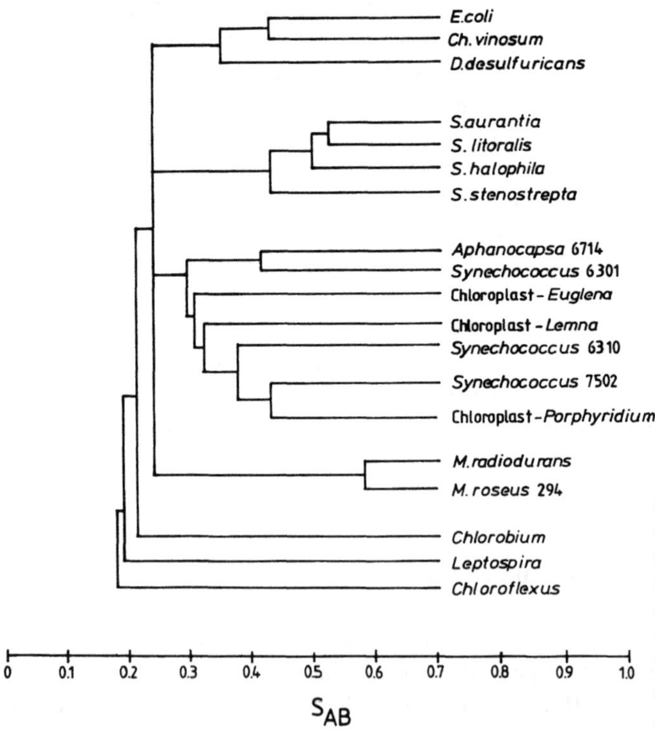

Fig. 1. Phylogenetic relationships between eubacteria and chloroplasts as deduced from coincidence coefficients (S_{AB}-values) derived from T_1-oligonucleotide catalogues of the respective 16S rRNAs. (Data from Fox et al. 1980)

nucleotides by polynucleotide kinase catalysed reactions after T_1-digestion of the respective rRNAs (Stackebrandt et al. 1981). Using this technique several plant 18S rRNA species have been analysed (Stöcklein et al. this volume). The use of cell cultures for labelling rRNA in vivo provides another alternative (for example see Zablen et al. 1975).

Unfortunately the association coefficients are related to the actual number (or percentage) of identical nucleotides of the respective 16SrRNAs in an unknown and nonlinear way. In particular they depend highly on the local and overall G content of the two sequences compared. A high G content shifts the T_1-oligomers into the short chain range and therefore reduces the number of G terminated oligomers with chain lengths $\geqq 6$. The statistical weight of the total and particularly of the identical oligomers counted is thereby reduced. On the other hand low G content or an uneven distribution of G residues by clustering often increases the number of very long oligomers (chain lengths > 12) which are more difficult for sequence analysis and which alone for their longer chain lengths have a higher probability for divergence. This lack of direct relationship between S_{AB}-values and actual sequence homology can easily be demonstrated by counting T_1-oligomers from known complete rRNA sequences and calculating S_{AB}-values for pairs of rRNA sequences according to the formula mentioned above. In Table 1 such nonexperimentally derived S_{AB}-values are listed alongside with the actual sequence homologies of rRNA sequences. For instance, for the pair of 16S rRNA sequences from *E. coli* and maize chloroplast an association coefficient of 0.17 was calculated. In contrast to this low value, which usually is interpreted of indicating low relatedness, the actual homoloy of the two sequences is 74.2% (Schwarz and Kössel 1980). The values calculated for most of the other rRNA sequence pairs listed in Table 1 are in a similar range. Only the sequence pairs of either 16S or 23S rRNAs from maize and tobacco chloroplasts show higher S_{AB}-values (0.76 and 0.63, respectively) and these correlate to sequence homologies of more than 90% (96% and 92%, respectively). It seems therefore that S_{AB}-values in the range of 0.25–0.90 correlate to actual sequence homologies in the range of 80%–98%. It is also noteworthy that in a few cases lowering of sequence homology is accompanied even by a slight increase of association coefficient and vise versa (compare for instance the 16S pairs from maize chloroplasts/*E. coli* and human and mouse mitochondria). This clearly demonstrates that association coefficients below 0.25 are not reliable as narrow quantitative taxonomic parameters, unless they are combined with other parameters as was done for establishing the two bacterial kingdoms (eubacteria and archaebacteria). This limitation appears not unexpected as the sample numbers of identical oligonucleotides (N_{AB}) underlying the S_{AB}-values below 0.25 are only in the range of 15 and below and such low numbers naturally lower the statistical significance of the corresponding S_{AB}-values considerably. It seems therefore appropriate to use association coefficient as quantitative taxonomic parameters only in the range above 0.25. Even in this range branch points of dendrograms which rest on S_{AB}-differences smaller than 0.05 should be interpreted with caution due to standard deviations related to the "sampling weight" as well as the errors in collecting and sequencing all the oligonucleotides of the respective fingerprints. This intrinsic limitation of the method is counterbalanced by its ease in practicability. The time necessary for working out several oligonucleotide catalogues in parallel ranges between 2–6 weeks. Its

Table 1. Sequence homology between pairs of rRNAs, rRNA leader regions, rRNA spacer regions and flanking tRNAs
Coincidence coefficients (S_{AB}-values) as calculated from T_1-oligonucleotides of the known rRNA (rDNA) sequences are included for comparison with the actual sequence homologies

Sequence pairs compared	Percentages of homology	S_{AB}-Values	References of the sequences compared
a) 16S rRNAs (rDNAs):			
Maize chloroplasts – tobacco chloroplasts	96.4	0.76	Schwarz and Kössel (1980) Tohdoh and Sugiura (1982)
Maize chloroplasts – *Euglena* chloroplasts	80.0	0.23	Graf et al. (1982)
Maize chloroplasts – *E. coli*	74.2	0.17	Schwarz and Kössel (1980)
Euglena chloroplasts – *E. coli*	72.4	0.19	Graf et al. (1982)
b) Mitochondrial 12S rRNAs (rDNAs):			
human-mouse	75.5	0.17	Van Etten et al. (1980) Eperon et al. (1980)
c) 23S rRNAs (rDNAs):			
Maize chloroplasts – tobacco chloroplasts	92.0	0.63	Edwards and Kössel (1981) Takaiwa and Sugiura (1982a)
Maize chloroplasts[a] – *E. coli:* total sequences	70.9	0.11	Edwards and Kössel (1981)
5′ halves	68.0	0.08	Edwards and Kössel (1981)
3′ halves[a]	73.8	0.13	Edwards and Kössel (1981)
d) Mitochondrial 16S rRNAs (rDNAs):			
human-mouse	73.0	0.23	Van Etten et al. (1980) Eperon et al. (1980)
e) 16S rRNA leader[b] from chloroplasts of maize – tobacco	83.2	–	Schwarz et al. (1981) Tohdoh et al. (1981)
f) 16S/23S spacer rDNA[c] from chloroplasts of maize – tobacco	92.9	–	Koch et al. (1981) Takaiwa and Sugiura (1982b)
g) tRNA genes[d]:			
$tRNA^{Val}$ genes from leader region of maize chloroplasts – tobacco chloroplasts	100.0	–	Schwarz et al. (1981) Tohdoh et al. (1981)

$tRNA^{Val}$ maize chloroplast leader – *E. coli* 2 b	66.7	–	Schwarz et al. (1981)
$tRNA^{Ile}$ from the 16S/23S spacer of			
maize chloroplasts – tobacco chloroplasts	100.0	–	Koch et al. (1981) Takaiwa and Sugiura (1982a, b)
maize chloroplasts – *Euglena* chloroplasts	83.4	–	Graf et al. (1980) Orozco et al. (1980)
maize chloroplasts – *E. coli*	80.6	–	Koch et al. (1981)
Euglena chloroplasts – *E. coli*	77.0	–	Graf et al. (1980) Orozco et al. (1980)
$tRNA^{Ala}$ from the 16S/23S spacer of			
maize chloroplasts – tobacco chloroplasts	98.7	–	Koch et al. (1981) Takaiwa and Sugiura (1982a, b)
maize chloroplasts – *Euglena* chloroplasts	90.3	–	Graf et al. (1980) Orozco et al. (1980)
maize chloroplasts – *E. coli*	76.7	–	Koch et al. (1981)
Euglena chloroplasts – *E. coli*	78.1	–	Graf et al. (1980) Orozco et al. (1980)

[a] The 4.5 S rRNA sequence is included in this comparison as it is a structural equivalent of the 3′ terminal region of bacterial 23S rRNA (see Sects. 3.2 and 3.3 of this chapter)
[b] Leader sequences from positions minus 1 to minus 477 (maize) were used for comparison; the $tRNA^{Val}$ genes (72 positions) contained in this region (see under g) were, however, not included as they show identical sequences
[c] Spacer tRNA genes (7% of the spacer sequences) are excluded in this comparison
[d] CCA ends of mature tRNAs are not included in this comparison as they are encoded only in the respective *E. coli* genes

application in the field of plant taxonomy is therefore likely to yield more and more important contributions in the future.

2.2 Restriction Endonuclease Mapping

Restriction endonucleases cleave DNA at specific sites. Mapping of such sites with the many restriction endonucleases known today is a well established method for characterization of cloned genes, of entire plasmid-DNAs and even of small genomes from phages or organelles. The underlying technique consists in separation of fragment mixtures derived from DNAs after cleavage with various restriction enzymes by one dimensional gel electrophoresis according to chain lengths. Clearly resolved patterns are possible only up to 20 or 30 fragments per individual gel run which limits the fragment numbers to be compared. A direct comparison of fragment patterns between two different DNA species is therefore of very limited taxonomic value especially as only slight sequence divergences (3%–10%) usually are correlated with much higher divergences of fragment patterns (40%–80%). Historically this method has been used to demonstrate the diversity of chloroplast DNA from higher plants, which show clearly different fragment patterns when DNAs from different species are digested with restriction endonuclease EcoRI (Atchison et al. 1976).

A reliable quantitation of similarity between less closely related DNA species would depend on a detailed mapping of very large numbers (> 100) of restriction sites which theoretically is possible; in view of the time required for such maps (many months or up to one year) and since restriction maps usually need confirmation by sequencing, restriction mapping certainly cannot be regarded as a routine method for taxonomy.

Comparative restriction maps with a limited number of restriction sites have been worked out for rRNA genes from several plants. Friedrich et al. (1979) have mapped the genes coding for cytoplasmic rRNA. This was facilitated by enrichment of rDNA directly from bulk DNA by density gradient centrifugation. This together with the universal occurrence and relatively high conservation of rDNA would qualify rDNA as an ideal candidate for taxonomic mapping. If direct isolation of rDNA is not feasible, cloning is a well established alternative since rRNA necessary as a hybridization probe for isolation of rDNA clones is easily accessible. Thus chloroplast rDNAs from several algae and higher plants have been cloned and mapped (for reviews see Bedbrook and Kolodner 1979, Rochaix 1981). In *Euglena* differences between restriction maps of rDNA from different strains have been observed (Wurtz and Buetow 1981). In Fig. 2 restriction maps of rDNAs from maize and mustard chloroplasts are compared (Link, Fritzsche and Kössel unpublished). Homology between individual restriction fragments of the two rDNAs was measured by cross hybridization as demonstrated in Fig. 3. This DNA/DNA cross hybridization is a valuable variant of the method originally developed by Southern (1975) for DNA/RNA hybridization; its application appears attractive whenever a detailed map or even a complete sequence of a cloned gene is available as reference – as it is the case for the maize chloroplast rRNA genes – which can then be used as a basis for selecting highly specific hybridization probes as outlined in Fig. 3. Only if such probes are available, assignment of the various DNA fragments to the respective gene regions – as for instance to a 16S rRNA gene or to its leader region – is possible, which in turn is a prerequisite for a taxonomic evaluation

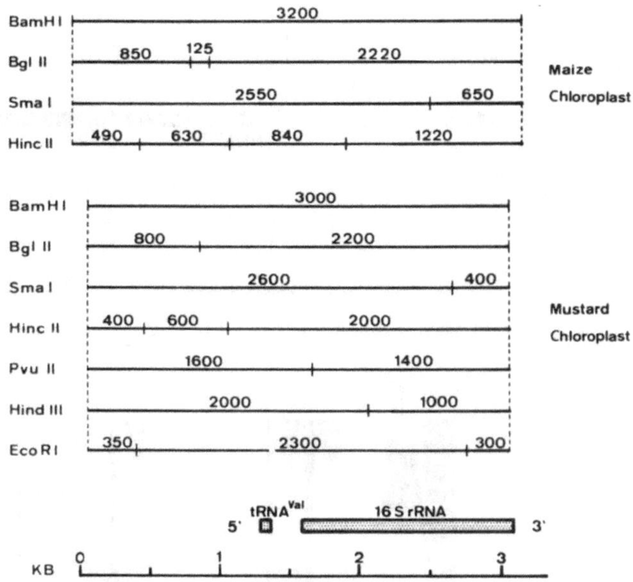

Fig. 2. Comparison of restriction maps from the 16S rDNA regions of maize and mustard chloroplasts (Fritzsche, Link and Kössel unpublished). Cleavage sites for the enzymes *PvuII, HindIII,* and *EcoRI* were observed only in the mustard but not in the maize DNA

of fragment positions and fragment sizes between two maps. Apart from the relatedness evident from the cross hybridization data shown in Fig. 3 a certain degree of relatedness is clearly evident also from a comparison of the two maps depicted in Fig. 2. However, numerical terms of expression of similarities between two maps – comparable to the association coefficients in oligonucleotide mapping (see Sect. 2.1) – have not been established as yet. Experimental error in fragment size determination is certainly one reason for this lack in quantitation. It is for instance difficult to judge from the respective gel patterns if fragment. BglII.800 from mustard is identical to fragment BlgII.850 from maize and similar uncertainty exists to greater or lesser extent for other fragment pairs (e.g. the entire BamHI fragments: 3,000 bp in mustard, 3,200 pb in maize, or the large BglII fragments: 2,200 bp in mustard, 2,220 bp in maize). Another reason is of course a relatively low sampling number – i.e. number of fragments compared – even if many more restriction enzymes than in Figs. 2 and 3 had been used. Finally, the percentage of conserved restriction sites – even if the sampling were of sufficient size – would reflect the actual sequence homology in an unknown and probably nonlinear way. For instance 16S rRNA from maize chloroplast and *E. coli* show 74% sequence homology but only less than one third of the restriction endonuclease cleavage sites are conserved between the two 16S rRNA species (Schwarz and Kössel 1980). Sequence conservation as high as 96.4% and 93% in the 16S (Schwarz and Kössel 1980, Tohdoh and Sugiura 1982) and in the spacer rDNA (Koch et al. 1981, Takaiwa and Sugiura 1982a,b) of maize and tobacco chloroplasts correlate to only 74% and 61% conservation of restriction sites. In summary the use of restriction endonuclease mapping so far appears of only limited value for taxonomic purposes except for providing a highly sensitive criteria for identity or near identity of two DNA species. Even for this limited purpose it should be kept in mind that identity of fragments between different DNA species can only be assigned to fragment

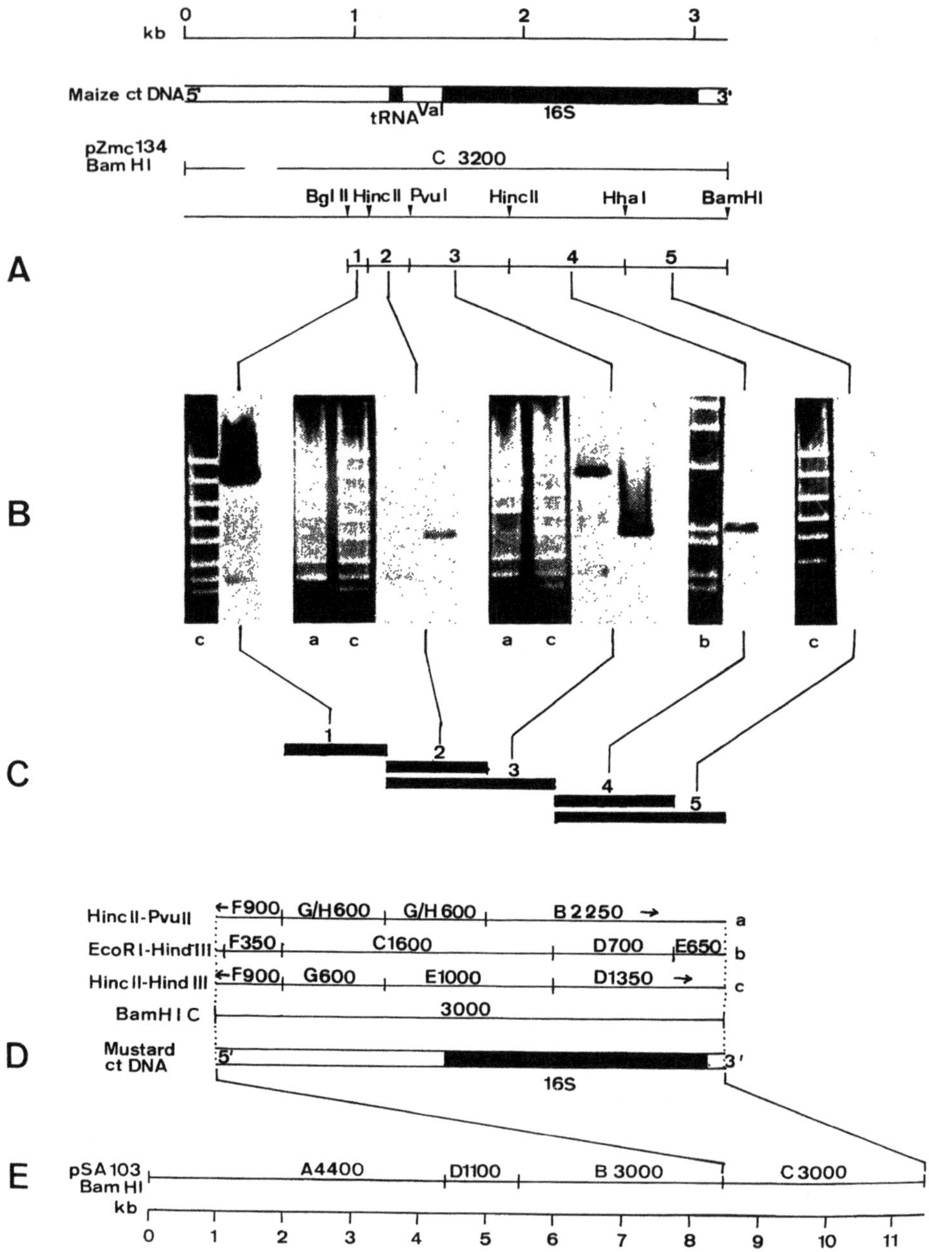

Fig. 3. Cross hybridization of ^{32}P labelled rDNA from maize chloroplast with rDNA from mustard chloroplasts. In **A** the maize rDNA fragments (1–5) used as hybridization probes are shown in correlation to 16S rDNA, the leader region and the tRNA^{Val} gene of maize. In **B** hybridization patterns are shown for fragments obtained from the mustard chloroplast 16S rDNA clone pSA 103 (Link, Fritzsche and Kössel unpublished) after cleavage with the enzyme mixtures HincII/PvuII (*a*), EcoRI/HindIII (*b*), and HincII/HindIII (*c*) followed by hybridization with the maize fragments 1–5 as radioactive probes. **B** Pictures with *white bands* on *black background* represent photographs of fragment patterns obtained after agarose gel electrophoresis and subsequent staining

sizes (with 5%–10% standard error) and to the very few positions of the enzyme recognition sequences but not to the sequences contained within the fragments. Homology between these sequences have to be confirmed either by cross hybridization or by sequencing.

2.3 DNA and RNA Sequence Analysis

Direct sequencing of large RNA species like 16S and 23S rRNA or mRNA for technical reasons is very difficult and therefore can not be regarded as a routine method In contrast to this, DNA sequencing techniques are now available which easily allow sequencing of entire operons or even small genomes. Thus, the majority of published nucleotide sequences are DNA sequences (ca. 1 Million bp. August 1983), and most of the RNA sequencing can now be reduced to sequencing of the corresponding DNAs (even in the case of tRNA and 5S rRNA sequences). However, direct RNA sequencing is indispensable for identification of modified bases and of termini of primary and mature transcripts (review by Sprinzl this volume).

In spite of this progress, DNA sequencing can still, practically speaking, not be recommended as a quick taxonomic method, since sequencing of a 1 kb DNA piece – especially of the final 5%–10% – often takes 6–12 months and more if cloning and restriction site mapping is included. Thus taxonomists will mainly depend on published sequences which have been and will be worked out also for other than taxonomic reasons. The advantages of nucleic acid sequence data – once they are worked out – as taxonomic tools are obvious. High resolution of subtle differences is readily achieved if long enough sequences are analysed. In principle a difference in 10 positions in a 1 kb DNA fragment would allow a 1% resolution (or in a 5 kb DNA fragment a 0.2% resolution), which is not possible if only small RNA sequences as from tRNAs or 5S rRNAs are compared. The selection of sequences can be adjusted to the individual taxonomic problem by using more rapidly diverging sequences ("faster going phylogenetic clocks") for comparison of more closely related species and more conserved DNA sequences for more distantly related species. Often such differential "clocks" are contained within one sequence unit; for instance third codon positions of protein coding sequences diverge more rapidly as compared to the first and second positions (Efstratiadis et al. 1980). Similarly, rRNA genes contain sequence domains of high and low divergence (Schwarz and Kössel 1980, Edwards and Kössel 1981). For general taxonomic purposes it is reasonable to use DNA sequences which code for products

Fig. 3. (continued)
with ethidium bromide. Pictures with *black bands* on *white background* are autoradiographs obtained after blotting the fragments onto nitrocellulose and hybridizing them with the radioactive probe DNA according to Southern (1975). *Probe 1* was derived from a subclone; besides the leader rDNA it contained pBR322 vector sequences (Zenke and Kössel unpublished) which lead to the strong band in the upper region of the autoradiographic picture (cross hybridization of vector DNA); only the lower weak band corresponding to the mustard fragment HincII/HindIII. G_{600} is due to chloroplast DNA cross hybridization as is the case for all autoradiographic bands of the other pictures. In C the fragments of mustard chloroplast rDNA hybridizing with the maize rDNA probes are summarized in correlation to the mustard rDNA map (D). The latter represents only the region of the fragment BamHI C_{3000}. Its relative position to the complete plasmid pSA103 and its BamHI cleavage sites is depicted in E

indispensible in a wide range of organisms. rRNA – especially the small ribosomal subunit RNA – meets this criteria, which is one reason why 16S rRNA sequences have already been used as a basis of the T_1-oligonucleotide mapping method (see Sect. 2.1). From the few examples of known complete rRNA genes and from the rRNA primary and secondary structures derived from these genes (see Sect. 3) a high taxonomic significance of any further rDNA sequences can easily be foreseen; especially since they contain more highly and less highly conserved domains, which may serve as differential parameters for relatedness. As comparison of related rRNA (rDNA) structures often gives valuable clues as to the function of individual rRNA domains it seems likely that the number of complete rDNA (rRNA) sequences, from different plant species, will increase substantially in the near future and that sooner or later the taxonomic significance of rDNA (rRNA) sequences will follow up or even supercede their protein counterparts.

Nucleotide sequences from protein coding genes of wide-spread occurrence such as the genes for the subunits of ribulose bisphosphate carboxylase (rubisco; see Wildman, this volume) of plants also appear well suited for taxonomic purposes. A comparison of the sequences coding for the large subunit of rubisco from maize and spinach chloroplasts (McIntosh et al. 1980, Zurawski et al. 1981) reveals a 10% divergence, and, as expected, a preferential divergence of third codon positions.

3 Structure and Comparison of rRNA Operons from Chloroplasts

A number of rRNA operons from chloroplasts of higher plants and algae have been mapped on the respective chloroplast DNAs. The operons from *Zea mays* and *Nicotiana tabacum* have been completely sequenced and selected regions from the *Chlamydomonas reinhardii* and *Euglena gracilis* rRNA operons have also been analyzed on the nucleotide level. In the following a comparison of chloroplast rRNA operons is presented in respect to

a) number and orientation of RNA operons per copy of plastid DNA;
b) nature and relative position of individual genes within the operon and in the flanking regions;
c) nucleotide sequences of individual genes;
d) secondary structures of rRNAs derived from the individual genes.

It will be demonstrated that useful taxonomic information can be gathered on all four levels.

3.1 Number and Orientation of rRNA Operons per Copy of Plastid DNA

Chloroplast DNAs from most higher plants contain rRNA genes within a pair of inverted repeat regions (Bedbrook and Kolodner 1979). For instance in maize chloroplasts two identical 22 kb inverted repeat regions are positioned on the plastid genome as depicted in Fig. 4a. Each inverted repeat includes within a 12 kb EcoRI fragment one copy of a 16S rRNA, 23S rRNA, 4.5S rRNA, and 5S rRNA gene and a 2.4 kb

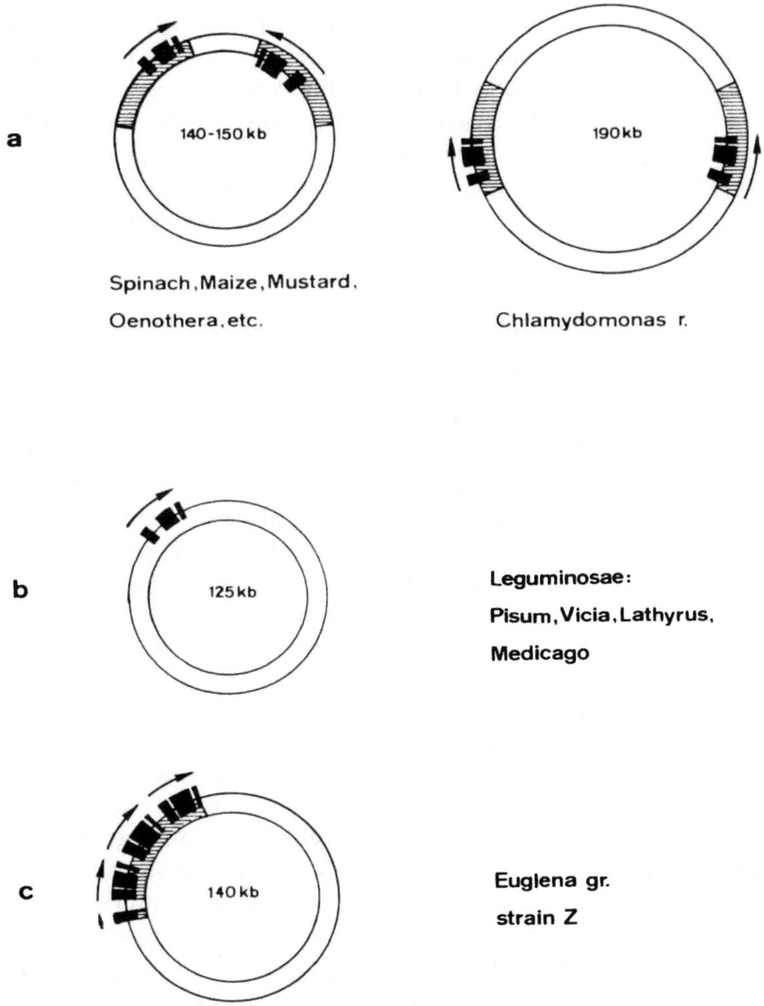

Fig. 4. Number, relative position, and orientation of rRNA operons on plastid DNA. In the center of the *circles* the sizes of plastid DNA is given in kb. Repeat regions (inverted, a, or in tandem array **c**) are marked by shading. rRNA genes are symbolized by *black blocks* (middle sized block for 16 S rDNA, *longer block* for 23 S rDNA, *small block* for 5 S rDNA. *Arrows* indicate the orientation of the rRNA operon in the direction of the genes coding for 16 S, 23 S, and 5 S. Note the extra 16 S rRNA in *Euglena* as symbolized by the first *(smaller) arrow*. For references see Bedbrook and Kolodner 1979, Palmer and Thompson 1982, Hallick 1982

spacer region between the 16S and 23S rRNA genes. A similar situation exists in chloroplast DNA from several other angiosperms (e.g. spinach, mustard, oenothera, tobacco) ferns and the alga *Chlamydomonas*. However, several notable exceptions from the inverted repeat situation have been found. One is the legume family, in which only a single copy of an rRNA operon is observed per plastid DNA (e.g. *Vicia faba:* Koller and Delius 1980, Fig. 4b). This situation has been confirmed for 10 species of this family and thus can be regarded as a useful phylogenetic marker (Palmer and Thompson 1982). A second exception is the alga *Euglena gracilis* which contains three copies of complete rRNA operons in tandem repeat and in certain strains one extra copy of a 16S rRNA gene (Fig. 4c). More recently an *Euglena* strain has been analyzed which deviates from this situation by having only one complete copy of an rRNA operon. Therefore, in *Euglena* the copy number of rRNA operons (together with variation in their restriction maps; see above) appears as a useful taxonomic marker even for strain differentiation (Wurtz and Buetow 1981, for review see Hallick 1982).

3.2 Nature and Relative Positions of Individual Genes Within the rRNA Operons and Their Flanking Regions

All of the chloroplast rRNA regions analyzed so far show the typically prokaryotic gene order of 16S, 23S, and 5S rDNA with a transcribed spacer between the 16S and 23S rRNA genes (Fig. 5). Also the existence of two spacer tRNA genes coding for $tRNA^{Ile}$ and $tRNA^{Ala}$ appears to be universal for chloroplast rDNA. However, superimposed on this basic skeleton, there appear elements characteristic for higher plant or algal rDNA. As indicated in Fig. 5 the region proximal to the 5′ end of the 16S rRNA gene contains a $tRNA^{Val}$ coding gene in maize (Schwarz et al. 1981), tobacco (Tohdoh et al. 1981), and mustard (Link, Fritzsche and Kössel unpublished) and a pseudo $tRNA^{Ile}$ gene in *Euglena* (Orozco et al. 1980). The $tRNA^{Val}$ sequences cannot be detected in the primary rRNA transcript (Kössel et al. 1982). However, their inclusion in the large precursor RNA followed by rapid processing or an inclusion during special regulatory states of the rRNA operon is difficult to exclude.

The spacer rDNAs from all higher plant chloroplasts are very long (1.6–2.4 kb), which is due to large intervening sequences in the two spacer tRNA genes (Koch et al. 1981, Takaiwa and Sugiura 1982b). The intervening sequences introduce a eukaryotic trait into the otherwise prokaryotic operons. However, the two spacer tRNA species as such which are also encoded in the *Euglena gracilis* spacer rDNA (but here without intervening sequences; see Fig. 5) are clearly of prokaryotic nature, as the same two species are observed in three of the seven *E. coli* rRNA operons (see Fig. 5), and in two of the ten *B. subtilis* rRNA operons (Loughney et al. 1982).

The 23S rRNA genes from chloroplasts phylogenetically show a tendency of being fragmented at their ends. For instance all ribosomes from higher plant chloroplasts contain within their large subunits an extra small rRNA species, termed 4.5S rRNA, in addition to 5S rRNA. From the position of the 4.5S rRNA genes within the operons and from sequence homology it is clear that 4.5S rRNA is a structural equivalent of the 3′ terminal region of prokaryotic 23S rRNA (Edwards et al. 1981; see Fig. 5). This 4.5S rRNA is, however, not present in chloroplast ribosomes from algae (*Euglena* and *Chlamydomonas*), liverworts, mosses and the fern *Adiantum* sp. (Bowman and Dyer 1979), but is again present in chloroplast ribosomes from the fern *Dryopteris acuminata* (Takaiwa et al. 1982). In *Chlamydomonas* chloroplast ribosomes other small rRNAs, a 3S and a 7S rRNA, are found. Again positional and sequence homology of the corresponding genes show that they represent equivalents of 5′ terminal regions of prokaryotic 23S rRNA (rDNA), (Rochaix 1981). It should be emphasized that all these small rRNAs are in all likelihood contained in the primary transcripts in which

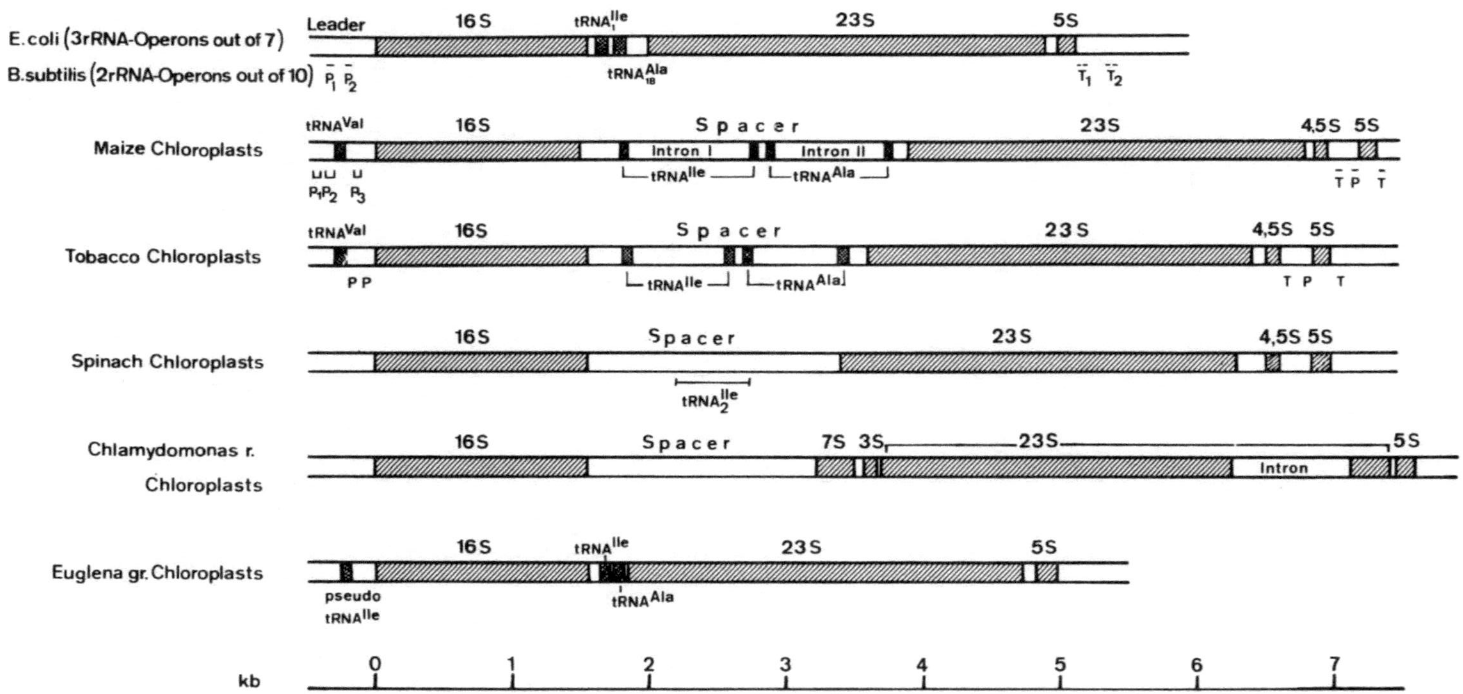

Fig. 5. Comparison of bacterial and plastid rRNA operons. The genes coding for the various rRNA species (16S, 23S, 5S, 3S, 7S, and 4.5S) are indicated. Promoter sites are designated by *P*. The chloroplast promoters should be regarded as tentative (Schwarz et al. 1981, Tohdoh et al. 1981). Terminator sites – also still tentative except for *E. coli* – are designated by *T*

they are therefore still connected with 23S rRNA in the form of a precursor rRNA. Thus fragmentation of plastid 23S rRNA genes occurs on the rRNA precursor level and probably is signalled by special sequences introduced during evolution (on the DNA level) between the 23S and the 4.5S and 7S rRNA genes respectively.

The 5S rRNA genes which are always positioned at or near the distal ends of plastid rRNA operons are probably separate transcription units as 5S rRNA sequences cannot be detected within the large primary transcription product (Hartley 1979, Kössel et al. 1982). This (but not the position of the 5S rRNA genes) is in contrast to the *E. coli* rRNA operons in which termination of the primary transcripts occurs distal to 5S rRNA genes. About 0.9 kb distal to the tobacco chloroplast 5S rRNA gene a $tRNA^{Asn}$ gene is observed (Kato et al. 1981). As many of the *E. coli* rRNA operons also contain "trailer" tRNA genes (one coding for a $tRNA^{Asp}$) it appears likely that such flanking tRNA genes are common to many chloroplast rRNA operons.

In summary, all the extra genetic elements of plastid rRNA operons such as the $tRNA^{Val}$ gene, the intervening sequences of the spacer tRNA genes, the small (4.5S, 3S, 7S) rRNA genes and perhaps trailer tRNA genes appear very suitable as taxonomic criteria of higher plants and algae even if their individual nucleotide sequences are not determined.

3.3 Nucleotide Sequences of Individual Genes

The availability of the complete rRNA operon sequences from maize (Schwarz and Kössel 1980, Koch et al. 1981, Edwards and Kössel 1981, Schwarz et al. 1981, Edwards et al. 1981) and from tobacco (Tohdoh et al. 1981, Tohdoh and Sugiura 1982, Takaiwa and Sugiura 1982a,b, Takaiwa and Sugiura 1980) chloroplasts together with sequences from *E. coli* rRNA operons have allowed a comparison of the individual rRNA and tRNA genes as well as leader and spacer regions with maximum resolution. Homologies in terms of percentage of identical nucleotide positions are listed in Table 1 for various regions of rRNA operon sequences. It is noteworthy that all the structural genes (rRNA genes and tRNA genes, the latter even in the split form of spacer tRNA genes) show maximum homology (92% and higher) between the two higher plant somewhat less homology (80%–90%) between the higher plants and the alga *Euglena gracilis* and even less (but still strong) homology (65%–77%) with the prokaryotic counterparts from *E. coli*. In Figs. 6 and 7 examples are depicted for homology between *E. coli* and maize chloroplast 16S rDNA sequences. The high conservation of rDNA and tDNA is not unexpected in view of the strong functional constraints resting on rRNA and tRNA structures. On the other hand, a more rapid divergence of intercistronic regions is also obvious from a comparison of the respective nucleotide sequences. In Fig. 7 for instance the abrupt transition from the highly conserved 3′ terminal region of 16S rDNA to the highly diverged 16S/23S spacer is shown for the maize chloroplast/*E. coli* sequence pair. In this case no homology between the two spacer sequences is apparent. However, the leader and 16S/23S spacer regions still show 83%–93% conservation if sequences between maize and tobacco are compared. Only very scarce homologies and large size variations are apparent from a comparison of algal and bacterial leader and spacer sequences with the higher plant leader and spacer sequences (the tRNA gene sequences contained in these regions have

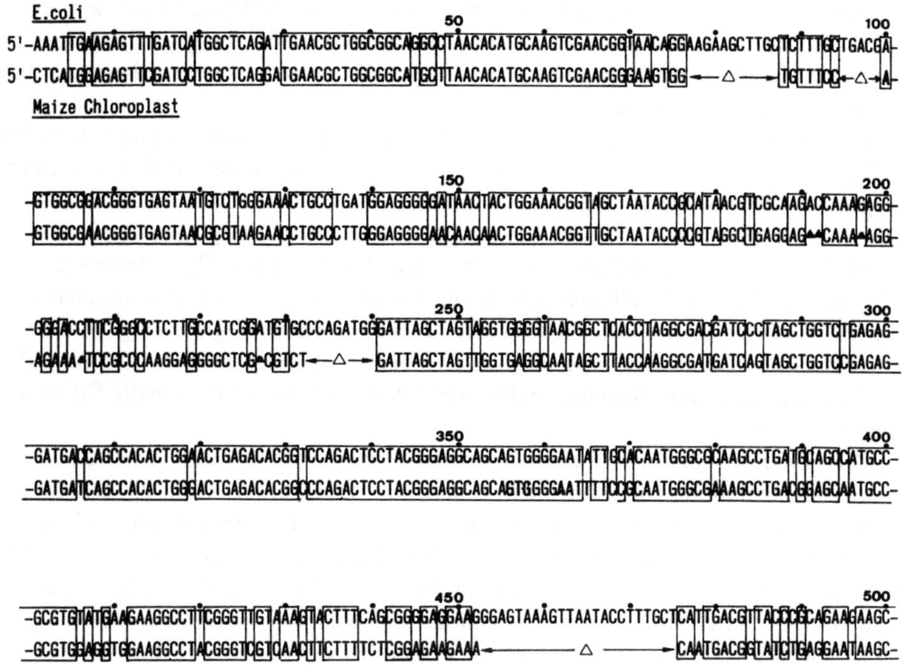

Fig. 6. Alignment of 16 S rDNA sequences from *E. coli* and *Zea mays* chloroplasts. Homologous sequences are framed. Single point deletions and larger deletions in the maize sequence are marked by ▲ and △, respectively. Only the 5′-terminal thirds of 16 S rDNA are shown. (Schwarz and Kössel 1980)

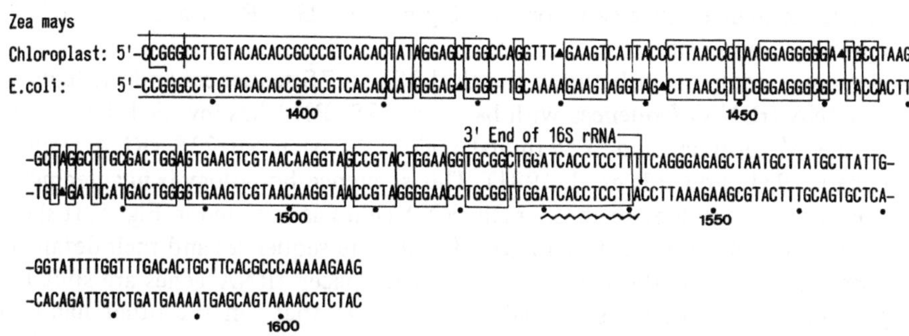

Fig. 7. Alignment of the 3′-terminal regions of 16 S rDNA and of the adjacent spacer rDNA from *E. coli* and maize chloroplasts. Homologous sequences are framed, single point deletions are marked by ▲. Sequences at the 3′ end of the 16 S rRNA gene, which are known to interact in the form of rRNA with translational start sites of mRNA are marked by ∼∼∼. (Schwarz and Kössel 1979)

to be excluded from the comparison as they represent structural genes). Thus within the plastid rRNA operons one could conveniently select regions of high and low divergence according to the taxonomic task at stake. In addition to the sequence from a monocotyledon (maize), a dicotyledon (tobacco, and an alga *(Euglena)*, it would be interesting to compare respective sequences from representatives of fern and gymnosperm chloroplast rDNAs. They would probably nicely fill the gap of apparent nonhomology which is now observed when leader and spacer sequences between higher plants and algae are compared.

Differential divergence is also observed within individual gene sequences. As listed in Table 1 the 5′ terminal half of maize 23S rDNA shows only 68% conservation in respect to the corresponding *E. coli* sequence, whereas the 3′ terminal halves from the two species show 73.8% homology (and 85% in the peptidyl transferase domain). Similarly in the 16S rDNAs the 5′ terminal thirds (especially between positions 70–250; see Fig. 6) in all cases show much lower conservation as compared to the central thirds or to the last 70 positions of the 3′ terminal regions (Fig. 7). The high conservation of the latter region was proposed to reflect its function for mRNA initiator site selection (Schwarz and Kössel 1980) which is characteristic for prokaryotic ribosomes.

Divergence of the rRNA genes is not only due to single base exchanges. Besides single base deletions (or insertions) which are often compensated by nearby single base insertions (or deletions) several larger deletions are observed when plastid 16S rDNAs are compared with *E. coli* 16S rDNA (Figs. 6 and 7). One of them, a 23 base pair deletion at positions 453–475 (Fig. 6), appears to be universal for chloroplast 16S rDNA (and rRNA), whereas others around positions 77–99 and 180–240 seem to be specific for algae or higher plants respectively (Graf et al. 1982). On the other hand, 23S rDNA from maize shows several larger insertions (25, 65, and 78 base pairs) when compared with *E. coli* 23S rDNA (Edwards and Kössel 1981). The 65 bp insertion seems to be specific for monocotyledons as it is not found in the tobacco 23S rDNA (Takaiwa and Sugiura 1982a). The 78 bp insertion separates the mature 23S rRNA coding region from the 4.5S rRNA coding region and apparently encodes the signal sequences necessary for splitting mature 23S rRNA and 4.5S rRNA from the primary transcript.

Sequence comparison between the algal 3S and 7S sequences and the higher plant 4.5S rRNA (rDNA) sequences with bacterial 23S rRNA has revealed their homology with the 5′ terminal and 3′ terminal regions of bacterial 23S rRNA respectively (Rochaix 1981, Edwards et al. 1981). This sequence homology is in agreement with the positional homology of these small rRNA genes as outlined in Fig. 5. It should be pointed out that only the availability of complete sequences and their detailed comparison have allowed the conclusion that the spacer tRNA genes are split by very large intervening sequences in higher plants; and that, on the other hand, strong homology exists between their exon regions and spacer tRNA genes from algal and bacterial rRNA operons.

A final evaluation of plastid (and perhaps cytoplasmic) rRNA operon sequences as a quantitative taxonomic tool must await more sequences – preferably 16S rDNA sequences – from other species. As alignment of any random sequences of similar base compositions would show 25% identical positions, it appears desirable to define homology more precisely either by subtracting this 25% noise level or by using numerical terms by which somehow a coupling with the environment of each sequence position is achieved. In other words, single points of homology between two sequences should contribute less to a homology parameter than positions embedded in long stretches of homology. In spite of these reservations, the comparison of the few known rDNA sequences from chloroplasts allow an optimistic view in regard to the taxonomic value of rDNA sequences in particular, and of DNA sequences in general.

3.4 Secondary Structures of rRNAs

Secondary structures have been proposed for both 16S (review: Stiegler et al. 1981) and 23S (Glotz et al. 1981, Branlant et al. 1981) ribosomal RNAs. Compensating base changes were used as one criteria which in many cases allow identical base paired structures between different rRNA species in spite of primary structure divergences (see Figs. 8 and 9). For this reason rRNA secondary structures are in general more highly conserved as compared to the underlying primary structures. In Figs. 8 and 9 the 5′ terminal domains of 16S and 23S rRNA secondary structures are depicted. In a few cases entire stem loop structures appear either deleted or newly introduced when the secondary structures of plastid rRNAs are compared with each other or with their bacterial counterparts. For instance the deletion of 23 bp from maize chloroplast 16S rDNA (positions 453–475 in the corresponding *E. coli* structure, see Fig. 6) "amputates" the secondary structure by one complete stem loop structure (Fig. 8). From other small deletions within the 5′ terminal region of 16S rDNA only size reductions of the corresponding stems are observed in both higher plant and algal 16S rRNA (deletion at position 77–99) or only in higher plant 16S rRNA (deletion between position 180–250). In the 23S rDNA from maize chloroplasts three insertions of 25, 65, and 78 bp are observed as compared to the *E. coli* sequence. All three cause the insertion of extra stem loop structures in the 23S rRNA secondary structure from maize chloroplasts. The 25 bp insertion is positioned within the 5′ terminal region which on the RNA secondary structure level corresponds to the first domain depicted in Fig. 9. The extra stem loop structure caused by the 25 bp insertion is depicted in the insert of this figure (top right). The 65 bp insertion (positioned in the central region of maize 23S rDNA; Edwards and Kössel 1981) and the corresponding stem loop structure is not present in the tobacco 23S rRNA structure (Takaiwa and Sugiura 1982a,b). The possibility exists that this stem loop structure is

Fig. 8. Secondary structure of 16S rRNA (5′-terminal domain). The principal sequence joined by *dots* is that of *E. coli* 16S rRNA (Brimacombe et al. 1983). This sequence is numbered every 10 bases from the 5′ end according to the DNA sequence depicted in Fig. 6. Base changes in the maize chloroplast 16S rRNA are indicated by nucleotides in *square* or *round boxes*; these latter sequences are numbered *(in brackets)* every 50 bases. Base changes in *square boxes* are those which are compensating or which enhance the secondary structure, whereas those in *round boxes* are in single-stranded regions or are non-compensating. *Dotted lines* indicate modified base-pairing in the maize sequence and a *bar* crossing out a base pair indicates that this base pair is not present in the maize sequence. A base change with an *arrow* pointing between two bases of the *E. coli* sequence is an insertion; single point deletions are marked by ▲. Major differences between the *E. coli* and maize structure are marked by *framing* (e.g. the 23 base deletion in the maize sequence) or are shown within *separate boxes*. Note that the corresponding primary structures of both *E. coli* and maize rDNA are depicted in Fig. 6

Fig. 9. Secondary structure of 23S rRNA (5′-terminal domain). For explanation of symbols see previous figure. A major difference between the *E. coli* and maize structure which is due to a 25 base insertion in the maize sequence is shown in the *boxed insert* on *top right*. (Glotz et al. 1981, Branlant et al. 1981)

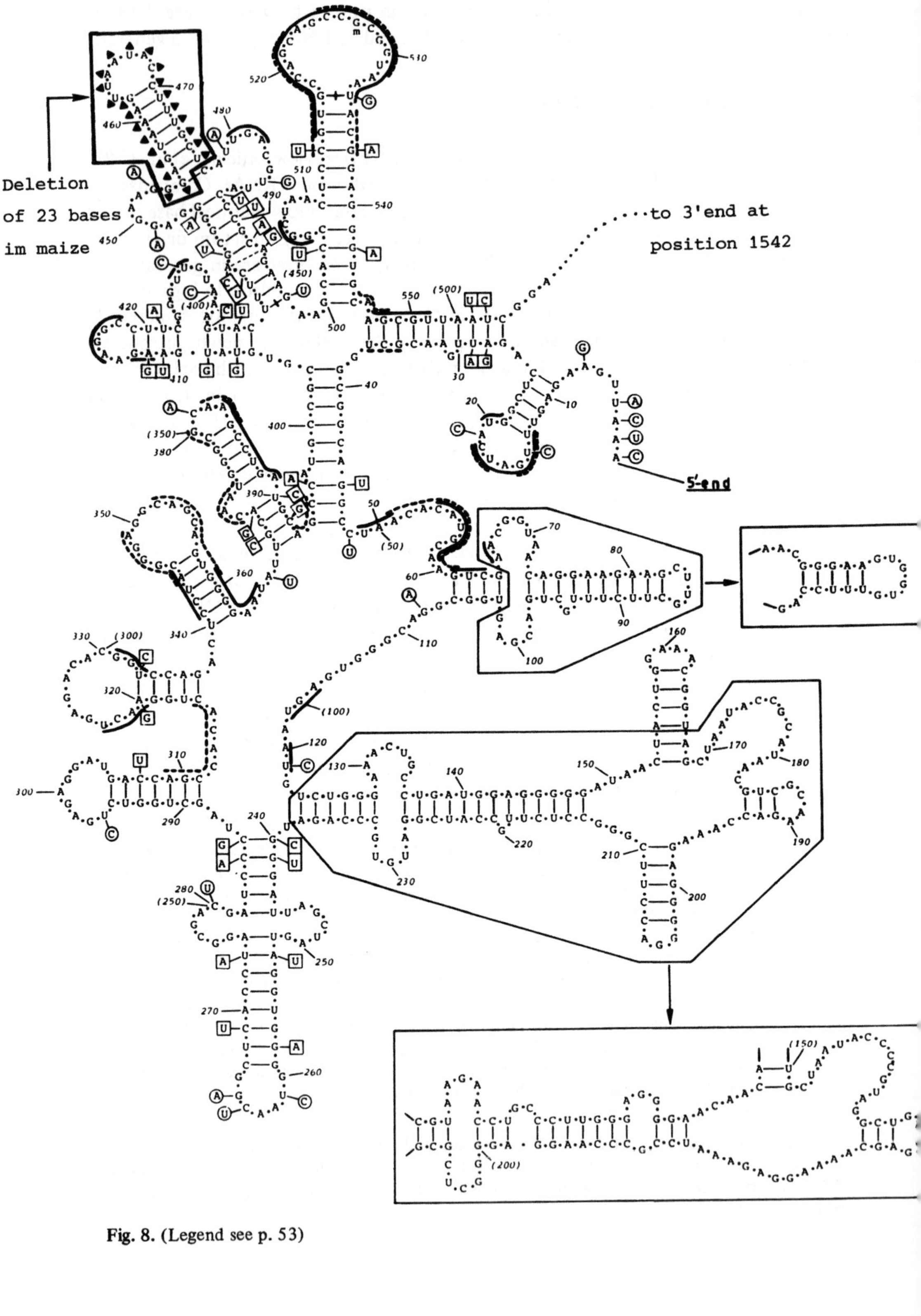

Fig. 8. (Legend see p. 53)

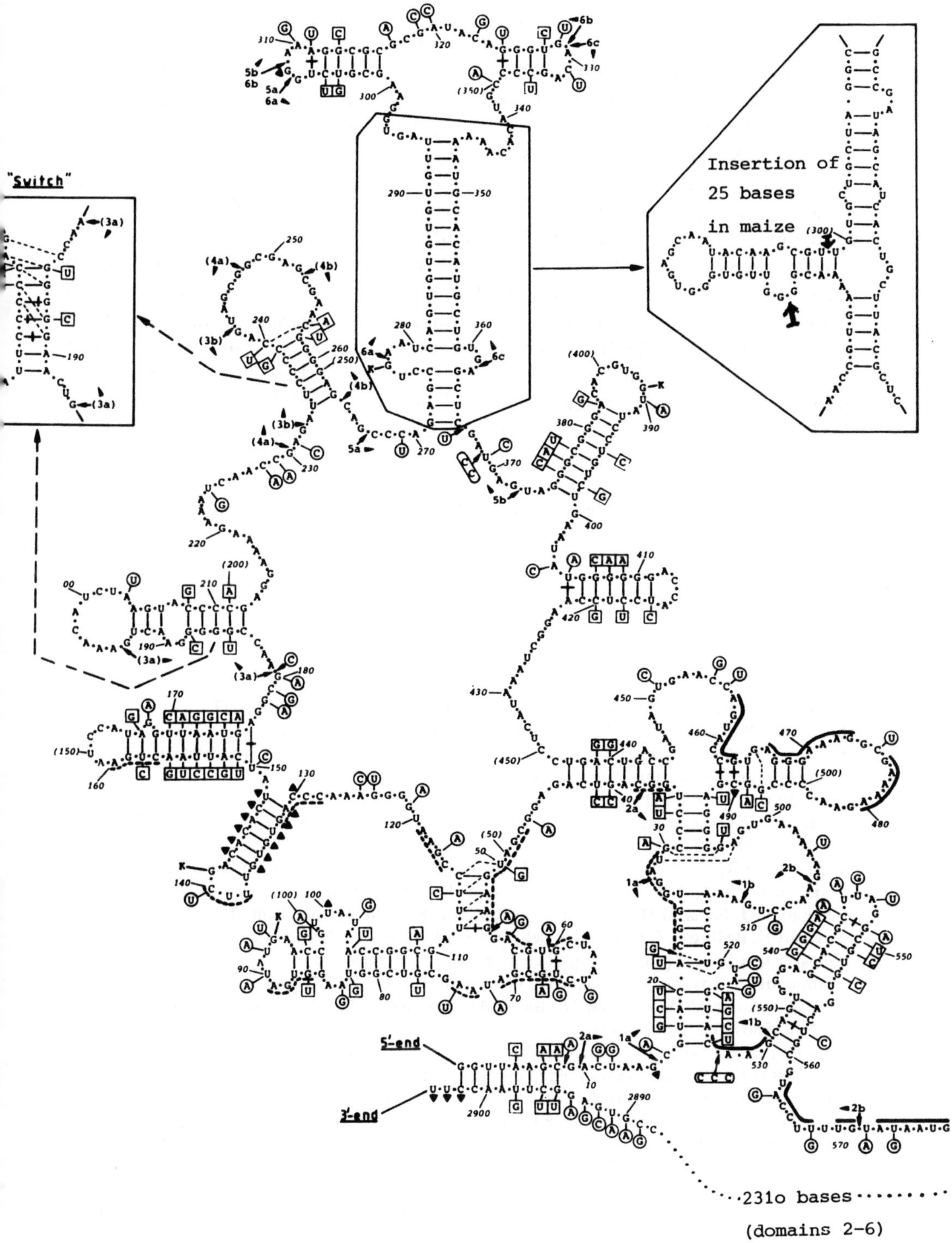

Fig. 9. (Legend see p. 53)

present in maize only in the precursor 23S rRNA and that it is spliced out during processing. It remains to be seen if this inserted stem loop structure (within the precursor or mature form) is also present in other monocotyledons and absent in other dicotyledons, or if it can be used as a taxonomic marker even between closer relatives. The 78 bp insert which is positioned near the 3′ end of 23S rDNA separates the 23S rRNA gene from the 4.5S rRNA gene. On the RNA level this insert and its secondary structure exists only as an intermediate within the primary transcript and is excised during maturation of the precursor to yield 23S and 4.5S rRNA.

Besides these larger insertions in 23S rRNA and deletions in 16S rRNA several smaller insertions and deletions, e.g. in 23S rRNA at positions 130–150 (see Fig. 9) are observed throughout the two molecules which probably can all serve as taxonomic markers. This view is supported by comparative studies between secondary structures of 16S rRNAs and 23S rRNAs from several species which have already been published (Stiegler et al. 1981, Glotz et al. 1981, Brimacombe et al. 1983). In conclusion it appears feasible that many of the "branches" and "leaves" of the secondary structure "trees" as depicted in Figs. 8 and 9 can serve as taxonomic markers in analogous ways to leaves and stems of plants on the macroscopic level.

Acknowledgements. The technical assistance of Mrs. E. Schiefermayr throughout our research reported herein is gratefully acknowledged. This work was supported by the Deutsche Forschungsgemeinschaft (SFB 46), and from the Fonds der Chemischen Industrie, Germany.

References

Atchison BA, Whitfeld PR, Bottomley W (1976) Mol Gen Genet 148:263–269

Bedbrook JR, Kolodner R (1979) Annu Rev Plant Physiol 30:593–620

Boulter D (1974) In: Bendz G, Santesson J (ed) Chemistry in botanical classification. Proc 25th Nobel Symp. Academic Press, London New York, pp 211–216

Bowman CM, Dyer TA (1979) Biochem J 183:605–613

Branlant C, Krol A, Machatt MA, Pouyet J, Ebel JP, Edwards K, Kössel H (1981) Nucleic Acids Res 9:4303–4324

Brimacombe R, Maly P, Zwieb C (1983) In: Cohn WE (ed) Progress in nucleic acid research and molecular biology. Academic Press, London New York, in press

Cedergren RJ, Samkoff D, LaRue B, Grosjean H (1981) Crit Rev Biochem 11:35–104

Edwards K, Bedbrook J, Dyer T, Kössel H (1981) Biochem Int 2:533–538

Edwards K, Kössel H (1981) Nucleic Acids Res 9:2853–2869

Efstratiadis A, Posakony JW, Maniatis T, Lawn RM, O'Connell C, Spritz RA, DeRiel JK, Forget BG, Weissman SM, Slightom JL, Blechl AE, Smithies O, Baralle FE, Shoulders CC, Proudfoot NJ (1980) Cell 21:653–668

Ehrendorfer F (1983) In: Jensen U, Fairbrothers DE (eds) Proteins and nucleic acids in plant systematics. Springer, Berlin Heidelberg New York

Eperon IC, Anderson S, Nierlich DP (1980) Nature (London) 286:460–467

Fox GE, Stackebrandt E, Hespell RB, Gibson J, Maniloff J, Dyer TA, Wolfe RS, Balch WE, Tanner RS, Magrum LJ, Zablen LB, Blakemore R, Gupta R, Bonen L, Lewis BJ, Stahl DA, Luehrsen KR, Chen KN, Woese CR (1980) Science 209:457–463

Friedrich HH, Hemleben V, Key JL (1979) Plant Syst Evol Suppl 2:73–88

Glotz C, Zwieb C, Brimacombe R, Edwards K, Kössel H (1981) Nucleic Acids Res 9:3287–3306

Graf L, Kössel H, Stutz E (1980) Nature (London) 286:908–910

Graf L, Roux E, Stutz E, Kössel H (1982) Nucleic Acids Res 10:6369–6381

Hallick RB (1982) In: Buetow DE (ed) Biology of Euglena 4. Academic Press, London New York, in press
Hartley MR (1979) Eur J Biochem 96:311–320
Jensen U (1981) In: Ellenberg H, Esser K, Kubitzki K, Schnepf E, Ziegler H (eds) Prog Bot 43:344–369
Kato A, Shimada H, Kusuda M, Sugiura M (1981) Nucleic Acids Res 9:5601–5607
Koch W, Edwards K, Kössel H (1981) Cell 25:203–213
Kössel H, Edwards K, Koch W, Langridge P, Schiefermayr E, Schwarz Zs, Strittmatter G, Zenke G (1982) Nucleic Acids Res Symp Ser 11:117–120
Koller B, Delius H (1980) Mol Gen Genet 178:261–269
Loughney K, Lund E, Dahlberg JE (1982) Nucleic Acids Res 10:1607–1624
McIntosh L, Poulsen C, Bogorad L (1980) Nature (London) 288:556–560
Orozco EM, Rushlow KE, Dodd JR, Hallick RB (1980) J Biol Chem 255:10997–11003
Osawa S, Hori H (1979) In: Chambliss G, Craven GR, Davies J, Davis K, Kahan L, Nomura M (eds) Ribosomes: Structure, function, and genetics. Univ Park Press, Baltimore, pp 333–355
Palmer JD, Thompson WF (1982) Cell 29:537–550
Rochaix JD (1981) Experientia 37:323–332
Schwartz RM, Dayhoff MO (1978) Science 199:395–403
Schwarz Zs, Kössel H (1979) Nature (London) 279:520–522
Schwarz Zs, Kössel H (1980) Nature (London) 283:739–742
Schwarz Zs, Kössel H, Schwarz E, Bogorad L (1981) Proc Natl Acad Sci USA 78:4748–4752
Southern EM (1975) J Mol Biol 98:503–517
Sprinzl M (1983) In: Jensen U, Fairbrothers DE (eds) Proteins and nucleic acids in plant systematics. Springer, Berlin Heidelberg New York
Stackebrandt E, Ludwig W, Schleifer KH, Gross HJ (1981) J Mol Evol 17:227–236
Stiegler P, Carbon P, Ebel JP, Ehresmann C (1981) Eur J Biochem 120:487–495
Stöcklein L, Ludwig W, Schleifer KH, Stackebrandt E (1983) In: Jensen U, Fairbrothers DE (eds) Proteins and nucleic acids in plant systematics. Springer, Berlin Heidelberg New York
Takaiwa F, Kusuda M, Sugiura M (1982) Nucleic Acids Res 10:2257–2260
Takaiwa F, Sugiura M (1980) Mol Gen Genet 180:1–4
Takaiwa F, Sugiura M (1982a) Eur J Biochem 124:13–19
Takaiwa F, Sugiura M (1982b) Nucleic Acids Res 10:2665–2676
Tohdoh N, Shinozaki K, Sugiura M (1981) Nucleic Acids Res 9:5399–5406
Tohdoh N, Sugiura M (1982) Gene 17:213–218
Van Etten RA, Walberg MW, Clayton DA (1980) Cell 22:157–170
Wurtz EA, Buetow DE (1981) Current Genet 3:181–187
Zablen LB, Kissil MS, Woese CR, Buetow DE (1975) Proc Natl Acad Sci USA 72:2418–2422
Zurawski G, Perrot B, Bottomley W, Whitfeld PR (1981) Nucleic Acids Res 9:3251–3270

Comparative Oligonucleotide Cataloguing of 18S Ribosomal RNA in Phylogenetic Studies of Eukaryotes

L. STÖCKLEIN, W. LUDWIG, K.H. SCHLEIFER, and E. STACKEBRANDT[1]

Abstract. The phylogenetic relationships of various plant species, and *Xenopus laevis, Saccharomyces cerevisiae*, and *Dictyostelium discoideum* were determined by comparing R Nase T_1-resistant oligonucleotides of their 18S ribosomal RNA. The results highly demonstrate the usefulness of this technique in phylogenetic studies of eukaryotes.

1 Introduction

Ribosomal ribonucleic acids fulfil all requirements of a reliable phylogenetic marker (see Kössel et al. this volume).

Especially the choice of the ribosomal RNA of the small subunit of ribosomes of prokaryotes (16S rRNA) has proved to be valuable, and the comparison of oligonucleotides of the size of hexamers and larger, generated by ribonuclease T_1 has proved to be informative. Their comparative analysis revealed data which provided new insights into our understanding of the early evolution of prokaryotes, and made it possible to arrange bacteria in a feasible natural system of relationships for the first time in the history of microbiology (Woese and Fox 1977, Fox et al. 1980, Stackebrandt and Woese 1981). So far, more than 350 bacterial species, but only a small number of eukaryotes, have been investigated by comparing the 16S and the 18S rRNA, respectively. The 18S rRNA of a few eukaryotes have been sequenced completely (Salim and Maden 1981, Rubtsov et al. 1980) and the oligonucleotide catalogue can be derived from the respective primary structure.

In this short communication we demonstrate the usefulness of the in-vitro labelling technique for sequencing oligonucleotides of the eukaryotic 18S rRNA by presenting preliminary data on the phylogenetic connections of some vascular plants and the slime mould *Dictyostelium discoideum*.

1 Lehrstuhl für Mikrobiologie, Technische Universität München, Arcisstraße 16, D-8000 München, FRG

Proteins and Nucleic Acids in Plant Systematic
ed. by U. Jensen and D.E. Fairbrothers

2 Material and Methods

All vascular plant species (Table 1) were cultivated heterotrophically as tissue cultures in batches of 300 ml[2)]. Cells were disrupted by passing through a precooled french pressure cell at 1,270 kg/cm^2. 18S rRNA was isolated following described methods (Stackebrandt et al. 1981), including one-dimensional polyacrylamide gel electrophoresis and electrophoretic elution.

18S rRNA was digested with RNase T_1 under controlled conditions and the 3′-terminal phosphates were removed by calf intestine phosphatase. Five μg of the digest was used for enzymic 5′-end group labelling, using ^{32}P-ATP and polynucleotide kinase (Silberklang et al. 1979, Stackebrandt et al. 1981). Labelled oligonucleotides were separated by high voltage electrophoresis on cellulose acetate strips in the first dimension (Sanger et al. 1965, Stackebrandt et al. 1982) and ammonium-formate gradient chromatography on DEAE cellulose plates in the second dimension (Silberklang et al. 1977, Stackebrandt et al. 1982). Sequence analysis of oligonucleotides of the size of heptamers and larger was performed by controlled alkaline hydrolysis, followed by a two dimensional separation of fragments by high voltage electrophoresis and homochromatography (Jay et al. 1974, Silberklang et al. 1979, Stackebrandt et al. 1981, 1982). The similarity coefficient (S_{AB*}-value) for each pair of oligonucleotide catalogue was calculated according to Fox et al. (1977). S_{AB*}-values are here defined as twice the number of nucleotides found in sequences, *heptamers* and larger, common to any pair of catalogues, A and B, divided by the total number of nucleotides in all sequences, heptamers and larger, in the two catalogues. They are designated S_{AB*}-values to distinguish them from the S_{AB}-values (see Kössel et al. this volume) which are based on oligonucleotides of the size of *hexamers* and larger. Average linkage cluster analysis (among the merged groups) was used to obtain the values required to construct a dendrogram.

3 Results and Discussion

The number of total bases of a 18S rRNA exceeds that of a 16S rRNA by about 300 to 350. Therefore, the number of RNase T_1 resistant oligonucleotides, used in the calculation of similarity coefficients, is also increased. Nevertheless, it could be shown that the in-vitro labelling technique, developed for the separation and sequencing of oligonucleotides of the 16S rRNA is also useful for the 18S rRNA without major modifications. However, the procedure is more time-consuming and laborious because of the larger number of oligonucleotides requiring more secondary and tertiary analyses. The only problem encountered concerns the exact sequence determination of oligonucleotides carrying posttranscriptional modifications, especially when the base residue is modified.

---

2 Plant tissue cultures and *D. discoideum* were a generous gift from Prof. M.H. Zenk (Institute of Pharmaceutical Biology, University of Munich, FRG) and Prof. G. Gerisch (Max Planck Institute of Biochemistry, Martinsried, FRG), respectively

Table 1. Similarity coefficients (S_{AB^*}-values) among organisms tested. To include *Xenopus laevis* and *Saccharomyces cerevisiae*, a RNase T_1 catalogue was generated from the primary structure of their 18S rRNA

Organisms	1	2	3	4	5	6	7	8	9	10
1 *Nicotiana tabacum*		0.89	0.87	0.85	0.74	0.67	0.60	0.33	0.26	0.17
2 *Galium mollugo*	0.89		0.87	0.85	0.73	0.66	0.60	0.33	0.26	0.18
3 *Catharanthus roseus*	0.87	0.87		0.84	0.77	0.66	0.63	0.35	0.26	0.18
4 *Ruta chalepensis*	0.85	0.85	0.84		0.77	0.66	0.60	0.32	0.25	0.17
5 *Dioscorea composita*	0.74	0.73	0.77	0.77		0.65	0.63	0.32	0.27	0.18
6 *Lemna minor*	0.67	0.66	0.66	0.66	0.65		0.54	0.31	0.25	0.18
7 *Abies grandis*	0.60	0.60	0.63	0.60	0.63	0.54		0.32	0.27	0.16
8 *Xenopus laevis*	0.33	0.33	0.35	0.32	0.32	0.31	0.32		0.27	0.16
9 *Saccharomyces cerevisiae*	0.26	0.26	0.26	0.25	0.27	0.25	0.27	0.27		0.15
10 *Dictyostelium discoideum*	0.17	0.18	0.18	0.17	0.18	0.18	0.16	0.16	0.15	

Table 1 shows the S_{AB^*}-values of the organisms investigated[3]. The oligonucleotide catalogues of the plants and of *D. discoideum* will be presented in a different context. Figure 1 is a dendrogram derived from the S_{AB^*}-values.

Although it is not possible to draw definitive phylogenetic deductions from this dendrogram, conceived only on the basis of 11 species randomly selected from plants, animals, and fungi, a preliminary systematic evaluation of the use of 18S rRNA data can be made.

All dicotyledons investigated form a coherent cluster ($S_{AB^*} > 0.83$), in which for the most closely related species *Nicotiana tabacum* and *N. silvestris* (S_{AB^*}-values not shown in Table 1) the most prominent similarity values have been detected ($S_{AB^*} = 1.0$).

The 18S rRNA branching pattern is in accord with proposed phylogenetic schemes of Ehrendorfer (1978) and Cronquist (1981), in which the Asteridae (all dicotyledons investigated, except *Ruta chalepensis*) are distinct from the Rosidae (represented here by *R. chalepensis*). The two species of the monocotyledons investigated, *Lemna minor* and *Dioscorea composita*, are not specifically related (S_{AB^*}-value of 0.65, Table 1). *Lemna minor* branches off earlier from the line of descent leading to the

3 The unpublished catalogue of *Lemna minor* 18S rRNA was kindly provided by Prof. C.R. Woese, University of Illinois, Urbana

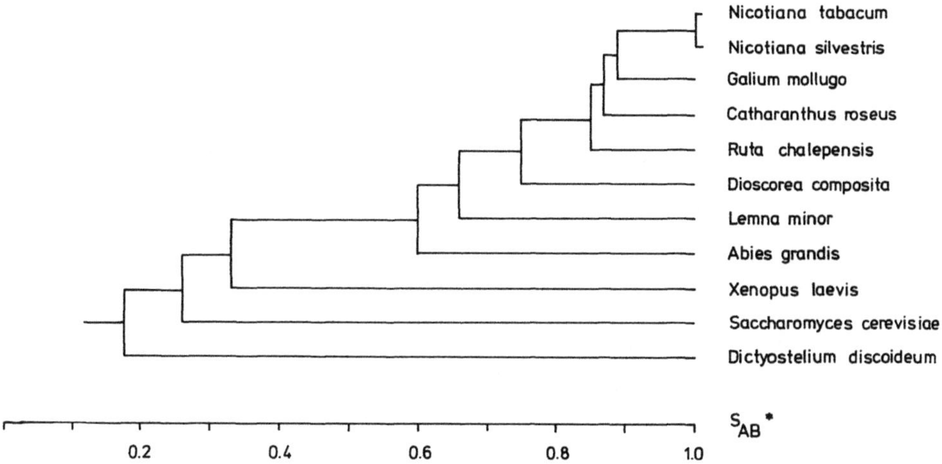

Fig. 1. Dendrogram of 18S rRNA similarities using 8 species of plants, one metazoan *(Xenopus laevis)*, one fungus *(Saccharomyces cerevisiae)*, and one slime mould *(Dictyostelium discoideum)*

dicotyledons than does *D. composita* which reveals higher S_{AB*}-values toward the dicotyledons (values ranging between 0.73 and 0.77). In this respect it is interesting to note that members of Dioscoreaceae have certain characteristics similar to those of dicotyledons, e.g. leaves with distinct stems and verticulate venation as well as a vestigial second cotyledon (Dahlgren and Clifford 1982).

The lowest S_{AB*}-values found among the plant representatives tested are those formed with *Abies grandis* (0.54–0.63), which corresponds with the distinct separation between the gymnosperms and the angiosperms.

Figure 1 also shows the branching points of those lines of descent leading to metazoa, fungi and the slime mould *Dictyostelium discoideum*, demonstrating once again the completely separated position of the fungi from the vascular plants. Due to the limited number of organisms investigated by our method and the even smaller number of organisms already investigated in respect to similarities of their 18S rRNA, it is not possible to definitely compare the phylogenetic tree of the 18S rRNA to the comprehensive ones based on cytochrome c (Fitch 1976, Ramshaw et al. 1974) and the 5S rRNA (Küntzel et al. 1981). However, a fairly good correspondence has been detected with the phylogenetic tree derived from the analysis of cytochrome c (Fitch 1976), while comparison with other trees, also derived from cytochrome c (Ramshaw et al. 1974), and from 5S rRNA (Küntzel et al. 1981) show differences, especially in the order of branching points leading to the various divisions. The inclusion of more representatives of these divisions together with those of algae and protozoa will allow more precise determination of the branching points, and will improve the evaluation of the RNA characters.

Acknowledgement. This study was supported by the Deutsche Forschungsgemeinschaft.

References

Cronquist A (1981) An integrated system of classification of flowering plants. Columbia Univ Press, New York

Dahlgren R, Clifford HT (1982) The monocotyledons. A comparative study. Academic Press, London New York

Ehrendorfer F (1978) In: von Denffer D, Ehrendorfer F, Mägdefrau K, Ziegler H (eds) Lehrbuch der Botanik. G Fischer, Stuttgart, pp 835–841

Fitch WM (1976) J Mol Evol 8:13–40

Fox G, Pechman KR, Woese CR (1977) Int J Syst Bacteriol 27:44–57

Fox GE, Stackebrandt E, Hespell RB, Gibson J, Maniloff J, Dyer TA, Wolfe RS, Balch WE, Tanner RS, Magrum LJ, Zablen LB, Blakemore R, Gupta R, Bonen L, Lewis BJ, Stahl DA, Luehrsen KR, Chen KN, Woese CR (1980) Science 209:457–463

Jay E, Bambara R, Padmanaghan R, Wu R (1974) Nucleic Acids Res 1:331–353

Kössel H, Edwards K, Fritzsche E, Koch W, Schwarz Zs (1983) In: Jensen U, Fairbrothers DE (eds) Proteins and nucleic acids in plant systematics. Springer, Berlin Heidelberg New York

Küntzel H, Heidrich M, Piechulla B (1981) Nucleic Acids Res 9:1451–1461

Ramshaw JAM, Peacock D, Meatyard B, Boulter D (1974) Phytochemistry 13:2783–2789

Rubtsov PM, Musakhanov MM, Zakharyev VM, Krayev AS, Skryabien KG, Bayev AA (1980) Nucleic Acids Res 18:5779–5794

Salim M, Maden BEH (1981) Nature (London) 291:205–208

Sanger F, Brownlee GG, Barrell BG (1965) J Mol Biol 13:373–398

Silberklang M, Gillum AM, RajBhandary UL (1977) Nucleic Acids Res 4:4091–4108

Silberklang M, Gillum AM, RajBhandary UL (1979) Methods. Enzymol 59G:58–109

Stackebrandt E, Woese CR (1981) In: Carlilie MJ, Collins JF, Moseley BEB (eds) Molecular and cellular aspects of microbial evolution. Soc Gen Microbiol Symp 32. Cambridge Univ Press, Cambridge, pp 1–31

Stackebrandt E, Ludwig W, Schleifer KH, Gross HJ (1981) J Mol Evol 17:227–236

Stackebrandt E, Seewaldt E, Ludwig W, Schleifer KH, Huser BA (1982) Zentralbl Bakteriol Hyg I Abt Orig C 3:90–100

Woese CR, Fox G (1977) Proc Natl Acad Sci USA 74:5088–5090

Zuckerkandl E, Pauling L (1965) J Theor Biol 8:357–366

Phylogenetic Information Derived from tRNA Sequence Data

M. SPRINZL[1]

Abstract. Comparison of 250 sequences of tRNAs revealed that the number of positions available for base replacements and thus the provision of information for phylogenetic evaluation were more limited than is generally assumed. The pattern of natural modification of tRNAs most probably reflects the development of the translation machinery in order to achieve higher precision of protein biosynthesis. It is demonstrated, in several experiments, that the introduction of modified nucleosides in tRNA affects the tRNA structure, tRNA: ribosome interaction, and also the fidelity of translation.

1 Introduction

By the introduction of new and rapid sequencing methods the number of known nucleic acid sequences have increased rapidly in recent years. With this development the comparison of nucleic acid sequences has become one of the prominent tools to investigate molecular evolution. The methods used in such phylogenetic studies rely in principle on a simple assumption that an evolutionary distance between two species is proportional to the amount of the nucleotide replacements in the DNA sequence when two homologous genes are compared. This approach was successfully used to derive phylogenetic trees and to study the evolutionary relationships between different cells and cell organells (Ninio 1982, Cedergren et al. 1981).

Direct comparison of two DNA sequences coding for homologous genes is only possible if the length of sequences to be compared is identical. If deletions or insertions occur resulting in different lengths of the DNA, the proper alignment and the identification of the inserted or deleted sequences may present a serious problem. Comparison of DNA sequences originating from different organisms and coding for a protein with homologous function is facilitated by the twenty-letter alphabet consisting of the codon-tripletts for twenty amino acids. In this case the alignment problem is the same as that for protein sequences. Functional RNA molecules as tRNAs and ribosomal RNAs and their genes consist only of a four letter alphabet of nucleic acids and therefore the alignment, identification of deletions and insertions, as well as the comparison of these sequences, is not possible without additional information. These sequences can only be aligned and compared since they possess universal structural features in their secondary structure, invariant and semiinvariant sequences, and modified nucleosides placed in typical positions of their secondary structure.

1 Lehrstuhl für Biochemie, Universität Bayreuth, Postfach 3008, D-8580 Bayreuth, FRG

Proteins and Nucleic Acids in Plant Systematics
ed. by U. Jensen and D.E. Fairbrothers

Common for all tRNAs and ribosomal RNAs is their participation in protein biosynthesis. During this complex process they interact with a variety of other macromolecular partners: proteins and nucleic acids. The evolution of this class of RNA molecules is therefore not independent, but reflects the evolution of the whole protein biosynthetic apparatus. The part of sequences of these RNAs, which interact with other macromolecules, are therefore either conserved or subjected to lower mutation rates than the sequences not involved in such interactions.

2 Conserved Features in tRNAs

Some invariances in the tRNA structure are so prominent that they became apparent soon after the elucidation of the first limited number of tRNA sequences. The first known sequence of tRNA, $tRNA^{Ala}$ from yeast, determined by Holley et al. (1965) was arranged in a proper cloverleaf structure. The next few sequences soon led to the recognition of the invariant parts and features of elongator tRNA. The elucidation of the tertiary structure of tRNA contributed to the identification of some additional common features in the tRNA sequences, and provided important hints for the understanding of the function of invariant or semiinvariant residues in the stabilization of the tertiary structure of tRNA (Rich and RajBhandary 1976, Clark 1978). The most important invariances in the tRNA structure are depicted in Fig. 1. The numbering of nucleotide residues used in Fig. 1 and allowing an alignment of all tRNA species was adopted at the Cold Spring Harbor Symposium on Transfer RNA in 1978 (Schimmel et al. 1979, Sprinzl and Gauss 1982). It is based on the comparison of the available tRNA sequences and on the known tertiary structure of $tRNA^{Phe}$ from yeast. This system can be applied to all tRNAs starting with species isolated from bacteria up to tRNAs originating from mammalian cells. The only exceptions are the tRNA species from mammalian mitochondria, some of which cannot be fitted into the generalized scheme shown in Fig. 1. This divergence is conferred to residues 8–26 and residues 44–60 in the generalized cloverleaf structure. Clearly the structure of the translational machinery in the mammalian mitochondrium is more divergent from that in its own cytoplasm as it is to that of the mammalian cytoplasm from the one in bacteria. The functional and phylogenetic consequences of this fact are not apparent, and will require detailed investigations of the mitochondrial translational machinery. It is interesting to note that the tRNAs isolated from mitochondria of lower eukaryotes and plants, although often being divergent in their nucleotide residues, which are considered to be invariant in other elongator tRNAs, can be well fitted into the generalized structure shown in Fig. 1. This fact implies a considerable divergence between the translational system in mammalian mitochondria and that of lower eukaryotes and plants.

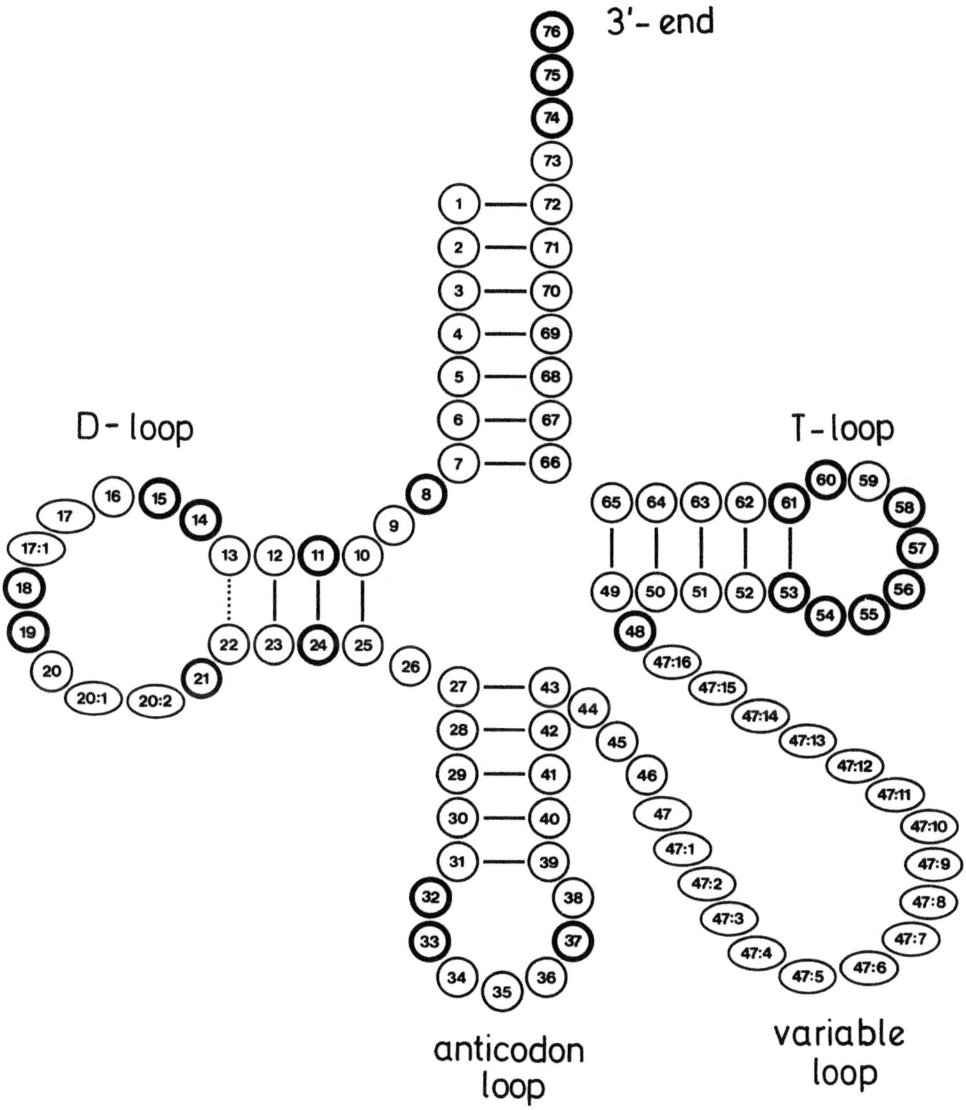

Fig. 1. Generalized cloverleaf structure of tRNA and the adopted numbering system for tRNAs (Schimmel et al. 1979). *Circles* represent nucleotides, which are always present; among these, the *thickedged circles* denote the residues, which are invariant or semi-invariant in the majority of elongator tRNA sequences. The residue 8 is always uridine or its derivative, the residue 48 is always a pyrimidine. For definition of other invariances see Tables 1 and 2. *Ovals* represent nucleotides, which are not present in all tRNAs: these are the nucleotides before the two constant guanosine residues (18, 19) in the D-loop, the nucleotides after these guanosine residues, and the nucleotides in the variable loop, which may be up to 17 nucleosides

3 Phylogenetic Studies on tRNA

A detailed review of previous research on the use of tRNA sequence data for phylogenetic studies has been recently published (Cedergren et al. 1981). Therefore, I will only present certain aspects and will especially discuss questions related to the content of information in tRNA sequences.

Although the alignment of tRNA sequences with the exception of mammalian mitochondrial tRNAs is possible, it is not easy to select the right tRNA species for phylogenetic comparison. This is due to the fact that in each cell for each amino acid a set of isoacceptor tRNAs is available. The differences in the sequences of tRNAs of different specificity originating from one organism are usually larger than the differences in the sequences of homologous tRNAs, e.g. $tRNA^{Phe}$ from different organisms. It is therefore obvious that only tRNA species of the same specificity, and thus homologous function, can be considered in phylogenetic studies. The isoacceptor problem complicates the phylogenetic consideration for two main reasons. (1) a single mutation can already cause a switch in tRNA specificity (Celis and Piper 1982) and (2) the selection of "homologous" tRNA from the set of isoacceptor is usually more a matter of experimental opportunism than of rational consideration. Although about 250 sequences are now available (Sprinzl and Gauss 1982), usually the most abundant tRNA species of given specificity were isolated from particular organisms, functionally characterized and sequenced. The sequence comparison of such arbitrarily chosen tRNA isoacceptor molecules from two species doesn't guarantee that the evolutionary, most closely related, and homologous species will be compared. Theoretically, for a selection of truly homologous isoacceptors all isoacceptor molecules of one group should be characterized and sequenced, or alternatively, the sequences of the DNA of all different tRNA isoacceptor genes in one species should be known. Due to the isoacceptor problem, most frequently the sequences of initiator tRNAs were compared for phylogenetic studies (Cedergren et al. 1980, 1981, Hasegawa et al. 1981, LaRue et al. 1979, 1981). These tRNAs with well defined and testable function in protein biosynthesis usually occur as one isoacceptor in the cell. Its derivatives differ only in the pattern of modification, not in the sequence of their genes. Other groups suitable for phylogenetic studies are composed of phenylalanine specific tRNAs (Eigen and Winkler-Oswatitsch 1981). Also $tRNA^{Gly}$ and $tRNA^{Val}$ were phylogenetically evaluated and it was shown that they have a common ancestor (Cedergren et al. 1980).

Let us ask the question, how reliable and accurate are the phylogenetic studies based on tRNA sequences? The accuracy of the ancestral sequence reconstitution, a crucial step in the derivation of the phylogenetic tree, depends on the number of available sequences and the amount of base residues available for mutations in a given molecule (Cedergren et al. 1981). The main problem which seriously limits the contents of information which can be expected from tRNA sequences, is the above mentioned multiple function of tRNA molecules. The tRNA interacts during its functional cycle with several enzymes, proteinous macromolecules, and ribosomes. In addition, it is involved in regulatory processes, and its biosynthesis is also dependent on specific interactions with processing enzymes (Schimmel et al. 1979). Mutations in tRNA must fit to this multifunctional system, or they must take place parallel with the changes in the interacting macromolecules. However, parallel mutations in two different genes are unlikely to occur. Therefore, in tRNAs parts of the sequences devoted to such interactions will be stable and the amount of nucleobases available for mutations will be limited. If the aligned tRNA sequences (Sprinzl and Gauss 1982) are evaluated as shown in Tables 1 and 2, it becomes apparent that the hot spots in the sequences available for mutations are much more limited, as expected from the presentation in Fig. 1, in which only the well recognized constant regions are depicted. Also the positions which were generally considered as variable (Clark 1978, Rich and RajBhandary 1976) are occupied, to a high degree by the same base or the same type of base. If homologous tRNAs, of the same amino acid specificity but originating from different organisms, are compared the number of positions in the sequence available for mutations is more reduced. As seen in Table 1, several structural features are not distributed randomly in the secondary structure of tRNA. These features are, for instance, the G–U (U–G) irregular base pairs, type of base pairs G–C (A–U), or polarity of the base pairs purine-pyrimidine (pyrimidine-purine). From Tables 1 and 2 it is obvious that in the tRNA structure there is a tendency to preserve the most suitable general structure, leading to a high content of semiinvariant nucleoside usage. Criteria such as polarity of base pairs, or type of base pairs in nucleic acids secondary structures, seem to be more important in the determination of the stability of the secondary and tertiary structures of nucleic acid than generally implicated (Borer et al. 1974, Mizuno and Sundaralingham 1978). It can be concluded that in addition to well known invariable sequences in tRNA, such as the GpTpΨpCpGp-region, 3′-CCA-end or guanosine-18 and -19 residues, the pressure to conserve the most suitable general tRNA structure additionally limits the number of positions which can be mutated. As a consequence, the homologous tRNA sequences of phylogenetically closely related organisms are very similar, or as in the case of mammalian tRNAs, often identical. For this reason, the phylogenetic analysis of tRNA sequences can only depict the main streams in the early evolutionary events, but the comparison of tRNA sequences is not suitable for the characterization of species within one closely related group of higher organisms.

More phylogenetic information, as compared to tRNA sequences, can be obtained by analysis of ribosomal RNAs (Kössel et al. this volume), which can be easily isolated, characterized, and sequenced. The isoacceptor problem hindering the comparison of homologous tRNA species to some extend does not exist in the case of ribosomal RNAs. The alignment of the sequences of 16S (Kössel et al. this volume) and 5S (Luehrsen and Fox 1981) ribosomal RNAs based on their secondary structures is possible. However, it can be expected that the limited mutation rates at particular positions of these RNAs, although due to a limited amount of sequence data not yet recognized, can also limit to some extent this sequence information for phylogenetic studies.

4 tRNA Modification

A remarkable feature of all tRNA molecules is the high extent of modification of their nucleotide residues. The structure of over 50 modified nucleosides originating

Table 1. Base composition of the stem regions of different classes of tRNA[a]: G | C / A | U

Aminoacylation stem

Base pair	Eubacteria 56 sequences		Archaebacteria 13 sequences		Plants cytoplasm 12 sequences		Plants chloropl. 16 sequences	
1–72	46 5 / 2 3	0 43 / 8 5	10 1 / 2 0	1 10 / 0 2	6 1 / 4 1	0 7 / 1 4	10 3 / 3 0	0 9 / 3 4
2–71	28 23 / 2 3	23 28 / 3 2	6 5 / 0 2	5 6 / 2 0	4 6 / 0 2	6 4 / 2 0	7 7 / 0 2	9 6 / 0 1
3–70	23 19 / 9 5	19 22 / 5 10	6 4 / 0 3	4 6 / 0 3	4 6 / 1 1	6 4 / 1 1	4 11 / 0 1	11 4 / 1 0
4–69	33 9 / 3 11	9 32 / 11 4	3 8 / 0 2	8 3 / 2 0	5 2 / 3 2	3 5 / 1 3	10 2 / 1 3	2 7 / 3 4
5–68	31 18 / 5 2	19 30 / 1 6	3 5 / 1 4	6 3 / 3 1	7 2 / 3 0	2 5 / 2 3	7 4 / 3 2	4 7 / 2 3
6–67	19 17 / 7 13	20 19 / 10 7	10 2 / 1 0	2 9 / 0 2	5 1 / 6 0	1 4 / 0 7	6 3 / 1 6	3 5 / 6 2
7–66	33 0 / 14 9	1 33 / 8 14	7 0 / 4 2	0 7 / 2 4	8 1 / 2 1	1 8 / 1 2	4 0 / 8 4	0 3 / 4 9

D-Stem

Base pair	Eubacteria		Archaebacteria		Plants cytoplasm		Plants chloropl.	
10–25	53 1 / 1 1	1 50 / 0 5	13 0 / 0 0	0 8 / 0 5	12 0 / 0 0	0 7 / 0 5	14 0 / 2 0	0 14 / 0 2
11–24	0 37 / 5 14	37 0 / 14 5	4 6 / 0 3	6 4 / 3 0	0 9 / 1 2	9 0 / 2 1	0 9 / 3 4	9 0 / 4 3
12–23	10 7 / 3 36	7 10 / 36 3	5 4 / 1 3	4 5 / 3 1	7 1 / 1 3	0 7 / 4 1	8 0 / 0 8	0 8 / 8 0
13–22	6 39 / 5 6	42 0 / 12 2	0 3 / 2 8	3 1 / 2 7	1 7 / 2 2	9 0 / 2 1	0 12 / 4 0	12 0 / 4 0

[a] Frequency of the occurrence of nucleotides in tRNAs was evaluated using tRNA sequence data (Sprinzl and Gauss 1982). For each position four numbers are given, representing the occurrence of guanosine *(upper left)*, cytidine *(upper right)*, adenosine *(lower left)*, and uridine *(lower right)*. The numbering of the positions is defined in Fig. 1. All known sequences belonging to the particular group and without selection for isoacceptors were evaluated. tRNA mutants were not considered. The modified nucleosides were replaced by their precursors as derived from the sequences of tRNA genes

Table 1. (continued)

Anticodon stem

Base pair	Eubacteria		Archaebacteria		Plants cytoplasm		Plants chloropl.	
27–43	9 28 7 12	29 6 11 10	2 8 0 3	8 2 3 0	4 1 1 6	1 4 6 1	3 6 2 5	7 2 4 3
28–42	8 33 7 8	33 8 8 7	2 5 1 5	6 2 4 1	0 5 0 7	5 0 7 0	3 4 5 4	5 3 3 5
29–41	21 7 15 13	7 21 13 15	6 2 3 2	2 6 2 3	5 1 6 0	1 5 0 6	10 1 3 2	1 10 2 3
30–40	41 13 0 2	13 40 2 1	7 6 0 0	6 6 0 1	11 1 0 0	1 10 0 1	11 5 0 0	5 11 0 0
31–39	8 26 19 3	26 8 3 19	4 6 3 0	6 4 0 3	5 2 5 0	2 5 0 5	4 1 5 6	3 3 4 6

TΨ-stem

Base pair	Eubacteria		Archaebacteria		Plants cytoplasm		Plants chloropl.	
49–65	39 5 9 3	6 26 2 22	6 7 0 0	7 4 0 2	2 6 4 0	6 2 0 4	4 2 9 1	2 3 1 10
50–64	25 18 6 7	19 17 6 14	5 5 1 2	5 5 2 1	2 9 0 1	9 2 1 0	1 8 2 5	8 1 5 2
51–63	22 15 11 8	16 22 7 11	6 5 1 1	5 6 1 1	6 0 5 1	1 6 0 5	4 7 3 2	7 2 2 5
52–62	49 0 7 0	0 49 0 7	6 1 5 1	2 6 0 5	10 0 0 2	0 10 2 0	12 0 4 0	0 12 0 4
53–61	56 0 0 0	0 56 0 0	13 0 0 0	0 13 0 0	12 0 0 0	0 12 0 0	16 0 0 0	0 16 0 0

Table 2. Base composition of the loop regions of different classes of tRNAs[a]: G | C / A | U

D-Loop

Base	Eubacteria 56 sequences				Archaebacteria 13 sequences				Plants cytoplasm 12 sequences				Plants chloropl. 16 sequences			
	G	C	A	U	G	C	A	U	G	C	A	U	G	C	A	U
14	0	0	56	0	0	0	13	0	0	0	12	0	0	0	16	0
15	43	0	13	0	10	0	3	0	9	0	3	0	11	0	5	0
16	1	18	3	34	0	6	0	7	0	3	0	9	0	4	0	12
17	0	9	0	27	0	8	0	2	0	1	0	5	0	1	0	13
17:1	0	1	0	3	0	2	4	4	0	0	0	1	0	1	1	4
18	54	0	0	2	13	0	0	0	12	0	0	0	16	0	0	0
19	54	0	0	2	13	0	0	0	12	0	0	0	16	0	0	0
20	6	10	2	38	0	8	2	3	3	0	4	5	0	1	0	15
20:1	0	3	3	15	1	7	0	1	0	0	3	0	0	1	4	1
20:2	2	2	3	0	0	2	3	1	0	0	0	0	0	0	0	0
21	3	0	52	1	0	0	12	1	2	0	10	0	4	0	12	0
26	42	0	12	2	7	1	4	1	10	0	2	0	8	0	6	2

[a] For definition of the group of evaluated tRNAs see footnote to Table 1. The variable (extra) loop of tRNA is not included in the table, but also tRNAs possessing a large variable loop were evaluated. Positions 17, 17:1, 20:1, 20:2 are not always occupied by a nucleotide (see also Fig. 1)

Table 2. (continued)

Anticodon loop

Base	Eubacteria	Archaebacteria	Plants cytoplasm	Plants chloropl.
32	0 36 / 0 20	0 6 / 0 7	0 11 / 0 1	0 11 / 1 4
33	0 0 / 0 56	0 0 / 0 13	0 2 / 0 10	0 0 / 0 16
34	24 12 / 0 20	7 1 / 0 5	5 7 / 0 0	4 7 / 0 5
35	6 16 / 22 12	2 2 / 4 5	0 1 / 10 1	2 1 / 13 0
36	8 19 / 13 16	4 3 / 2 4	1 2 / 4 5	3 1 / 6 6
37	2 2 / 52 0	8 0 / 5 0	5 0 / 7 0	0 0 / 16 0
38	3 8 / 38 7	2 2 / 7 2	0 1 / 9 2	0 3 / 12 1

TΨ-loop

Base	Eubacteria	Archaebacteria	Plants cytoplasm	Plants chloropl.
54	0 0 / 0 56	0 0 / 0 13	0 0 / 4 8	0 0 / 0 16
55	2 0 / 0 54	0 0 / 0 13	0 0 / 0 12	0 0 / 0 16
56	0 56 / 0 0	0 13 / 0 0	0 12 / 0 0	0 16 / 0 0
57	42 0 / 14 0	13 0 / 0 0	9 0 / 3 0	5 0 / 11 0
58	0 0 / 56 0	0 0 / 13 0	0 0 / 12 0	0 0 / 16 0
59	18 2 / 20 16	0 0 / 13 0	2 0 / 5 5	6 0 / 10 0
60	0 7 / 0 49	0 0 / 0 13	0 3 / 4 5	0 0 / 0 16

from tRNA was determined and several more modified nucleosides in tRNA were detected, the structure of which remains to be elucidated (Sprinzl and Gauss 1982). The types of modification range from simple methylation, either on the nucleobase or on the 2′-hydroxyl group of the ribose residue, up to the introduction of hypermodified bases with complicated side chains (Schimmel et al. 1979). The modified bases in tRNA are not distributed randomly. Some positions in the sequence are almost always modified and a significant regularity in the type of modification at some positions of the sequence is obvious. Nucleosides in other positions are never or very seldomly modified. The frequency of the modification of the particular position in tRNA is evaluated in Fig. 2. Typical examples for the frequent sequence-specific occurrence of modified nucleosides is e.g. the 4-thiouridine (s^4U) in position 8, dihydrouridine (D) in position 16, 20 and 47, 2,2-dimethylguanosine (m_2^2G) in position 26, 7-methylguanosine (m^7G) in position 46, 5-methylcytidine (m^5C) in positions 48, 49, and 50, ribothymidine (T) in position 54 and pseudouridine (Ψ) in position 55. Types of modification at some other positions in the tRNA, although frequently present, can be very variable. This situation is observed especially in the anticodon loop of tRNAs. Position 32 and in high degree the position 34 (the first letter of the anticodon) and 37 (3′-neighbor to the anticodon) are almost always occupied with a variety of types of modified nucleosides. These modifications are most probably necessary for the fine structural tuning of the anticodon loop structures in order to maintain the high precision of the codon recognition. This aspect of tRNA modification is discussed in detail by Cedergren et al. (1981). Due to the variety of anticodons it is not surprising that the types of modification at these regions vary as well depending on tRNA isoacceptor. On the other hand the high conservation of the type of modification in other regions of the sequence points to some kind of universal function of this particular segment of the molecule. The positions 8, 16, 20, 26, 48, 54, 55, and 58 are typical candidates for this type of considerations (Fig. 2).

No evidence for a simple yes or no function of tRNA modification exists. Perhaps the only exception to this rule is the suggested role of pseudouridine in $tRNA^{His}$ from *S. typhimurium*. In this case modifications at positions 38 and 39 were found to be related to the regulation of histidine operon (Turnbough et al. 1979). All other reports in the literature point to the role of tRNA modification as a way to modulate the tRNA structure in order to effect a more efficient participation of these molecules in a particular function. For instance, the modification of the nucleosides in the anticodon of tRNA or the nucleoside residues adjacent to the anticodon were shown to be important for the modulation of the codon-anticodon interaction (Cedergren et al. 1981). So the removal of the hypermodified Y-base from the position 37 of the $tRNA^{Phe}$ from yeast severely effects its activity in the translation. Similarly the presence of the modified base Q in the anticodon of several mammalian tRNAs is related to the cell differentiation (Dingermann et al. 1981) and is also found to be different in normal and cancer cells (Okada et al. 1978, Shindo-Okada et al. 1981). The 2-thiouridine in position 34 of tRNA is imposing a restriction for the recognition of the wobble base (Sebiya et al. 1969) during the codon-anticodon interaction, and the 2-thiocytidine residue in position 32 of $tRNA^{Arg}$ from *E. coli* modulates the structure of the anticodon loop and contributes to the recognition of the correct codon (Kruse et al. 1978) and to the fidelity of translation (Baumann and Sprinzl in preparation).

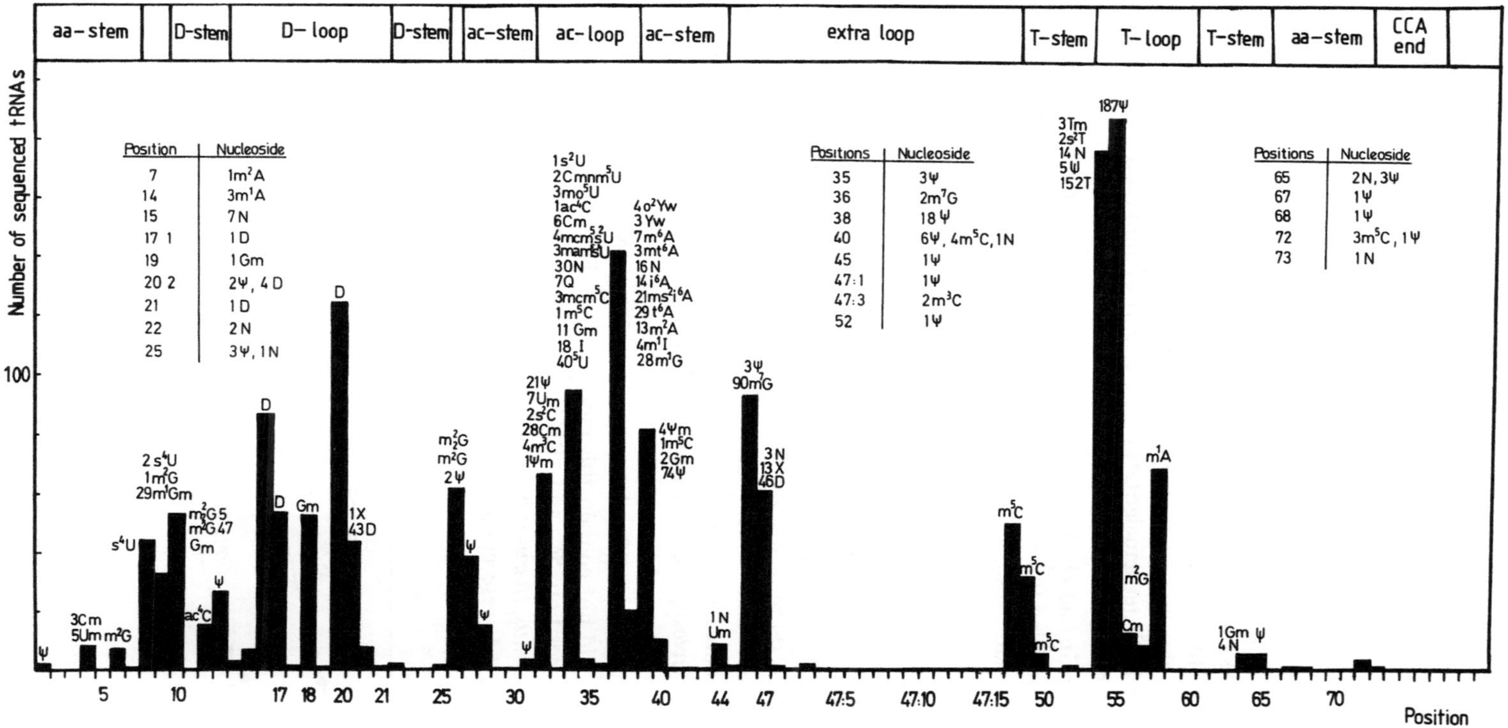

Fig. 2. Positions, frequency and the occurrence of modified nucleosides in 206 sequences evaluated from the data of Sprinzl and Gauss (1982). Numbers are given only for those columns containing different nucleosides. tRNA mutants were omitted in the compilation. The numbering system defined in Fig. 1 is used

Special attention was paid to the elucidation of the role of modification in the position 54 of the tRNA, which is present in the majority of sequences of elongation tRNAs from different sources (Sprinzl and Gauss 1982). Ribothymidine-54 or its derivatives are part of the highly conservative sequence GpTpΨpCpGp, which was suggested to be a recognition site of tRNA involved in its binding to ribosomes (Sprinzl et al. 1976). An example for an experiment leading to this suggestion is shown in Fig. 3 (Watanabe and Sprinzl in preparation). Oligonucleotides TpΨpCpGp and s^2TpΨpCpGp originating from the sequences 54 to 57 were excised from different types of tRNAs and their ability to bind to 70S ribosomes was compared. It was expected that these tetranucleotides would bind to the tRNA-binding site of the ribosomes and thus inhibit their normal function. Indeed an inhibition was observed in the case of both tetranucleotides. The extent of inhibition was, however, dependent on the modification of the 5′-terminal uridine residue in the tetranucleotides. The s^2TpΨpCpGp had a higher inhibitory activity than the tetranucleotide TpΨpCpGp or the corresponding unmodified oligonucleotide. The chemical modification of the 2-thioribothymidine residue in s^2TpΨpCpGp with cyanogen bromide or its desulphuration lead as expected to a lowering of the inhibitory activity. These experiments demonstrate that the efficiency of interaction of the oligonucleotide with the ribosome is dependent on the presence of the modified 5′-nucleoside. Similar experiments with intact tRNA species, modified on the residue-54 to different extents, provide similar results (Dingermann et al. 1980, Kersten et al. 1981).

The influence of the modification at position 54 of the tRNA on its overall structure can be directly measured by ^{1}H-nuclear magnetic resonance spectroscopy (NMR). In a high field region of the NMR spectrum of tRNA the methyl and methylene groups of modified nucleosides can serve as suitable spectroscopic reporter signals for structural investigations. These signals are especially sensitive to the changes in the tertiary and secondary structure of tRNA. In order to elucidate the structural consequences of the modification at position 54, the NMR spectra of $tRNA^{fMet}$ from *E. coli*, which did not contain this modification, $tRNAf^{Met}$ from *E. coli* having ribothymidine in position 54, and $tRNA^{fMet}$ isolated from an extreme thermophilic bacterium *Thermus thermophilus*, which contained 2-thioribothymidine instead of ribothymidine 54 were compared (Davanloo et al. 1979a). The temperature stability of these tRNAs was found to be strongly dependent upon the modification of the

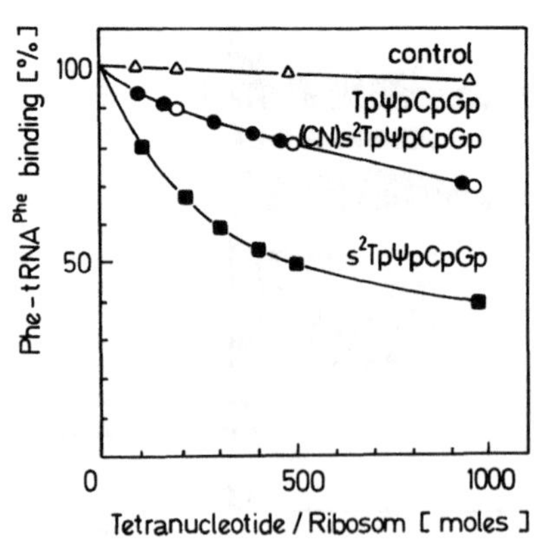

Fig. 3. Inhibition of the template-dependent binding of aminoacyl-tRNAs to *E. coli* 70 S ribosomes by tetranucleotides containing different modified nucleosides in their sequence. For details see Sprinzl et al. (1976)

residue 54. The tRNA containing 2-thioribothymidine was most stable, followed by the ribothymidine-containing tRNA^{fMet} from *E. coli*. The most unstable was the tRNA^{fMet} possessing no modified U-54 in its sequence. The most interesting result from these experiments, is that the structural changes leading to higher stability of tRNA is not confined to the local region, where the modification was introduced, i.e. to the position 54 and its surrounding, but is transmitted over a long distance also to other regions of the molecule. This could be concluded from the NMR experiments, where the changes of the spectroscopic signals originating from the residue D-20 were monitored and found to be dependent on the modification of the distant position 54. Similar, long range, and modification-dependent effects in the tRNA were detected by the NMR technique also in other regions of the molecule (Davanloo et al. 1979b). For instance, if the tRNA is aminoacylated on the residue 76, the T-54 NMR signal changes its chemical shift. In a similar manner, changes in T-54 NMR resonances were observed by involvement of tRNA in codon-anticodon interactions (Fig. 4), demonstrating that the modifications of tRNA on distinct positions of its sequence can effectively influence the conformations on other parts of the molecule. These results support the general idea, that the introduction of modified nucleosides in tRNA, by modifying enzymes, serves primarily to modulate the tRNA structure into a shape, in which the best structure necessary for its given function is formed.

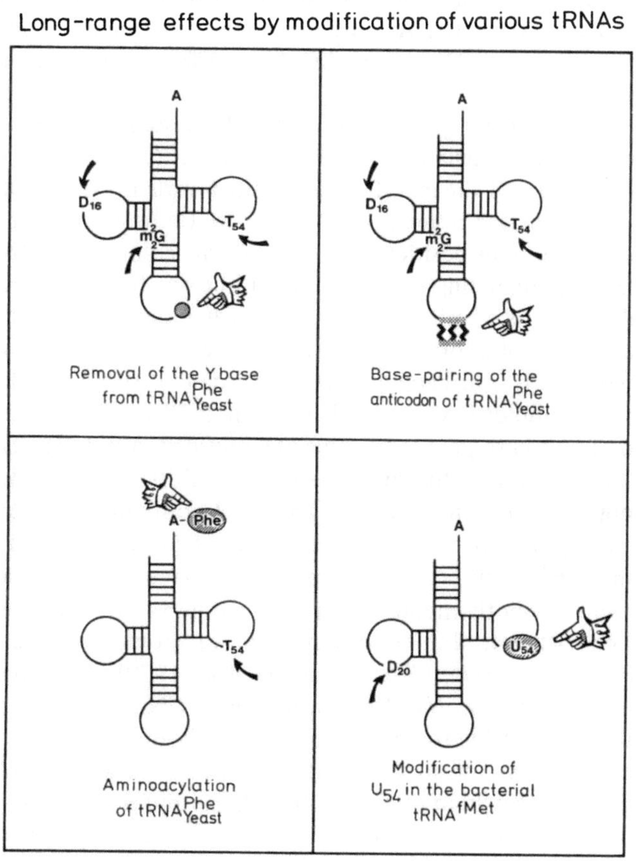

Fig. 4. Long-range effects caused by enzymatic or chemical modification of particular residues in the tRNA as determined by nuclear magnetic resonance spectroscopy (Davanloo et al. 1979a,b). The *arrows* indicate the modified nucleosides, in which changes of NMR signals were observed

5 Evolution of tRNA Modification

The tRNA modifications are to a considerable degree origin specific. In general, the amount of residues modified in tRNAs from the cytoplasm of eukaryotic cells is significantly higher than that of prokaryotic tRNAs or tRNAs from organells of eukaryotic cells, mitochondria or chloroplasts. For instance, the tRNA species isolated from the cytoplasm of mammalian cells are modified on 37 different positions and usually to a relatively higher extent than tRNAs from eubacteria, which are modified only in 23 positions (Fig. 5). However, some positions, which are very frequently modified in tRNAs from eubacteria, are not modified in cytoplasmic tRNAs from higher eukaryotes. These are positions 8 and 18, where in eubacteria 4-thiouridine and 2-methylguanosine, respectively, are often present. More typical is the reverse case, when cytoplasmatic tRNA from higher eukaryotes are modified on positions which the eubacterial tRNAs never carry a modification. Such positions are especially 4, 6, 10, 12, 14, 17:1, 25, 26, 27, 48, 49, and 50. For eukaryotic tRNA typical modifications are located especially in regions joining two stems; around positions 10, 26, and 48. It is significant that tRNAs from archaebacteria and from cytoplasm of eukaryotic cells share a common modification pattern, especially in these regions, whereas eubacteria are not so modified. The similarity between the archaebacteria and eukaryotic tRNAs does not apply to all modifications. Some positions in these two classes of tRNA are modified in a different way, which is typical for the particular group.

tRNAs from plant chloroplasts share some common features with tRNAs from eubacteria in their modification. These are: (1) the low amount of nucleoside residues modified, (2) the modification of the invariant U-8 residue, and (3) the lack of modification of the residues 48, 49, and 50. Contrastingly the occurrence of modified nucleosides in the positions 25, 26, 27 of the chloroplast tRNAs is typical for eukaryotic tRNAs originating from cytoplasm but not for eubacterial tRNA species. Mitochondrial tRNAs of primitive eukaryotes carry their modifications on the positions, which are typical also for eukaryotic cytoplasmic samples. However, the number of positions modified is considerably lower in the case of tRNAs originating from mitochondria.

Specific differences in tRNAs isolated from different classes of organisms also are observed in respect to the type of modification at particular positions of the sequence. Some examples for such specific modification of tRNAs originating from different organisms are presented in Table 3. Ribothymidine, one of the most frequently occurring modified nucleosides is present in tRNAs from all sources, except that of mammalian mitochondria (Sprinzl and Gauss 1982) and archaebacteria. Some derivatives of ribothymidine occur in position 54 in tRNAs only from certain organisms. For instance, 2-thioribothymidine was found only in thermophilic eubacterium *Thermus thermophilus* (Watanabe et al. 1979), 1-methylpseudouridine is present in archaebacterium *Halobacterium volcanii* (Pang et al. 1982), other archaebacteria having 2′-O-methyluridine *(Sulfobolus acidocaldarius)*, pseudouridine *(Thermoplasma acidophilum)* or an unidentified derivative of uridine *(Halcoccus mourhuae)* at this position (Kuchino et al. 1982). 2′-O-methyl-ribothymidine was found in position 54 only in tRNAs from the cytoplasm of mammalian cells (Sprinzl and Gauss 1982).

Table 3. Positions of occurrence of some modified nucleosides in tRNA[a]

Modified Nucleosides	Positions in the sequence of tRNA from					
	Eubacteria	Archaebacteria	Plants		Mammals	
			Cytoplasma	Chloroplasts	Cytoplasma	Mitochondria
Ribothymidine	54	–	54	54	54	–
2-Thioribothymidine	54	–	–	–	–	–
1-Methylpseudouridine	–	54	–	–	–	–
2′-O-Methylribothymidine	–	–	–	–	54	–
Acp^3-Uridine	47	–	–	47	20:1	–
4-Thiouridine	8	–	–	–	–	–
Dihydrouridine	16, 17, 20, 47	–	16, 17, 20, 47	16, 17, 20, 47	16, 17, 20, 20:1, 47	16, 17
2′-O-Methylcytidine	32	32, 56	4, 13, 32	34	4, 32, 34	–
2-Thiocytidine	32	–	–	–	–	–
5-Methylcytidine	48	39, 40, 48, 49, 56	38, 48, 49, 50	–	38, 48, 49, 50, 72	49
1-Methyladenosine	22, 58	58	58	–	14, 58	9, 58
7-Methylguanosine	46	–	46	36, 46	46	–
2-Methylguanosine	–	10, 26, 57	10, 26	26	6, 7, 10, 26	10
2,2,-Dimethylguanosine	–	10, 26	26	26	26	–
1-Methyladenosine	22, 58	58	58	–	14, 58	9, 58

[a] The data were taken from the compilation of tRNA sequences (Sprinzl and Gauss 1982)

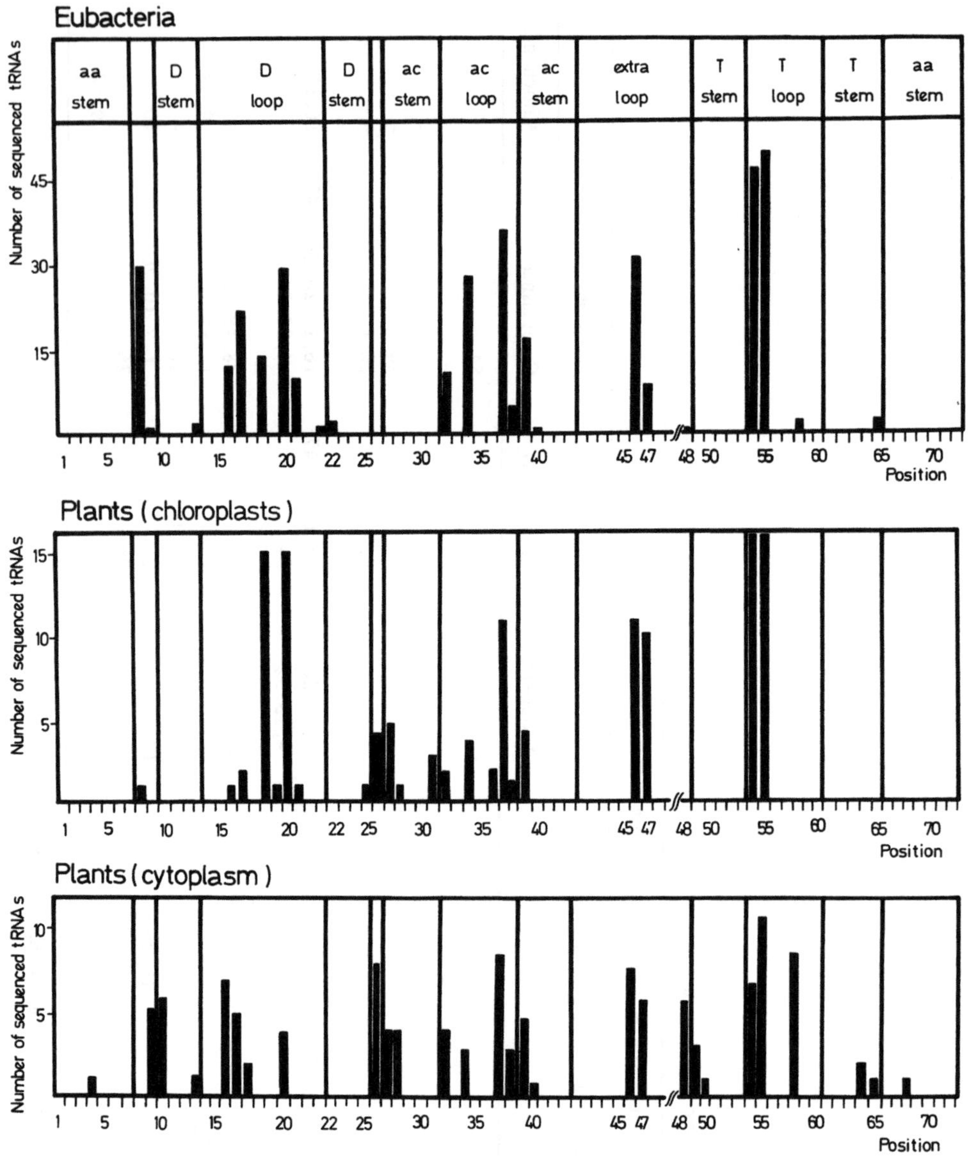

Fig. 5a,b. Frequency of the occurrence of modified nucleosides in tRNAs isolated from **A** eubacteria (56 sequences), plant chloroplasts (17 sequences), plant cytoplasm (12 sequences). **B** Archaebacteria (14 sequences), mitochondria of mycophyta (20 sequences), cytoplasm of mycophyta (38 sequences). Sequence data from Sprinzl and Gauss (1982). tRNA mutants were omitted. Also structurally unidentified modified nucleosides were considered

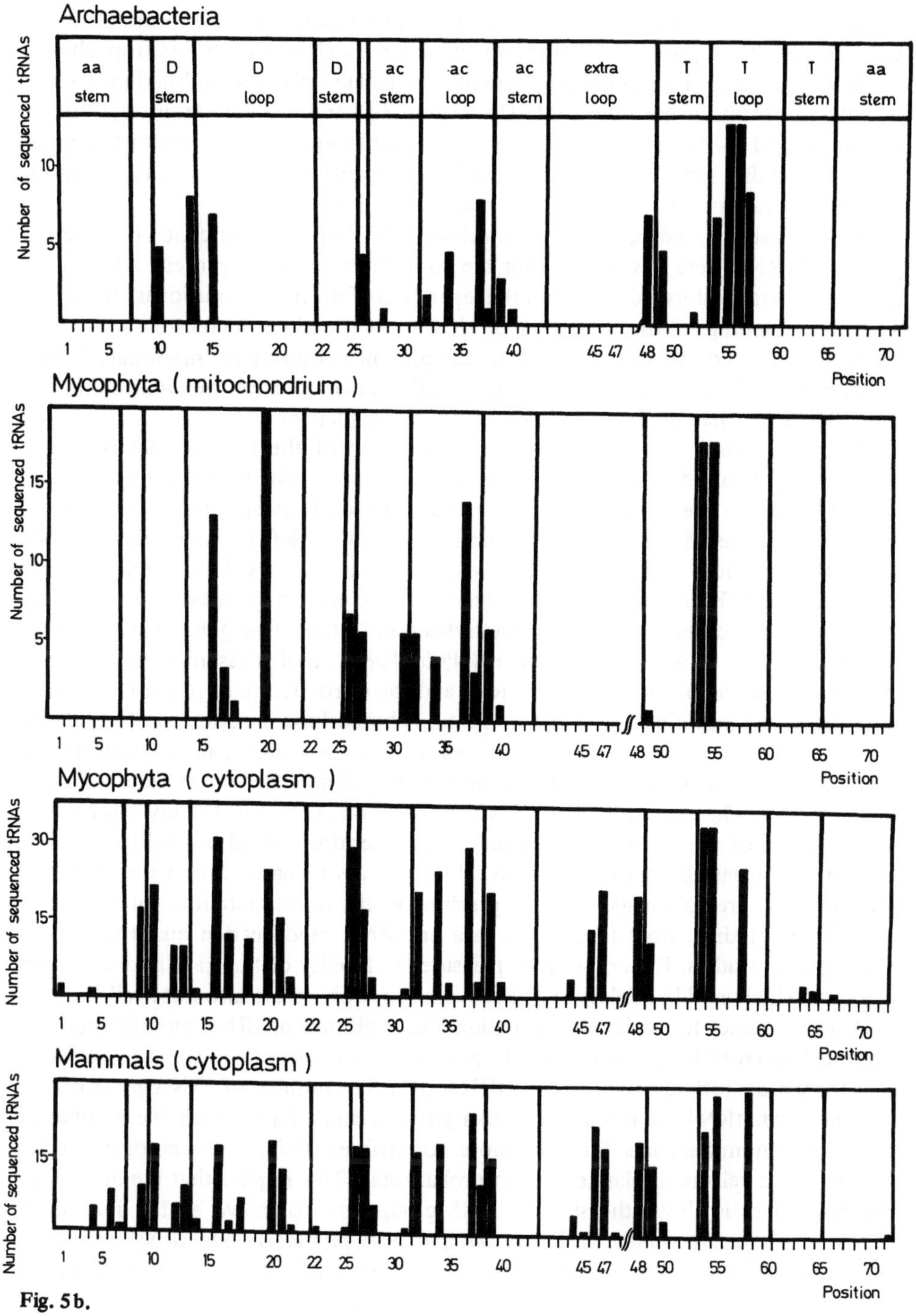
Archaebacteria
aa stem
D stem
D loop
D stem
ac stem
ac loop
ac stem
extra loop
T stem
T loop
T stem
aa stem
Number of sequenced tRNAs
Position
Mycophyta (mitochondrium)
Number of sequenced tRNAs
Position
Mycophyta (cytoplasm)
Number of sequenced tRNAs
Position
Mammals (cytoplasm)
Number of sequenced tRNAs
Position

Fig. 5b.

Another example is the derivative of uridine, 3-(3-amino-3-carboxy-propyl)uridine, which is present in position 47 of several eubacterial tRNAs and tRNAs from chloroplasts. In tRNAs isolated from cytoplasm of mammalian cells, this modified nucleoside occurs exclusively in position 20:1. Up to now 3-(3-amino-3-carboxypropyl)uridine was not found in tRNAs from archaebacteria, plant cytoplasm or mammalian mitochondria. Similar significant uneven distribution in different groups of tRNA also is observed in the case of other modified nucleosides (Table 3).

Data in Table 3 indicate that the mechanism leading to modification of different groups of tRNA were developed after the basic features of the protein synthetising apparatus were established, and after the separation of the main evolutionary branches. The understanding of the evolution of tRNA modifications therefore can be very useful for the study of the evolution of the protein biosynthetic "machinery". How did tRNA modification evolve and what was the reason to drive the evolution in this direction rather than use base replacements in the sequence?

As stated above the late stages in the evolution of the translational apparatus mostly were involved in increasing its precision of processing the information coded by mRNA. Since the sequence of large parts of tRNA is determined by the general mechanism of the translation, and the number of positions left for mutations in tRNA is rather low, there must have been a way to bypass this restriction and allow the development of tRNA toward higher precision of the protein synthesis. The modification of the nucleoside residues by methylases and other modifying enzymes could represent a way to solve this problem (Ninio 1982). If this assumption is correct, then the modification of the nucleic acids and the introduction of minor nucleosides into the RNAs participating in the translation should have an effect on the fidelity and precision of protein synthesis. Indeed as discussed above, most experimental evidence dealing with the problem of the function of modified nucleosides points to this direction. The consequence of this consideration is, that in the investigation of the evolution of the translational system more attention should be paid to the question of its increasing fidelity. However, although it is recognized that the modifications of RNA are connected with the precision of the translation, it is not clear at all at the present time, how to use the data on tRNA modification quantitatively in phylogenetic studies. Experiments to measure the fidelity of the translational system in a way, which could be relevant to the evolution problem are also not available.

Another aspect to explain the function and evolution of tRNA modification was suggested recently by Kubli (1980). In genomes of prokaryotic as well as eukaryotic cells there are multiple copies for one tRNA gene. The number of tRNA genes coding for one given tRNA isoacceptor in eukaryotes is much higher than the number of these genes in prokaryotes. Still the amount of different tRNA isoacceptor molecules in eukaryotic cells is similar to that in prokaryotes. This implies that the number of different genes is about the same in both groups, but as shown by the analysis of tRNA genes in *Drosophila melanogaster* and *Xenopus laevis* one tRNA gene in a eukaryotic cell may occur in multiple copies of several hundreds. If a gene for particular tRNA is present in multiple copies in one genome, a correction mechanism keeping the tDNA sequences constant is required. Such a correcting mechanism will protect the tDNA from mutations and allow a fast synthesis of tRNA species with identical sequence. However, mutational adaptability of tRNAs is lost in this system.

Again, the way out of this dilemma could be the modification of the primary transcripts of tRNA (Kubli 1980). By modifying enzymes, the tRNAs can be gradually modified to achieve the best adaptation for the given function. Extreme variability can be achieved in this way. Remarkable high stability of tRNA sequences in higher organisms (Sprinzl and Gauss 1982) connected with a high extent of modification of these tRNAs support this hypothesis.

6 Conclusions

The general structure of tRNA, as shown above, and its sequence was to a large extent conserved during the evolution of the species. This remarkable feature of tRNA structure implies that the evolution of the mechanism of protein synthetizing, translational apparatus, i.e. the evolution of the shape of participating macromolecules or principles of their interactions was essentially completed before the "main streams" in living organisms diverged. Since the same type of an adaptor RNA (tRNA) is found in all living cells, the general shape of tRNA and consequently that of other components of protein synthetizing machinery had to be established before evolutionary divergence between the various kingdoms of living organisms took place. Using this general mechanistic principle, the translational "machinery", had to develop later to a form which is capable of process the information of higher and higher complexity (Woese 1979). To fulfill this task a more sophisticated regulation mechanism had to arise, and particularly the precision of the processing of information had to increase. Thus, in the main time of evolution the translational machinery was developing from less regulative, not very accurate system to a well regulated system of high fidelity. The molecular principles in the translation "machinery" remained constant.

It seems that the tRNA modification through evolution replaced, at least partly, its adaptation by mutations of the sequence. This change, from mutational adaptation to the evolution by the action of modifying enzymes, took place in very early stages of the evolution, and in part before the "main streams" of living organisms separated. The structural and functional patterns of the tRNA modification is presently not understood well enough to allow a phylogenetic evaluation. Nevertheless, further study of the function of tRNA modification will lead to a better understanding of the translational "machinery". The simple comparison of the tRNA sequences alone is not enough to provide additional information. Therefore attention must be turned more strongly toward the study of the evolution of the precision of translation, and in this context to the study of the modification of nucleic acids. Regardless that modern tRNA sequencing methods allow rapid sequence determinations, in the case of the study of the evolution of the translation apparatus, we should not overlook the details. One such detail is that the RNA molecules are modified.

References

Borer P, Dengler B, Tinoco I, Uhlenbeck O (1974) J Mol Biol 86:843–853
Cedergren RJ, LaRue B, Sankoff D, Lapalme G, Grosjean H (1980) Proc Natl Acad Sci USA 77:2791–2795
Cedergren RJ, LaRue B, Sankoff D, Grosjean H (1981) CRC Grit Rev Biochem 11:35–104
Celis JE, Piper PW (1982) Nucleic Acids Res 10:r83–r91
Clark BFC (1978) In: Altmann S (ed) Transfer RNA. MIT Press, Cambridge
Davanloo P, Sprinzl M, Watanabe K, Albani M, Kersten H (1979a) Nucleic Acids Res 6:1571–1581
Davanloo P, Sprinzl M, Cramer F (1979b) Biochemistry 15:3189–3199
Dingermann T, Pistel F, Kersten H (1980) Eur J Biochem 104:33–40
Dingermann T, Ogilvie A, Pistel F, Mühldorfer W, Kersten H (1981) Z Physiol Chem 362:763–773
Eigen M, Winkler-Oswatitsch R (1981) Naturwissenschaften 68:217–228
Hasegawa M, Yano TA, Miyata T (1981) Precam Res 14:81–98
Holley RW, Apgar J, Everet GA, Madison J, Marquisee M, Merril SH, Penswick JR, Zamir A (1965) Science 147:1462–1465
Kersten H, Albani M, Mannlein E, Praisler R, Wermbach P, Nierhaus K-H (1981) Eur J Biochem 114:451–456
Kössel H, Edwards K, Fritzsche E, Koch W, Schwarz Zs (1983) In: Jensen U, Fairbrothers DE (eds) Proteins and nucleic acids in plant systematics. Springer, Berlin Heidelberg New York
Kruse TA, Clark BFC, Sprinzl M (1978) Nucleic Acids Res 5:879–891
Kubli E (1980) Trends Biochem Sci, pp 90–91
Kuchino Y, Ihara M, Yabusaki Y, Nishimura S (1982) Nature (London) 298:684–685
LaRue B, Cedergren RJ, Sankoff D, Grosjean H (1979) J Mol Evol 14:287–300
LaRue B, Newhouse N, Nicoghosian K, Cedergren RJ (1981) J Biol Chem 256:1539–1543
Luehrsen KR, Fox GE (1981) Proc Natl Acad Sci USA 78:2150–2154
Mizuno H, Sundaralingham M (1978) Nucleic Acids Res 5:4451–4461
Ninio J (1982) Molecular approaches to evolution. Pitmann Books Ltd
Okada N, Shindo-Okada N, Sato S, Hoh YH, Oda KI, Nishimura S (1978) Proc Natl Acad Sci USA 75:4247–4251
Pang H, Ihara M, Kuchino Y, Nishimura S, Gupta R, Woese CR, McCloskey JA (1982) J Biol Chem 257:3589–3592
Rich A, RajBhandary UL (1976) Annu Rev Biochem 45:805–860
Schimmel PR, Söll D, Abelson JN (1979) Transfer RNA: Structure, properties, and recognition. Cold Spring Harbor Lab, USA
Sebiya T, Takeishi K, Ukita T (1969) Biochem Biophys Acta 182:411–426
Shindo-Okada N, Terada M, Nishimura S (1981) Eur J Biochem 115:423–428
Sprinzl M, Gauss DH (1982) Nucleic Acids Res 10:r1–r55
Sprinzl M, Wagner T, Lorenz S, Erdmann VA (1976) Biochemistry 15:3031–3039
Turnbough CL, Neill RJ, Landsberg R, Ames BN (1979) J Biol Chem 254:5111–5119
Watanabe K, Kuchino Y, Yamaizumi Z, Kato M, Oshima T, Nishimura S (1979) J Biochem (Tokyo) 86:893–905
Woese CR (1979) In: Chambliss G, Craven GR, Davies J, Davis K, Kahan L, Nomura M (eds) Ribosomes: Structure, function, and genetics. Univ Park Press, Baltimore

Proteins: Structural Properties

From Genes to Proteins: Genotypic and Phenotypic Analysis of DNA Sequences by Protein Sequencing

K. BEYREUTHER[1], K. STÜBER, B. BIESELER, J. BOVENS, R. DILDROP, T. GESKE, I. TRIESCH, K. TRINKS, S. ZAISS, and R. EHRING

Abstract. Sequences of genes, and gene products such as proteins, provide a detailed and largely self-consistent context for discussion of molecular phylogenies and taxonomic relationships. Analysis at the DNA level, an alternative and often more rapid approach for protein sequence analysis, requires identification of codogenic regions. We describe how protein sequencing, sequencing of peptide mixtures, and amino acid analysis can be used to complement DNA sequencing. We demonstrate the usefulness of protein structural studies for the assignment of initiation codons to protein genes, of exons and introns to DNA sequences and for the identification of sequence internal residues such as active site residues and post-translational modifications. In addition, we show that partial protein sequencing is very useful for proving the existence of a gene product if neither specific antisera nor any functional assay exists. For proteins belonging to a known family, sequence comparison by indirect methods such as the comparison of amino acid compositions, may permit an accurate, inexpensive, and rapid description of phylogenies and taxonomic relationships. Finally, we discuss the characteristics of naturally occurring DNA sequences and the codon distribution of known reading frames in prokaryotic and eukaryotic genes using a nucleotide sequence data base of 600,000 base pairs.

1 Introduction

A gene is defined by its products, either a protein or one of several kinds of RNA molecules.

The relation between a gene and its product is generally a colinear but not necessarily a contiguous one, and the DNA sequence of a gene is ordinarily as useful for genetic and genealogical studies as the sequence of the gene product itself (Perler et al. 1980, Efstratiadis et al. 1980).

Since the analysis of DNA has been markedly improved by the development of powerful techniques for cloning and sequencing of DNA molecules (Abelson and Butz 1980, Maxam and Gilbert 1977, Sanger et al. 1977) DNA sequence analysis became the method of choice to establish protein and RNA sequences. However, the identification of noncoding regions and transcribed and translated regions in DNA sequences requires at least partial sequence information of the gene product itself. DNA sequencing, together with partial protein sequencing, has been very useful in establishing the sequences of many polypeptide products including those synthesized in vivo in minute amounts and those not suited for direct protein sequencing – as for

[1] Institut für Genetik, Universität zu Köln, Weyertal 121, D-5000 Köln 41, FRG

Proteins and Nucleic Acids in Plant Systematics
ed. by U. Jensen and D.E. Fairbrothers

instance, integral membrane proteins (Büchel et al. 1980, Ehring et al. 1980, Higgins et al. 1982). The direct analysis of the gene product is also required for active-site analysis and phenotypic analysis. It is obvious that postsynthetic modifications of RNA or protein molecules cannot be estimated accurately from DNA sequences due to the complexity of the modification reactions, even if their "code" (sequence specificity and species specificity) could be known. The many functional groups in proteins that are synthesized by posttranslational modifications are presumably present because of their special functions and therefore are considered to contribute to the phenotype (Uy and Wold 1977). Posttranslational modifications of proteins include proteolytical processing as well as more than 140 alterations of sidechains of almost all twenty encoded amino acids (Uy and Wold 1977).

It remains a task to the protein chemist to find out whether or not a gene product exists, whose putative sequence was deduced from the DNA sequence, whether post-translational alterations of the primary translational product occur, which amino acids are modified (including the identification of the cysteine residues forming S-S bridges in the native protein), and which residues constitute the active site.

In this paper we compare some of the techniques for protein analysis which are not only relevant for the interpretation of DNA sequences but also for establishing molecular phylogenies and taxonomic relationships.

2 The Information Content of Amino Acid Analyses

It is sometimes desirable to study the possible relationship of two proteins, two peptides or of a protein and a DNA sequence when protein sequence data are unavailable. Few methods are useful for such an analysis. Comparisons based on charge, molecular size, functional criteria, genetic localization, end group analyses, and fingerprinting of proteolytic digests all have serious shortcomings. These indirect techniques for sequence comparisons, in contrast to actual sequencing methods, are based on genotypic as well as phenotypic properties of proteins. Direct techniques as protein and DNA or RNA sequencing provide information on the genotype, i.e. the order of the twenty amino acids that are coded for in protein synthesis. The indirect analyses include the twenty encoded amino acids, but also the many that are synthesized by posttranslational modifications (Uy and Wold 1977). Therefore, the afore-mentioned indirect methods are not only unreliable if the two compared proteins differ due to proteolytical processing or posttranslational modifications of side chains, but also if minor sequence differences do exist (amino acid substitutions, short deletions or insertions). For the same reasons these methods should only be applied with a note of caution for comparison of in vitro synthesized proteins (primary translational products) with mature forms of the same protein.

Possible useful criteria are amino acid composition (Cornish-Bowden 1980), immunological cross-reactivity (Wilson et al. 1977) and under some circumstances electrophoretic mobility of digests of proteins (Cleveland et al. 1977). Immunological and electrophoretic criteria may be useful as long as cross-reactivity is found, but such cross-reactivity is frequently one of the earliest common features to be lost.

Amino acid analysis is the most suitable indirect method for sequence comparison. Compositions can be used to obtain phylogenetic trees of related proteins. Sequence trees are better than composition trees if the most distantly related proteins in a group differ by at least 10–15 substitutions. Otherwise, the extra labour required for sequence measurements does not give better phylogenetic results than those one could expect to get from the corresponding compositions (Cornish-Bowden 1980). Relatedness of groups of two proteins can be readily ascertained from amino acid compositions, for instance by calculating the correlation coefficients of Reisner and Westwood (1982) without other than compositional information of the two proteins. Very important for the application of amino acid residue data is that the compositions have been accurately established, a requirement which is not easy to fulfill with present technology. Amino acid analyses may be performed with less than 10^{-10} mole of protein (Fig. 1) with an accuracy of about 5%. Such amounts of pure protein are easily obtained from complex mixtures of several hundred different proteins by a single two-dimensional electrophoretic separation on polyacrylamide gels (O'Farrell 1975). The compositions can than be used for calculation of the molecular weight of the polypeptides, for comparison of the actual amino acid residue data with values deduced from DNA sequences, and for a classification of the proteins by comparison with amino acid compositions of a data bank (Reisner and Westwood 1982, and cited therein). The latter application gives information on whether a protein of similar or the same composition has been isolated before and whether it can be used for phylogenetic studies. Amino acid compositions are relatively insensitive to charge or size

Method	Sensitivity information	
Quantitative amino acid analysis	0.5 to 5 μg protein	Accurate mol. weight, composition, comparison with DNA reveals posttranslational modifications, evolutionary relations
Peptide mapping	0.1 to 10 nmol (radioactive proteins: pmol to fmol)	Relationship of preform, proform to mature protein, mutationally altered proteins
Limited proteolysis	0.1 to 5 μg	Domains (structural and functional) relationship to mature protein evolution
Radio immune assay/elisa	0.1 μg	Homologous proteins evolution
N-terminal sequence analysis of protein or internal peptides (for blocked proteins)	50 to 1,000 pmol (0.5 to 10 μg of protein of M 10)	Correlation to gene (DNA sequence): for gene isolation and antibody isolation, evolution

Fig. 1. Sensitivity and information content of indirect and direct methods for sequence comparisons of proteins

differences brought about by posttranslational modifications if they are done with acid hydrolysates.

An example of the comparison of a protein composition with data derived from a DNA sequence is given in Fig. 2. Here, the amino acid analysis provided the information that the integral membrane protein lactose permease of the bacteria *Escherichia coli* does not undergo C-terminal proteolytic processing (mature form). The values for those residues, occurring twice in the C-terminal sequence of permease, are in agreement with the composition of the primary translational product deduced from the known DNA sequence of the permease gene (Büchel et al. 1980). Even if one assumes an experimental error of 5% for the determination of the amino acid composition determined for lactose permease isolated from the cytoplasmic membrane of *Escherichia coli*, the values for the residues included in the sequence presented in Fig. 2 are still in agreement with the assumption that C-terminal processing does not occur. This finding was substantiated by direct sequencing (Beyreuther and Wright to be published).

Sequence comparison by compositional analysis is a valuable method for establishing evolutionary relations of pairs of proteins, for the comparison of unknown proteins with proteins of known function, for the correlation of genes and their products, and for probing the presence or absence of some posttranslational modifications (proteolytical processing) as shown here. The relatively high experimental error of conventional amino acid analysis, however, limits the deduction of phylogenetic trees from compositional data to small proteins (100–200 residues). In practice, the usual error of 5% in determining amino acid compositions of proteins does not influence the results if the most frequent residue does not occur more than ten to twenty times in a chain. This is only the case for proteins of the aforementioned chain length. As a rule, amino acid compositions of individual proteins are kept, during evolution, as near to the genetic code frequencies as possible within the limits set by function. The consequence is that an amino acid, such as leucine, for which six different triplets exist

	AAA	DNA-Sequence
Asp	21.0	22
Thr	19.1	19
Ser	27.2	29
Glu	21.8	22
Pro	11.7	12
Cys	6.6	8
Gly	35.3	36
Ala	34.4	35
Val	28.7	29
Met	13.5	14
Ile	28.7	33
Leu	49.8	54
Tyr	11.7	14
Phe	35.4	56
His	4.3	4
Lys	12.0	12
Arg	12.6	12
Trp	4.7	6
		417

L - S - G - P - G - P - L - S - L -
400

L - R - R - Q - V - N - E - V - A - COOH
410 417

Fig. 2. Comparison of the amino acid compositions of lactose permease of *Escherichia coli* deduced from the DNA-sequence of the permease gene (Büchel et al. 1980: *right column*) with the composition of the protein determined by amino acid analysis (AAA, *left column*). *Bottom line:* the C-terminal sequence 400–417 is given as deduced from the DNA-sequence. Residues which occur twice in this sequence are *underlined* in the compositions and the sequence except for leucine (leucine and phenylalanine values are too low, even after 96 h of acid hydrolysis, and are therefore not included in the comparison). Q (Gln) and E (Glu) are measured as E after hydrolysis

(about 10% of all code words) does account for about 10% of all amino acid residues in proteins. This means, proteins of 100 to 200 residues usually contain 10 to 20 leucine residues.

As already mentioned, composition trees and sequence trees of slowly evolving proteins are of comparable quality (10 to 15 substitutions separate the most distantly related proteins). Thus, closely related taxa may be distinguished with the same accuracy by comparison of protein sequences or amino acid compositions of proteins belonging to a family, provided the differences are low. It does not matter by which method numbers are determined that are not significantly different (for instance one, two or three amino acid substitutions). Only for numbers that are different according to statistical consideration sequences produce nearly perfect results, whereas appreciably worse ones are obtained for compositions. For closely related proteins distinguished by a few amino acid replacements the compositions are very similar and errors due to the hydrolysis procedure are more or less the same for similar proteins. Distantly related proteins are expected to have compositions differing substantially also in those residues which are acid labile such as serine, threonine, cysteine, and tryptophan residues. They may also differ in the number of peptide bonds resistant to acid hydrolysis under standard conditions. Under these conditions compositions even of small proteins cannot be obtained with the required precision.

3 Identification of a Gene by Protein Sequencing

3.1 General Considerations

The recent development of DNA cloning and sequencing methods has made available structural information of genetic material from a wide variety of sources (see Sect. 7). Whilst nucleotide sequencing alone is able to define the approximate length of coding regions from open reading frames and comparison of wild-type with mutant genes or alleles, amino acid sequence analysis is still important for several reasons. Protein sequence information of in vitro synthesized proteins corresponding to the primary translational products encoded by the DNA is used to identify the initiation codon for the gene. In vitro transcription-translation and translation systems suited for this purpose are available for DNA or RNA from prokaryotes and eukaryotes (Zubay systems, wheat germ system, reticulocyte system, oocyte injection system). Furthermore, protein sequencing provides a valuable extension and examination of DNA sequences, especially for regions of the DNA which are difficult to sequence directly or which are not included in cDNA clones. Protein sequencing is also important to establish the status of the active protein in order to decide whether the already mentioned posttranslational alterations modify the primary translational product. This is of special interest to secreted proteins, intrinsic membrane proteins, and proteins transported into intracellular compartments such as chloroplasts, mitochondria, lysosomes, vacuoles or peroxisomes. These proteins are often subject to proteolytical processing during transport and to glycosylation during export. Secreted proteins, in contrast to cytoplasmic proteins, often contain S-S bridges ("Hartley rule"). The covalently linked cysteines of cystines have to be identified by protein chemical techniques at least for the first member of a protein family. N-linked glycosylation is exclusively found at asparagine residues occupying the N-terminal position of internal tripeptides with C-terminal (third position) serine or threonine residues (Hubbard and Ivatt 1981). As a rule this is the acceptor site for N-linked glycosylation of secreted proteins whose verification in nature has to be proved by protein sequencing for each individual new polypeptide of a species.

Protein sequence analysis is still indispensable for the identification and subsequent localization of active site regions of proteins and of accessible regions ("surface residues", antigenic determinants) all of which are definable by chemical modifications or mutational alterations.

The amount of protein needed for N-terminal sequencing may be less than 10^{-9} mole with direct identification of the derivatives released during Edman degradation by high performance liquid chromatography (Fig. 3). Amounts as low as 10^{-15} mole are required for radiolabel protein sequencing techniques (Fig. 3).

3.2 Proof of a Gene Product Encoded by an Insertion Element

The smallest mobile genetic elements are insertion sequences (IS). Several copies of four different IS sequences (IS1, IS2, IS3, and IS4) have been found in the chromosome of *Escherichia coli* (Starlinger 1980). Since all bacterial IS elements sequenced thus far have open reading frames (Klaer et al. 1981, and cited therein) it was attempted to find out whether proteins are detectable corresponding to one of the two open reading frames of IS4. The search was successful for the protein corresponding to the large open reading frame of IS4 (Trinks et al. 1981). A protein of Mr 47,000 encoded by IS4 is synthesized in minicells in the presence of a multicopy plasmid carrying IS4 but not in its absence. The protein was radioactively labeled in vivo and separated from contaminating proteins by sodium dodecylsulfate polyacrylamide gel electrophoresis and subsequently eluted. N-terminal sequencing of homogeneous Mr 47,000 protein labeled with radioactive methionine and leucine showed that these residues occupy the same positions as predicted for the putative gene product of the large open reading frame of IS4 (Fig. 4). The conformity of the aminoterminal sequences suggests that we deal with a polypeptide encoded by IS4. The molecular weight of 47,000 of the isolated IS4 protein is unexpected since the predicted molecular weight is 54,000 (Klaer et al. 1981). A protein of Mr 54,000 has not been detected on gels. Proteolytic processing effecting on C-terminal residues, an erroneous DNA

Protein sequencing method		Amount of protein
Phenylisothiocyanate Method:	Manual procedure	1 nmol/10 μg[a]
(HPLC-detection of PTH--derivatives)	Automated procedures	0.05 to 1 nmol/0.5 to 10 μg[a]
(Radiolabel sequencing)	Automated procedures	0.001 to 1 pmol

[a] Calculated for proteins of molecular weight of 10,000

Fig. 3. Sensitivity of actual protein sequencing methods

IS-ELEMENT IS 4

LARGE READING FRAME: 442 CODONS

RESULTING MOLECULAR WEIGHT OF PROTEIN: 54,000

DNA SEQUENCE:	TT TTT ATG CAC ATT GGA CAG GCT CTT GAT CTG GTA TCC CGT TAC GAT
DEDUCED PROTEIN:	MET HIS ILE GLY GLN ALA LEU ASP LEU VAL SER ARG TYR ASP
PROTEIN SEQUENCE:	MET ——————— LEU ·—— LEU —— ———————
	1 7 14
DNA SEQUENCE:	TCT CTG CGT AAC CCA CTG ACT TCT CTG GGG GAT TAC CTC GAC
DEDUCED PROTEIN:	SER LEU ARG ASN PRO LEU THR SER LEU GLY ASP TYR LEU ASP
PROTEIN SEQUENCE:	—— LEU ——————— LEU ·· ——— LEU ——— ——— LEU ——
	16 20 28

Fig. 4. Co-ordination of protein and DNA sequencing results of IS4. The DNA sequence (Klaer et al. 1981) was used for deduction of a protein sequence (shown is the first methionine residue and the following residues of the large open reading frame). The protein sequence was determined with a Mr 47,000 protein radiolabelled in minicells with ^{35}S-methionine and ^{3}H-leucine (Trinks et al. 1981) using an automated spinning cup sequencer

sequence or an abnormal electrophoretic behaviour of the protein in the presence of sodium dodecylsulfate could account for the molecular weight discrepancy. Which of the three possibilities does hold true will possibly be shown with the help of overproducer strains for IS4 protein.

The work with the IS4 protein indicates that gene protein correlation can be studied even for proteins being made in very small amounts. Less than 10^{-14} mole of radiolabelled protein was employed for the sequence analysis results which are presented in Fig. 4, for which neither specific antiserum nor any functional assay exists, and which had previously not even been known to exist.

Similar assignments of putative gene products to its coding DNA-sequences were also performed in this laboratory with proteins isolated from plants and eukaryotic cell lines.

4 Assignment of Initiation Codons to Protein Genes

4.1 Proteins of the lac Operon of Escherichia coli

The lactose operon of *Escherichia coli* encodes by *lacI* the repressor (tetramer with subunit Mr of 39,000), by *lacZ* the enzyme β-galactosidase (tetramer with subunit Mr of 116,000), by *lacY* the transport protein lactose permease (Mr 46,000 for the subunit, aggregational state unknown), and by *lacA* the enzyme thiogalactosid transacetylase (dimer with subunit Mr of 25,000), respectively (Fig. 5). The complete nu-

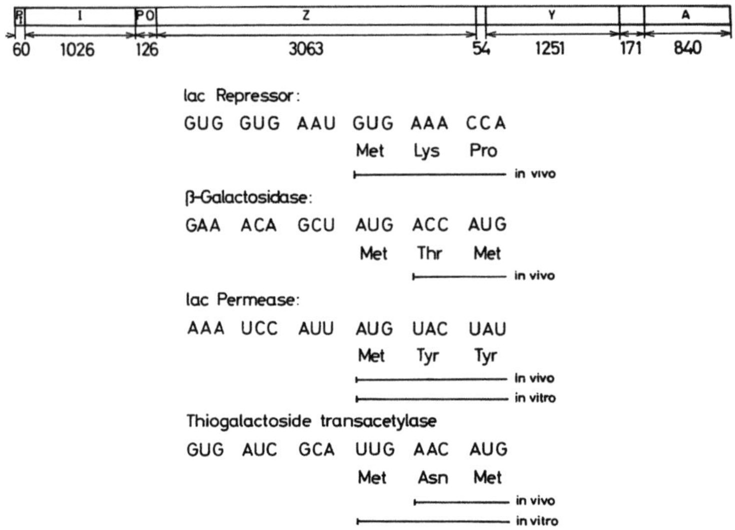

Fig. 5. The lactose operon of *Escherichia coli*. The length of the genes and intergenic regions is given in base pairs at the *top*. The RNA sequences given are deduced from the known DNA sequence of the corresponding regions. Proteins which were isolated from bacteria and used for sequencing are named as in vivo proteins. Sequences of proteins synthesized in cell-free systems are indicated as in vitro products. Only the N-terminal 3 residues are included for clarity; the *vertical bar* indicates the N-terminal amino acid determined

cleotide sequences and protein sequences of all these genes and proteins are known, except for parts of the transacetylase (Beyreuther et al. 1973, Beyreuther 1978, Fowler and Zabin 1977, Zabin and Fowler 1978, Farabaugh 1978, Büchel et al. 1980, Ehring et al. 1980, Kalnins et al. 1983). The sequence of the aminoterminal region of the repressor already indicates that the initiation codon for *lacI* is not the normal AUG but GUG (Fig. 6). Restart proteins in *lacI* also use GUG and UUG as initiation signals (Weber and Geisler, 1978). The translation of β-galactosidase and permease is started at an AUG codon whereas *lacA* uses UUG as the start triplet (Fig. 6).

The identification of the start sites for *lacI* and *lacZ* was based on conventional protein sequence data which were available before the nucleic acid sequence (Fig. 5). The sequences of the amino terminal regions of lactose permease and lactose transacetylase were determined by conventional protein sequencing of material isolated from the cytoplasm or the cytoplasmic membrane of *Escherichia coli* strains but also by radiolabel microsequencing of in vitro synthesized proteins (Beyreuther et al. 1979, Ehring et al. 1980, Zabin and Fowler 1978, Beyreuther, Triesch and Ehring to be published). The in vitro synthesis of the two proteins was done in a plasmid-DNA directed cell-free transcription-translation system (Ehring et al. 1980) in the presence of radioactive amino acids. The predominant radiolabelled product obtained in the presence of the plasmid pGM21 DNA (*lacY*; Teather et al. 1980) is lactose permease, which can be pelleted by centrifugation at 100,000 x g. Total translation mixture or the sedimented biosynthetically labelled permease can be directly used for protein sequencing studies after dialysis against dilute formic acid (Beyreuther et al. 1979). No

GENE	PRODUCT	INITIATION CODON
I	LAC REPRESSOR (1-360)	GUG
	RESTART 23 (23 - 360)	GUG
	RESTART 62 (62 - 360)	UUG
Z	ß-GALACTOSIDASE	AUG
Y	LAC PERMEASE	AUG
A	THIOGALACTOSIDE TRANSACETYLASE	UUG

Fig. 6. The initiation codons in the lactose operon

further purification is required. For the in vitro synthesis of lactose transacetylase a derivative of pGM21 was used which carried the *lacY* and *lacA* genes. The two major products obtained from this plasmid in the in vitro system, thiogalactoside transacetylase, and permease, were separated on sodium dodecylsulfate polyacrylamide gels by electrophoresis and subsequently eluted from the gels or simply by centrifugation at 100,000 x g. The dialyzed supernatant of the centrifugation step gave the same aminoterminal sequence as the Mr 25,000 protein band isolated from gels. Figure 7 shows an experiment performed with ^{35}S-methionine labelled thiogalactoside transacetylase of the supernatant fraction.

The examples presented in this chapter show that the utilization of proper recombinant plasmids renders unnecessary, laborious, and often less involved protein purifications. The sequencing results do indicate that both polypeptides, transacetylase, and permease, are initiated at a single site (Fig. 5), and that the primary translation products have the same N-terminal residue as the mature proteins except for β-galactosidase and transacetylase. The latter two proteins are subjected to demethionylation (Fig. 5).

4.2 Genes may contain more than one used Initiation Site

Initiation codons being not in phase allow the synthesis of different polypeptide products from the same stretch of DNA. Such overlapping genes have been shown to exist in DNA and RNA viruses (Barrell et al. 1976, Tooze 1980). Overlapping genes enable these viruses to code for more proteins than expected from the nucleotide content.

In *Escherichia coli* potential initiation sites for translation may be predicted for open reading frames by searching for Shine-Delgarno ribosome binding sequences

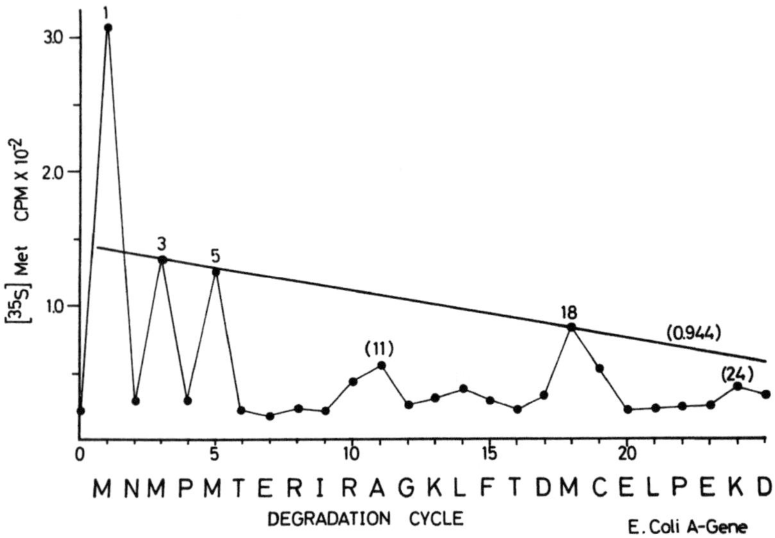

Fig. 7. Amino acid sequence analysis of thiogalactoside transacetylase. The protein was synthesized in a cell-free protein synthesis system programmed with lacY, lacA plasmid DNA in the presence of ^{35}S-methionine. The supernatant obtained after centrifugation at 100,000 *g* (60 min) was dialyzed against 50 mM NH_4HCO_3, and 1 M formic acid and degraded in a spinning cup sequencer. The radioactivity in the fraction obtained after each cycle of Edman degradation is shown. The N-terminal sequence of transacetylase at the *bottom* is according to Zabin and Fowler (1978) and Büchel et al. (1980). The repetitive yield of the degradation is 94.4% for each step. Lactose permease has methionine residues at positions 1, 11, and 23

(Steitz 1979). A 5′ untranslated ribosome binding site is, however, not essential for the initiation of protein synthesis. This has been demonstrated for the CI mRNA transcribed from the promoter P_{RM} of the *Escherichia coli* phage lambda (Walz et al. 1976, Beyreuther and Gronenborn 1976). This mRNA begins immediately with the codon for the N-terminal methionine residue at the 5′ end. The mature protein, lambda repressor, lacks this N-terminal methionine residue and starts at a serine residue encoded by codon two of the CI mRNA (Fig. 8).

The simultaneous utilization of two AUG initiation codons has also been demonstrated by protein-nucleic acid comparison (Meyer et al. 1980). Amino acid

C I PROTEIN:[1] H_2N SER THR LYS LYS LYS PRO LEU THR GLN GLU
(1 ... 10)

m RNA:[2] ppp AUG AGC ACA AAA AAG AAA CCA UUA ACA CAA GAG

Fig. 8. Comparison of the protein sequence of lambda repressor with the sequence of the CI messenger RNA started at the promoter P_{RM} reveals the absence of a leader sequence for the RNA. (*1*) The protein sequence is according to Beyreuther and Gronenborn (1976); (*2*) the RNA sequence was determined by Walz et al. (1976)

sequencing of radiolabelled bacteriophage fd gene 2 protein revealed that the synthesis was initiated at two distinct AUG codons close to the ribosome binding site. The two initiation codons are in phase, separated by two triplets encoding isoleucine and aspartic acid and are used in vivo. The two resulting translational products were found to begin at the first AUG signal for 90% of the chains and at the second AUG codon for 10% of the sequenced molecules. The predominant use of a first or second initiation codon could be of regulatory or functional significance.

5 Identification of Sequence Internal Residues by Radiolabel Sequencing of Peptide Mixtures

The method for identification of sequence internal residues of proteins of known sequence has been worked out with lactose permease of *Escherichia coli* (Fig. 5). It is of general use and allows the localization of residues constituting the active site and of "surface residues" provided the sequence of the protein is known. This reference sequence can either be a DNA, RNA or protein sequence.

Lactose permease is a strongly hydrophobic protein consisting of 417 residues of which 71% are nonpolar (Fig. 2). It contains one essential cysteine residue which is specifically labelled in the absence of substrate using the assay of Fox and Kennedy (1965). This cysteine residue was identified by radiolabel sequencing of peptide mixtures of specifically, at cysteine residues, radioalkylated lactose permease (Beyreuther et al. 1982). The assignment of the cysteine residue at position 148 of permease as the essential one is based on the release of the radioactive cysteine derivative after Edman degradation of peptide mixtures (Fig. 9) and on the knowledge of the DNA sequence of the *lacY* gene (Büchel et al. 1980). Chemical or enzymatic cleavage of lactose permease yields fragments with cysteines at positions predictable from the known sequence of the protein (Beyreuther et al. 1982). The combination of two cleavage reactions for the creation of peptide mixtures (secondary peptide mixtures) provides a test for the prediction of cysteine positions (Fig. 9). The shift of the radioactivity released at positions 4, 6, and 13 by Edman degradations of tryptic digests of lactose permease labelled at the essential cysteine residue to position 3 in secondary peptide mixtures obtained after trypsin plus cyanogen bromide cleavage, is only consistent with cysteine residue 148 carrying the essential SH-group (Fig. 9). An advantage of the method is its independence of pure proteins. The experiments described in Fig. 9 are performed with membrane vesicles enriched in lactose permease (Teather et al. 1980) containing over 100 different polypeptides aside from the permease. The procedure depends only on a radiolabelling procedure of high specificity for the protein to be studied but not on pure proteins.

The same technique was successfully employed in this laboratory for the identification of some "surface residues" of detergent solubilized and membrane-bound lactose permease (Beyreuther et al. 1982). Recently, radiolabel sequencing of peptide mixtures of the same protein has allowed us to probe its orientation in the lipid membrane (Bieseler, Kisters, and Beyreuther in preparation).

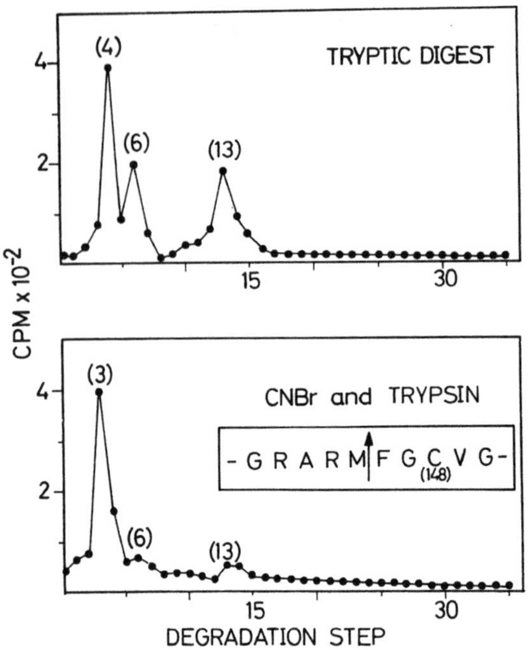

Fig. 9. Radioactivity recovered by automated Edman degradation of specifically radiolabelled lactose permease treated with trypsin *(upper part)* or trypsin followed by cyanogen bromide (CNBr) treatment *(lower part)*. Membrane bound permease was pretreated with N-ethylmaleimide (NEM) and the substrate thiodigalactoside (TDG) and then reacted with ^{14}C-NEM in the absence of TDG. The *insert* in the lower panel shows the sequence around cysteine residue 148. Trypsin cleaves after arginine (R) 142 and 144, giving rise to the release of radioactivity at positions 4 and 6 *(upper panel)*. The tryptic site at position 142 is preceded by a site at position 135 (not shown). Cleavage at this site accounts for the radioactivity release at degradation step 13 *(upper panel)*. CNBr cleaves very efficiently at methionine residues and leads to a shift of all radioactivity to degradation step 3 proving that cysteine 148 is the essential residue

Another application of radioactive sequencing of unfractionated peptide mixtures has been published by Kitamura et al. (1980). Protein and RNA sequence studies allowed the genetic mapping of the genome-linked protein of picornaviruses which is a constituent of a polyprotein product of the RNA genome.

6 Protein Sequencing Complementary to DNA Sequencing

Work in this laboratory is also concerned with the analysis of gene – protein correlations of immunoglobulin heavy chains. The emphasis has been on variant immunoglobulins, in particular on class switch variants and V region variants (Beyreuther et al. 1981, Dildrop and Beyreuther 1981, Dildrop et al. 1982, Zaiss and Beyreuther 1983).

Immunoglobulin genes are composed of individual genetic elements encoding signal sequence, variable region including D- and J-region, constant domains, hinge-region, and membrane segment (only for surface immunoglobulin) of light and heavy chains, respectively (Early and Hood 1981). Which of the exons, discovered as open reading frames on germ-line DNA sequences, constitute the immunoglobulin chains expressed in an immune response of a B-cell can be derived from cDNA sequences or protein sequences. The assignment of cysteine residues to cystines and of actual glycosylation sites remains to the protein chemist.

We have sequenced several murine heavy chains of class switch variants and were thus able to rule out amino acid substitutions (somatic mutations) as accompanying phenomenon of class switching (Beyreuther et al. 1981, Dildrop and Beyreuther 1981, Zaiss and Beyreuther 1983, Bovens et al. to be submitted). One of the murine class switch variants analysed in this laboratory was an IgD. The amino acid sequence of the heavy chain of IgD was partially known from a cDNA sequence when we started our work (Tucker et al. 1980). The DNA derived sequence data did not include the N-terminal sequence of the first domain of the IgD heavy chain and predicted two possible exons for the C-terminal part. The variable region of the heavy chain was also not covered by the sequenced cDNA clone. Conventional protein sequence analysis of the IgD heavy chain (isolated from the hybridoma cell line B1-8delta 1) solved the question which of the C-terminal exons is expressed in secreted mouse IgD (Fig. 10). We were also able to co-ordinate cysteines to cystines (Fig. 11) and to identify the glycosylation sites at asparagine residues. One of the 7 predicted glycosylation sites (Tucker et al. 1980) is outside the actual used exons, the NPT-site carries no oligosaccharide side chain, the five other sites are glycosylated (Fig. 11) (Dildrop and Beyreuther 1981, Dildrop et al. 1982).

Co-ordination of protein and DNA sequencing as exemplified here provides insights into structure-function relationships of proteins which neither of the two sequencing techniques alone, protein or DNA sequencing, is able to accomplish. Genetic mapping is another powerful product resulting from combined efforts of protein and DNA sequencing. The major advantage is however the most efficient way for sequence

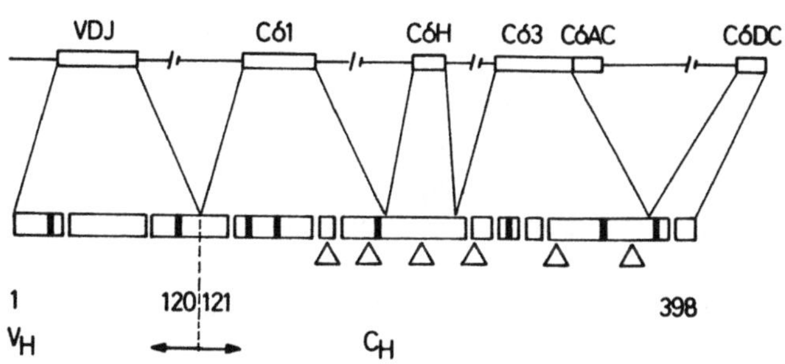

Fig. 10. Schematic confrontation of exons for IgD and of the ordered cyanogen bromide peptides isolated from the heavy chain of IgD. *Connecting lines* indicate which exons are actually used. *Solid bars* indicate positions of cysteines in the sequence and *open triangles* potential glycosylation sites as postulated from the DNA sequence of Tucker et al. (1980). The AC exon is not expressed (Dildrop and Beyreuther 1981). The VDJ region and parts of the first constant domains are not included in the reference DNA sequence

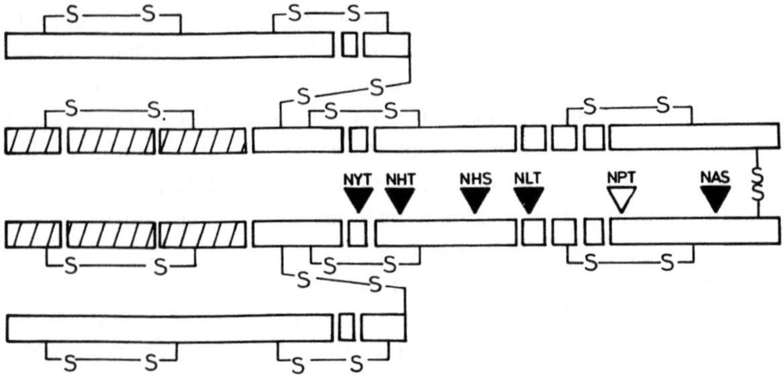

Fig. 11. The complete IgD molecule with schematically indicated cystines and glycosylation sites. *Dark triangles* are the actually glycosylated carbohydrate attachment sites. The sequence NPT (Asn-Pro-Thr) is not glycosylated *(open triangle)*. Light chains and heavy chains are of B1–8 delta 1 (Dildrop et al. 1982). The CNBr peptide pattern of both chains are given. The variable region of the heavy chain is *hatched*

determination, active site localization, gene and gene product identification, establishment of molecular phylogenies and of evolutionary relations is provided by the combination of the present protein and nucleic acid sequencing techniques. This combination has become essential for finding genes in complementary DNA libraries and in genomic libraries by employing mixed oligonucleotide probes synthesized on the information obtained of partial protein sequences (Agarwal et al. 1981).

7 Evaluation of DNA Sequences

The naturally occurring DNA sequences differ much from the expected randomness. Nussinov (1980) took a sample of 11 prokaryotic sequences and 33 eukaryotic sequences, each longer than 450 base pairs and counted the dinucleotides. She found 3,175 times GT, 2,194 times TG, 2,406 times GC, and 956 times CG. According to statistical expectations the occurrence of GT should be similar to the occurrence of TG, and the same applies to the GC/CG pair. Grantham et al. (1980, 1981) investigated a sample of 161 DNA sequences for the frequencies of the 64 different codons in known reading frames. Again strikingly nonrandom distributions were found, which appeared to be specific for the organism investigated and the type of gene. A different distribution was found for highly transcribed genes versus genes that are expressed at a lower level. Table 1 shows the dimer and trimer analyses from the sample of about 600,000 bases contained in the EMBL nucleotide sequence data base (Hamm and Stüber 1982). Table 2 gives the codon distribution of known reading frames in prokaryotic and eukaryotic genes from the same sample of DNA sequences. Almost all of these frequencies prove to be significantly different from those frequencies expected from the composition of the DNA when tested statistically.

Table 1. Occurrence of dinucleotides and trinucleotides in the codon catalog (sequence data base: 600,000 base pairs; Hamm and Stüber 1982)

A: Dimers and trimers counted in eukaryotes (sample size: 218,477 base pairs)

AA	17,431	AC	11,726	AG	15,317	AT	13,885
CA	15,531	CC	13,417	CG	5,746	CT	15,827
GA	14,346	GC	11,963	GG	14,198	GT	11,324
TA	11,047	TC	13,440	TG	16,532	TT	16,344
AAA	5,686	AAC	3,203	AAG	4,584	AAT	3,939
ACA	3,848	ACC	3,020	ACG	1,348	ACT	3,494
AGA	4,579	AGC	3,610	AGG	4,184	AGT	2,921
ATA	3,095	ATC	3,011	ATG	3,883	ATT	3,882
CAA	4,220	CAC	3,216	CAG	4,500	CAT	3,570
CCA	4,013	CCC	3,611	CCG	1,670	CCT	4,089
CGA	1,371	CGC	1,469	CGG	1,613	CGT	1,280
CTA	2,602	CTC	3,841	CTG	5,366	CTT	3,980
GAA	4,301	GAC	2,889	GAG	4,162	GAT	2,968
GCA	3,576	GCC	3,162	GCG	1,359	GCT	3,839
GGA	4,043	GGC	3,344	GGG	3,604	GGT	3,174
GTA	2,140	GTC	2,649	GTG	3,681	GTT	2,838
TAA	3,183	TAC	2,399	TAG	2,048	TAT	3,400
TCA	4,076	TCC	3,606	TCG	1,347	TCT	4,368
TGA	4,307	TGC	3,521	TGG	4,747	TGT	3,935
TTA	3,196	TTC	3,913	TTG	3,580	TTT	5,622

B: Dimers and trimers counted in a prokaryote *(E. coli)* (sample size: 52,817 base pairs)

AA	4,178	AC	2,938	AG	2,858	AT	3,149
CA	3,149	CC	2,881	CG	3,837	CT	3,003
GA	3,383	GC	4,061	GG	3,269	GT	3,231
TA	2,383	TC	2,994	TG	3,977	TT	3,493
AAA	1,380	AAC	995	AAG	985	AAT	814
ACA	686	ACC	803	ACG	790	ACT	658
AGA	719	AGC	928	AGG	621	AGT	588
ATA	504	ATC	905	ATG	900	ATT	815
CAA	813	CAC	660	CAG	974	CAT	700
CCA	726	CCC	526	CCG	1,007	CCT	619
CGA	894	CGC	1,083	CGG	903	CGT	953
CTA	397	CTC	545	CTG	1,416	CTT	643
GAA	1,209	GAC	703	GAG	583	GAT	882
GCA	951	GCC	895	GCG	1,246	GCT	966
GGA	642	GGC	1,021	GGG	635	GGT	966
GTA	732	GTC	672	GTG	851	GTT	971
TAA	768	TAC	575	TAG	312	TAT	726
TCA	784	TCC	654	TCG	790	TCT	759
TGA	1,119	TGC	1,026	TGG	1,106	TGT	723
TTA	750	TTC	872	TTG	806	TTT	1,062

Table 2. Codon distribution of known reading frames in the nucleotide sequence data base of Hamm and Stüber (1982)

A: Codon frequencies of eukaryotes

TTT	444	TCT	556	TAT	371	TGT	374
TTC	693	TCC	541	TAC	515	TGC	377
TTA	197	TCA	375	TAA	0	TGA	0
TTG	423	TCG	139	TAG	0	TGG	549
CTT	339	CCT	499	CAT	393	CGT	188
CTC	572	CCC	401	CAC	443	CGC	191
CTA	233	CCA	534	CAA	536	CGA	125
CTG	947	CCG	142	CAG	703	CGG	134
ATT	377	ACT	517	AAT	402	AGT	343
ATC	605	ACC	665	AAC	632	AGC	519
ATA	225	ACA	470	AAA	707	AGA	501
ATG	603	ACG	175	AAG	1,029	AGG	384
GTT	375	GCT	682	GAT	544	GGT	580
GTC	492	GCC	663	GAC	623	GGC	555
GTA	226	GCA	419	GAA	691	GGA	459
GTG	756	GCG	151	GAG	782	GGG	303

B: Codon frequencies of a prokaryote *(E. coli)*

TTT	189	TTC	215	TTA	97	TTG	110
TCT	134	TCC	126	TCA	69	TCG	80
TAT	122	TAC	143	TAA	0	TAG	0
TGT	54	TGC	70	TGA	0	TGG	109
CTT	104	CTC	99	CTA	32	CTG	598
CCT	76	CCC	53	CCA	78	CCG	230
CAT	98	CAC	116	CAA	137	CAG	301
CGT	267	CGC	221	CGA	60	CGG	60
ATT	241	ATC	327	ATA	24	ATG	280
ACT	124	ACC	211	ACA	66	ACG	106
AAT	131	AAC	248	AAA	394	AAG	147
AGT	53	AGC	139	AGA	31	AGG	17
GTT	256	GTC	140	GTA	164	GTG	225
GCT	254	GCC	231	GCA	236	GCG	317
GAT	275	GAC	250	GAA	424	GAG	195
GGT	338	GGC	283	GGA	54	GGG	90

The functional meaning of this has yet to be explained. It is attempted below to substantiate some hypotheses on the evolutionary value of these observed distributions, but the understanding is far from being complete.

The dimer distribution found shows great differences between eukaryotes and prokaryotes. The preference for GC towards CG might be explained assuming a general genetic necessity (methylation?) because this is so often observed in eukaryotes.

The choice between different isocoding codons is suspected to be governed by the need for translation security. Modiano et al. (1981) have investigated a sample of wild type and mutant globin genes observed in the human population, and concluded that preferentially such codons are used which are non-vulnerable to mutations i.e. more than one mutation is necessary to convert a codon to a nonsense codon. This trend towards translational security can also be observed in the genetic code itself, where similar amino acids (hydrophobic or hydrophilic) often have similar codons. Many mutations are thus exchanges between similar and isofunctional amino acids.

Grantham et al. (1980, 1981) found that proteins which are expressed at different levels have different codon frequencies such that often used proteins avoid unusual codons. This might be explained by the distribution of the different isoaccepting tRNA's.

The nonrandom distribution of codons can be used to discern coding regions from noncoding regions (Fickett 1983, Shulman et al. 1981). It has to be remembered that DNA, apart from simply coding for proteins, also has many other functions to specify. The expression of genes has to be regulated by regulatory signals, the DNA itself must be replicated in a regulated fashion again involving special sequences, the mRNA transcribed from DNA has to be stable at physiological temperature, and has to be spliced in an ordered fashion which cannot be achieved without the occurrence of special directing subsequences between the other signals.

DNA can thus be envisaged as a complex carrier of information for many diverse and even contradicting needs. This is schematically outlined in Fig. 12. The proportions given to the different information segments in Fig. 12 are of course arbitrarily chosen and will differ for each species and gene which is investigated.

There remain parts of sequences, which obviously carry no information where the sequence might be varied at random without disrupting function. These sequences, like introns, intergenic regions or repetitive sequences, still have dimer and trimer distributions which are far from random. This reflects a general tendency of unused DNA to accumulate redundancy and might possibly be traced back to the mechanisms of DNA replication and repair.

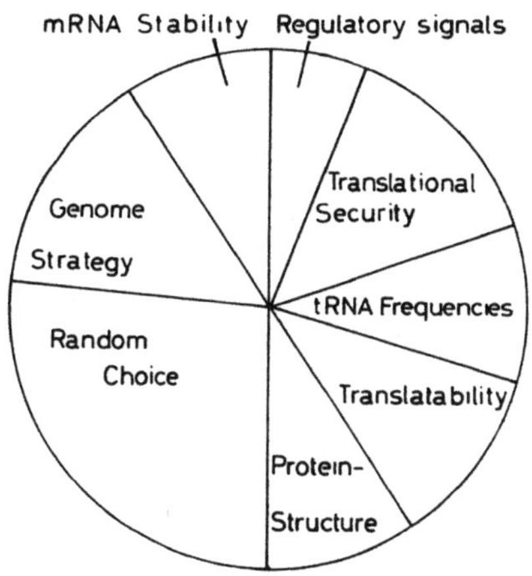

Fig. 12. Information content of genomic DNA (genetic information) divided into arbitrary units which may differ for each species. The distribution shown applies for a typical eukaryote like mouse nuclear DNA. In prokaryotes, mitochondria, chloroplasts, bacteriophages, and viruses protein structure may account for more than 90% of the DNA content and random choice may approach zero

8 Conclusions

For genetic and genealogical studies DNA cloning and sequencing, together with at least partial protein sequencing, has been very successfully employed in the past four years.

Protein sequencing is still complementary to DNA sequencing. It is indispensable for the interpretation of DNA sequences providing information on codogenic regions, introns, exons, the number of gene products encoded by a stretch of DNA (coding capacity), the modification of the primary gene product and the functional assignment of individual parts of a gene.

Comparison, of DNA sequences of genes or exons, reveals two kinds of differences, those resulting in amino acid substitutions and those leading to silent substitutions. The amino acid sequence comparison's fail to account accurately for the latter kind of alterations, even when employing mathematical correction methods. However, silent substitutions may be of limited use for revealing genealogical relations, and for measuring evolutionary time. They are of value, but only for recently diverged genes. The same holds true for the evaluation of substitutions of noncoding sequences (Efstratiadis et al. 1980). Thus, codogenic regions of DNA sequences have to be determined either by protein sequencing, or by one of the nuclease mapping techniques, or by the comparison of germ line DNA sequences with cDNA sequences. Within a given gene family, this has to be done only once within a species but repeated for each separate species. Number and length of exons and introns may differ even for genes encoding functional homologous proteins from species to species (Perler et al. 1980, Efstratiadis et al. 1980).

Comparison of DNA and protein sequences for construction of phylogenies does not necessarily yield better trees, as for instance, comparison of amino acid compositions of related proteins. Closely related proteins of well known families and their genes need not be sequenced for the establishment of taxonomic relationships if they differ only by a few substitutions. The precision of amino acid analysis, however, sets limits to the size of proteins being compared. Proteins up to a molecular weight of about 25,000 daltons are well suited for sequence comparison by the indirect method of amino acid analysis. Larger proteins are much less suitable, due to the relatively high error of the determination of amino acid compositions.

Acknowledgements. This work was supported by grants from the DFG through SFB 74, from the Minister für Wissenschaft und Forschung des Landes Nordrhein-Westfalen, from the Bundesministerium für Forschung und Technologie, and from the Fonds der Chemischen Industrie to K.B.

References

Abelson J, Butz E (1980) Science 209:1317–1438

Agarwal KL, Brunstedt J, Noyes BE (1981) J Biol Chem 256:1023–1028

Barrell BG, Air GM, Hutchinson CA (1976) Nature (London) 264:34–40

Beyreuther K (1978) In: Miller JH, Reznikoff WS (eds) The operon. Cold Spring Harbor Lab, Cold Spring Harbor, pp 123

Beyreuther K, Gronenborn B (1976) Mol Gen Genet 147:115–117
Beyreuther K, Adler K, Geisler N, Klemm A (1973) Proc Natl Acad Sci USA 70:3576–3580
Beyreuther K, Ehring R, Overath P, Wright JK (1979) In: Birr C (ed) Methods in peptide and protein sequence analysis. Elsevier/North-Holland, Amsterdam New York Oxford, pp 199
Beyreuther K, Bovens J, Dildrop R, Dorff H, Geske T, Liesegang B, Müller C, Neuberger MS, Radbruch A, Rajewsky K, Sablitzky F, Schreier PH, Zaiss S (1981) In: Janeway C, Sercarz EE, Wigzell H, Fox CF (eds) Immunoglobulin idiotypes and their expression. Academic Press, London New York, pp 229
Beyreuther K, Bieseler B, Ehring R, Müller-Hill B (1982) In: Elzinga M (ed) Methods in protein sequence analysis. Humana Press, Clifton, NJ, pp 139
Büchel DE, Gronenborn B, Müller-Hill B (1980) Nature (London) 283:541–545
Cleveland DW, Fischer SG, Kirschner MW, Laemmli UK (1977) J Biol Chem 253:1102–1106
Cornish-Bowden A (1980) Biochem J 191:349–354
Dildrop R, Beyreuther K (1981) Nature (London) 292:61–63
Dildrop R, Brüggemann M, Radbruch A, Rajewsky K, Beyreuther K (1982) EMBO 1:635–640
Early P, Hood L (1981) Cell 24:1–3
Efstratiadis A, Posakony JW, Maniatis T, Lawn RM, O'Connell C, Spritz RA, De Riel JK, Forget BG, Weissman SM, Slightom JL, Blechl AE, Smithies O, Baralle FE, Shoulders CC, Proudfoot NJ (1980) Cell 21:653–668
Ehring R, Beyreuther K, Wright JK, Overath P (1980) Nature (London) 283:537–540
Farabaugh PJ (1978) Nature (London) 274:765–769
Fickett JW (1983) Nature (London) (in press)
Fowler AV, Zabin I (1977) Proc Natl Acad Sci USA 74:1507–1510
Fox CF, Kennedy EP (1965) Proc Natl Acad Sci USA 54:891–899
Grantham R, Gautier C, Gouy M, Mercier R, Pave A (1980) Nucleic Acids Res 8:49–62
Grantham R, Gautier C, Gouy M, Jacobzone M, Mercier R (1981) Nucleic Acids Res 9:43–74
Hamm G, Stüber K (1982) Nucleotide sequence data library news. Eur Mol Biol Lab Heidelberg, I
Higgins CF, Haag PD, Nikaido K, Areshir F, Garcia G, Ferro-Luzzi Ames G (1982) Nature (London) 298:723–727
Hubbard SC, Ivatt RJ (1981) Annu Rev Biochem 50:555–583
Kalnins A, Otto K, Rüther U, Müller-Hill B (1983) EMBO J 2:593–597
Kitamura N, Adler CJ, Rothberg PG, Martinko J, Nathenson SG, Wimmer E (1980) Cell 21:295 to 302
Klaer R, Kühn S, Tillmann E, Fritz H-J, Starlinger P (1981) Mol Gen Genet 181:169–175
Maxam AM, Gilbert W (1977) Proc Natl Acad Sci USA 74:560–564
Meyer TF, Beyreuther K, Geider K (1980) Mol Gen Genet 180:489–494
Modiano G, Battistuzzi G, Motulsky AG (1981) Proc Natl Acad Sci USA 78:1110–1114
Nussinov R (1980) Nucleic Acids Res 19:4545–4562
O'Farrell PH (1975) J Biol Chem 250:4007–4021
Perler F, Efstratiadis A, Lomedico P, Gilbert W, Kolodner R, Dodgson J (1980) Cell 20:555–565
Reisner AH, Westwood NH (1982) J Mol Evol 18:240–250
Sanger F, Nicklen S, Coulson AR (1977) Proc Natl Acad Sci USA 74:5463–5467
Shulman MJ, Steinberg CM, Westmoreland N (1981) J Theor Biol 88:409–420
Starlinger P (1980) Plasmid 3:241–259
Steitz JA (1979) In: Goldberger RF (ed) Biological regulation and development. Plenum Press, New York London, pp 349
Teather RM, Bramhall J, Riede I, Wright JK, Fürst M, Aichele G, Wilhelm U, Overath P (1980) Eur J Biochem 108:223–231
Tooze J (ed) (1980) The molecular biology of tumour viruses, 2nd edn. Cold Spring Harbor Lab, Cold Spring Harbor
Trinks K, Habermann P, Beyreuther K, Starlinger P, Ehring R (1981) Mol Gen Genet 182:183–188
Tucker PW, Liu CP, Mushinski JF, Blattner FR (1980) Science 209:1353–1356
Uy R, Wold F (1977) Science 198:890–896
Walz A, Pirrotta V, Ineichen K (1976) Nature (London) 262:665–669

Weber K, Geisler N (1978) In: Miller JH, Reznikoff WS (eds) The operon. Cold Spring Harbor Lab, Cold Spring Harbor, pp 155
Wilson AC, Carlson SS, White TJ (1977) Annu Rev Biochem 46:573–639
Zabin I, Fowler AV (1978) In: Miller JH, Reznikoff WH (eds) The operon. Cold Spring Harbor Lab, Cold Spring Harbor, pp 89
Zaiss S, Beyreuther K (1983) Eur J Immunol 13:508–513

Protein Characters and their Systematic Value

P. v. SENGBUSCH[1]

Abstract. Much knowledge has been accumulated during the past two decades with regard to the application of protein sequences for the construction of genealogies. In the present paper the evolution of protein structures is evaluated in view of natural selection. Special attention has been called to the influence of the intracellular environment (being to a large extent under control of the cellular genome) on the selection of protein structures, and to the importance of the evolution of redundancy (redundant information, redundant structures).

1 Introduction

One of the pertinent problems in the fields of taxonomy and evolution is the construction of genealogies (phylogenetic trees), which should represent the ascendency of any particular taxon by tracing historic events as accurately as possible. The traditional approach is based on the comparison of morphological characters of living specimens in addition to the study of fossils. However, there are no general rules which guide the choice of the most suitable morphological entities. For most groups of taxa different sets of characters have been applied. This is partly due to the circumstance that nearly all of them are adaptive, i.e. they evolved for optimal fitness under natural selection. Often it is difficult to decide whether a character is a primitive or a derived trait, and due to selection forces divergent and convergent lines of evolution arise.

2 Proteins as Probes for Reconstruction of Phylogenies

The congeniality of two taxa is based on the relatedness of their genomes. Therefore, two species which recently diverged from a common ancestor are closer related to one another than species which have branched off in earlier times.

To obtain an unbiased record of the changes which accumulated in both lines of descent, it would be ideal to have complete information about the genomes of taxa which are under consideration. For practical and technical reasons, however, this ap-

[1] Institut für Allgemeine Botanik und Botanischer Garten, Universität Hamburg, 2000 Hamburg 52, FRG

Proteins and Nucleic Acids in Plant Systematics
ed. by U. Jensen and D.E. Fairbrothers

proach is inaccessible at the present time. Therefore one is forced to apply approximation methods.

Since morphological characters are end products of complex interconnected biochemical pathways which are under the control of a number of genes and external signals, it is obvious to analyze proteins which are known to be primary gene products. The framework behind this notion is the classical one gene – one enzyme (polypeptide) hypothesis (Beadle and Tatum 1941, Horowitz 1948). When methods became available during the late fifties and sixties for the analysis of protein structures, it was feasible to comprehend changes in proteins for a better understanding of genes. During that period one of the highlights was the deciphering of the genetic code.

It also became conceivable to study proteins in order to solve problems in phylogeny. Similarities and dissimilarities were believed to reflect evolutionary changes. Evidence for this was furnished as early as 1963 from a comparison of cytochrome c – sequences of seven species (man, horse, pig, rabbit, chicken, tuna, and baker's yeast) (Margoliash 1963).

These data led to the concept of an evolutionary (molecular) clock (Zuckerkandl and Pauling 1965) and to the neutralist hypothesis (theory of non-darwinian evolution) (Kimura 1968, King and Jukes 1969, Kimura and Otha 1971, 1974). Both ideas are based on the assumption that the genetic information which is encoded in a linear array of nucleotide bases in DNA is translated into a colinear sequence of amino acids, forming a polypeptide chain. Substitutions of amino acids were thought to reflect spontaneous alterations (point mutations) in the DNA which occur as stochastic events and accumulate in the genome. Most of these events are disadvantageous and therefore are discarded by natural selection. A few of them, however, are neutral or near neutral and therefore are accepted as amino acid substitutions in the protein. The mutant residue thus would be equivalent to the residue which is replaced, as far as function is concerned. The comparison of the number of amino acid substitutions (replacements) in homologous proteins, from species whose time of evolutionary divergence is known from other sets of data (e.g. fossil record), allowed the calibration of rates at which mutations accumulated in the genes. The rates are expressed quantitatively as the number of accepted point mutations per 100 amino acid residues per 100 million years.

It soon became apparent that the rate of change (the "molecular clock") differs in proteins, having different functions (Jukes and Holmquist 1972). Cytochrome c, e.g. evolved consistently slower than fibrinopeptides, but much more rapidly than histon IV. Rates were found to vary over three orders of magnitude. Analytical data on a number of proteins from many species have been collected and discussed (Dayhoff 1972, Boulter 1973, Cronquist 1976, Wilson et al. 1977, Goodman 1981, Jensen 1981, Scogin 1981, Ramshaw 1982).

Protein evolution thus seemed to open a new way of constructing phylogenetic trees which are independent of morphological evidence (Fitch and Margoliash 1967). The data turned out to be highly suitable for evaluation by sophisticated mathematical procedures. With knowledge of the genetic code it was possible to attribute a particular amino acid substitution to a minimal number (1, 2 or 3) of nucleotide replacements in the corresponding codon. It also was possible to calculate frequencies of amino acid replacements and to compare these predictions with actually determined data.

Computer simulation of evolutionary changes allowed an estimation of the proportion of expected error due to convergencies and backmutations (Dayhoff 1969, Fitch 1971, Peacock and Boulter 1975, Fitch and Langley 1976).

Using immunological techniques as analytical tools, it has been shown that albumin (and other monomeric serum proteins) are excellent markers for tracing protein evolution in several vertebrate classes (Wilson et al. 1974a,b, Prager and Wilson 1976). In an extensive study evidence was provided that protein evolution proceeds at a rate independent of that of chromosomal and organismal evolution, although the phylogenies in general agree with respect to the branching order. In the same space of time one group of organisms (e.g. mammals) have evolved in great morphological variety, while another group (e.g. frogs) morphologically remained essentially unchanged. However, the proteins of frogs generally have shown the same rate of sequence change as those of mammals.

In contrast, rates of gene rearrangement (chromosome alterations) turned out to be unusually frequent in groups with both high rates of phenotypic evolution and speciation. Within the group of frogs, Maxson and Wilson (1975) found that the genus *Arcis* on the protein level is more similar to some species of the genus *Hyla* than the latter are to other *Hyla* species. Yet *Arcis* strikingly differs from *Hyla* in anatomy and the way of life. Evidently the *Arcis* lineage underwent unusually rapid organismal evolution.

Compatible results were obtained (King and Wilson 1975) by comparing proteins and DNA of humans and African apes (chimpanzees and gorillas). The overall protein composition differs by an average of 0.8% in the amino acid sequences and by about 1.1% in the unique sequences of DNA. There are, however, changes of chromosome number (2n = 46 in humans, 2n = 48 in African apes) and various re-arrangements and inversions of chromosome sections, which are believed to be responsible for the rapid evolution and speciation of humans. These alterations most certainly are closely connected with protein evolution. However, they do not primarily change sequences, but rather the regulation of synthesis and the activity of proteins, thereby generating new intracellular environments, which subsequently may select new sequences.

3 Natural Selection Acting on Proteins

Despite the results presented in the preceeding section and the apparent evidence for a clocklike behavior of proteins in evolution, overwhelming evidence accumulated over the years is in favour of a selectionist or darwinian view of protein evolution (Goodman et al. 1975). In the most stringent sense the proponents of this alternative claimed that one particular amino acid residue will be optimum at a given site in each protein. Substitutions will be accepted only if they improve its structure and function.

To understand the arguments which support the impact of natural selection it is necessary to realize that each protein has a particular, specific function in the cell or organism, and natural selection preserves anything that is essential for survival. In

1971 Eigen pointed out that the two classes of macromolecules (nucleic acids and proteins, encode a folding device for adapting a state of lower free energy (as com- but are incapable of self-replication without a high rate of error. They are capable of instructing the formation of protein molecules. The specific amino acid sequences of proteins, encode a folding advice for adapting a state of lower free energy (as compared to the unfolded molecule), thereby spontaneously forming unique three-dimensional structures. In this distinct conformation the protein is capable of performing a specific catalytic function. Most likely the original function of a primitive ancient protein (enzyme) was catalysis to ensure maintenance of the genetic information (instructing this protein), and to minimize the error rate by proof-reading during nucleic acid replication. A duplication (multiplication) of genetic information was the first step to give rise to more genes and thus was a prerequisite for protein evolution. While the original function could have been conserved, the gene product(s) of the newly acquired gene(s) were free to accumulate mutations, thereby evolving new functions and specificities.

When discussing protein (enzyme) evolution, it is important to distinguish between protein differentiation and protein speciation. Differentiation, thereby, is the process which leads to functional diversification of homologous proteins. Descendents of the ancestral protein also are termed paralogous. Speciated proteins (orthologous proteins) are those which fulfill the same function in different organisms. They can be used, as has been demonstrated, to estimate the genealogy of organisms.

3.1 Evolution (Differentiation) of Enzymes

Koshland (1976) proposed that enzyme differentiation occurred on four evolutionary levels, (1) catalytic power, (2) specificity, (3) regulation, and (4) co-operation.

3.1.1 Evolution of Catalytic Power

To obtain catalytic power a specific three-dimensional structure has to evolve. This structure – and not particular amino acid residues – is most stringently preserved. All cytochrome c's, e.g. in eukaryotic cells, have exactly the same topology and arrangement of the polypeptide chain, but only a few amino acid residues, being located in the active site, are invariable. In prokaryotes cytochrome c function is accomplished by related proteins which have been shown to have a three-dimensional structure very similar to that of cytochrome c (Dickerson 1971, 1980). These structures proved to be much better guides to the study of distantly related proteins than were amino acids alone (topological relatedness: Schulz 1977, Schulz and Schirmer 1979).

3.1.2 Evolution of Specificity

Enzymes evolved in families (Dayhoff et al. 1975), e.g. dehydrogenases use NAD as coenzyme. The polypeptide-chains are folded to give rise to two (or more) independent structural domains (folding units) (Rossmann et al. 1975). One of these domains (the NAD-binding domain which also is responsible for catalytic function) is the same in all of these enzymes. Depending on the type of dehydrogenase it is joined to (at least) a second one, which ensures specificity, i.e. it discriminates between the substrates as glutaraldehyde, lactate, ethanol etc. The enzymes thus have evolved on a basis of a modular system. By the fusion of two pieces of genetic information new information originated, which is capable of instructing the synthesis of a polypeptide chain with two structural domains as modules.

In bacterial populations, change of specificity towards a substrate has been induced under selective laboratory conditions (Betz et al. 1974, Wills 1976, Hartley 1980). By subsequent analysis of the corresponding enzymes it was shown that a change of substrate specificity was caused by single amino acid substitutions at the expense of the original substrate specificity. This kind of evolutionary strategy is frequent in fast replicating species, which have large population numbers. The genetic information is lost in the individual and its descendents, but it is retained in the gene pool of the population.

Changes of specificity by loss and new arrangement of original information also is known in eukaryotic species, as illustrated by the following example. In plants, especially legumes, specific carbohydrate binding proteins (lectins) exist, e.g. concanavalin A in *Canavalia ensiformis*, which has the capacity to bind to α-D-mannosyl- and/or α-D-glucosyl-residues. The amino acid sequence and the three-dimensional structure of this protein is known. A single polypeptide chain contains 237 amino acid residues. In *Vicia faba* another lectin (favin) is present. Surprisingly, a comparison of the two proteins revealed a circular permutation of extensive homologous sequences. Favin is made up of two polypeptides; the α-chain is homologous to residues 70–119 of concanavalin A, and the β-chain to the remaining sequences 120–237 followed by 1–69. It is a fusion product of the two terminal sections of concanavalin A. Since the sugar binding specificity remained unchanged, it was assumed that the three-dimensional structures of both lectins also are the same. There is additional evidence that similar three-dimensional structures are present in the lectins from soybean, peanut, and red kidney bean. These, however, are known to have different sugar binding specificities (Cunningham et al. 1979).

3.1.3 Evolution of Regulation

Regularity arose by the evolution of flexibility of the protein structure and the capacity to bind small molecules, other than the substrate, thereby inducing conformational changes (induced fit) of the active site. These structural changes alter the affinity of the enzyme to its substrate.

3.1.4 Evolution of Co-operation

Co-operation is based on the capacity of two or more polypeptide chains (subunits) to form allosteric complexes, in which information about binding the first substrate molecule to one of the subunits is transmitted to the other(s), thereby tuning its (their) sensitivity towards the binding of additional substrate molecules (Monod et al. 1963, Perutz and TenEyck 1972). The evolution of the tetrameric haemoglobin in vertebrates is a well studied example for the evolution of co-operative effects (Goodman and Moore 1973). Mutations, which have improved haemoglobin function, were accepted at an accelerated rate in early vertebrate evolution. During this period gene duplications occurred, which were the basis for the independent evolution of the myoglobin and the α- and β-chains of haemoglobin. During the transition of monomeric haemoglobin into a tetramer (with a sigmoid oxygen equilibrium curve) the residue positions (in the α- and in the β-chain) which acquired co-operative function changed more rapidly than other positions. Mutation rates decelerated (by a factor of 10) after the function had been optimized and both chains were mutually co-adapted. Possibilities for improvements were great in the beginning, but the chances gradually declined. After the function was perfected, evolution slowed since the majority of amino acid substitutions were detrimental and therefore were discarded through stabilizing selection. The tetrameric form is much more efficient in transporting oxygen from an oxygen-rich environment to the tissues and organs than is a monomer. This step was very likely one of the major prerequisites of the evolution of fast and large bodied fish, and the subsequent evolution of terrestric vertebrates.

3.2 Functional and Structural Constraints of Structural Proteins

The sequence data on histon IV (along with those on histons III, IIa, and IIb) express extensive conservation. Histon IV sequences have remained virtually unchanged since before the divergence of animals and plants. Only 2% amino acid substitutions were found by comparing pea and calf histon IV (DeLange et al. 1969a,b). There is, however, a 22% difference, if *Tetrahymena* (a ciliat) is included in this comparison. Histon I is characterized by a fairly large degree of variance (DeLange 1980). Each two molecules of histones IV, III, IIa, and IIb are known to form the core of a nucleosome (Kornberg 1977), around which 140 base pairs of DNA are supercoiled. Histon I is not involved in this multimolecular (supramolecular) structure. Instead, it is reversibly attached to regions of spacer DNA, thereby being under less functional constraints. The nucleosome structure is characteristic for all eukaryotic cells. It is essential for the structural organisation of the genetic material and its accurate distribution during cell division. Handling of genetic information (storage, transmission, replication) is the most vital set of activities in a cell, thus being under extremely high stabilizing selection pressure. Optimization of this supramolecular unit was achieved during early eukaryote evolution. The differences found in present day *Tetrahymena*, are thus a relic of still imperfect sequences. It is noteworthy also to recall that chromosome organization and the mechanism of mitosis are present, but not yet optimized, in the most primitive group of eukaryotes, the protists.

Just another example of a conservative multimolecular unit is the ribosome which is involved in translation, an array of functions which are essentially as important as DNA-handling. Both proteins and ribosomal RNA (rRNA) are essential for functional integrity. It has been shown that the rRNA has evolved for conserving a specific folding pattern, allowing us to trace back the evolution of very primitive cells as well as of kingdoms of organisms (see Kössel et al. this volume).

Less stringent conditions with respect to amino acid substitutions have been elucidated by comparing the coat proteins of four tobacco mosaic virus (TMV) strains.

The coat protein has the inherent property to aggregate into rodlike helices which enclose the genetic material (a single stranded RNA) of the virus. In all four strains a few amino acid positions are invariant, and it was assumed that these residues were involved in contacts with the RNA (Wittmann et al. 1969). This notion was confirmed by comparison of sequence data and the three-dimensional structure of the coat protein (Bloomer et al. 1978) (Figures 1 and 2). In most other positions, 2, 3 or even 4 different amino acids were accepted which indicates that the conservation of this particular three-dimensional structure and the subunit → subunit bonds can be maintained by using sets of different amino acids. Nothing is known about the origin of this virus. The analyzed sequences do not provide convincing clues about the genealogy of the strains.

A number of mutants of one of the wildtypes (*vulgare*) were isolated and the amino acid sequences of their proteins were determined (Wittmann and Wittmann-Liebold 1966). The mutants differ from the wildtype by one or two amino acid substitutions each. Using rabbit antiserum to wildtype virus, it was possible to identify some, but not all, of the amino acid substitutions (v. Sengbusch 1965, van Regenmortel 1967). In contrast to the monomeric albumin, TMV coat protein is polymerized and thus only a small portion of the polypeptide chain is exposed at the surface of intact virus particles. It was predicted that only amino acid residues which alter this region can be monitored using antiserum. When the experimentally determined three-dimensional structure became known, this prediction was verified (Figures 3 and 4).

4 Intra- and Extracellular Proteins

4.1 Constraints by the Cellular Milieu

Cells are considered as units of life. This assertion obviously leads to the assumption that cellular functions have an impact on proteins and their evolution. Cells contain numerous different enzymes. The majority of them are oligomeric, i.e. they are built from several subunits. These are involved in catalysis and regulation of a variety of interconnected metabolic pathways. The amount and the efficiency of each particular enzyme is specific and depends on the enzyme itself, its rate of synthesis, and the supply of metabolites. The concentrations of many enzymes in corresponding cells of closely related species may vary by an order of magnitude. The adaptive significance of these concentration differences is understood only in a few cases (Wilson et al. 1977).

The formation of enzymes is instructed exclusively by the genome of a given cell. Since enzymes within a cell have to co-operate in many aspects, it is evident that the requirements which they have to fulfill are dictated primarily by the presence of other proteins. Consequently the selective force, to which cellular proteins are exposed, is controlled by the genome of the cell more than by the environment. In contrast, secondary metabolites and morphological entities evolve mainly in response to external stimuli. Therefore it is not unexpected to find that protein evolution and organismal evolution proceed at different rates.

```
                                    10                  20
vulgare         ac S Y S I T T P S Q F V F L S S A W A D P I E L I N L C
dahlemense                   S                   V               L   V
U 2             -  P   T   N             Y         Y       V
HRG                    N     N S N   Y Q Y F A A V     E   T P M L   Q
--------------------------------------------------------------------------
                      30                  40                  50
vulgare            T N A L G N Q F[Q T Q]Q A R T V V Q R Q F S Q V W K P
dahlemense           S S                         T     Q       E
U 2                                              T     Q     A D A
HRG                V S     S Q S Y       A G   D     R Q     A N L L S T
--------------------------------------------------------------------------
                            60                  70                  80
vulgare            S P Q V T V R F P D S D F K V Y R Y N A V L D P L V T
dahlemense         F     S           G D V Y                         I
U 2                    V M           A       Y           S T         I
HRG                I V A P A V         T G   R     V N S     I R     Y E
--------------------------------------------------------------------------
                                  90                 100
vulgare            A L L G A[F D T R N R I I]E V E N Q A N P T T A E T L
dahlemense                 T                           Q S
U 2                      N S                 Z   B B Z   B         V(T,P,
HRG                    M R S                 Q T   E   S R   S A S Q V A
--------------------------------------------------------------------------
                   110                 120                 130
vulgare            D A T R[R V D D A T V A I R]S A I N N L I V E L I R G
dahlemense                                                 V N   V
U 2                B,I)Z Z                     A S         A N   V
HRG                N A   Q                       Q   Q L   L N     S N H
--------------------------------------------------------------------------
                         140                 150             158
Vulgare            T G S Y N R S S F E S S S G L V W T S G P A T
dahlemense             L     Q N T       M               A     S
U 2                    M F   Q A G     T A           T T
HRG                G   Y M     A E    /-,A,I,-/  P     T A
--------------------------------------------------------------------------
```

Fig. 1. Aligned amino acid sequences of four wildstrains of TMV coat protein. The sequence for the strain *vulgare* is written out in full. For the three other strains only amino acid positions and the nature of amino acid substitutions which are different from those in *vulgare* are listed. Runs of invariable positions exist between residues 36–38, 87–94, and 113–122. These regions were thought to contain amino acid residues which make contact to RNA. (Wittmann et al. 1969)

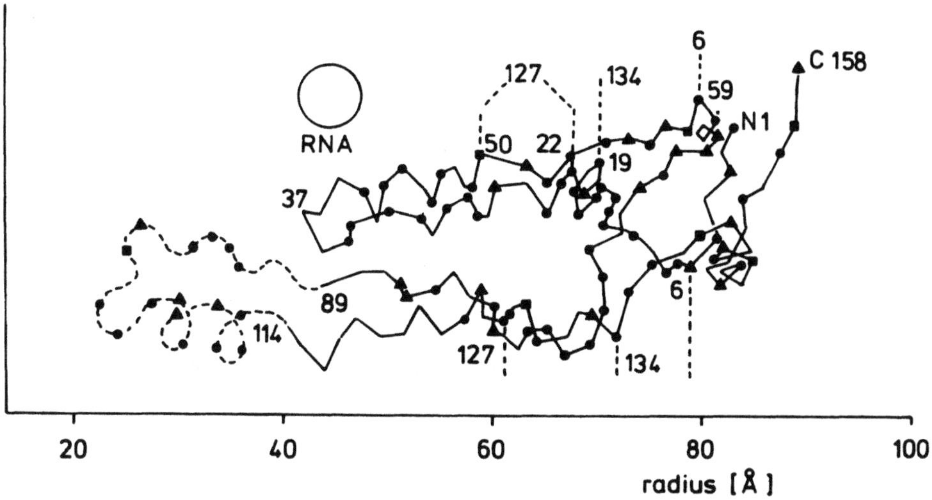

Fig. 2. Folding pattern (three-dimensional structure) of the TMV polypeptide chain (redrawn from Bloomer et al. 1978). Inserted are positions in which alternative amino acid residues may occur in four wildtypes (see Fig. 1). The number of different amino acids in a given position is indicated by ● = 2, ▲ = 3, and ■ = 4. *Dotted lines* represent subunit-subunit interactions. Invariable amino acid sequence sections are involved in interaction with RNA

Certainly there are fundamental as well as gradual differences with respect to the degree of contribution of genome and environment as selective forces in cell types of various origin. The development of multicellular animals (metazoa) e.g. proceeds according to a rigid genetic program which instructs the successive formation of gene products that give rise to an algorithm of (usually irreversible) differentiation steps. This does not apply in such a stringent mode to plants. Development in plants to a large extent is controlled by external signals (as light and/or temperature). These factors implement differentiation processes by inducing transcription of gene sets which are selected from corresponding alternatives. Consequently, the composition of the cell, as e.g. where proteins are concerned, will vary. Due to the presence of alternative protein compositions a particular intracellular protein will be exposed to variable micro-environments. Therefore one is likely to assume that this kind of variance may enhance the accumulation rate of amino acid substitutions in intracellular plant proteins, as compared to corresponding proteins from animal cells.

4.2 Experimental Constraints

By far the most speciated proteins which have been studied, are those of vertebrate origin. They were chosen mainly because they are fairly easy to purify and process. These restrictions, however, imply that such a sample of proteins is not representative for all cellular proteins, and thus it does not adequately reflect the evolution of the genome.

```
                     10                  20                  30
ac S Y S I T T P S Q F V F L S S A W A D P I E L I N L C T N A L G N Q F
                                         L V       S               S
           40                  50                  60                  70
   Q T Q Q A R T V V Q R Q F S Q V W K P S P Q V T V R F P D S D F K V Y
                       G                         I           G G
                     80                  90                  100
   R Y N A V L D P L V T A L L G A F D T R N R I I E V E N Q A N P T T A
                       A                               G   R
           110                 120                 130                 140
   E T L D A T R R V D D A T V A I R S A I N N L I V E L I R G T G S Y N
     M                                     S     T/V           I   F   S
                   150               158
   R S S F E S S S G L V W T S G P A T
                 F               L   -
```

Fig. 3. The amino acid sequence of TMV *vulgare*. Amino acid substitution in mutants are summarized in the line *below*. Underlined positions are those which may be identified (working with undenatured virus particles) with an antiserum directed to TMV *vulgare*. (v. Sengbusch 1965, van Regenmortel 1967)

It is striking that very few intracellular enzymes have been included in such investigations. The majority of proteins which were used for genealogies are either

a) extracellular (e.g. albumin, immunoglobulins, and transferrin), or
b) being located in intramembrane spaces, or membrane-surrounded vesicles (e.g. cytochrome c in mitochondrial intramembrane space, plastocyanin in thylacoids of chloroplasts), or
c) intracellular in highly specialized, metabolically essentially inactive cells (e.g. haemoglobin in erythrocytes).

These proteins, at least in part, are exposed to selective forces which differ from those inside the cytosol or the matrices of mitochondria or chloroplasts. Some common features, however, should not be disregarded. Like a cytosolic enzyme, cytochrome c, e.g. constitutes a link in a biochemical pathway (respiratory chain). It accepts electrons from a reductase and transfers them to a cytochrome oxidase. Although there is very little known about the structural premises at the outer surface of the inner mitochondrial membrane, there is some evidence for a stochiometric existence of the constituents of the respiratory chain. The micro-environment of cytochrome c has been shown to differ in various classes of organisms. This conclusion was drawn from the observation that distinct sets of amino acid substitutions accumulate in mammals as compared to insects, fungi and/or higher plants, and that the rate of fixation of mutants varies in different branches of descent (Margoliash 1980).

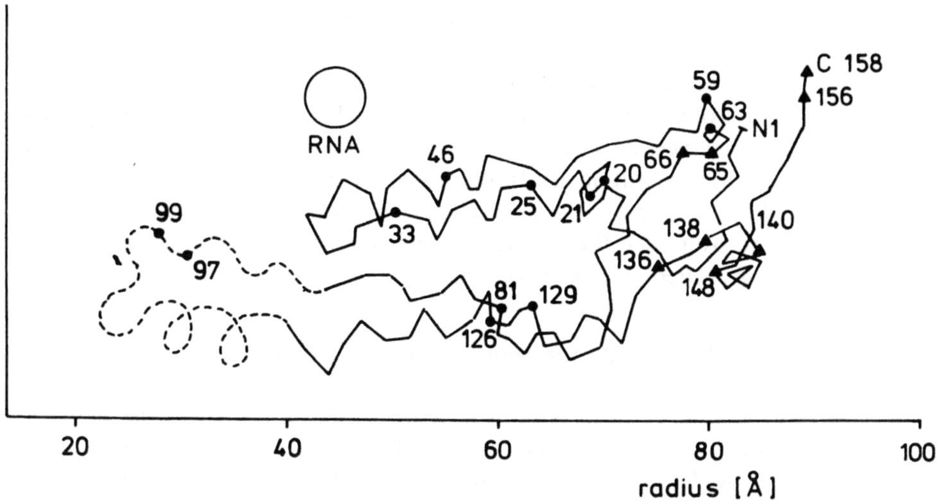

Fig. 4. Folding pattern of the polypeptide chain of TMV *vulgare*. (Redrawn from Bloomer et al. 1978). Inserted are positions in which, in mutants, single amino acid substitutions occur. ▲ = difference is detectable using an antiserum to TMV *vulgare*, ● = no difference detectable. This scheme illustrates that only amino acid replacements which alter the conformation of the surface of the virus particle contribute to the serological properties. Replacements in the interior of the polypeptide chain remain undetected

It still is of eminent importance, for the understanding of protein and organismal evolution, to obtain at least estimates on the degree of variation of intracellular proteins, especially of enzymes. Using gel electrophoresis as a method to trace changes in proteins Selander and Kaufman (1973) and Selander (1976) determined the degree of polymorphism and heterozygosity. Between 30% and 80% of all gene loci were shown to be polymorph, and 5%–20% of all genes in sexually reproducing organisms, to be heterozygous. Quantitative differences concerning the degree of this variance exist in different organismal classes, and in part they may be attributed to the population structure (population size, elapsed time since the population was founded, breeding behavior, and inbreeding or outbreeding). Valentine (1976) showed, for marine invertebrates, that there is a positive correlation between stability of food resources and degrees of genetic variation in the gene pool. In response to changing environments natural selection does not favour changes in particular protein structures, but prefers the alternative expression of equivalent or near equivalent proteins, which are available as products of alleles or independent gene loci.

5 Redundancy

For a period of a decade the proponents of the neutralist hypothesis and those of the selectionist hypothesis of protein evolution were trying to obtain evidence for their views. As data accumulated the foundations, on which the neutralist hypothesis was based, began to disintegrate. There is no doubt anymore that proteins, like any other biological characters, are subjected to natural selection. The essence of the problem is, however, how does natural selection really work? It is also adequate to ask for the arguments in favour of a particular amino acid residue in any position of any protein. On many levels of molecular and cellular organization structure and function have been shown to be highly redundant. Information theory (Wiener 1948) implies that messages are liable to accumulate errors during information transfer, and redundancy is the only way to overcome these problems and to reconstruct the original message from impaired transmitted message fragments.

5.1 The Genetic Code Evolved to Minimize Changes in Proteins

A considerable number of nucleotide replacements are termed silent, because - of the degeneracy of the code - they are not translated into amino acid substitutions. Another, also fairly large number of nucleotide replacements causes conservative amino acid substitutions (e.g. hydrophobic → hydrophobic; polar → polar). The chemical character of the amino acid side chain thereby remains virtually unaltered. Only a small portion of nucleotide replacements causes radical amino acid substitutions (as hydrophobic → polar or vice versa). Natural selection often prevents their fixation in the gene pool of a species.

Often conservative (and occasionally radical) amino acid substitutions in many positions of a polypeptide chain do not affect the function of the protein, and therefore may (at rates being specific for a given protein and a given line of descent) accumulate as time elapses. These types of substitutions have been termed neutral or near neutral, and their occurrence was and is valuable to taxonomists. As pointed out, however, these apparently neutral replacements are by no means neutral with respect to natural selection. Any protein is part of a complex system, like a cell or a multicellular organism, which evolved and gained fitness by acquisition of redundancy. Natural selection is not assaying for the most appropriate amino acid in any position of any protein. If this was the case, complex systems would not be fit to survive. Instead, selection is favouring integrated functional units (phenotypes), which resist perturbances of the environment and of its own genome. Thus protein molecules, and within them particular amino acids, have to act as buffers which are well adapted, as long as their function fits into the network of biochemical pathways and their action is not rate limiting for the need of the cell or organism.

6 Conclusions and Perspectives

Research on proteins and the application in the construction of phylogenetic trees has more than 20 years of tradition. Serious controversies arose about the value of this procedure and about the evaluation of obtained data. When such research started it was believed to be the most direct approach to obtain information about genes and changes within genes. Indeed, a vast amount of knowledge has been accumulated. Much was learned about the mechanism of protein differentiation and speciation, variant and invariant residues, strategies of evolution at the molecular level, evolution of co-operation and evolution of biochemical pathways. It became clear how new functions evolved as a consequence of successive gene duplications and re-arrangements, and it also became evident how redundant information is acquired, and to what extent it is used. It was important to demonstrate that protein evolution proceeded at a different rate than organismal evolution, and that protein evolution is not the cause of the organismal evolution.

One original aim, however, the construction of genealogies of taxa was partially successful only because of numerous difficulties. In retrospect it was discovered that selected proteins were not always representative markers of the genome, and thus were not appropriate for the construction of trees which agree with those based on morphological evidence.

The major drawbacks to application for taxonomic purposes have been (and still are): (1) difficulty in obtaining pure starting material in sufficiently large quantities, and (2) the laborious and extremely expensive procedures required to determine amino acid sequences.

Powerful methods for analyzing nucleic acids have recently been developed. They are in all aspects actually superior to the corresponding methods in analytical protein chemistry. Without risk it is predictable, just by extrapolation from present research activities, that sequencing genes and other genome sections will give valuable additional information about congenialities of sequences, the mode of recombination and gene transfer, re-arrangements and expression of genes, as well as about their potential to evolve.

In view of this progress in research, for a taxonomist a take home lesson might be this – that just because of the large experimental expenditures, looking on proteins might not be the most straightforward procedure to obtain information about genes and genomes, and their value for reconstructing phylogenetic trees. However, studying proteins will continue to be important in enhancing our understanding of how cellular functions evolved, and how advantage was taken to improve the fitness of the organisms as opposed to natural selection.

References

Beadle WG, Tatum EL (1941) Proc Natl Acad Sci USA 27:499–506

Betz JL, Brown PR, Smyth MJ, Clarke PH (1974) Nature (London) 247:261–264

Bloomer AC, Champness JJ, Bricogne G, Staden R, Klug A (1978) Nature (London) 276:362–368

Boulter D (1973) In: Bendz G, Santesson J (eds) Chemistry in botanical classification. Academic Press, London New York, pp 211–216

Cronquist A (1976) Brittonia 28:1–27

Cunningham BA, Hemperley JJ, Hopp HP, Edelman GM (1979) Proc Natl Acad Sci USA 76: 3218–3222

Dayhoff MO (1969) Sci Am July, 86–95

Dayhoff MO (1972) Atlas of protein sequence and structure, vol V. Nat Biochem Res Found, Washington DC, (1973) Suppl 1, (1976) Suppl 2, (1978) Suppl 3

Dayhoff MO, McLaughlin PJ, Barker WC, Hunt LT (1975) Naturwissenschaften 62:154–161

DeLange RJ (1980) In: Sigman DS, Brazier MAB (eds) The evolution of protein structure and function. Academic Press, London New York, pp 151–158

DeLange RJ, Fambrough DM, Smith EL, Bonner J (1969a) J Biol Chem 244:319–334
DeLange RJ, Fambrough DM, Smith EL, Bonner J (1969b) J Biol Chem 244:5661–5679
Dickerson RE (1971) J Mol Evol 1:26–45
Dickerson RE (1980) In: Sigman DS, Brazier MAB (eds) The evolution of protein structure and function. Academic Press, London New York, pp 173–202
Eigen M (1971) Naturwissenschaften 58:465–523
Fitch WM (1971) Syst Zool 20:406–416
Fitch WM, Langley CH (1976) Fed Proc 35:2092–2097
Fitch WM, Margoliash E (1967) Science 155:279–284
Goodman M (1981) Prog Biophys Mol Biol 37:105–164
Goodman M, Moore GW (1973) Syst Zool 22:508–532
Goodman M, Moore GW, Matsuda G (1975) Nature (London) 253:603–608
Hartley BS (1980) In: Sigman DS, Brazier MAB (eds) The evolution of protein structure and function. Academic Press, London New York, pp 39–48
Horowitz NH (1948) Genetics 33:612–613
Jensen U (1981) In: Ellenberg H, Esser K, Kubitzki K, Schnepf E, Ziegler H (eds) Prog Bot 43:344–369
Jukes TH, Holmquist R (1972) Science 177:530–532
Kimura M (1968) Nature (London) 217:624–626
Kimura M, Ohta I (1971) Nature (London) 229:467–469
Kimura M, Ohta I (1974) Proc Natl Acad Sci USA 71:2848–2852
King JL, Jukes TH (1969) Science 164:788–798
King MC, Wilson AC (1975) Science 188:107–117
Kornberg R (1977) Annu Rev Biochem 46:931–954
Koshland DE (1976) Fed Proc 35:2104–2111
Margoliash E (1963) Proc Natl Acad Sci USA 50:672–679
Margoliash E (1980) In: Sigman DS, Brazier MAB (eds) The evolution of protein structure and function. Academic Press, London New York, pp 299–321
Maxson LR, Wilson AC (1975) Syst Zool 24:1–15
Monod J, Changeux JP, Jacob F (1963) J Mol Evol 6:306–329
Peacock D, Boulter D (1975) J Mol Biol 95:513–527
Perutz MF, TenEyck LF (1972) Cold Spring Harbor Symp Quant Biol 36:295–310
Prager EM, Wilson AC (1976) J Mol Evol 9:45–57
Ramshaw JAM (1982) In: Boulter D, Parthier B (eds) Nucleic acids and proteins in plants, vol I. Springer, Berlin Heidelberg New York, pp 229–290
Regenmortel van MHV (1967) Virology 31:467–480
Rossmann MG, Liljas A, Bränden C-I, Banaszak LJ (1975) Enzymes 11:61–102
Schulz GE (1977) J Mol Evol 9:339–342
Schulz GE, Schirmer RH (1979) Principles of protein structure. Springer, Berlin Heidelberg New York
Scogin R (1981) In: Young DA, Seigler DS (eds) Phytochemistry and angiosperm phylogeny. Praeger, New York, pp 19–42
Selander RK (1976) In: Ayala FJ (ed) Molecular evolution. Sinauer, Sunderland, Mass, pp 21–45
Selander RK, Kaufman DW (1973) Proc Natl Acad Sci USA 70:1875–1877
Sengbusch v P (1965) Z Vererbungsl 96:364–386
Valentine JW (1976) In: Ayala FJ (ed) Molecular evolution. Sinauer, Sunderland, Mass, pp 78–94
Wiener N (1948) Cybernetics or control and communication in the animal and the machine. Actualites scientifiques et industrielles 1053, Hermann and Cie, Paris
Wills C (1976) Fed Proc 35:2098–2101
Wilson AC, Maxson LR, Sarich VM (1974a) Proc Natl Acad Sci USA 71:2843–2847
Wilson AC, Maxson LR, Sarich VM (1974b) Proc Natl Acad Sci USA 71:3028–3030
Wilson AC, Carlson SS, White TJ (1977) Annu Rev Biochem 46:573–639
Wittmann HG, Wittmann-Liebold B (1966) Cold Spring Harbor Symp Quant Biol 31:163–172
Wittmann HG, Hindennach I, Wittmann-Liebold B (1969) Z Naturforsch 24b:877–885
Zuckerkandl E, Pauling L (1965) In: Bryson V, Vogel JH (eds) Evolving genes and proteins. Academic Press, London New York, pp 97–166

Plant Protein Sequence Data Revisited

D. BOULTER[1]

Abstract. A short historical introduction setting the record straight, is followed by a review of the present and future position for protein sequence data in phylogenetic studies. It is concluded that more amino acid sequence data are needed in order to help elucidate flowering plant relationships, but this must be considered with all other available data, both biological and molecular. As this becomes available and particularly nucleotide sequence data, which can give a more refined picture of deduced relationships, interpretation of the sequence data set should prove to be more effective.

1 Introduction

In 1970, only six complete amino acid sequences of plant proteins had been determined, three of which were from cytochrome c (Ramshaw 1982). In view of the fact that relatively few animal vertebrate cytochrome c sequences had been successfully used to construct a phylogenetic tree which corresponded to the fossil data, Boulter et al. (1970) concluded at that time that "if the present potential of the method continues to be realised it is hoped to establish the major branches of a phylogenetic tree for the plant kingdom in the future".

By 1972, sufficient plant cytochrome c data had been established to allow the construction of a unique tree using the ancestral sequence method (Boulter et al. 1972). This tree gave the first indications that the molecular data did not support the general ideas of the Cronquist (1968) and Takhtajan (1969) schema. Even so, it was concluded by Boulter et al. that "we do not consider phylogenetic speculation to be profitable at this time, we are hopeful that work on future sequences now in progress, will confirm and develop the ideas put forward here". These ideas were that, (1) buckwheat (Polygonaceae) and spinach (Chenopodiaceae) diverged early from the other flowering plant families investigated; (2) that the Compositae were not so advanced a family as suggested by Cronquist; and (3) that the Asteridae were not necessarily derived from ancestors referable to the Rosidae. We did not draw detailed conclusions from this unique tree, since we pointed out that it was constructed on various assumptions, thus phylogenetic speculation should be avoided.

In 1973, Boulter presented a tree relating the cytochrome c sequence of 25 species constructed using the ancestral amino acid sequence method. This tree, with its earliest point in time deduced from the fossil record and fixed by the *Ginkgo biloba* sequence,

[1] University of Durham, Department of Botany, South Road, Durham DH1 3LE, UK

Proteins and Nucleic Acids in Plant Systematics
ed. by U. Jensen and D.E. Fairbrothers

showed for the flowering plants investigated, that buckwheat and spinach diverged first, and "then at approximately the same point in time" a group consisting of some monocotyledons and dicotyledons and two other dicotyledonous groups. It was concluded from these results "that present-day angiosperms represent a relic taxa (group)" and that the four flowering plant groups mentioned above have had a long separate evolutionary history", i.e. that the tree is a "shrub"; it was pointed out to that extent, the molecular data did not agree with either the Takhtajan or Cronquist schema.

The Durham group subsequently established some complete and many partial sequences of the photosynthetic protein plastocyanin and also several ferredoxin sequences, in the hope of supplementing and confirming the suggestions inherent in the cytochrome c data. In the event, more or less unique, stable trees could not be assembled from these additional data on the two proteins and consequently this was not possible. In 1979 therefore, Boulter et al. concluded that "at this stage in the investigation it is not possible to say to what extent amino acid sequence data give new insights into the phylogeny of the Flowering Plants since 'random' evolutionary amino acid substitutions lead to distortions when 'trees' are constructed from amino acid sequence data using the present data handling methods" (Boulter et al. 1979).

2 Present and Future Position

What can be said therefore in 1982 of the present and future situation for the use of higher plant amino acid sequence data for phylogenetic reconstructions? Apart from occasional exceptions which may have a technical explanation, whenever species of the same genus have been examined, their amino acid sequences have been shown to be identical or very similar. For example, *Brassica oleracea* (cauliflower) and *B. napus* (rape) cytochrome c (Boulter et al. 1972), plastocyanin of two *Heracleum* species (Boulter et al. 1979), and the ferredoxin sequences of species of *Equisetum* (Matsubara et al. 1978). At a higher level, sequences of closely related genera also show great similarity. Thus there were great similarities between *Brassica* and *Cucurbita* cytochrome c, between *Triticum* and *Aegilops* ferredoxin (Shin et al. 1979), and between *Solanum tuberosum* and *Lycopersicon esculentum* plastocyanin. Occasionally even at this taxonomic level anomalies occur, for example the considerable difference between the plastocyanin sequences of two species of *Senecio* (Boulter et al. 1979).

At the family level, the cytochrome and plastocyanin plant data sets show many plant families have distinct sequences, members resembling one another to a much greater extent than those of other families, but again there are exceptions, for example the plastocyanin of *Phaseolus* not fitting in with the rest of the legume plastocyanins. In view of the general taxonomic "sense" which has been clearly demonstrated in both the animal (Penny et al. 1982) and plant data sets through to the family level, how are we to explain such anomalous results? Surely not by "a wave of the hand" that amino acid sequence data do not have potential in phylogeny (Kubitzki 1977), but rather that the amino acid sequence data can give useful phylogenetic information, but only when a relatively large number of sequences has been determined for each particular problem. Thus, the indications are that the data from flowering plants will contain a relatively high proportion of neutral and parallel substitutions and that "polymorphism" exists among proteins. Both ferredoxin and plastocyanin can occur in multiple forms and usually only one sequence has been determined and the isoform(s) not identified. Amphiploidy and gene duplication have occurred repeatedly in angiosperm evolution making it often difficult to interpret sequence data. The criticism therefore that proteins are difficult taxonomic characters to use because of the amount of research required, is admissable, but this must be set against the fact that there is no evidence that other characters have been useful in phylogenetic speculations.

Recently Dahlgren (1975 and personal communication 1980) has constructed an evolutionary "shrub" relating many flowering plant families. In contrast to the earlier schema mentioned, the protein data lend support to several main features of the Dahlgren scheme, e.g. (1) the "shrub"-like nature of the relationships, (2) the early separation of both the Chenopodiaceae and Polygonaceae, (3) the close relationship of Brassicaceae and Cucurbitaceae, (4) the non-relatedness of Cronquist's Rosidae and the Compositae, (5) the early origin of the Compositae, (6) the lack of evidence of the Ranalian type flower being ancestral to other flowering plant families.

With the advent of gene cloning and quick accurate nucleotide sequence determination methods, it is now possible, if a cloned gene or copy gene is available together with its protein product, to correlate gene and amino acid sequences. Studies of this kind have shown that eukaryote genes are split, i.e. contain intervening sequences, introns, which although transcribed into heterogeneous nuclear RNA (hnRNA) do not form part of the coding mRNA; those parts of the gene which give rise to coding sequences are called exons. Thus protein amino acid sequence data can help predict exons and trace evolutionary events at the gene level and reciprocally DNA structures can aid in studying protein evolution.

Recently part of the DNA sequence of the cDNA for the 7S storage protein, phaseolin, of *Phaseolus vulgaris* has been published (Sun et al. 1981). We have recently determined the amino acid sequence of one of the subunits of the 7S protein vicilin of peas (Hirano et al. 1982). Although such amino acid sequence data are not yet available for phaseolin, the cDNA data allow a corresponding amino acid sequence to be deduced and this shows that the two 7S proteins are homologous (Hirano et al. 1982).

Another recent example of the reciprocity of DNA and protein sequence data relates to leghaemoglobin. Amino acid sequence data for plant leghaemoglobins, the protein of which is coded by the host flowering plant in *Rhizobium* interactions, show homology with animal myoglobin and haemoglobin (Hunt et al. 1978). Since leghaemoglobin is not found elsewhere in the plant kingdom, and the animal and plant lines of descent diverged some thousand million or more years ago (Dayhoff 1972), several possibilities could account for this finding:

a) Leghaemoglobin genes occur elsewhere in the flowering plants but are silent and never expressed.
b) Lateral gene transfer has taken place relatively recently.
c) The similarity of leghaemoglobin to animal globins is due to convergence.

Whilst at the moment there is no evidence to decide between these possibilities, I favour the last one for the following reasons:

Go (1981) has suggested that the central exonic regions of the haemoglobin gene consist of two fused exons with a division somewhere between residue position 66 and 71. Leghaemoglobin gene has four exons rather than the three found in haemoglobin gene and these are 1–32, 33–68, 69–103, and 106 to the C-terminus, suggesting that leghaemoglobin gene is a primitive gene from which the haemoglobin gene evolved (Blake 1981, Jensen et al. 1981, Hyldig-Nielson et al. 1982).

Haemoglobin consists of four polypeptide chains which exhibit cooperative action in order to fulfil its physiological function and fusion of the exons presumably took place under strong selection pressure. We can speculate that a similar pressure at the time of the evolution of nitrogen fixation caused exons from other haem binding protein genes to be combined to form the leghaemoglobin gene with its own specific function. To that extent we can consider leghaemoglobin and haemoglobin speculatively to be products of convergent evolution, namely the combination of exons from other genes. Work with animal genes (Jeffreys 1981) shows that there are many cryptic changes in nucleotide sequences, i.e. silent changes which do not lead to a change in the amino acid sequence of the protein product; clearly, nucleotide sequences of homologous genes will give a more refined picture of deduced relationships.

3 Conclusions

Whilst it may be said that more amino acid sequence data are urgently needed in order to help elucidate flowering plant relationships, it is also true generally that the existing information from amino acid sequence data has been underutilised so far by systematists and taxonomists (see Jensen 1981). Amino acid sequence data must be considered with all other available data; as more data, e.g. serological data, accumulate we can expect a more effective interpretation of the amino acid sequence data set to become possible.

References

Blake CCF (1981) Nature (London) 291:616

Boulter D (1973) In: Bendz G, Santesson J (eds) Chemistry in botanical classification. Nobel Foundation, Stockholm. Academic Press, London New York, pp 211–216

Boulter D, Laycock MV, Ramshaw JAM, Thompson EW (1970) Taxon 19:561–564

Boulter D, Ramshaw JAM, Thompson EW, Richardson M, Brown RH (1972) Proc R Soc London Ser B 181:441–455

Boulter D, Peacock D, Guise A, Gleaves JT, Estabrook G (1979) Phytochemistry 18:603–608

Cronquist A (1968) The evolution and classification of flowering plants. Nelson, London Edinburgh

Dahlgren R (1975) Bot Not 128:119–147

Dayhoff MO (1972) Atlas of protein sequence and structure, vol V. Nat Biomed Res Found, Washington DC

Go M (1981) Nature (London) 291:90–92

Hirano L, Gatehouse JA, Boulter D (1982) FEBS letters 145:99–102

Hunt TL, Hurst-Calderone S, Dayhoff MO (1978) In: Dayhoff MO (ed) Atlas of protein sequence and structure, vol V, Suppl 3. Nat Biomed Res Found, Washington DC, pp 229

Hyldig-Nielson JJ, Jensen EÖ, Paludan K, Wiborg O, Garrett R, Jörgensen P, Marcker KA (1982) Nucleic Acids Res 10:689–701
Jensen EÖ, Paludan K, Hyldig-Nielson JJ, Jörgensen P, Marcker KA (1981) Nature (London) 291:677–679
Jensen U (1981) In: Ellenberg H, Esser K, Kubitzki K, Schnepf E, Ziegler H (eds) Progress in botany, vol 43. Springer, Berlin Heidelberg New York, pp 344–369
Kubitzki K (1977) In: Ellenberg H, Esser K, Merxmüller H, Schnepf E, Ziegler H (eds) Progress in botany. Springer, Berlin Heidelberg New York
Matsubara H, Hase T, Wakabayashi S, Wada K (1978) In: Matsubara H, Yamanaka T (eds) Evolution of protein molecules. Jpn Sci Soc Press, Tokyo, pp 209–220
Penny D, Foulds LR, Hendry MD (1982) Nature (London) 297:197–200
Ramshaw JAM (1982) In: Boulter D, Parthier B (eds) Encyc Plant Physiol, New Ser, vol 14A. Springer, Berlin Heidelberg New York, pp 229–290
Shin M, Yokoyama Z, Abe A, Fukusawa H (1979) J Biochem (Tokyo) 85:1075–1081
Sun SM, Slighton JL, Hall TC (1981) Nature (London) 289:37–41
Takhtajan A (1969) Flowering plants – origin and dispersal. Oliver Boyd, Edinburgh

Discrimination among Infraspecific Taxa using Electrophoretic Data

H. STEGEMANN[1]

Abstract. Discrimination among taxa can be done only by using a diversity of physical parameters of the constituents. It is shown that different electrophoretic techniques meet this prerequisite for proteins. Drawing conclusions based on data from one method alone is not advisable.

Presently it is common practice in taxonomy and genetics to seek biochemical information in addition to the more traditional characteristics. Detecting electrophoretic discrimination of infraspecific taxa is no longer a problem for many species; however, for some species it remains unsolved. The virtue of biochemical methods lies, at the moment, mainly in this technique, and the practical applications are increasing rapidly. However, gaining new insights into relationships is much more complicated, and often too far-reaching conclusions are being drawn from biochemical data, especially when data from only one technique forms the basis for such conclusions. Biochemical changes occur within each single plant during development, and applying only one methodological approach could form wrong impressions concerning diversity and relationships.

Let us compare the principles of the two static and the two dynamic electrophoretic procedures developed for separating proteins. Dynamic non-equilibrium separations are obtained from using PAGE[2] and SDS-PAGE[2]. PAGE will respond to any change of charge, size and configuration, while SDS-PAGE responds mainly to size of the uncoiled protomer and to changes of a carbohydrate moiety in a glycoprotein. PAGIF[2] or porosity gradient polyacrylamide electrophoresis (PoroPAGE[2]) is more or less static in respect to the endpoint of the separation, governed by the isoelectric points of the proteins or by their molecular weight. The chance for discriminating species using static procedures is lower than when using dynamic procedures. Immunological methods will mirror properties of the surface of a molecule (charge and configuration), but will not reveal all the aspects. The combination of these techniques with 2-dimensional separations, e.g. the combination of PAGIF and immunoelectrophoresis or PAGIF with PAGE, PoroPAGE or SDS-PAGE (Stegemann 1979b, Stegemann and Pietsch 1983), will reveal almost all the information about physical parameters of the proteins.

---

[1] Institut für Biochemie, Biologische Bundesanstalt, D-3300 Braunschweig, Messeweg 11, FRG

[2] Abbreviations:
PAGE = Poly-Acrylamide-Gel-Electrophoresis
SDS = Sodium Dodecyl-Sulfate
Poro = Porosity-Gradient
PAGIF = Isoelectric Focusing in Poly-Acrylamide Gels

Proteins and Nucleic Acids in Plant Systematics
ed. by U. Jensen and D.E. Fairbrothers

The first example includes proteins extracted from tubers of *Solanum tuberosum* used for differentiating cultivars and determining primitive cultivars, grown for centuries by natives of Latin America. The protomer size in potato proteins has not changed during the evolution of different varieties (Fig. 1D). The protomer, that is the protein subunit formed from the native protein by SDS and mercapto-ethanol, becomes randomly coiled and loaded with SDS. The same proteins separated by size in native conformation by PoroPAGE indicated that the proteins, not disassembled, are somewhat more characteristic for each cultivar (Fig. 1 C). Likewise, separating the native proteins according to their isoelectric points by PAGIF also yields quite characteristic patterns (Fig. 1E). However, the easiest and most economic way to compare extracts of tuber proteins is to apply the standard PAGE which provides a detailed differentiation (Fig. 1A and 1B). When this technique is extended to esterases and, with caution because of some environmental influence, for peroxidases, a more accurate identification is obtained (Fig. 1E to I). The position and intensity of tuber storage proteins and esterases are not influenced by environmental factors or virus infection of the plant.

A second example has been obtained from *Zea mays* experiments. The different extracts (water, buffer, propan-2-ol, buffer with SDS) are not equally suitable for taxonomy or experiments in biochemical genetics. All extracts reveal that the relation between the individual proteins is almost exclusively governed by the genetic background, but it has been discovered that buffer extracts serve well for the discrimination of cultivars. For this reason SDS-PAGE in two different buffers is used. The individual migration rates which form the characteristic electrophoretic patterns are partly due to a different carbohydrate content of the glycoproteins. These pick up different amounts of SDS leading to different migration rates in SDS-PAGE. Otherwise one could not understand the strong influence of the buffer in producing different protein patterns from the same extracts (Stegemann 1979a,b). Furthermore it is essential that there is no change from season to season, a prerequisite which cannot be proven by one-dimensional techniques only. Such proof comes from high-resolution two-dimensional methods as introduced by us (Macko and Stegemann 1969, Stegemann et al. 1973, Fig. 2).

For a third example data from legumes are discussed. The patterns obtained using a tris-borate buffer pH 7.1 with 0.1% SDS, also used for cereals, revealed a good differentiation among cultivars and species of tested legumes (Hamza and Stegemann in preparation). Some examples are demonstrated in Fig. 3. The use of the same buffer is essential to allow comparisons among all seed proteins from different origins. From these data only a preliminary evaluation is possible since one-dimensional SDS-PAGE was used. The influence of the buffer composition on the protein patterns is demonstrated.

The discrimination and identification of infraspecific taxa has become a very important aspect of practical applications. For trade and processing, the electrophoretic analyses of potatoes and cereals has largely replaced former identification methods, which were usually based on the morphology of the sprout. The protein patterns can be obtained within hours and are very reliable. They have been used e.g. for eliminating duplicates in the collection of about 13,000 primitive cultivars of potatoes, for checking imported crops, and for protecting consumers from purchasing mislabeled culti-

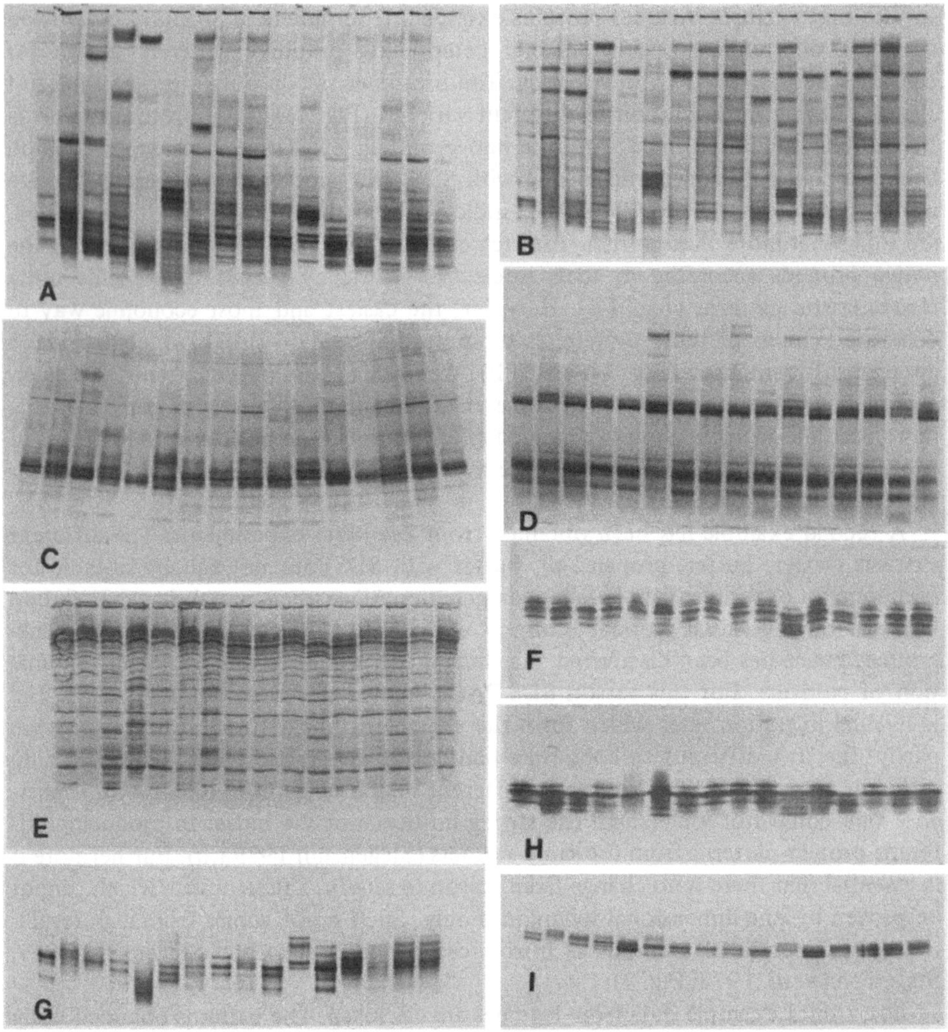

Fig. 1 A–I. Potato proteins (from cultivars registered in 1982 in West Germany), separated by different procedures to show the importance for choosing the best fitting method for differentiation. The choice should combine pattern characteristic, speed, and economy.
A PAGE in Tris-borate buffer pH 8.9 (storage proteins)
B PAGE in Tris-borate buffer pH 7.9 (storage proteins)
C PoroPAGE in Tris-borate buffer pH 8.9 (storage proteins)
D SDS-PAGE in Tris-borate buffer pH 8.9 (storage proteins)
E PAGIF in Servalyt T, pH 4–9 (storage proteins)
F same as E. 2,500 Vh, run for 3.5 h (esterases)
G same as A. (esterases)
H same as E. 2,500 Vh, run for 17 h (esterases)
I same as C (esterases)

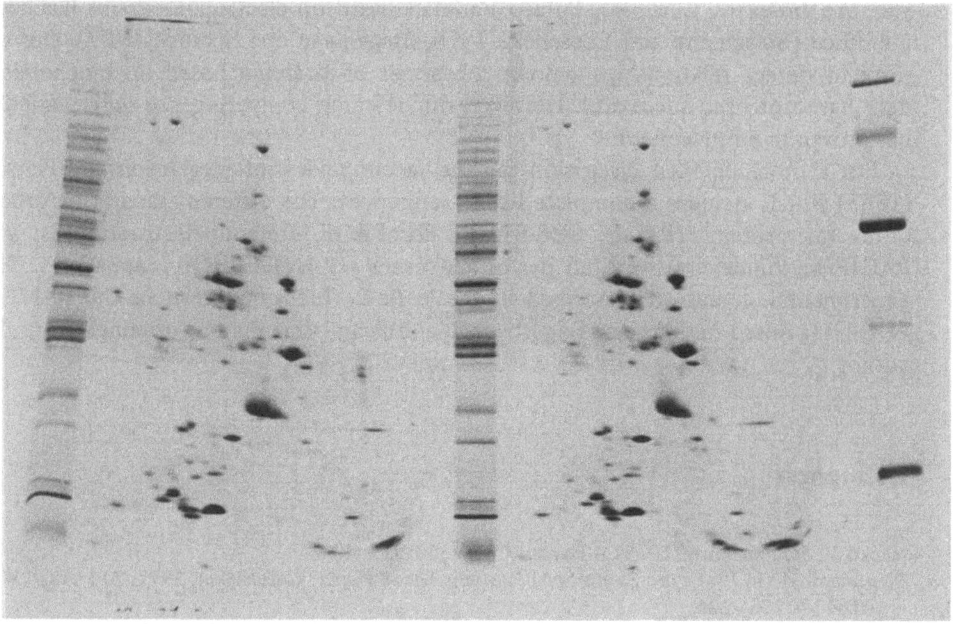

Fig. 2. One-dimensional separation of protomers (water-soluble, SDS-loaded maize proteins) in the *vertical lanes* compared with mapping (first PAGIF in rods with 6 M urea and 4% Servalyt pH 5–9 placed *horizontally* for second dimension in PoroPAGE with 10 to 28% PAA and SDS-Tris-borate buffer pH 8.9). Even sharp bands in one-dimensional SDS-PoroPAGE *(vertical)* are further resolved by mapping. Moreover, nearly identical patterns are seen on the *left* (seeds harvested in 1978) and on the *right* (same cultivar, but harvested in 1977). MW-markers: *far right.* The two mappings were done in one gel

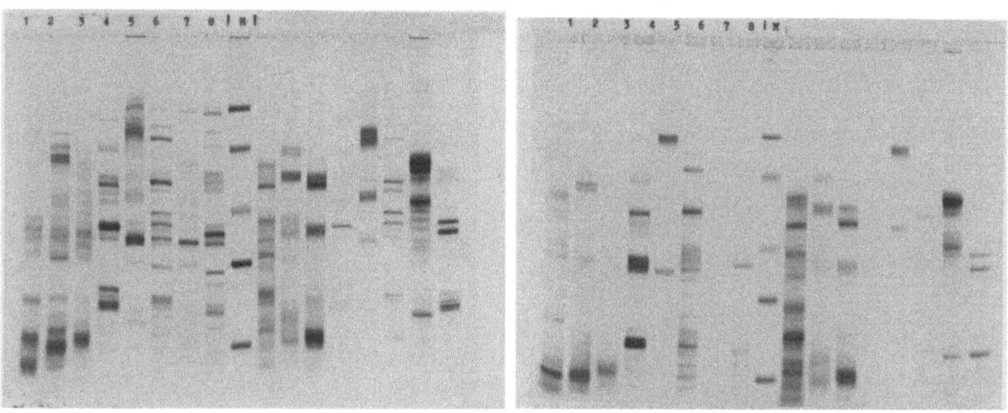

Fig. 3. Cereal and legume proteins, compared as protomers by SDS-PAGE in Tris-borate-buffer at pH 7.1 *(left figure)* and at pH 8.9 *(right)*. Water extracts are shown on the *left* side (slot *1* to *8*), buffer extracts (Tris-borate pH 8.9 after water extraction) on the *right* (slot *10* to *17*), markers for the molecular weight in the *middle* with 14.3, 25.7, 37, 67, 97.4 kD, respectively.
Zea mays, var. 1087 waxy (slot *1* and *10*); *Triticum aestivum*, var. Sacha 8 (slot *2, 11*); *Oryza sativa*, var. S 258 (slot *3, 12*); *Vicia faba*, var. Express (slot *4, 13*); *Phaseolus vulgaris*, var. Dopp. holl. Prinzess (slot *5, 14*); *Pisum sativum*, var. Kleine Rheinländerin (slot *6, 15*); *Lupinus albus*, commercial var. (slot *7, 16*); *Cicer arietinum*, var. Giza 2 (slot *8, 17*)

vars. An Index of European Potato Varieties based on electropherograms has been published (Stegemann and Loeschcke 1976, Stegemann and Schnick 1982). Our efforts to detect relationships among subspecies of *Solanum* based on biochemical data have not been successful. However, our research comparing *Zea mays* varieties has proven more successful.

For a more detailed discussion see the recent review of Stegemann and Pietsch (1983) which includes a complete list of references. The different separation procedures for proteins (PAGE, SDS-PAGE, PoroPAGE, Immunoelectrophoresis, and PAGIF in thin-layers) were all performed using a PANTA-PHOR[3] apparatus. The experimental details are described in the leaflet: "Electrophoresis in the PANTA-PHOR" (printed in German, English or Spanish) and this may be obtained from the author.

References

Macko V, Stegemann H (1969) Z Physiol Chem 350:917–918

Stegemann H (1979a) Proc Symp Seed Proteins Dicot Plants, Gatersleben 1977. Abh Akad Wiss DDR N4:215–224

Stegemann H (1979b) In: Righetti PG, v Oss CJ, Vanderhoff JW (eds) Electrokinetic separation methods. Elsevier, Amsterdam, pp 313–336

Stegemann H, Loeschcke V (1976) Mitt Biol Bundesanst 168:1–360

Stegemann H, Pietsch G (1983) In: Gottschalk W, Müller HP (eds) Seed proteins – biochemistry, genetics, nutritional value. Nijhoff, Den Haag, pp 45–75

Stegemann H, Schnick D (1982) Mitt Biol Bundesanst 211:1–220

Stegemann H, Francksen H, Macko V (1973) Z Naturforsch 28c:722–732

[3] Available from Labor-Müller, D-3510 Hannoversch-Münden

Monoclonal Antibodies: Scope and Limitations in Phylogenetic Studies

S. FAZEKAS de ST. GROTH[1]

Abstract. The principles and practice of producing monoclonal antibodies is surveyed, with special attention to recent improvements and simplifications that make the method generally accessible.

After discussing the special properties of monoclonals (many of which are averaged out or masked in conventional antisera), their use as preparative, analytical or diagnostic reagents is outlined.

The special problems attending their application as criteria of classification and, generally, as tools in phylogenetic studies are pointed out, together with ways of ensuring that their great potential is not wasted.

1 The Immune Response

When an animal comes up against something alien, it will respond to it. If such an alien substance enters its interior, so that the animal cannot run away or tear it apart, the adequate response is immunological. One way of doing this is to generate cells specialized in dealing with intruders; but *cellular immunity* is not our subject today. Another way, *humoral immunity*, consists in producing antibodies, molecules specialized in dealing with intruders.

A humoral response is never simple: many kinds of antibody are made against even the simplest, chemically pure antigen. The reason for this is twofold. First, the immune system evolved in a way to meet all contingencies, and meet them fast. This can be done only if all kinds of antibody-producing cells are preformed, so that a response amounts to the expansion of preexisting subpopulations. A system like this is redundant as any antigen can be complemented in more than one way and will therefore also stimulate more than one kind of responder. As a result, with each cell making its own kind of antibody, a large family of antibodies is raised, each of its members more or less fitting the single stimulating antigen.

The second reason for heterogeneity is that what we call an antigen is, immunologically speaking, not homogeneous – any one molecule may have several surface regions (antigenic determinants or *epitopes*) which might stimulate antibodies of different specificities. I do not even mention the fact that what the chemist is pleased to call a pure substance will always carry minor impurities which are still immunogenic. With biological material "minor" becomes a euphemism, covering contaminants that are by no means negligible in the immunological sense.

[1] Basel Institute for Immunology, CH-4005 Basel, Switzerland

Proteins and Nucleic Acids in Plant Systematics
ed. by U. Jensen and D.E. Fairbrothers

No wonder, then, that when an animal responds to all these epitopes with a variety of antibodies against each, it will produce a unique mixture: no two antisera are the same, indeed, even two bleeds from the same animal contain different assortments of antibodies. This miscellany of exquisite specificities is a nuisance, and a great deal of effort has gone into making antisera more homogeneous. There are several ways of doing this, all quite cumbersome and none quite satisfactory. The final solution came from an idea of Georges Köhler, and was realized by Köhler and Milstein (1975).

2 Production of Monoclonal Antibodies

2.1 Strategy

The idea was to separate antibody-producing cells right at the outset and then grow them in vitro into clones of a single specificity. The trouble is that antibody-producing cells are terminally differentiated, they will not divide and will live for a week at most. So these ephemeral cells were fused with myeloma cells which are tumors of the immune system, effectively immortal and readily grown in culture.

Fusion of two cells is a rare event; fusion of the two nuclei in the resulting heterokaryon even rarer. And such hybrid cells, initially tetraploid, will not survive unless they reduce their chromosome complement to a mitotically manageable number, retain whatever genetic mechanism ensures immortality, and at the same time eliminate whatever prevented the antibody-producing partner from multiplying.

Isolation of these rare survivors from the overwhelming majority of vigorously growing myeloma cells was the second problem. This was solved by using mutant myeloma cells which could produce purines only through the folic acid pathway. Such cells die in the presence of aminopterine, while normal cells survive by virtue of the hypoxanthine escape pathway. So another rare event militates against the success of fusion: only hybrids retaining the gene for ribosyltransferase will survive. The product of all these probabilities comes to about 10^{-5}. But this still leaves a large number of viable hybrids as a mouse spleen contains more than 10^8 cells and we can therefore expect about a thousand survivors.

Viable hybrids are not necessarily useful hybrids. For that they should also carry the two chromosomes coding for the heavy and the light chain of the specific antibody we are hoping to obtain. This reduces the probability by another order of magnitude.

2.2 Tactics

Apart from problems of principle there were also problems of practice; these, too, have all been solved by now. We could follow the route of progress by comparing the original method with the present state of the art, but this is not the occasion for going into esoteric technical detail – that we have done before (Fazekas de St. Groth and Scheidegger 1980). Here we shall just look at the sequence of steps and look for the rationale behind particular modifications (Fig. 1).

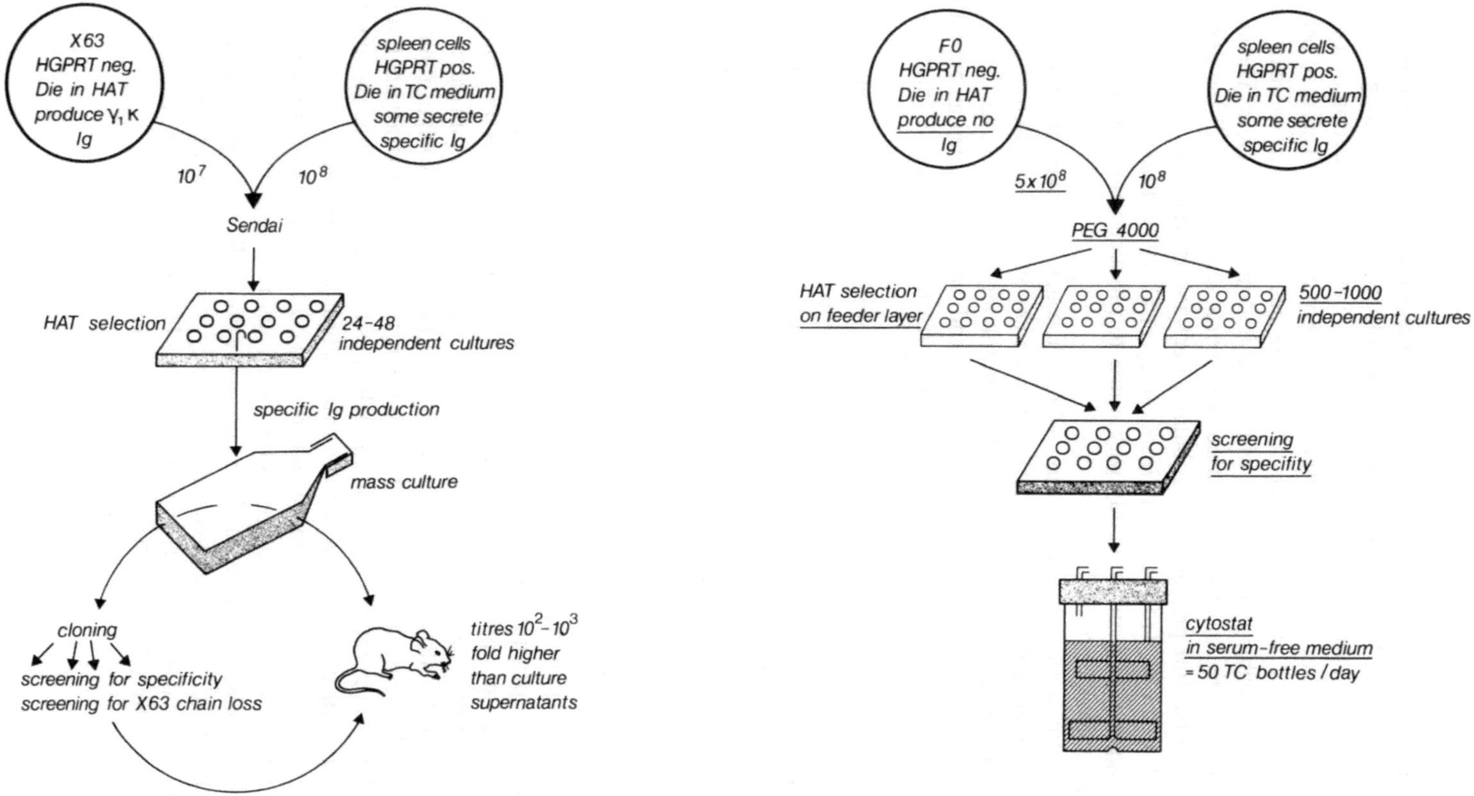

Fig. 1. Production of monoclonal antibodies in 1975 and in 1982. The improvements are underlined. *X63, FO* = myeloma lines; *HGPRT* = hypoxanthine-guanine phosphoribosyl transferase; *HAT* = hypoxanthine-aminopterine-thymidine medium; *PEG* = polyethyleneglycol; *TC* = tissue culture

2.2.1 Choice of the Partners in Fusion

The classical myeloma line, X63, had all the right properties, except that it produced a globulin of its own. If this cell is fused with a spleen cell that is making antibody, i.e. another globulin, we have the curious and greatly annoying situation of a monoclonal culture of hybrid cells still secreting a mixture of antibodies. This is so because the pairs of light and heavy chain made by the same cell will associate more or less at random and all possible combinations will be released into the medium. The answer to this problem is to find or manufacture a myeloma line that has all the desirable properties of X63, but produces no globulin. There are several such available today (Shulman et al. 1978, Trowbridge 1978, Kearney et al. 1979, Fazekas de St.Groth and Scheidegger 1980), and anybody who wants to produce truly monoclonal antibodies will not look beyond them.

There is also a second consideration when choosing the myeloma partner. All hybrid cells are aneuploid and may lose chromosomes even after prolonged passage. It is an advantage therefore to have rapidly multiplying hybrids, so that their chain-loss-mutants will not outgrow them. With this in mind we have selected an exceptionally fast-growing myeloma line that also transmits this property to many of its hybrids.

When choosing the other partner in fusion it is well to remember that all useful myeloma lines originated in the BALB/c strain of mice. This strain (including its Fl offspring) is therefore the natural choice for contributing the immunized cells. Fusion between histoincompatible partners is feasible but requires certain precautions. Indeed, successful mouse-rat and mouse-human fusions have been made, but the rate of success is lower, as is the stability of the hybrids.

2.2.2 Quantitative Aspects

Originally (and in a few laboratories even now) 10^8 spleen cells and 10^7 myeloma cells were fused. One does not have to be a mathematician to see that these proportions favour self-fusion of spleen cells, which is useless. So we shall have to change the proportions. We shall have to change the numbers, too, as such fusion mixtures, when distributed over 24–48 culture cups, will give multiple hybrids in each of them. Separating multiple clones means not only an inordinate amount of extra work but also constant disappointment as some of the initially positive cultures will be swamped by rapidly multiplying irrelevant hybrids and eventually score as negative.

Since setting up a thousand extra culture cups on the day of fusion means about an hour's extra work, while cloning even 100 mixed cultures might take several months of solid effort, it is reasonable to aim at not more than 10%–20% positive cultures. Under such conditions cloning becomes effectively superfluous as less than 1 culture in 10 will have two clones in it, and less than 1 in 100 more than two. Reducing the spleen cell input by a factor of 50 will give the desired result.

2.2.3 Fusion and Selection

The original fusing agent, a paramyxovirus, has been replaced by the more readily available and controllable polyethylene glycol (Galfrè et al. 1979). As fusion occurs in a narrow range, with failure of fusion on one side and damage to the cells on the other, a number of elaborate rituals have been proposed for this step. Non-toxic preparations that are isotonic at the effective concentration (e.g. PEG 4,000, gas chromatography grade) will give reproducible results, even without observing any of the strictures.

The selective sequence can also be simplified. Instead of gradual increase of the aminopterine concentration by changing the medium every day, the fusion mixtures can be directly exposed to the final concentration and then left alone for the rest of the week. Apart from saving labour, this treatment yields also more hybrids.

2.2.4 Maintenance, Screening, Preservation

After a week of selection against the myeloma cells, the aminopterine is removed from the medium. By this time about a quarter of the eventual clones can be detected, and twice as many again will appear over the next week. The positive cultures are then compacted in transfer trays, over a feeder layer of macrophages which provide intermediate metabolites missing from even the most complex media. Screening for specificity, chain composition, etc., is done on these secondary cultures and those found positive can be stored permanently in liquid nitrogen.

Screening is a common stumbling block. If the overnight yield of, say, 10^4 cells is to be tested, only methods detecting fractions of a picomole are of any use. The average hybrid puts out about 1,000 antibody molecules per second, i.e., some 10^{11} molecules/ml will accumulate in a culture cup in a day. The popular radioimmunoassays or enzyme-linked immunoassays will detect such quantities, but they come dangerously close to their limit of sensitivity. The danger lies in the need for extensive washing of the antigen-antibody complex: antibodies of lower affinity dissociate at this stage, and otherwise useful hybrids will score as negative. Agglutination of antigen-coated red cells is at least as sensitive and does not suffer from self-stultifying technical requirements. The objection that antigen-coated red cells were unstable and had to be freshly prepared for each test, has been recently overcome (Fazekas de St. Groth 1983). While this objection still holds for plaque assays, they are unquestionably the most sensitive as they will detect a single antibody-producing cell. The vagaries of day-to-day variation can be minimized, and methods have been developed for simultaneous testing of 96 microcultures (Bankert et al. 1980).

The choice, then, is one of personal preference: the impatient hybridizer will opt for plaque assays, the fashion-conscious for RIA or ELISA, while passive agglutination is left for those preferring comfort and reproducibility.

2.2.5 Production

A properly designed fusion test will yield about 100 specific hybrids per mouse spleen, 90 of them already growing in separate cultures. There is no need to retain and waste effort on cloning the few mixed cultures unless the interest lies in the study of the antibody repertoire rather than in the practical application of monoclonal antibodies.

For practical application monoclonal antibodies are needed in quantity. Originally there were two alternatives: either injecting the hybrids into mice and collecting the ascitic fluid after a week or two, or scaling up the in vitro production. The former method yields a small volume of concentrated but rather impure product, the latter is expensive both in materials and labour.

Nowadays we do not need either. All hybrid lines tested to date could be grown in suspension cultures, in serum-free medium. If such a suspension culture is combined with an overflow arrangement and the input of medium exactly matched to the growth rate of the cells, we have a *cytostat* with a constant number of exponentially growing cells and a constant output of their products (Fazekas de St.Groth 1983). Since the monoclonal antibodies are the only globulins in the effluent, they are readily isolated. In fact, a cytostat can be directly coupled to an affinity column, so that all that is left for us is to dissociate and collect pure antibodies every day or two.

3 The Uses of Monoclonal Antibodies

What all this amounts to is that if a mouse can make an antibody against a particular epitope, we can make that antibody-producing cell immortal and collect its produce in quantity. The method has some obvious advantages over conventional antisera: the hybrids produce antibodies of a single specificity, even very rare specificities can be isolated, impure antigens lead to pure antibody reagents, and the supply is unlimited.

The only obvious disadvantage is an embarras de richesse. We shall therefore look at the antibody response more closely, see what kind of monoclonal antibodies we can hope to get and what guidelines we can lay down for their use.

3.1 Consequences of Heterogeneity

3.1.1 Cross Reactivity

An epitope (a small area with a distinct van der Waals contour, schematically represented in Fig. 2 by a fork with prongs of various length and shape) can be complemented in several ways by the combining areas, *paratopes*, of antibody molecules. There may be one that fits perfectly, and these are rare – we found only 7 among 793 monoclonal antibodies reacting with a single epitope of an influenza virus. The rest will fit imperfectly or fit only in part.

What happens then when such a family of antibodies is brought together with a related antigen that differs minimally from the immunogen, say, by having a leucine instead of a valine at one point in its epitope? The situation is like trying to fit a glove

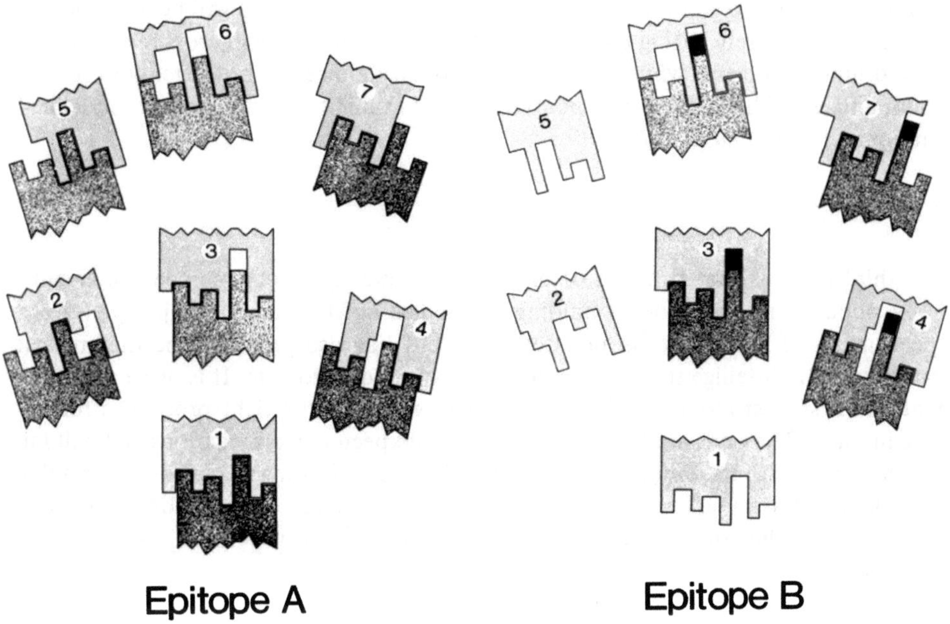

Fig. 2. Reactions and cross reactions of an antibody population raised against epitope A. Representative paratopes are labelled *1–7*

to a hand with a swollen finger. All antibodies that fitted perfectly at the VAL locus will be sterically hindered by the LEU from making effective contact with the mutant epitope and will thus be excluded from cross reactions. The rest of the antibody family will react with both epitopes.

3.1.2 Heterocliticity

We have also a special case. Antibody No. 3 will fit better to the mutant antigen than to the immunizing antigen, it is said to be *heteroclitic*. There might exist, in theory at least, epitopes that bind any cross-reactive antibody better than does the immunogen, and we should be prepared for this when using monoclonal antibodies. The problem hardly arises with standard sera, as in any one test the small fraction of antibodies heteroclitic for the test antigen is masked by reactions of the majority.

3.1.3 Seniority

What happens then in the reverse situation, when epitope A is brought together with a family of antibodies made against epitope B? The situation is like fitting a large glove to a small hand: any paratope fitting LEU will also accomodate the smaller VAL. To give it a name, we shall call the epitope with the bulkier component *senior*. (Senior is used here in the military sense, as a young lieutenant outranks a sergeant-major old enough to be his father.) It follows that cross reactions will be asymmetric: even the most closely fitting antibody against a senior antigen will cross react with all related junior antigens, while well fitting junior antibodies will not react with related senior antigens. This is the second point to be kept in mind when choosing monoclonal antibodies for particular use.

3.1.4 Affinity

The third point is that firmness of binding varies greatly among members of an antibody population, with poorly binding antibodies making up the majority. Such a spectrum of affinities has its practical uses, but has also proved to be the greatest liability to the intelligent use of monoclonal antibodies to date. It is naive to expect that the first best monoclonal we come upon will have the right properties for the task in hand. The odds are that it will exhibit unexpected cross reactions and will fail to react even with its immunogen in certain situations. For many purposes, including most kinds of assay and comparisons of related substances, only antibodies of high affinity should be considered, and that usually means not more than 1 in 10 of the isolated clones.

3.1.5 Isotypy

There is yet another aspect of heterogeneity to be considered. The same paratopes may occur on different classes of immunoglobulins. Their specificity, i.e., the variable regions of their two peptide chains, is identical but the rest of the heavy chains is characteristically different and serves also different functions. Thus the same paratopes may come in the form of pentavalent macroglobulins or several kinds of smaller molecules, some which bind to Protein A of *Staphylococcus aureus* and are thus very easily purified, some which fix complement and are thus especially suited for plaque assays, some which are readily reduced to the monovalent form and thus applicable where aggregation is to be avoided. Only by collecting and characterizing a battery of monoclonal antibodies against any one antigen can the full potential of the method be realized.

3.2 Purification of Antigens

While preparations used for immunization need not be pure, assays and especially quantitative assays demand reagents of greater or at least of known purity. This is one of the areas where monoclonal antibodies stand on their own: there is no method for the purification of macromolecules which would compare in general applicability and ease of application with what monoclonal antibodies can do.

Optimally, one would choose for this purpose antibodies of middling affinity, high enough to bind the antigen even in dilute systems and low enough to allow dissociation without denaturation. But *any* monoclonal antibody can be used by adjusting concentrations appropriately, and will be found equal to or better than available alternatives. There are several well-tried methods of coupling globulins to suitable carriers, and such immunosorbents can then be used and reused either in a batch process or in the form of columns. Some of the applications are quite dramatic: in the case of interferon, for instance, the very laborious and expensive purification procedure could be replaced by a single passage through a column of monoclonal anti-interferon, with striking improvements in both yield and purity.

3.3 Antigenic Analysis

Most larger molecules, whether proteins, carbohydrates or lipids, are antigenic. Since even single molecules may have several different epitopes, an organism is a vast collection of them. Antigenic comparison of organisms implies therefore selection, i.e., dealing only with a small handful of molecules that, for some reason, are deemed to be of interest. As a preliminary to any evolutionary comparison, these representative molecules have to be analysed first.

The analysis has three purposes. First, the number of distinct epitopes has to be determined. In practice this means reacting pairs of monoclonals with the test substance and seeing which interfere with binding of the other.

The second purpose is to define the nature of particular epitopes. This is done by comparing homologous substances from several species. Some monoclonals will be found to bind to many of them: they are directed against highly conserved epitopes. Others will exhibit narrower cross reactivity, restricted perhaps to a single species: these would mark distinct, but evolutionarily equally acceptable structural solutions. And there will be some which distinguish individuals or groups of individuals within a species: these may define epitopes at the most refined level of systematics, and are of interest in delimiting the structural and conformational freedom of the test substances. Epitopes of all three kinds could be defined on an influenza antigen, where monoclonal antibodies have been used extensively and the exact shape and composition of several mutant proteins is known. The correlation of antigenic variation with evolutionary and chemical changes was striking.

The third purpose of antigenic analysis is to evaluate the affinity and cross reactivity of the panel of monoclonals. This information is a prerequisite for their proper use in all kinds of tests, but requires no additional effort as the experiments on competition between paratopes and on reactivity with various epitopes define these properties, too.

3.4 Use in Systematics

In principle, monoclonal antibodies could be applied to either agglomerative or divisive classification. But their strength lies in the recognition of minute areas on the surface of molecules, and this marks them as special tools for subdividing rather than compounding existing units. It is in this area that considerations of seniority and affinity become paramount. When looking for close relationships, such as single amino acid differences in a group of proteins, an antibody against the senior member will not recognize differences within the test population. Antibody against junior members, on the other hand, will suggest absolute differences as it will fail to react with any of the more senior antigens. I cannot urge strongly enough the use of several monoclonal antibodies for such purposes. Provided they have been shown on the junior antigen to compete against each other, a set of hierarchically arranged paratopes will unequivocally define a set of hierarchically arranged epitopes. And, since a single point mutation always leads to asymmetric cross reactions, even the number of mutational steps can often be guessed from reactions of a small panel of monoclonals.

The hierarchy of epitopes is, of course, unknown at the beginning of a study. If luck is on our side, we may pick as immunogen the junior member of a set and get also all more senior antibodies from a single immunization, as the response will contain the required heteroclitics.

If fortune does not favour us and we immunize with a senior antigen, there will be no specific anti-junior antibodies in our collection of monoclonals. This does not mean that they will not react with junior epitopes, only that they will not discriminate between them. In such a situation it is mandatory to use at least one of the presumably junior homologues and generate a second set of monoclonals.

But whether fortune smiles on us or not, we should always work with high affinity antibodies; low affinity antibodies, and that means the majority of monoclonals, will only confuse classifications. Low affinity means a small contact area between epitope and paratope, and the smaller the area the less characteristic it is and the more likely to occur also as part of other epitopes either on the same or on unrelated molecules. Cautionary examples of downright misclassification, of distinguishing spurious "public" and "private" specificities abound in the literature and there is certainly no need for multiplying them.

Documenting similarities between distant taxa does still not relax the requirement for high affinity test antibodies. A rational approach here would rather rely on choosing a suitable test antigen, one of the more highly conserved framework epitopes. Firmly fitting monoclonals will detect single amino acid substitutions, and following the accumulation of these should be as good an indicator of evolutionary distances as any.

3.5 Use as Reagents

Even before the advent of monoclonal antibodies, a wide variety of assays and diagnostic tests relied an the specificity of antisera. Most of these are being gradually replaced by the more reproducible and often also simpler methods employing monoclonal antibodies. The trend will continue, with the predictable flood of literature and mushrooming of associated industries. What is needed here is not an apology for monoclonals but rather a word of caution.

A monoclonal antibody, whether produced in our own laboratory or coming from a commercial source, remains an unknown reagent until we have defined its specificity (i.e., the epitope it recognizes, *not* just the name of the antigenic molecule of which the epitope forms a small part), its affinity and the range of its cross reactions. This is the information needed for effective and economic use; without it their remarkable powers of discrimination remain unexploited. It goes without saying that in a field where closely related objects are to be characterized and delicate differences recognized, only well-characterized reagents will be of any help.

To finish, there is one more area where monoclonal reagents are proving useful, in fact, indispensable. Location of a gene, a DNA sequence, always starts with isolating the corresponding messenger RNA. This is usually the most difficult step, possibly the only uncertain one, in the sequence of manufacturing the required probes and eventually isolating and analysing the heritable message. Monoclonals can be selected which recognize and precipitate even nascent peptide chains. Such precipitates will contain also the ribosomes on which the peptide is made, as well as the messenger RNA from which the peptide is being translated. Recovery of the messenger is then a matter of routine. Monoclonal antibodies helped in this way over some of the difficult steps that led to the unravelling of their own complex inheritance and of the molecular basis of their astounding variety.

References

Bankert RB, DesSoye DE, Powers L (1980) J Immunol Methods 35:23–32
Fazekas de St.Groth S (1983) J Immunol Methods 57:121–136
Fazekas de St.Groth S, Scheidegger D (1980) J Immunol Methods 35:1–21
Galfrè G, Milstein C, Wright B (1979) Nature (London) 277:131–133
Kearney J, Radbruch A, Liesegang B, Rajewsky K (1979) J Immunol 123:1548–1550
Köhler G, Milstein C (1975) Nature (London) 256:495–497
Shulman M, Wilde CD, Köhler G (1978) Nature (London) 276:269–270
Trowbridge IS (1978) J Exp Med 148:313–323

Particular Proteins Contributing to Phylogeny and Taxonomy

Phycobiliproteins and Phycobiliprotein Organization in the Photosynthetic Apparatus of Cyanobacteria, Red Algae, and Cryptophytes

W. WEHRMEYER[1]

Abstract. Amino acid sequence analyses of phycobiliproteins of cyanobacteria, red algae, and cryptophytes and new immunological results of these pigments are summarized and compared. The similarities are evaluated under functional and evolutionary aspects regarding both infraspecific and transspecific categories. The results obtained by these different methods coincide. Cryptomonad biliproteins show immunological connections with each other and with cyanobacterial and rhodophytan biliproteins transgressing spectroscopic classes. Similar results are obtained for cyanobacterial phycoerythrocyanin.

Abbreviations. APC allophycocyanin; C-, R-PC ("cyanobacterial", "red algal") phycocyanin; PEC phycoerythrocyanin; C-, B-, R-PE ("cyanobacterial", "bangiophycidean", "red algal") phycoerythrin; PCB phycocyanobilin; PEB phycoerythrobilin; PUB phycourobilin; PXB undetermined phycobiliviolin type.

1 Introduction

Antennae pigments in the photosynthetic apparatus of plants function in light absorption and transfer of absorbed light energy to the reaction centres of the photosystems, particularly chlorophyll a. In the evolution of pigment systems that optimize light harvesting and enhance the survival of the organism in specific ecological environments, the development of the phycobiliproteins has played a decisive role. Phycobiliproteins are characteristic antennae pigments of the prokaryotic cyanobacteria (Cyanophyta) and the eukaryotic rhodophytes and cryptophytes. The presence of phycobiliproteins in the photosynthetic apparatus of such diverse taxonomic groups has raised various phylogenetic questions. Phycobiliproteins are involved in oxygenic and facultative anoxygenic photosynthesis of cyanobacteria (Padan 1979), an ability which is reminiscent of the corresponding step from anaerobic to aerobic life in evolution of photosynthetic organisms in the Precambrium era (Schopf 1974). Phycobiliproteins also successfully have survived the evolutionary transition from pro- to eukaryotic cells which is demonstrated by their distribution in red algae and cryptophytes. Considering that this step is a transgression of "the largest evolutionary discontinuity among contemporary organisms" (Stanier 1977), the biliproteins may become markers of specific interest. Lastly, the cryptophytes are unicells with two subequal apical flagella. Present day rhodophytes predominantly include the most

[1] Fachbereich Biologie, Universität Marburg, D-3550 Marburg, FRG

Proteins and Nucleic Acids in Plant Systematics
ed. by U. Jensen and D.E. Fairbrothers

highly differentiated forms of algae, in addition to unicells with complete absence of flagella in all members and developmental phases including gametes, which is apparently a primary condition. Therefore, can biliproteins provide evidence of a possible correlation between Aconta (Rhodophyta) and Contophora (all other eukaryotic algae) as designated in Christensen's 1962 classification of algae?

Biliproteins are appropriate marker proteins because of their age as living fossils dating from the Precambrium period; their disadvantage is the lack of universality since they are restricted to the above mentioned phyla. (Phytochromes, i.e. non-photosynthetic "biliproteins" which are found in many organisms and function in very small amounts as photoreversible pigments in photomorphogenetic reactions are not included.) To answer these questions three lines of evidence are pursued: (1) the analysis of amino acid sequences of phycobiliproteins, (2) the analysis of immunological cross-reactivity of phycobiliproteins, and (3) analysis of the biliprotein organization of the photosynthetic apparatus in cyanobacteria, red algae and cryptophytes, because the molecular structure of the polypeptide chains of the biliproteins tend to aggregate and, in part, give rise to supramolecular forms of biliprotein organization (= phycobilisomes).

2 Phycobiliprotein Organization in the Photosynthetic Apparatus of Cyanobacteria, Red Algae, and Cryptophytes

Phycobiliproteins are intensively coloured, highly fluorescent, water soluble chromoproteins. They are composed of at least two dissimilar polypeptides, α and β subunit, in a molar ratio of 1:1. These both carry prosthetic groups, covalently bound by cysteinyl thioether linkages, closely related in structure to bile pigments. Depending upon the quality and number of chromophores in correlation with the protein environment and state of aggregation the light absorption of phycobiliproteins range from 500 to 671 nm: 500–570 nm for phycoerythrins, 610–640 nm for phycocyanins, and 650–671 nm for allophycocyanins (Table 1).

Among the physicochemical properties of biliproteins the ability to aggregate is particularly important with respect to the structural and functional organization of the antenna. The general phenomena of aggregation in successive steps can be demonstrated by *Mastigocladus* phycobiliproteins (Table 2). Starting with the polypeptide chains of the α and β subunits of allophycocyanin, phycocyanin, and phycoerythrocyanin as a general pattern protomeric $(\alpha\beta)$, trimeric $(\alpha\beta)_3$, and hexameric $(\alpha\beta)_6$ aggregates can be derived. The latter are the constituent elements of the phycobilisomes, the supramolecular aggregation forms of biliproteins in cyanobacteria and red algae. Studies on the hemidiscoidal phycobilisomes of red algae (Mörschel et al. 1977, Wehrmeyer unpublished results) and numerous cyanobacteria (Bryant et al. 1979, Williams et al. 1980, Yamanaka et al. 1980, Nies and Wehrmeyer 1980, 1981, Rosinski et al. 1981) convincingly demonstrate the structural similarity of the phycobilisomes of this type regardless of the taxonomic classification. The exchange of hexameric biliprotein components of the peripheral rods of the phycobilisomes following chromatic adaptation in green or red light indicates that these biliprotein structures of

Table 1. Distribution, subunit structure, and spectroscopic properties of phycobiliproteins of cyanobacteria and red algae. Data from Glazer (1977), Muckle and Rüdiger (1977), Glazer and Hixson (1977), Nies and Wehrmeyer (1980), and Bryant (1982)

Biliprotein	Distri-bution	Visible absorption max (nm)	Fluorescence emission maximum (nm)	Subunit structure of stable assembly form	Chromophore content of the protomer (αβ)
Allophycocyanin B	C R	671 > 618	673	$(\alpha\beta)_3$	2 PCB
Allophycocyanin	C R	650	660	$(\alpha\beta)_3$	2 PCB
C-phycocyanin	C R	620	640	$(\alpha\beta)_3$; $(\alpha\beta)_6$	3 PCB
R-phycocyanin	R	617 > 555	636	$(\alpha\beta)_3$; $(\alpha\beta)_6$	2 PCB; 1 PEB
Phycoerythrocyanin	C	570 > 595(s)	625	$(\alpha\beta)_3$; $(\alpha\beta)_6$	2 PCB; 1 PXB
C-phycoerythrin	C	560	577	$(\alpha\beta)_3$; $(\alpha\beta)_6$	5-6 PEB
b-phycoerythrin	R	545 > 563(s)	570	$(\alpha\beta)_n$	6 PEB
B-phycoerythrin	R	545 > 563 > 498(s)	575	$(\alpha\beta)_6\gamma$	6 PEB
R-phycoerythrin	C R	565 > 540 > 498	578	$(\alpha\beta)_6\gamma$	undetermined

C cyanobacteria
R red algae
(s) shoulder
PCB phycocyanobilin
PEB phycoerythrobilin
PXB phycobiliviolin-type chromophore of undetermined structure

phycoerythrin and phycocyanin must be compatible with each other. A review discussing the organization and composition of cyanobacterial and rhodophycean phycobilisomes recently has been completed (Wehrmeyer 1983). These supramolecular similarities in the phycobilisomes and phycobilisome constituents of cyanobacteria and red algae can be interpreted as an indication of a common origin. Disregarding the principal differences of prokaryotic and eukaryotic cells, there are striking similarities between the photosynthetic apparatus of these two phyla beyond their biliprotein

Table 2. Physicochemical characteristics of subunits and assembly forms of allophycocyanin, C-phycocyanin, and phycoerythrocyanin of *Mastigocladus laminosus*. Data from 1) Gysi and Zuber (1974), 2) Frank et al. (1978), 3) Fueglistaller et al. (1981), and Nies and Wehrmeyer (1981)

state of aggregation	allophycocyanin		C-phycocyanin		phycoerythrocyanin	
	mol wt	sed. const.	mol wt	sed. const.	mol wt	sed. const.
○ α	17 200 [1]		18 000 [2]		19 100 [3]	
○ β	17 200		19 400		20 300	
$(\alpha\beta)$	34 400		37 400		39 400	
$(\alpha\beta)_3$	103 200	6 S	112 200	6 S	118 200	6 S
$(\alpha\beta)_3$ 2×	206 000	11 S	224 000	11 S	236 000	11 S
$(\alpha\beta)_6$ 2×			448 000	17 S		

organization (Table 3). The cryptophytes, in contrast, deviate in many aspects in spite of their eukaryotic cell organization. The main differences in comparison to the red algal chloroplast are (1) the periplastidal ER cisterna in addition to the chloroplast envelope, (2) the pairing of the thylakoids forming 2- or 3-thylakoid bands, (3) the presence of chlorophyll c_2 in addition to chlorophyll a, and (4) the prevailing of a carotene over β carotene.

Most striking differences come from the cryptophytan biliprotein antenna which is organized within the intrathylakoidal space and appears as electron-dense material. Cryptophytes do not capture the light by a collective unit of different phycobiliproteins as it is the case in cyanobacteria as well as in red algae, but possess only one single biliprotein which is either phycocyanin or phycoerythrin and lack allophycocyanin. This is surprising since allophycocyanin is thought to be the most essential link in the energy transfer to membrane bound chlorophyll. Furthermore, the biliproteins of cryptophytes show a reduced molecular weight of the dissimilar subunits with values of 9–12 kD for a and 16–19 kD for β which occur with a 1:1 stoichiometry. Molecular weight determinations of "native" cryptophytan biliproteins made by gel filtration reveal values which suggest that these pigments exist in a dimeric state of aggregation. Considering these differences the basis of taxonomic correlations between cryptophytes and rhodophytes appears scanty. Therefore amino acid sequence analyses, and the study of immunochemical reactivity of these pigments seem desirable and promising.

Table 3. Comparison of the photosynthetic apparatus of Cyanobacteria, Rhodophyta, and Cryptophyta

	Cyanobacteria (-phyta)	Rhodophyta	Cryptophyta
energy supply (exeptions)	oxygenic & anoxygenic photosynthesis	oxygenic photosynthesis (heterotrophic parasites)	oxygenic photosynthesis (some heterotrophs).
photosynthetic apparatus	single thylakoids within prokaryotic cells	single thylakoids within chloroplasts	paired thylakoids within chloroplasts
additional envelope	—	absent	periplastidal ER
ribosomes	cytoplasmic 70 S type	chloroplastic 70 S type	chloroplastic 70 S type
RNA sequence homology	cytoplasmic 16S rRNA	chloroplastic 16S rRNA	?
membrane bound pigments	chlorophyll a β carotene myxoxanthophyll, zeaxanthin, echinenone	chlorophyll a β (α) carotene zea-, cryptoxanthin	chlorophyll a, c_2 α (β) carotene allo-, crocoxanthin
phycobiliproteins	APC, C-PC, C-(R) PE PEC	APC, C-, R-PC, B-, R-PE	PC types, PE types
organization of biliproteins	phycobilisomes (bundle shaped, hemidiscoidal type)	phycobilisomes (hemidiscoidal, hemiellipsoidal type)	disperse, locally in regular array
collective unit of phycobiliproteins	at least 2 different	at least 2 different	one biliprotein per species
localization	on the outer thylakoidal surface	on the outer thylakoid surface	intrathylakoidal
basic aggregates	$(\alpha\beta)_3$, $(\alpha\beta)_6$ α β	$(\alpha\beta)_3$, $(\alpha\beta)_6$, $(\alpha\beta)_6\gamma$ α β γ	$(\alpha\beta)_2$, $\alpha\alpha'\beta_2$ α β
mol wt of subunits	15-19 kD 18-22 kD	15-19 kD 18-22 kD 29-30 kD	9-12 kD 16.-19 kD
chromophores	PCP, PEB, PXB, (PUB)	PCB, PEB, PUB	unidentified
linker polypeptides	present in different sizes	present in different sizes	unknown
photosynthetic products	amylopectin-like glucan	floridean starch	starch
localization		outside of chloroplasts	outside of chloroplasts

3 Amino Acid Sequence Analyses of Phycobiliproteins

3.1 Methods

In spite of the increasing interest in amino acid sequence analyses and the high number (about 1,100) of completely analyzed proteins (Walsh et al. 1981), there are only 6 biliproteins with 11 polypeptide chains known in detail. However, numerous sequences of phycobiliproteins have partially been determined at their N terminus. This is the basis of the discussions in the following sections, which is admittedly small, but perhaps will stimulate further research.

The amino acid sequences were written using one-letter abbreviations (IUPAC-IUB Commission 1969) as proposed by Zuber (1978). To achieve maximum homology between all of the subunits which were compared some deletions had to be introduced into the primary structure of the polypeptide chains. For *Mastigocladus laminosus* allophycocyanin Sidler et al. (1981) demonstrated, that the maximum homology with the subunits of C-phycocyanin can be obtained by omitting the amino acid positions 1 + 2, 82 + 83, 151–160 in the α chain and the positions 64, 78 + 79, and 151 to 160 in the β chain, whereas the α subunit of C-phycocyanin has a deletion in position 149–160 and the β chain at positions 73 + 74. This framework consequently was used for all other sequence data of biliproteins and has proven to be a valuable tool for comparison.

3.2 Infraspecific Amino Acid Sequence Analyses of Biliproteins

Presently the filamentous heterocystous cyanobacteria *Mastigocladus laminosus* Cohn (= *Fischerella laminosa* according to Rippka et al. 1979) is the best known organism with respect to its antennae pigments. The primary structure of both subunits of allophycocyanin, C-phycocyanin, and phycoerythrocyanin has been completely analyzed (Frank et al. 1978, Sidler et al. 1981, Fueglistaller et al. 1983). The results of the comparison of the amino acid sequences of the biliproteins are summarized in Fig. 1. When the complete amino acid sequences of α and β subunits of

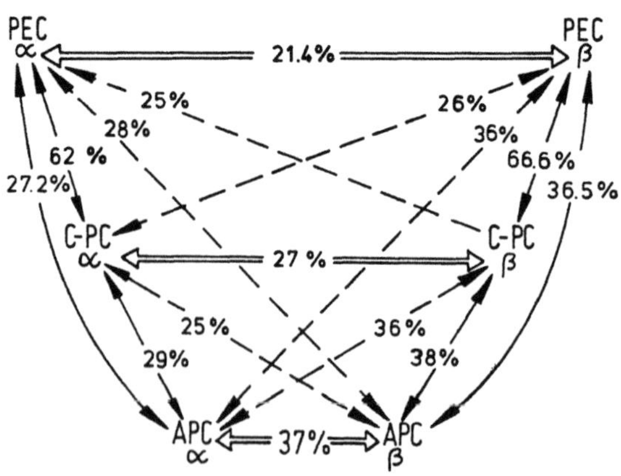

Fig. 1. Homologies of the 3 α and 3β subunits of the biliproteins of *Mastigocladus laminosus* which are completely analyzed in their amino acid sequence: C-phycocyanin (Frank et al. 1978), allophycocyanin (Sidler et al. 1981), phycoerythrocyanin (Fueglistaller et al. 1983)

M. laminosus allophycocyanin, phycocyanin, and phycoerythrocyanin are compared after alignement, there is a low and significantly decreasing degree of homology between the subunits (broad horizontal arrows). The homology between the two allophycocyanin polypeptide chains (37%) is significantly higher than that between α and β phycocyanin (27%), and between α and β phycoerythrocyanin (21,4%). This implies that the differentiation of a common ancestral chain into the α and β chain is a rather early event in the evolution of the phycobiliproteins. Tracing back beyond this hypothetical ancestor, a reasonable explanation for the observed homology between α and β subunits is the postulation that gene duplication occurred in the early history of biliprotein evolution. The same explanation already was presented when only 15–20 amino acids of the N-terminus of the subunits of biliproteins were analyzed (Williams et al. 1974, Troxler et al. 1975). This mode of protein evolution recently has been reviewed (Dayhoff 1978) and shown to occur in not less than 40 other proteins, and is thus well established. Comparing exclusively α or β subunits from different biliproteins within this species (Fig. 1, continuous arrows), the degree of homology is either similar to that just described above for α and β subunits of single biliproteins as far as allophycocyanin-phycocyanin or allophycocyanin-phycoerythrocyanin relationships are concerned, or significantly higher. The 62% homology of the α and the 66.6% homology of the β subunits of C-phycocyanin and phycoerythrocyanin indicate a special correspondence between these pigments, and indicate a separate evolutionary development of α and β subunits in general. From the actual distribution of phycoerythrocyanin in derived orders of predominantly heterocystous filamentous cyanobacteria (Bryant 1982), it can be concluded that phycoerythrocyanin probably originated from C-phycocyanin which is common in all cyanobacterial orders including the basic Chroococcales.

The developmental connection between β C-phycocyanin and β allophycocyanin, however, remains uncertain. The lack of the second chromophore at position 155 in β allophycocyanin together with 10 amino acids at position 151–160 can be interpreted as being caused by a deletion in the gene of the β subunit of C-phycocyanin. However, it is also reasonable to assume that an insertion in the gene of β allophycocyanin caused the phylogenetic transition from allophycocyanin to C-phycocyanin.

Lastly, a relatively low degree of homology is observed for α and β subunits of different biliproteins (Fig. 1, dashed arrows). Among these the 36% homology between α allophycocyanin and β C-phycocyanin in comparison to 38% homology of β allophycocyanin to β C-phycocyanin is remarkable. Whether an α or β type allophycocyanin subunit was the ancestral protein from which phycocyanin arose is difficult to determine, although the β subunit is a slightly favorable candidate.

3.3 Transspecific Amino Acid Sequence Analyses of Biliproteins Belonging to the Same Spectral Class

With respect to this taxonomic question special attention must be paid to the similarities of biliproteins from different species. Unfortunately, completely analyzed subunits of phycobiliproteins belonging to different orders and classes are still lacking. In particular, the biliproteins of red algae and cryptophytes have not yet been intensively investigated. Nevertheless, some guiding principles are emerging and will be outlined.

With respect to the presently available data the biliproteins of cyanobacteria and red algae are presented first, following the spectral classification into allophycocyanins, phycocyanins, and phycoerythrins; separately the exceptional position of phycoerythrocyanin is considered. Subsequently, the biliproteins belonging to different spectral classes are compared. The biliproteins of cryptophytes, which still stand apart, are dealt with only briefly because of the lack of adequate information about these pigments.

3.3.1 Allophycocyanin

There are only three complete amino acid sequences of subunits of allophycocyanin available: α and β allophycocyanin of *Mastigocladus laminosus* and the β subunit of *Anabaena variabilis* (Sidler et al. 1981, DeLange et al. 1981). However, these data are completed by amino acid sequences of the N terminus of several other allophycocyanins from cyanobacteria and red algae published by Troxler et al. (1980). From these the α and β subunit of *Nostoc* sp. and *Cyanidium caldarium* which are analyzed up to position 30 of the polypeptide chain are presented in Table 4.

It is well known from the description of the phycobilisomes, that the allophycocyanin containing center is the most conservative part of the phycobilisome. Allophycocyanin is the most decisive link in the energy transfer chain from the peripheral biliproteins, especially phycocyanin, to the membrane-bound chlorophyll *a* being the reaction centers of photosystem II. This essential functional role can only be sustained if the tertiary structure of the polypeptide chains as well as the quaternary structure of the whole protomer are conserved, which is accomplished by a high degree of constancy in the primary structure. The observed 90% homology between the β subunit of allophycocyanin of *Mastigocladus laminosus* and that of *Anabaena variabilis* is an excellent confirmation (Table 4). The same tendency becomes apparent in the partially sequenced polypeptide chains of allophycocyanins from *Nostoc* and *Cyanidium*. The N terminus of β allophycocyanin of *Nostoc* showed a 67% or 77% homology to β allophycocyanin of *Mastigocladus laminosus* or *Anabaena variabilis* (all cyanobacteria), respectively. Comparing β allophycocyanin of *Mastigocladus laminosus* or *Anabaena variabilis* with that of *Cyanidium caldarium*, which species is suggested to be a red alga (see below), exhibited 73% homology within the sequence of 30 amino acids of the N terminus. The degree of homology of partially sequenced α subunits of allophycocyanin was found to be ever higher, with 93% for α allophycocyanin of *C. caldarium* to α allophycocyanin of *M. laminosus* (and *Nostoc*) and 97% of α allophycocyanin of *M. laminosus* with that of *Nostoc* sp. This value, however, is reduced to 82% when the complete amino acid sequence (160 amino acids) of the α subunits of allophycocyanin of *C. caldarium* is considered (Offner and Troxler 1982).

The exceptionally high degree of homology between the subunits of allophycocyanin from cyanobacteria and a red alga is certainly due to functional reasons which will be discussed later. This conservatism, on the one hand, makes allophycocyanin an ideal candidate to demonstrate the close relationship of cyanobacterial and red algal biliproteins. On the other hand, this pigment actually is unsuitable for further taxonomic differentiation within the different classes.

Table 4. Complete amino acid sequence of α and β-subunits of allophycocyanin from *Mastigocladus laminosus* (Sidler et al. 1981) and *Anabaena variabilis* (De Lange et al. 1981) supplemented by partial sequences of allophycocyanin from *Cyanidium caldarium* and *Nostoc* sp. (Troxler et al. 1980)

N.	α	APC	—SIVTKSIVNADAEARYLSPGELDRIKSFVT
M.l.	α	APC	—SIVTKSIVNADAEARYLSPGELDRIKSFVSSGEKRLRIAQILTDNRERIVKQAGDQLF
C.c.	α	APC	—MIVTKSIVNADAEARYLSPGELDRIKSFVL
C.c.	β	APC	MQDAITAVINTADVQGKYLDSSXIZKLKGY
N.	β	APC	AQDAITAVINAADVGQKYLDATALSKLKAY
A.v.	β	APC	AQDAITAVINSADVQGKYLDTAALEKLKAYFSTGELRVRAATTISANAAAIVKEAVAKSL
M.l.	β	APC	MQDAITAVINSSDVQGKYLDTAALEKLKSYFSTGELRVRAATTIAANAAAIVKEAVAKLT
A.v.	β	APC	LYS—DITRDGGNMYTTR—RYAACIRDLDYYLRYATYAMLAGDPSILDERVLNGLKETYN (PCB at C)
M.l.	β	APC	LYS—DITLPGGDMYTTR—RYAACIRDLDYYLRYATYAMLAGDPSILDERVLNGLKETYN (PCB at C)
A.v.	β	APC	SLGVPVGATVQAIQAIKEVTASLVGADAGK——EMGIYLDYISSGLS
M.l.	β	APC	SLGVPISATVQAIQAMKEVTASLVGPDAGK——EMGVYFDYICSGLS

3.3.2 Phycocyanin

Most of the research on amino acid sequences has been done with phycocyanins. Both subunits of C-phycocyanin from the following species were completely analyzed (Table 5): *Synechococcus* 6301 (= *Anacystis nidulans*) belonging to the cyanobacterian order of Chroococcales (α-subunit: Walsh et al. 1980, β-subunit: Freidenreich et al. 1978), *Mastigocladus laminosus* from the cyanobacterian Stigonematales (α and β-subunits: Frank et al. 1978), and *Cyanidium caldarium*, an eukaryotic alga presently of uncertain affinity (α-subunit: Offner et al. 1981, β-subunit: Troxler et al. 1981). The taxonomic position of *Cyanidium* is still under discussion; thus this organism may be considered a red alga (Chapman et al. 1968, Seckbach et al. 1981) or, as proposed recently, an endosymbiontic form, arising from a symbiosis between cyanobacteria and apochlorotic eukaryotic host (Kremer et al. 1978, Kremer 1982).

In Fig. 2 the results of amino acid sequence analyses after alignment are compared. It becomes apparent that α and β-subunits of the C-phycocyanins exhibit only low sequence homology of 28%–31% (broad horizontal arrows). On the contrary, comparison within α or β chains, respectively, of different C-phycocyanins exhibits a considerably high homology. This ranges from 73%–79% within the β, and from 68%–81% within the α subunits (continuous arrows). The high degree of homology of the amino acid sequences of the α and β-subunits of *Cyanidium* with those of members of the cyanobacteria may be interpreted as evidence for the close relationship and may underline the cyanelle character. It is conceivable that a comparison with subunits of C-phycocyanin from *Porphyridium aerugineum* which is certainly a red alga, may show an equivalent percentage of homology. Such analyses are desirable.

Also in this case of transspecific comparison, the similarities among all α and among all β subunits sequenced convincingly demonstrate and elucidate the separated lines of polypeptide evolution. The homology between α and β-subunits of C-phycocyanins, on the other hand, does not exceed 27% (dashed arrows) respectively 31%. The high degree of homology observed within the α and β polypeptide chains restricts the use of C-phycocyanins for evaluating taxonomic transspecific differentiation within

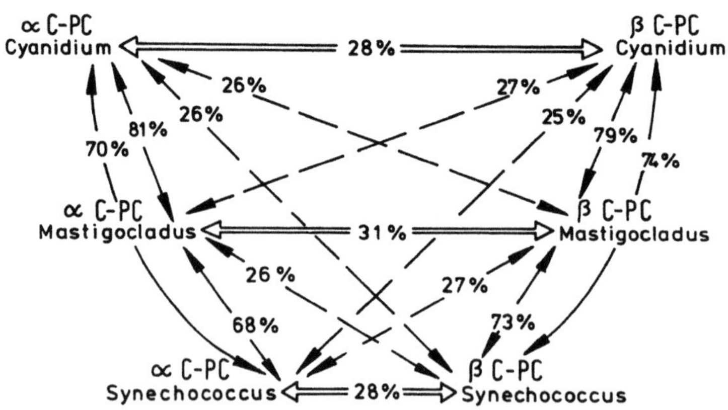

Fig. 2. Homology values of completely analyzed subunits of C-phycocyanin from different sources according to Table 5

Table 5. Complete amino acid sequences of α and β-subunits of C-phycocyanin from different organisms (*Mastigocladus laminosus*, both subunits: Frank et al. 1978, *Synechococcus* α C-PC: Walsh et al. 1980, *Synechococcus* β C-PC: Freidenreich et al. 1978, *Cyanidium caldarium* α C-PC: Offner et al. 1981, *Cyanidium caldarium* β C-PC: Troxler et al. 1981)

Organism	Subunit		Sequence (residues 1–60)
C.c.	α	C-PC	MKTPITEAIAAADNQGRFLSNTELQAVNGRYQRAAASLEAARSLTSNAERLINGAAQAVY
S.	α	C-PC	SKTPLTEAVAAADSQGRFLSSTELQVAFGRFRQAASGLAAAKALANNADSLVDGAANAVY
M.l.	α	C-PC	VKTPITDAIAAADTQGRFLSNTELQAVNGRYQRAAASLEAARALTANAQRLIDGAAQAVY
C.c.	β	C-PC	MLDAFAKVVAQADARGEFLSNTQLDALSKMVSEGNKRLDVVNRITSNASAIVTNAARALF
M.l.	β	C-PC	AYDVFTKVVSQADSRGEFLSNEQLDALANVVKEGNKRLDVVNRITSNASTIVTNAARALF
S.	β	C-PC	TFDAFTKVVAQADARGEFLSDAQLDASLRLVAEGNKRIDTVNRITGNASSIVANAARALF

Organism	Subunit		Sequence (residues 61–120)
			PCB (above α rows, at the Cys following residue 80)
C.c.	α	C-PC	SKFPYTSQMPGPQYASSAVGKAKCARDIGYYLRMVTYCLVVGGTGPMDEYLIAGLEEINR
S.	α	C-PC	SKFPYTTSTPGNNFASTPEGKAKCARDIGYYLRIVTYALVAGGTGPIDEYLIAGLDEINT
M.l.	α	C-PC	QKFPYLIQTSGPNYAADARGKSKCARDIGHYLRIITYSLVAGGTGPLDEYLIAGLNEIND
C.c.	β	C-PC	SEQPQLIQPGGTAYTNR—RMAACLRDMEIILRYVSYAIIAGDSSILDDRCLNGLRETYQ
M.l.	β	C-PC	EEQPQLIAPGGS—ATRNGTMAACLRDMEIILRYITYAILAGDASILDDRCLNGLRETYQ
S.	β	C-PC	AEQPSLIAPGGNAYTN——RMAACLRDMEIILRYVTYAVFTGDASILDDRCLDGLRETYL
			PCB (below β rows, at the Cys)

Organism	Subunit		Sequence (residues 121–172)
C.c.	α	C-PC	TFDLSPSWYVEALNYIKANHGLSGQAAN——EANTYIDYAINALS
S.	α	C-PC	KFDLAPSWYVEALKYIKANHGLSGDSRD——EANSYIDYILNALS
M.l.	α	C-PC	AFELSPSWYIEALKYIKANHGLSGQAAN——EANTYIDYVINALS
C.c.	β	C-PC	ALGVPGASVAVGIEKMKDSAIAIANDPSGITTGDCSALMAEVGTYFDRAATAVQ
M.l.	β	C-PC	ALGTPGSSVAVGIQKMKEAAINIANDPNGITKGDCSALISEVASYFDRAAAAVA
S.	β	C-PC	ASGVPGALVAEGVRKMKDAAVAIVSDRNGITQGDCSA–ISELGSYFDKAAAAVA
			PCB (below β rows, at the Cys)

the cyanobacteria. However, from the taxonomic point of view the differentiation between C- and R-phycocyanin would be of specific interest. C-phycocyanin is a characteristic pigment for probably all cyanobacteria and many Porphyridiales among the red algae, whereas R-phycocyanin predominates among all higher orders of this division. At present, there are only 21 amino acids of the α and 19 of the β subunit of R-phycocyanin at the N-terminus and about 28 of β R-phycocyanin at the C-terminus around the chromophore bearing cysteinyl residue of *Porphyridium cruentum* analyzed (Glazer et al. 1976, Bryant et al. 1978). For these sequences a homology of 76%, 68%, and above 50% with the corresponding subunits of C-phycocyanin of *Mastigocladus laminosus* could be estimated. This relatively high homology corresponds with the fact that C- and R-phycocyanin are representing the same spectral class of biliproteins.

3.3.3 Phycoerythrin

Among the cyanobacterial and red algal biliproteins the lack of knowledge in the spectral class of phycoerythrins is particularly obvious. The scant data have been summarized by Glazer and Hixson (1977). In addition, the amino acid sequences of α and β C-phycoerythrin of *Fremyella diplosiphon* were kindly provided by W. Sidler (unpublished results).

The homology values of the N terminal sequences of the α-subunits of B- and C-phycoerythrin and of the β-subunits of B- and C-phycoerythrin of *Porphyridium cruentum* and *Fremyella diplosiphon* were 78% and 63%, respectively. These values may not be overestimated as only 18 or 19 amino acids of the α and β-subunits of B-phycoerythrin of *P. cruentum* were available. They are, however, in the order of magnitude comparable with observations for other biliproteins belonging to a distinct spectral class.

3.3.4 Phycoerythrocyanin

This only recently discovered pigment is dealt with separately because of its exceptional position. It is characteristically distributed within filamentous cyanobacteria, namely in non-heterocystous Oscillatoriales *(Pseudanabaena)*, heterocystous Nostocales *(Anabaena, Cylindrospermum, Nostoc, Aphanizomenon, Scytonema, Calothrix)*, and in Stigonematales *(Fischerella, Chlorogloeopsis)* which exhibit true branching (Bryant 1982). With the exception of *Chroococcidiopsis* it has not been found in either Chroococcales or Pleurocapsales. Very recently, both subunits of phycoerythrocyanin of *Mastigocladus laminosus* were completely sequenced (Fueglistaller et al. 1983). In addition, 21 amino acids of the N terminus of the α subunit of phycoerythrocyanin of *Anabaena variabilis* and 19 amino acids of the β-subunit were analyzed (Glazer et al. 1976); further 38 amino acids of the C terminus of the β phycoerythrocyanin are known (Bryant et al. 1978). There is 80% homology within the α and 68% within the β-subunits at the N terminus and 81% in the β-subunits at the C terminus of phycoerythrocyanin (Table 6).

Table 6. Amino acid sequences of α and β-subunits of phycoerythrocyanin from different sources in comparison with *Mastigocladus laminosus* C-phycocyanin (Frank et al. 1978) (*Mastigocladus laminosus*, both subunits of PEC: Fueglistaller et al. 1983, *Anabaena variabilis* α and β PEC, N-terminus: Glazer et al. 1976, *Anabaena variabilis* β PEC, C-terminus: Bryant et al. 1978)

			Sequence (positions 1–60; numbered 10, 20, 30, 40, 50, 60)
A.v.	α	PEC	MKTPLTEAIGAADVRGXYLXN
M.l.	α	PEC	MKTPLTEAIAAADLRGSYLSNTELQAVFGRFNRARAGLEAARAFANNGKKWAEAAANHVY
M.l.	α	C-PC	VKTPITDAIAAADTQGRFLSNTELQAVNGRYQRAAASLEAARALTANAQRLIDGAAQAVY
A.v.	β	PEC	MLNAFXKVVEQANRKGNYL
M.l.	β	PEC	MLDAFSRVVEQADKKGAYLSNDEINALQAIVADSNKRLDVVNRLTSNASSIVANAYRALV
M.l.	β	C-PC	AYDVFTKVVSQADSRGEFLSNEQLDALANVVKEGNKRLDVVNRITSNASTIVTNAARALF

			Sequence (positions 61–120; numbered 70, 80, 90, 100, 110, 120)
			PCB (at C, position 84, α PEC)
M.l.	α	PEC	QKFPYTTQMQGPQYASTPEGKAKCVRDIDHYLRTISYCCVVGGTGPLDDYVVAGLKEFNS
M.l.	α	C-PC	QKFPYLIQTSGPNYAADARGKSKCARDIGHYLRIITYSLVAGGTGPLDEYLIAGLNEIND
M.l.	β	PEC	AERPQVFNPGGP—CFHHRNQAACIRDLGFILRYVTYSVLAGDTSVMDDRCLNGLRETYQ
M.l.	β	C-PC	EEQPQLIAPGGS—ATRNGTMAACLRDMEIILRYITYAILAGDASILDDRCLNGLRETYQ
			PCB (at C, β C-PC)

			Sequence (positions 121–; numbered 130, 140, 150, 160, 170)
M.l.	α	PEC	ALGLSPSWYIAALEFVRDNHG—LTGDVAG————EANTYINYAINALS
M.l.	α	C-PC	AFELSPSWYIEALKYIKANHGLSGQAAN————EANTYIDYVINALS
A.v.	β	PEC	KDAAIAIANDSKGITKGDCSQLIAELASYFDRAASAVV
M.l.	β	PEC	ALGTPGDAVASGIKKMKEAAIALANDPNGITKGDCSQLMSELASYFDRAAAAV
M.l.	β	C-PC	ALGTPGSSVAVGIQKMKEAAINIANDPNGITKGDCSALISEVASYFDRAAAAVA
			PCB (at C of GDCS, β C-PC)

3.4 Comparison of Biliproteins Belonging to Different Spectral Classes

3.4.1 Allophycocyanin - Phycocyanin

In Fig. 3 the values of homology between the completely analyzed *a* (and *β*) subunits of allophycocyanin of *Mastigocladus laminosus* (+ *β* APC of *Anabaena variabilis*) and the *a* (and *β*) subunits of completely analyzed C-phycocyanins from the same and different species are shown. In this case it can convincingly be demonstrated that the homology between subunits belonging to different spectral classes of biliproteins is – as a rule – low; however the values of the heterologous *β*-subunits with 37%–40% are significantly higher than those of the *a*-subunits with 28%–29%. Thus it can be concluded that the *β*-subunits are functionally more important and are therefore conserved to a higher extent than the *a*-subunits.

3.4.2 C-phycocyanin - C-phycoerythrin

The alignment of 49 residues from the N terminus of *a* C-phycoerythrin of *Fremyella diplosiphon* with *a* C-phycocyanin of *Mastigocladus laminosus*, and 60 residues from *β* C-phycoerythrin with *β* C-phycocyanin of the same species resulted in a homology of 49% or 43%, respectively (Table 7). This is high, as C-phycocyanin and C-phycoerythrin belong to different spectral classes. The homology values of *a* and *β* C-phycoerythrin compared to *a* and *β* phycoerythrocyanin of *M. laminosus* in this restricted part of the N terminus was 45% and 52%. These calculations are preliminary and the percentage of homology will change as soon as the sequences of the pigments are completed, but, there is good evidence from the immunological research that the tendency observed will be sustained.

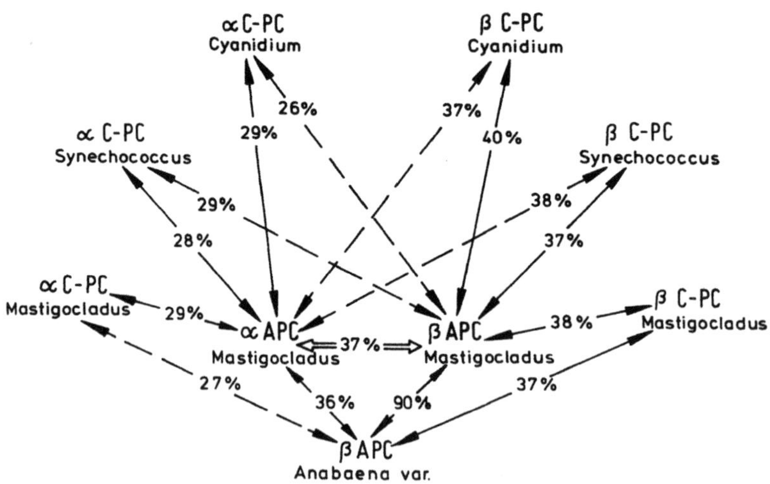

Fig. 3. Homology values of completely analyzed subunits of C-phycocyanin from different sources in comparison with *a* and *β* allophycocyanin from *Mastigocladus laminosus* and *β* allophycocyanin from *Anabaena variabilis*

Table 7. Amino acid sequences of cyanobacterial and red algal phycoerythrins in comparison with *Mastigocladus laminosus* C-phycocyanin (*Porphyridium cruentum* α and β B-PE: Glazer and Hixson 1977, *Fremyella diplosiphon* α and β C-PE: Sidler et al. unpublished results)

			1 10 20 30 40 50 60
P.c.	α	B-PE	MKSVIXTVVSAADAAGRF
F.d.	α	C-PE	MKSVVTTVIAAADAAGRFPSTSDLESVQGSIQXAAARLEAAEKLANNID
M.l.	α	C-PC	VKTPITDAIAAADTQGRFLSNTELQAVNGRYQRAAASLEAARALTANAQRLIDGAAQAVY
F.d.	β	C-PE	MLDAFSRAVVSADASGSTVSDIA—ALRAFVASGNRRLDAVNAIASNASRMVVDAVRGMI
M.l.	β	C-PC	AYDVFTKVVSQADSRGEFLSNEQLDALANVVKEGNKRLDVVNRITSNASTIVTNAARALF
P.c.	β	B-PE	MLDAFTRVVVNADAKAAYV

3.4.3 C-phycocyanin – Phycoerythrocyanin

The high degree of homology of α and β-subunits of completely sequenced C-phycocyanin and phycoerythrocyanin from *Mastigocladus laminosus* with 62% and 67% has already been mentioned above (Fig. 1, Table 6). This homology indicates an otherwise unknown relatedness of biliproteins belonging to different spectral classes. It does not seem to be species-dependent, however, as the sequences of phycoerythrocyanin of *Anabaena variabilis* in Table 6 reach a 57% homology for α and 42% and 78% homology for the N and C terminus of the β-subunit with C-phycocyanin of *Mastigocladus laminosus*.

3.5 Amino Acid Sequence Analyses of Cryptophytan Biliproteins

Complete amino acid sequences of cryptophytan biliproteins are still lacking, only 26 amino acids from the N terminus of the α and 19 from the β-subunit of *Hemiselmis virescens* phycocyanin have been analyzed (Glazer and Apell 1977). Comparing these sequences with corresponding composite sequences derived from studies of cyanobacterial and red algal biliproteins, a characteristic pattern of invariant residues within the β-subunits could be demonstrated. The β-subunits of *Hemiselmis virescens* phycocyanin, particularly, resembled most closely that of *Porphyridium cruentum* B-phycoerythrin with fifteen identities in the sequence of nineteen residues. Surprisingly Glazer and Apell (1977) detected no homology between the α type of *H. virescens* phycocyanin and any other α-subunits of biliproteins tested.

Sidler et al. (1983) succeeded in separating the subunits of *Chroomonas* phycocyanin-645. As was demonstrated earlier (Mörschel and Wehrmeyer 1975) *Chroomonas* phycocyanin-645 is composed of three subunits α, α', and β with molecular weights of 9.2, 10.4, and 15.5 kD, respectively. A high degree of homology of the β phycocyanin-645 with the cyanobacterial phycocyanin was detected in confirmation of the above mentioned results. Also the α chains showed sequence homologies with known cyanobacterial biliproteins. The homology started at position 62 of the polypeptide chain, demonstrating a loss of the first 61 amino acids in cryptophytan biliproteins.

These are convincing evidences for an intimate connexion of biliproteins from either cyanobacteria, red algae, and cryptophytes. Thus there is no support for an independent biliprotein evolution in the cryptophytes without connections to the other biliprotein containing phyla. Thus, the postulation can be made that the phycocyanins of *Hemiselmis virescens* and *Chroomonas* are derived from the same ancestral gene as those of cyanobacterial and rhodophytan biliproteins. A reduced molecular weight of the α chains in comparison to that of cyanobacterial and red algal biliproteins is frequently found in cryptophytes (Mörschel and Wehrmeyer 1977, Gantt 1979). It may be that the length of the lost polypeptide chain varies to a certain degree depending on species and pigment. The complete study of cryptophytan phycocyanin and its homologies with amino acid sequences of the cyanobacterial biliproteins – allophycocyanin, C-phycocyanin, and phycoerythrocyanin – will certainly produce far reaching consequences: either the phycocyanin-645 is the evolutionary remnant of only the

C-phycocyanin of its ancestor, or it represents more than one biliprotein component and thus must be looked upon as a specific type of cryptophytan antennae pigment system.

4 Immunological Analyses of Phycobiliproteins

4.1 Methods

The general procedures for the immunological assays include isolation and purification of the biliproteins, and the immunization and antisera production followed by double immunodiffusion tests (Ouchterlony 1949). In tabulating the results of immunodiffusion experiments no distinction has been made between reactions of identity and reactions with spurring indicating only partial identity. As starting material we exclusively used isolated phycobilisomes and highly purified biliproteins obtained from these fractions after ultracentrifugation of dissociated phycobilisomes and gel electrophoreses.

4.2 Immunological Cross-Reactions with Biliproteins of Cyanobacteria and Rhodophytes of the Same Spectral Class

A historical survey of the immunological research has been published by MacColl and Berns (1979). Table 8 summarizes some of the most reliable immunological data obtained from the literature, only those experiments using purified biliproteins are included. There is convincing evidence that all phycobiliproteins belonging to the same spectral class (namely allophycocyanins, phycocyanins, and phycoerythrins) of cyanobacteria and red algae, regardless of the taxonomic position of the organism, cross-react with each other and are immunologically similar (Berns 1967, Glazer et al. 1971). This is in accord with a 82% homology for α and a 90% homology for β-subunits of amino acid sequences of allophycocyanin of different species (p. 150). Similarly, a homology of 68%–81% for α and 73%–79% for β-subunits within completely analyzed C-phycocyanins was found (Fig. 2); only the 63%–78% homology observed for phycoerythrins (p. 154) refers to partially sequenced polypeptide chains of subunits Allophycocyanin exhibited by far the greatest antigenic similarities independent of algal source, and consequently the highest degree of homology in the amino acid sequences of subunits.

Contrastingly no cross-reactions can be detected by double diffusion between allophycocyanins, phycocyanins, and phycoerythrins, and there is no immunological similarity among these three biliprotein classes, even when the immunizing antigens are derived from the same organism (Glazer et al. 1971). This is in parallel with the low degree of homology in amino acid sequences between subunits belonging to different spectral classes, e.g. allophycocyanin and C-phycocyanin with 26%–29% for α and 37%–40% for β-subunits (Fig. 3). Since phycocyanin and allophycocyanin have the same chromophore their failure to cross-react immunologically shows that the

Table 8. Cross-reactivity of antisera of biliproteins belonging to different spectral classes from cyanobacteria and red algae, with different homologous and heterologous biliproteins as antigens. Data from (1) Cohen-Bazire et al. (1977), (2) Glazer et al. (1971), (3) Kao et al. (1973), (4) Berns (1967), (5) Eder et al. (1978), (6) MacColl et al. (1976)

	Antiserum to	APC Synechococcus 6312	APC Synechococcus 63o1	APC Aphanocapsa 67o1	APC Fremyella diplosiphon	APC Cyanidium caldarium	C-PC Synechococcus 6312	C-PC Synechococcus 63o1	C-PC Synechococcus lividus	C-PC Coccochloris elabens	C-PC Aphanocapsa 67o1	C-PC Plectonema calothricoides	C-PC Phormidium luridium	C-PC Fremyella diplosiphon	C-PC Cyanidium caldarium	C-PC Porphyridium aerugineum	C-PC Pseudanabaena	C-PE Aphanocapsa 67o1	C-PE Pseudanabaena W 1173	C-PE Pseudanabaena catenata	C-PE Fremyella diplosiphon	C-PE Tolypothrix tenuis	C-PE Phormidium persicinum	C-PE Calothrix membranacea	B-PE Porphyridium cruentum	R-PE Ceramium rubrum			
1	APC Synechococcus 6312	+																											
2	" Synechococcus 63o1		+	+	+	+		−			−			−				−			−								
2	" Aphanocapsa 67o1		+	+	+	+		−			−			−				−			−								
2	" Fremyella diplosiphon		+	+	+													−			−								
1	C-PC Synechococcus 6312						+																						
2	" Synechococcus 63o1		−	−	−			+			+			+	+			−			−								
3.4	" Synechococcus lividus								+	+		+	+		+	+													
2	" Aphanocapsa 67o1		−	−	−			+			+			+	+			−			−								
3.4	" Plectonema calothricoides								+	+		+	+		+	+													
4	" Phormidium luridum								+			+	+		+	+													
2	" Fremyella diplosiphon		−	−	−			+		+				+				−			−								
4	" Cyanidium caldarium								+		+	+			+	+													
2	C-PE Aphanocapsa 67o1		−	−	−			−			−			−				+			+								
5	" Pseudanabaena W1173																−	+	+	+	+								
2	" Fremyella diplosiphon		−	−	−			−			−			−				+	+		+								
4	" Phormidium persicinum																	+				+			+	+			
4	" Tolypothrix tenuis																					+			+	+			
4.6	B-PE Porphyridium cruentum																					+	+	+	+	+			

phycocyanobilin does not function as a determinant, and presumably lies in a fold, or at least not exposed on the surface of the quaternary protein structure.

The above mentioned conclusions derived from immunological studies on phycobiliproteins similarly have been presented by Berns (1967) and Glazer et al. (1971) and were accepted by Stanier (1974) and MacColl and Berns (1979). However, our recent research modify these generalizations. In particular, cross-reactions of biliproteins belonging to different spectral classes earn attention.

4.3 Immunological Cross-Reactions of Cyanobacterial Phycoerythrocyanins

An exception among cyanobacterial phycobiliproteins is phycoerythrocyanin, recently discovered by Bryant et al. (1976). It is also a constituent biliprotein of phycobilisomes of *Mastigocladus laminosus* (Nies and Wehrmeyer 1980). Antisera directed against purified phycoerythrocyanin isolated from this species reacted with the homologous phycoerythrocyanin antigen, plus cross-reacted with the heterologous C-phycocyanin from *M. laminosus* and from *Fremyella diplosiphon* (Morisset 1983).

These observations confirm and extend earlier investigations of Bryant et al. (1976) with antiserum raised against phycoerythrocyanin of *Anabaena* which also cross-reacted with *Anabaena* phycoerythrocyanin as well as with *Anabaena* phycocyanin. No immunological reaction was observed with phycoerythrocyanin antiserum from *M. laminosus* or *Anabaena* using allophycocyanin as antigen. In Table 9 these results are summarized. Considering the high degree of homology of amino acid sequences of α phycocyanin of *Mastigocladus* and of the corresponding β-subunits of 60% and 64%, respectively, these immunoreactions are not surprising.
In addition, antiserum to phycoerythrocyanin of *M. laminosus* also cross-reacts with all spectral types of phycoerythrin, i.e. C-phycoerythrin from the cyanobacterium *Fremyella diplosiphon*, B-phycoerythrin from the bangiophycidean red alga *Porphyridium cruentum* and R-phycoerythrin from the florideophycidean red alga *Delesseria sanguinea*. It can be concluded from these experiments that the antiserum to *M. laminosus* phycoerythrocyanin shares basic common determinants with C-phycocyanins and C-, B-, and R-phycoerythrins, and thus is the first example of an immunological cross-reaction of biliproteins transgressing spectral classes. The thorough analyses of reactions of partial identity (spurring) may probably reveal the degree of relatedness of the different biliproteins and the number of determinants.

Supplementary studies performed with antisera against *Mastigocladus laminosus* C-phycocyanin likewise gave cross-reactions with C-phycocyanins of *Mastigocladus* and *Fremyella* as expected, but also with phycoerythrocyanin of *M. laminosus* and C-, B-, and R-phycoerythrins of *Fremyella diplosiphon, Porphyridium cruentum,* and *Delesseria sanguinea*, respectively (Table 9). These results suggest that there is a partial identity between C-phycocyanin of *M. laminosus* and the three types of phycoerythrin C-, B-, and R-phycoerythrin. This is, in part, confirmed as far as C-phycoerythrin is concerned by amino acid sequence analyses with a partial homology of C-phycocyanin and C-phycoerythrin of 49% and 43% for α and β-subunits, respectively. Hitherto it has not been substantiated for the other phycoerythrins.

Table 9. Cross-reactivity of antisera raised against purified phycoerythrocyanin and C-phycocyanin from *Mastigocladus laminosus* with cyanobacterial and red algal biliproteins as antigens. [a] Data from Bryant et al. (1976)

	Cyanobacteria									Rhodophyta				
Antiserum to	PEC Anabaena variabilis	C-PC Anabaena variabilis	PEC Mastigocladus lam.	C-PC Mastigocladus lam.	C-PC Fremyella diplosiphon	APC Mastigocladus lam.	APC Anabaena variabilis	C-PE Fremyella diplosiphon	C-PE Aphanocapsa 6701	B-PE Porphyridium cruentum	R-PE Delesseria sanguinea			
PEC Anabaena sp.[a]	+	+					−		−					
PEC Anabaena sp. 6411[a]	+	+												
PEC Mastigocladus lam.			+	+	+	−		+		+	+			
C-PC Mastigocladus lam			+	+	+	−		+		+	+			

4.4 Immunological Cross-Reactions of Cryptophytan Phycobiliproteins

In the earlier biliprotein immunological investigations the cyanobacterial and red algal pigments have played the central role, and the cryptophytan biliproteins were subordinate. In agreement with the taxonomic uncertainty of the cryptophytes, separated from all other algal phyla, their biliproteins also seem to deviate. Phycocyanins and phycoerythrins of cryptophytes revealed no immunological cross-reaction with any phycobiliproteins of rhodophytes or cyanobacteria, and they were suggested to be immunologically unrelated (Berns 1967, Glazer et al. 1971). This was questioned by MacColl et al. (1976) after repeating an earlier experiment conducted by Berns (1967); they obtained interactions of antisera prepared against phycoerythrin of *Rhodomonas lens* with B-phycoerythrin of the red alga *Porphyridium cruentum* as antigen. In addition, cryptomonad phycoerythrins and phycocyanins cross-reacted with each other with only little spurring, demonstrating that they were immunologically very similar. These results could be confirmed and extended (Table 10). Antisera raised against purified phycoerythrin-545 and phycoerythrin-555 from *Cryptomonas maculata* and *Hemiselmis rufescens* cross-reacted with cryptophytan phycoerythrins and with phycocyanin-645 of *Chroomonas*. Likewise, antiserum against phycocyanin-645 cross-reacted with phycocyanin-645 and both phycoerythrins as well. This is a further example of immunological cross-reactions transgressing spectral biliprotein classes, which seems to be a characteristic feature of biliproteins of this phylum. The following is a possible explanation for this behavior. The loss of a considerable part of the polypeptide chain at the N terminus – e.g. 61 residues of α phycocyanin-645 of *Chroomonas* in comparison to a phycocyanin of *Mastigocladus laminosus* as mentioned above – causes a loss of protection of regions with a high degree of homology in the amino acid sequences. These are thus capable of functioning as additional antigenic sites common to several phycobiliproteins during the antisera production.

Table 10. Cross-reactivity of antisera raised against purified cryptophytan phycoerythrin and phycocyanin with cryptophytan and red algal biliproteins as antigens.
[a] Data from MacColl et al. (1976)

Antiserum to	Cryptophyta						Rhodophyta					
	PE Rhodomonas lens	PE Cryptomonas	PC Chroomonas	PE_{545} Cryptomonas	PE_{555} Hemiselmis rufescens	PC_{645} Chroomonas		B-PE Porphyridium cruentum	R-PE Delesseria sanguinea			
PE Rhodomonas lens[a]	+	+	+					+				
PE Cryptomonas ovata[a]	+	+	+									
PC Chroomonas[a]		+	+									
PE_{545} Cryptomonas				+	+	+						
PE_{555} Hemiselmis				+	+	+						
PC_{645} Chroomonas				+	+	+			+			
B-PE Porphyridium cr.[a]	+	+	+					+				

An antiserum raised against phycoerythrin of *Rhodomonas lens* was capable of recognizing B-phycoerythrin from *Porphyridium cruentum* (MacColl et al. 1976), another antiserum produced against phycocyanin-645 of *Chroomonas* cross-reacts with R-phycoerythrin of *Delesseria sanguinea* and other higher red algae (Morisset 1983; see Table 10). Again this demonstrates that basic ancestral genes are common to all biliprotein producing phyla, and that the cryptophytes have a closer phylogenetic relationship to the red algae than had previously been assumed. Additionally it demonstrates that under certain circumstances, probably for the same reasons, cross-reactions of cryptophytan phycocyanin and red algal phycoerythrins beyond spectral biliprotein classes are possible.

5 Conclusions

1. Cyanophytan and rhodophytan phycobiliproteins belonging to the *same* spectral class (namely, allophycocyanins, phycocyanins or phycoerythrins) are similar. They show a high cross-reactivity in immunochemical tests and a high degree of structure homology with 80%–90% for complete sequences of allophycocyanin, 70%–80% for complete sequences of phycocyanin subunits and 60%–80% for partially analyzed subunits of phycoerythrin. The amino acid sequence analyses are thus in accord with and confirm the results obtained by immunochemical methods.

2. The usual lack of cross-reactivity of biliproteins of cyanobacterial or rhodophycean origin belonging to *different* spectral classes coincides with a low degree of homology, below 30% for α, and below 40% for β-subunits (comparing allophycocyanin of *Mastigocladus* with different phycocyanins). Since antibody-antigen interactions involve principally the surface of the antigen, it can be postulated that congruent amino acid sections lie buried in the interior of the protein conformation or are otherwise not available to antibody reactions. This is, particularly, suggested for some regions C-terminal to the chromophore binding sites of α and β-subunits which show high homology values according to Zuber (1978) and Sidler et al. (1981). This may also be due to parts of the N termini which show a high homology of amino acids regardless of the biliprotein class and do not display any cross-reactivity (Eder et al. 1978).

3. The generally accepted statement (Berns 1967; Glazer et al. 1971, MacColl and Berns 1979) that no serological cross-reactions can be obtained between biliproteins belonging to different spectral classes, must now be revised. There are some distinct cross-reactions of antisera directed against purified *Mastigocladus* phycoerythrocyanin with heterologous C-phycocyanin of *M. laminosus* and C-phycoerythrin of *Fremyella diplosiphon* (Morisset 1983). This is in accord with amino acid homology values of α phycoerythrocyanin to α C-phycocyanin of 62% and α C-phycoerythrin of 45% and of β phycoerythrocyanin to β C-phycocyanin of 66.6% (Fueglistaller et al. 1983) and β C-phycoerythrin of 52% (determined from Sidler et al., unpublished results), respectively. Moreover, distinct antiserum directed against purified C-phycocyanin of *M. laminosus* cross-reacted with all types (C-, B-, R-) of heterologous phycoerythrin and phycoerythrocyanin (Morisset 1983).

These are the first examples of immunochemical cross-reactions of biliproteins transgressing spectral classes. The statement, that members of each phycobiliprotein class, provided that they are of cyanobacterial or rhodophytan origin, share a common set of major antigenic determinants (Glazer et al. 1971) must be extended. Phycoerythrocyanins share some determinants with cyanobacterial phycocyanins and cyanobacterial and rhodophytan phycoerythrins and, likewise, cyanobacterial phycocyanin (from *Mastigocladus laminosus*) shares some determinants with cyanobacterial and rhodophytan phycoerythrins.

4. The detected persistence of high sequence homology among biliproteins can only be understood by means of a strong functional requirement, vital for maintenance of the functional properties of biliproteins (Fueglistaller et al. 1981). Among these functionally essential features, as far as cyanobacterial and red algal biliproteins are considered, the following must be accounted for: (a) the fixed position of at least a part of the chromophoric groups in a more or less oriented position within the proteinic environment to enable optimal energy transfer; (b) the specific organization of aggregates with probably different types of bonding, i.e. between α and β-subunits to form protomers ($\alpha\beta$), between protomers to build up trimers $(\alpha\beta)_3$, which in their turn, give rise to hexameric aggregates $(\alpha\beta)_6$ by face to face association; (c) specific regions at the surface of the aggregates for cooperation with the "linker" polypeptides to bind to and organize sequential association of hexamers to create peripheral rods and the allophycocyanin-containing phycobilisome center. Particularly allophycocyanins function in establishing specific contact among themselves in the phycobilisome center, with the peripheral rods and with the supporting thylakoid membrane. It is for these reasons, that allophycocyanins show such a high degree of homologies in their amino acid sequences and such strong cross-reactivity.

The high degree of homology of allophycocyanin subunits, and the proteinchemical conservatism and function of biliproteins in general restrict an extended evaluation of sequences for systematic purposes to the highest taxonomic levels.

5. The high degree of proteinchemical similarity among phycobiliproteins (amino acid sequences, mol. wt. of subunits, tertiary and quarternary structure, immunological and aggregation behaviour) is expressed in the phycobilisomes, complex supramolecular organization forms of biliproteins in the cell of cyanobacteria and red algae. The hemidiscoidal type of phycobilisomes is predominantly present in cyanobacteria as well as in many bangiophycidean red algae with nearly identical structural architecture and can thus be looked upon as an evidence of relationships between pro- and eukaryotic organisms.

6. The hypothesis that cryptophytan phycocyanins and phycoerythrins do not cross-react immunologically with any red algal or cyanobacterial phycobiliprotein and are hence unrelated (Stanier 1974) must be revised. Antisera produced against phycoerythrin from *Rhodomonas lens* and phycocyanin-645 from *Chroomonas* cross-react with B-phycoerythrin of *Porphyridium cruentum* and R-phycoerythrin of *Delesseria sanguinea*, respectively (MacColl et al. 1976, Morisset 1983).

Comparative sequence analyses of phycocyanin-645 from *Chroomonas* and C-phycocyanin from *Mastigocladus laminosus* (Sidler et al. 1983) showed a high amino acid homology of the β-subunits and a lower homology in the α-subunits; the latter comparison was only possible when the N terminus was started at position 62 of the known cyanobacterial biliprotein chain. This gives convincing evidence that the cryptophytes might have a closer phylogenetic relationship to the other biliprotein containing phyla than had previously been assumed. Regarding the high homology values of either cyanobacterial or red algal phycobiliproteins it is difficult to decide to which phyla the narrowest phylogenetic affinities of cryptophytan biliprotein exist. The question is whether both the cryptophytes and the red algae independently originated from cyanobacteria (Leedale 1974) or whether there is a phylogenetic line from cyanobacteria to the cryptophytes from which the rhodophytes diverged long ago (Ragan and Chapman 1978). Thus recent observations on a nucleomorph, i.e. a nucleus-like organelle in the compartment between the chloroplast envelope and the periplastidal ER cisterna in Cryptophytes, provide evidence for the possible derivation of a cryptophytan chloroplast from an eukaryotic symbiotic organism (Gibbs 1981).

7. Antiserum directed against cryptophytan phycocyanin-645 from *Chroomonas* cross-reacted with phycoerythrin-545 of *Cryptomonas* as well as with phycoerythrin-555 of *Hemiselmis*; likewise, antisera against these cryptophytan phycoerythrins gave positive reactions with *Chroomonas* phycocyanin-645 antigen (Morisset 1983). This is a second example of immunochemical cross-reaction of biliproteins transgressing spectral classes. This increased cross-reactivity of antisera may originate from amino acid homologies in the region C terminal to the chromophoric group in position 84 of the polypeptide chain; they may be freely exposed in cryptophytan biliprotein after omission of 61 amino acids at the N terminus of the α-subunit and are, on the contrary, protected in cyanobacterial and red algal biliproteins.

Deviating from cyanobacteria and red algae the biliproteins of cryptophytes are localized within the thylakoid lumen, show a reduced molecular weight and a lower native state of aggregation.

8. It is hazardous to construct any type of phylogeny on the basis of a single trait and with the admittedly sparse knowledge available at present, but from the amino acid sequence analyses and the immunological data it can be deduced that basic ancestral genes are common to all biliprotein containing phyla, and the cryptophytes have a closer phylogenetic relationship to the red algae than had been previously assumed.

Acknowledgements. The author thanks Mrs. R. Kort for excellent photographic work, Dipl.-Biol. W. Morisset for making unpublished results accessible, and Dr. A.M. Roberton, University of Auckland, New Zealand, for improving the English manuscript. The research was supported by a grant from the Deutsche Forschungsgemeinschaft (SFB 103).

Dedicated to Prof. Dr. H.A. von Stosch, Marburg, on the occasion of his 75th birthday 1983.

References

Berns DS (1967) Plant Physiol 42:1569–1586
Bryant DA (1982) J Gen Microbiol 128:835–844
Bryant DA, Glazer AN, Eiserling FA (1976) Arch Microbiol 110:61–75
Bryant DA, Hixson CS, Glazer AN (1978) J Biol Chem 253:220–225

Bryant DA, Guglielmi G, Tandeau de Marsac N, Castets AM, Cohen-Bazire G (1979) Arch Microbiol 123:113–127
Chapman DJ, Cole WJ, Siegelman HW (1968) Am J Bot 55:314–322
Christensen T (1962) In: Böcher TW, Lange M, Sörensen T (eds) Systematisk Botanik 2. Munksgaard, Copenhagen, pp 1–178
Cohen-Bazire S, Beguin S, Rimon S, Glazer AN, Brown DM (1977) Arch Microbiol 111:225–238
Dayhoff MO (1978) In: Dayhoff MO (ed) Atlas of protein sequence, vol V, Suppl 3. Natl Biomed Res Found, Washington DC
DeLange RJ, Williams LC, Glazer AN (1981) J Biol Chem 256:9558–9566
Eder J, Wagenmann R, Rüdiger W (1978) Immunochemistry 15:315–321
Frank G, Sidler W, Widmer H, Zuber H (1978) Z Physiol Chem 359:1491–1507
Freidenreich P, Apell GS, Glazer AN (1978) J Biol Chem 253:212–219
Fueglistaller P, Widmer H, Sidler W, Frank G, Zuber H (1981) Arch Microbiol 129:268–274
Fueglistaller P, Suter F, Zuber H (1983) Ber Dtsch Bot Ges 95 (in press)
Gantt E (1979) In: Levandowsky M, Hutner SA (eds) Biochemistry and physiology of Protozoa, vol I, 2nd edn. Academic Press, London New York, pp 121–137
Gibbs SP (1981) Ann NY Acad Sci 361:193–208
Glazer AN (1977) Mol Cell Biochem 18:125–140
Glazer AN, Apell GS (1977) FEMS-Lett 1:113–116
Glazer AN, Hixson (1977) J Biol Chem 252:32–42
Glazer AN, Cohen-Bazire G, Stanier RY (1971) Proc Natl Acad Sci USA 68:3005–3008
Glazer AN, Apell GS, Hixson CS, Bryant DA, Rimon S, Brown DM (1976) Proc Natl Acad Sci USA 73:428–431
Gysi J, Zuber H (1974) FEBS Lett 48:209–213
IUPAC-IUB Commission on Biochemical Nomenclature (1969) Z Physiol Chem 350:793–797
Kao OHW, Berns DS, Town WR (1973) Biochem J 131:39–50
Kremer BP (1982) Br Phycol J 17:51–61
Kremer BP, Feige GB, Schneider HAW (1978) Naturwissenschaften 65:157
Leedale GF (1974) Taxon 23:261–270
MacColl R, Berns DS (1979) TIBS 4:44–47
MacColl R, Berns DS, Gibbons O (1976) Arch Biochem Biophys 177:265–275
Mörschel E, Wehrmeyer W (1975) Arch Microbiol 105:153–158
Mörschel E, Wehrmeyer W (1977) Arch Microbiol 113:83–89
Mörschel E, Koller KP, Wehrmeyer W, Schneider H (1977) Cytobiology 16:118–129
Morisset W (1983) Thesis, Fachbereich Biol, Univ Marburg FRG
Muckle G, Rüdiger W (1977) Z Naturforsch 32c:957–962
Nies M, Wehrmeyer W (1980) Planta 150:330–337
Nies M, Wehrmeyer W (1981) Arch Microbiol 129:374–379
Offner GD, Troxler RF (1982) Fed Proc 41(4):1179
Offner GD, Brownmas AS, Ehrhardt MM, Troxler RF (1981) J Biol Chem 256:12167–12175
Ouchterlony Ö (1949) Acta Pathol Microbiol Scand 26:507–515
Padan E (1979) Annu Rev Plant Physiol 30:27–40
Ragan MA, Chapman DJ (1978) A biochemical phylogeny of the protists. Academic Press, London New York, pp 317
Rippka R, Deruelles J, Waterbury JB, Herdman M, Stanier RY (1979) J Gen Microbiol 111:1–61
Rosinski J, Hainfeld JF, Rigbi M, Siegelman HW (1981) Ann Bot (London) 47:1–12
Schopf JW (1974) Evol Biol 7:1–43
Seckbach J, Hammerman IS, Hanania J (1981) Ann NY Acad Sci 361:409–424
Sidler W, Gysi J, Isker E, Zuber H (1981) Z Physiol Chem 362:611–628
Sidler W, Kumpf B, Frank G, Morisset W, Wehrmeyer W, Zuber H (1983) Ber Dtsch Bot Ges 95 (in press)
Stanier RY (1974) Symp Soc Gen Microbiol 24:219–240
Stanier RY (1977) Carlsberg Res Commun 42:77–98
Troxler RF, Foster JA, Brownmas AS, Franzblau C (1975) Biochemistry 14:268–274
Troxler RF, Greenwald LS, Zilinskas BA (1980) J Biol Chem 255:9380–9387

Troxler RF, Ehrhardt MM, Brownmas AS, Offner GD (1981) J Biol Chem 256:12176–12184
Walsh KA, Ericsson LH, Parmelee DC, Titani K (1981) Annu Rev Biochem 50:261–284
Walsh RG, Wingfield P, Glazer AN, DeLange RJ (1980) Fed Proc 39:1998
Wehrmeyer W (1983) In: Papageorgiou GC, Packer L (eds) Photosynthetic procaryotes: Cell differentiation and function. Elsevier/North Holland, New York, pp 1–22
Williams RC, Gingrich JC, Glazer AN (1980) J Cell Biol 85:558–566
Williams VP, Freidenreich P, Glazer AN (1974) Biochem Biophys Res Commun 59:462–466
Yamanaka G, Glazer AN, Williams RC (1980) J Biol Chem 255:11004–11010
Zuber H (1978) Ber Dtsch Bot Ges 91:459–475

Phylogenetic Consideration of Ferredoxin Sequences in Plants, Particularly Algae

H. MATSUBARA and T. HASE [1]

Abstract. We have compared 38 amino acid sequences of low potential [2Fe-2S] ferredoxins isolated from angiosperms, a fern, horsetails, Chlorophyta, Chrysophyta, Rhodophyta, and Cyanophyta. Some species had two ferredoxins and the correspondence between gene duplication and speciation was considered. Amino acid differences among ferredoxins indicated that algae were close to each other and remote from higher plants; green algae showed nearly equal distances from all other tested taxa; horsetails were very divergent from all other organisms. Species of higher plants belonging to the same family or order did not necessarily show higher homology to each other than to members of different families or orders, indicating that differences in ferredoxin structures did not reflect suggested taxonomic relationship among higher plants. However, the phylogenetic relationships among algae were reasonable when considered on the basis of the ferredoxin dendrogram. When a deletion was assigned a special value for comparing amino acid sequences the systematic relationship of algae becames clearer. Several unique amino acids or deletions were found in the sequences and these reflected phylogenetic characteristics.

1 Introduction

Taxonomy was the first discipline developed for classifying, systematizing, and arranging the special features of individual organisms and taxa. Later systematic phylogeny was introduced to correlate various biological characteristics with biological evolutionary aspects. Morphology, ecology, physiology, genetics, developmental and differentiation biology, biochemistry, paleontology, and other fields of knowledge are used in combination in studies of systematic phylogeny, but of these morphology is still regarded as the most significant.

Since the time when the amino acid sequence of insulin was first determined, the sequences of many proteins have been reported and the comparison of these sequences has made it possible to study biological evolution at a molecular level (Bryson and Vogel 1965, Fitch and Margoliash 1967). Three dimensional structural studies have also been introduced in investigating the molecular evolution of proteins (Dickerson 1980). Recent striking progress in structural studies on nucleic acids has enlarged the field of molecular evolution, leading to a new phase in which there is a hope that combined studies will provide a clue to the mechanism of biological evolution.

We have investigated the molecular evolution of plant ferredoxins (Matsubara et al. 1978, Matsubara et al. 1980), because ferredoxins are widely distributed in plants, and because their small molecular size facilitates studies on their structure.

[1] Department of Biology, Faculty of Science, Osaka University, Toyonaka, Osaka 560, Japan

Proteins and Nucleic Acids in Plant Systematics
ed. by U. Jensen and D.E. Fairbrothers

Ferredoxins are a group of acidic proteins with equimolar amounts of non-heme iron and inorganic sulfur ranging from each 2 to 8 atoms. Their redox potentials range from about −500 to +350 mV. They have characteristic absorption and electron paramagnetic resonance spectra and unique amino acid sequences. They function in diverse redox systems as important electron carriers (Lovenberg 1973, 1977).

We have established several sequences of ferredoxin from higher plants, algae, and bacteria (Matsubara et al. 1980) and also the three-dimensional structure of one [2Fe–2S] ferredoxin from *Spirulina platensis* (Tsukihara et al. 1981).

By comparison of these sequences we have arranged these ferredoxins in a phylogenetic tree and obtained information on various phenomena, such as gene duplication, which have been used for taxonomic classification purpose. We have also correlated the sequences with the three-dimensional structure in terms of conserved sites and variable sites on the molecule during evolution.

This paper is focussed on chloroplast-type [2Fe-2S] ferredoxins; 38 sequences are compared, and their evidence for the systematic phylogeny of plants is discussed.

2 Sequence Comparison of Chloroplast-Type Ferredoxins

We have now 37 complete amino acid sequences of chloroplast-type ferredoxins isolated from angiosperms, ferns, horsetails, Chlorophyta, Chrysophyta, Rhodophyta, and Cyanophyta. They are aligned as shown in Fig. 1 by inserting gaps at the amino- and carboxyl-termini and 6 positions, namely 4, 11, 12, 16, 37, and 61, from the amino-terminus, giving a total of 102 comparable sites with high homology. Figure 1 also includes the partial amino-terminal sequence of *Rhodymenia palmata* (Rhodophyta). Two species of *Phytolacca* two of *Equisetum, Dunaliella salina, Aphanothece sacrum, Nostoc muscorum,* and *Nostoc* strain MAC each contain two molecular species of ferredoxin in the same organism. All of these iso-ferredoxins have been sequenced except one of the two detected in *N. muscorum*. These iso-ferredoxins have very different amino acid sequences from each other, suggesting gene duplications that occurred early in evolution (Matsubara et al. 1978, 1980). Some heterogeneities have also been noted; for example, in *Leucaena glauca* ferredoxin (Benson and Yasunobu 1969), *Arctium lappa* ferredoxin (Takruri et al. 1982), *Phytolacca esculenta* ferredoxin I (Wakabayashi et al. 1980), and *Dunaliella salina* ferredoxin I (Hase et al. 1980). In these cases, two different amino acids were found at several sites in ferredoxin preparations, suggesting the presence of allelic genes. However, in Fig. 1 only one of the two is arbitrarily used for comparison.

Eighteen sites are occupied by common amino acids (nonvariant sites) in all ferredoxins. These sites are residues 40, 42, 43, 46, 48, 49, 51, 53, 58, 68, 69, 73, 77, 81, 82, 84, 94, and 97. These sites include the 4 cysteine residues, Cys-43, -48, -51, and -82, chelating the two iron atoms in the [2Fe-2S] cluster. The other sites can be classified into conservative sites, occupied by chemically similar amino acids, and random or variable sites, occupied by chemically different amino acids. The common and conservative sites are located near the cysteine residues constructing the [2Fe-2S] cluster and also in the areas surrounding the cluster in the three-dimensional structure

```
                                                                                                      1
           1         2         3         4         5         6         7         8         9         0
  1        0         0         0         0         0         0         0         0         0         0
(1)  --A-FKVKLLT-PDG-PKEFECPDDVYILDQAEELGIDLPYSCRAGSCSSCAGKLVEGDL-DQSDQSFLDDEQIEEGWVLTCAAYPRSDVVIETHKEEELTG
(2)  -AA-YKVTLVT-PTG-NVEFQCPDDVYILDAAEEEGIDLPYSCRAGSCSSCAGKLKTGSL-NQDDQSFLDDDQIDEGWVLTCAAYPVSDVTIETHKEEELTA
(3)  -AS-YKVKLVT-PEG-TQEFECPDDVYILDHAEEEGIVLPYSCRAGSCSSCAGKVAAGEV-NQSDGSFLDDDQIEEGWVLTCVAYAKSDVTIETHKEEELTA
(4)  -AS-YKVKLIT-PDG-PQEFECPDDVYILEHAEELGIDIPYSCRAGSCSSCAGKLVAGSV-DOSDOSFLDDEQIEEGWVLTCVAYPKSDVTIETHKEEELTA
(5)  -AT-YNVKLIT-PDG-EVEFKCDDDVYVLDQAEEEGIDIPYSCRAGSCSSCAGKVVSGSI-DQSDQSFLDDEQMDAGYVLTCHAYPTSDVVIETHKEEEIV-
(6)  -AT-YKVKFIT-PEG-EQEVECDDDVYVLDAAEEAGIDLPYSCRAGSCSSCAGKVVSGFV-DQSDESFLDDDQIAEGFVLTCAAYPTSDVTIETHKEEELV-
(7)  -AT-YKVTLIT-PEG-KQEFEVPDDVYILDHAAEEVGDLPYSCRAGSCSSCAGKVTAGSV-DQSDGSFLDDDQMEAGWVLTCVAYPTSDVTIETHKEEELTA
(8)  -AT-YKVKLVT-PEG-EVELEVPDDVYILDQAEEEGIDLPYSCRAGSCSSCAGKLVSGEI-DQSDQSFLDDDQMEAGWVLTCHAYPKSDIVIETHKEEELTA
(9)  -AT-YKVKLVT-PSG-QQEFQCPDDVYILDQAEEVGIDLPYSCRAGSCSSCAGKVKVGDV-DQSDGSFLDDEQIGEGWVLTCVAYPVSDGTIETHKEEELTA
(10) -AT-YKVTLVT-PSG-TQTIDCPDDTYVLDAAEEAGLDLPYSCRAGSCSSCTGKVTAGTV-DQEDQSFLDDDQIEAGFVLTCVAFPKGDVTIETHKEEDIV-
(11) AAS-YKVTFVT-PSG-TNTITCPADTYVLDAAEESGLDLPYSCRAGACSSCAGKVTAGAV-NQEDGSFLEEEQMEAGWVLTCVAYPTSDVTIETHKEEDLTA
(12) -AT-YKVTLVT-PSG-TQTIDCPDDTYVLDAAEEAGLDLPYSCRAGSCSSCTGKVTAGTV-DQEDQSFLDDDQIEAGFVLTCVAYPKGDVTIETHKEEDIA-
(13) AAS-YKVTFVT-PSG-TKTITCPADTYVLDAAEDTGLDLPYSCRAGACSSCAGKVTAGSV-NQEDGSFLDEEQMEAGWVLTCVAYPTSDVTIETHKEEDLSA
(14) -AI-FKVKFLT-PDG-ERTIEVPDDKFILDAGEEAGLDLPYSCRAGACSSCTGKLLDGRV-DQSEQSFLDDDQMAEGFVLTCVAYPAGDITIETHAEEKL--
(15) --A-YKTVLKT-PSG-EFTLDVPEGTTILDAAEEAGYDLPFSCRAGACSSCLGKVVSGSV-DQSEGSFLDDGQMEEGFVLTCIAIPESDLVIETHKEEELF-
(16) --A-YKVTLKT-PDG-DITFDVEPGERLIDIASEKA-DLPLSCQAGACSTCLGKIVSGTV-DQSEGSFLDDEQIEQGYVLTCIAIPESDVVIETHKEDEL--
(17) --A-YKTVLKT-PSG-EFTLDVPEGTTILDAAEEAGYDLPFSCRAGACSSCLGKVVSGSV-DESEGSFLDDGQMEEGFVLTCIAIPESDLVIETHKEEELF-
(18) --A-YKVTLKT-PDG-DITFDVEPGERLIDIGSEKA-DLPLSCQAGACSTCLGKIVSGTV-DQSEGSFLDDEQIEQGYVLTCIAIPESDVVIETHKEDEL--
(19) -AT-YKVTLKT-PSG-DQTIECPDDTYILDAAEEAGLDLPYSCRAGACSSCAGKVEAGTV-DQSDQSFLDDSQMDGGFVLTCVAYPTSDCTIATHKEEDLF-
(20) --S-YMVTLKT-PSG-EQKVEVSPDSYILDAAEEAGVDLPYSCRAGSCSSCAGKVESGTV-DQSDQSFLDDDQMDSGFVLTCVAYATSDCTIVTHQEENLY-
(21) --A-YKVTLKT-PSG-DQTIEVSPDAYILDAAEEAGLDLPYSCRAGACSSCAGKVEAGTI-DQSDQSFLDDDQQGRGFVLTCVAYATSDCTISTHQEESLY-
(22) -AD-YKIHLVSKEEGIDVTFDCSEDTYILDAAEEEGIELPYSCRAGACSTCAGKVTEGTV-DQSDQSFLDDEQMLKGYVLTCIAYPESDCTILTHVEQELY-
(23) -AS-YKIHLVNKDQGIDETIECPDDQYILDAAEEQGLDLPYSCRAGACSTCAGKLLEGEV-DQSDQSFLDDDQVKAGFVLTCVAYPTSNATILTHQEESLY-
(24) -AVKYTVTLST-PGG-VEEIEGDETTYVLDSAED
(25) -AT-YSVTLVNEEKNINAVIKCPDDQFILDAAEEQGIELPYSCRAGACSTCAGKVLSGTI-DQSEQSFLDDDQMGAGFLLTCVAYPTSDCKVQTHAEDDLY-
(26) -AS-YTVKLIT-PDG-ENSIECSDDTYILDAAEEAGLDLPYSCRAGACSTCAGKITAGSV-DQSDQSFLDDDQIEAGYVLTCVAYPTSDCTIETHKEEDLY-
(27) -AS-YKVTLKT-PDG-DNVITVPDDEYILDVAEEEGLDLPYSCRAGACSTCAGKLVSGPA-PDEDQSFLDDDQIQAGYILTCVAYPTGDCVIETHKEEALY-
(28) -AT-YKVTLVR-PDGSETTIDVPEDEYILDVAEEQGLDLPFSCRAGACSTCAGKLLEGEV-DQSDQSFLDDDQIEKGFVLTCVAYPRSDCKILTNQEEELY-
(29) -AT-YKVTLI-DAEGTTTTIDCPDDTYILDAAEEAGLDLPYSCRAGACSTCAGKLVTGTI-DQSDQSFLDDDQVEAGYVLTCVAYPTSDVTIETHKEEDLY-
(30) -AT-YKVTLINDAEGLNQTIEVDDDTYILDAAEEAGLDLPYSCRAGACSTCAGKIKSGTV-DQSDQSFLDDDQIEAGYVLTCVAYPTSDCTIETHKEEELY-
(31) -AT-YKVTLINEAEGLNKTIEVPDDQYILDAAEEAGIDLPYSCRAGACSTCAGKLISGTV-DQSDQSFLDDDQIEAGYVLTCVAYPTSDCVIETHKEEELY-
(32) -AT-YKVTLINEAEGINETIDCDDDTYILDAAEEAGLDLPYSCRAGACSTCAGTITSGTI-DQSDQSFLDDDQIEAGYVLTCVAYPTSDCTIKTHQEEGLY-
(33) -AS-YKVTLINEEMGLNETIEVPDDEYILDVAEEEGIDLPYSCRAGACSTCAGKIKEGEI-DQSDQSFLDDDQIEAGYVLTCVAYPASDCTIITHQEEELY-
(34) -AT-YKVTLISEAEGINETIDCDDDTYILDAAEEAGLDLPYSCRAGACSTCAGKITSGSI-DQSDQSFLDDDQIEAGYVLTCVAYPTSDCTIQTHQEEGLY-
(35) -AT-FKVTLINEAEGTKHEIEVPDDEYILDAAEEEGYDLPFSCRAGACSTCAGKLVSGTV-DQSDQSFLDDDQIEAGYVLTCVAYPTSDVVIQTHKEEDLY-
(36) -AT-YKVRLFNAAEGLDETIEVPDDEYILDAAEEAGLDLPFSCRSGSCSSCNGILKKGTV-DQSDQNFLDDDQIAAGNVLTCVAYPTSNCEIETHREDAIA-
(37) -AT-YKVTLINEEEGINAILEVADDQTILDAGEEAGLDLPSSCRAGSCSTCAGKLVSGAAPNQDDQAFLDDDQLAAGWVMTCVAYPTGDCTIMTHQESEVL-
(38) -ATVYKVTLV-DQEGTETTIDVPDDEYILDIAEDQGLDLPYSCRAGACSTCAGKIVSGTV-DQSDQSFLDDDQIEKGYVLTCVAYPTSDLKIETHKEEDLY-
```

Fig. 1. (Legend see p. 171)

(Fukuyama et al. 1980, Matsubara et al. 1980). The hypervariable sites are located in regions remote from the cluster and in outer regions of the molecule.

3 Amino Acid Difference Matrix

The matrix showing amino acid differences among 37 ferredoxin sequences is given in Fig. 2, and a deletion is counted as one difference. These amino acid differences were then recalculated for seven groups as shown in Fig. 3. There are 30–33 differences among ferredoxins of algae including Rhodophyta, Chrysophyta, Cyanophyta, but not Chlorophyta. This is fewer than the 34 to 50 differences between the ferredoxins of these algae and ferredoxins of angiosperms, a fern and horsetails. This indicates that all these algae, except Chlorophyta, are relatively related. Chlorophyta ferredoxins have 30–35 differences from other algal ferredoxins, which is about the same as their 31–38 differences from the sequences of higher plant ferredoxins. Horsetails have larger numbers of difference in amino acids from those of ferredoxins of other organisms. They have 30–50, which are the greatest number of differences detected.

The numbers of differences among ferredoxins of organisms belonging to the same group are 2–37 for angiosperms, 1–31 for horsetails, 15–20 for Chlorophyta, 28 for

Fig. 1. Comparison of sequences of [2Fe-2S] ferredoxins from higher plants and algae. Amino acids are written by one letter abbreviations: *A* alanine; *C* cysteine; *D* aspartic acid; *E* glutamic acid; *F* phenylalanine; *G* glycine; *H* histidine; *I* isoleucine; *K* lysine; *L* leucine; *M* methionine; *N* asparagine; *P* proline; *Q* glutamine; *R* arginine; *S* serine; *T* threonine; *V* valine; *W* tryptophan; *Y* tyrosine. Several gaps are inserted to give high homology among ferredoxin sequences. When micro-heterogeneity of amino acids at a certain position was observed, only one amino acid, chosen arbitrarily, is shown in this figure. When leucine and valine were found at position *9* in *L. glauca* ferredoxin, it is expressed as 9 L/V after the scientific name below, and this expression is applied to other positions. Residues ambiguously identified as Asx and Glx are defined as either their acid or amide depending on which gives higher homology with other sequences. Organisms for ferredoxins are: *(1) Leucaena glauca* 9 L/V, 17 P/A, 38 D/E, 102 G/A; *(2) Spinacia oleracea*; *(3) Medicago sativa*; *(4) Sambucus nigra*; *(5) Petroselinum sativum*; *(6) Brassica napus*; *(7) Arctium lappa* 68 F/Y; *(8) Triticum aestivum*; *(9) Colocasia esculenta*; *(10) Phytolacca americana* (ferredoxin I); *(11) P. americana* (II); *(12) P. esculenta* (I) 52 T/A, 101 A/V; *(13) P. esculenta* (II); *(14) Gleichenia japonica*; *(15) Equisetum telmateia* (I); *(16) E. telmateia* (II); *(17) E. arvense* (I); *(18) E. arvense* (II); *(19) Scenedesmus quadricauda*; *(20) Dunaliella salina* (I) 6 M/Q, 37 V/L, 55 V/L; *(21) D. salina* (II); *(22) Porphyra umbilicalis*; *(23) Cyanidium caldarium*; *(24) Rhodymenia palmata*, Ser-26 originally reported is revised to be Thr-26 in the present study. *(25) Bumilleriopsis filiformis*; *(26) Synechocystis* 6714; *(27) Aphanothece sacrum* (I); *(28) Synechococcus sp.*; *(29) Aphanizomenon flos-aquae*; *(30) Chlorogloeopsis fritschii*; *(31) Mastigocladus laminosus*; *(32) Spirulina platensis*; *(33) Aphanothece halophitica*; *(34) Spirulina maxima*; *(35) Nostoc muscorum*; *(36) Nostoc* strain MAC (II); *(37) Aphanothece sacrum* (II); *(38) Nostoc* strain MAC (I).

References are: (1)–(3), (9)–(11), (15)–(19), (22), (23), (27), (31), (32), (34), (35), and (37), listed in Hase et al. (1978); (4), Takruri and Boulter (1979a); (5), Nakano et al. (1981); (6), Takruri and Boulter (1980); (7), Takruri et al. (1982); (8), Takruri and Boulter (1979b); (12) and (13), Wakabayashi et al. (1980); (14), Hase et al. (1982); (20) and (21), Hase et al. (1980); (24), Andrew et al. (1981a); (30), Takahashi et al. (1981); (25), (26), (28), (29), (33), (36), and (38), Hase et al. (unpublished data)

	1	2	3	4	5	6	7	8	9	10	11	12	13	14	15	16	17	18	19	20	21	22	23	25	26	27	28	29	30	31	32	33	34	35	36	37	38
1 *Leucaena glauca*	0	20	23	15	25	25	27	20	20	35	37	34	37	34	36	39	37	40	33	37	36	40	39	47	31	35	35	34	36	31	40	35	39	32	43	47	38
2 *Spinacia oleracea*	20	0	19	20	27	25	23	21	18	30	31	29	31	37	39	45	40	46	30	35	35	37	37	42	31	33	36	31	32	31	36	32	34	34	41	40	37
3 *Medicago sativa*	23	19	0	14	30	22	16	19	16	27	27	26	28	37	39	45	40	46	29	32	33	37	36	44	28	36	36	33	32	32	38	32	37	34	42	45	36
4 *Sambucus nigra*	15	20	14	0	23	22	18	20	17	29	31	28	31	34	38	42	39	43	28	34	35	39	36	46	24	34	35	31	31	30	37	32	35	32	41	44	36
5 *Petroselinum sativum*	25	27	30	23	0	20	26	19	26	31	37	31	36	40	38	42	39	43	31	34	38	37	42	40	28	34	40	32	33	32	35	36	33	33	43	45	37
6 *Brassica napus*	25	25	22	22	20	0	23	23	22	25	33	25	33	30	34	43	35	44	25	27	31	38	36	41	25	36	36	29	25	27	31	35	30	30	37	39	33
7 *Arctium lappa*	27	23	16	18	26	23	0	19	20	27	27	26	27	36	36	41	37	42	25	30	31	38	37	42	26	34	35	29	26	27	33	31	31	27	38	40	33
8 *Triticum aestivum*	20	21	19	20	19	23	19	0	21	31	34	30	35	32	33	43	34	44	30	32	34	37	35	39	29	30	31	29	29	25	34	29	33	27	37	38	30
9 *Colocasia esculenta*	20	18	16	17	26	22	20	21	0	28	29	27	30	36	36	43	37	44	26	33	33	36	37	41	30	37	35	34	31	32	37	34	36	36	40	46	35
10 *Phytolacca americana* I	35	30	27	29	31	25	27	31	28	0	23	2	23	31	35	43	36	44	19	31	28	38	34	39	24	32	32	23	26	29	28	35	28	32	34	42	30
11 *Phytolacca americana* II	37	31	27	31	37	33	27	34	29	23	0	22	6	38	39	47	40	48	25	37	34	40	37	44	27	34	39	30	34	36	36	39	35	38	46	48	37
12 *Phytolacca esculenta* I	34	29	26	28	31	25	26	30	27	2	22	0	22	30	35	43	36	44	18	30	27	37	33	38	23	31	31	22	25	28	27	34	27	31	32	41	29
13 *Phytolacca esculenta* II	37	31	28	31	36	33	27	35	30	23	6	22	0	38	38	47	39	48	25	37	34	40	37	44	27	35	39	30	34	35	36	39	34	38	46	49	35
14 *Gleichenia japonica*	34	37	37	34	40	30	36	32	36	31	38	30	38	0	35	46	36	45	28	34	32	44	34	40	29	35	33	34	33	33	38	36	37	36	38	42	37
15 *Equisetum telmateia* I	36	39	39	38	38	34	36	33	36	35	39	35	38	35	0	29	1	30	30	34	34	39	44	47	36	41	36	38	37	36	42	43	40	38	46	50	38
16 *Equisetum telmateia* II	39	45	45	42	42	43	41	43	43	43	47	43	47	46	29	0	30	1	42	44	41	43	51	52	41	40	41	42	39	39	44	43	44	40	50	56	38
17 *Equisetum arvense* I	37	40	40	39	39	35	37	34	37	36	40	36	39	36	1	30	0	31	31	35	35	40	45	48	37	41	37	39	38	37	43	44	41	39	47	51	39
18 *Equisetum arvense* II	40	46	46	43	43	44	42	44	44	44	48	44	48	45	30	1	31	0	43	45	42	44	52	53	42	41	42	43	40	40	45	44	45	41	51	55	39
19 *Scenedesmus quadricauda*	33	30	29	28	31	25	25	30	26	19	25	18	25	28	30	42	31	43	0	20	16	31	27	32	20	29	29	22	21	24	25	30	25	28	32	40	28
20 *Dunaliella salina* I	37	35	32	34	34	27	30	32	33	31	37	30	37	34	34	44	35	45	20	0	15	37	34	38	26	34	33	34	28	30	32	33	32	34	39	41	35
21 *Dunaliella salina* II	36	35	33	35	38	31	31	34	33	28	34	27	34	32	34	41	35	42	16	15	0	35	29	36	25	31	30	28	25	28	27	29	27	33	36	41	33
22 *Porphyra umbilicalis*	40	37	37	39	37	38	38	37	36	38	40	37	40	44	39	43	40	44	31	37	35	0	28	35	32	41	32	31	30	31	29	31	28	36	43	45	34
23 *Cyanidium caldarium*	39	37	36	36	42	36	37	35	37	34	37	33	37	34	44	51	45	52	27	34	29	28	0	31	27	33	25	25	26	25	25	24	25	29	32	38	30
25 *Bumilleriopsis filiformis*	47	42	44	46	40	41	42	39	41	39	44	38	44	40	47	52	48	53	32	38	36	35	31	0	35	38	35	31	31	27	28	31	28	31	42	39	33
26 *Synechocystis* 6714	31	31	28	24	28	25	26	29	30	24	27	23	27	29	36	41	37	42	20	26	25	32	27	35	0	24	28	20	17	21	20	24	18	25	35	39	25
27 *Aphanothece sacrum* I	35	33	36	34	34	36	34	30	37	32	34	31	35	35	41	40	41	41	29	34	31	41	33	38	24	0	29	28	27	25	31	27	30	26	37	38	28
28 *Synechococcus* sp.	35	36	36	35	40	36	35	31	35	32	39	31	39	33	36	41	37	42	29	33	30	32	25	35	28	29	0	27	26	25	28	23	27	27	35	41	21
29 *Aphanizominon flos-aquae*	34	31	33	31	32	29	29	29	34	23	30	22	30	34	38	42	39	43	22	34	28	31	25	31	20	28	27	0	14	16	15	24	15	17	31	35	16
30 *Chlorogloeopsis fritschii*	36	32	32	31	33	25	26	29	31	26	34	25	34	33	37	39	38	40	21	28	25	30	26	31	17	27	26	14	0	8	11	16	12	17	25	30	18
31 *Mastigocladus laminosus*	31	31	32	30	32	27	27	25	32	29	36	28	35	33	36	39	37	40	24	30	28	31	25	27	21	25	25	16	8	0	15	16	16	13	25	29	20

32 *Spirulina platensis*	40	36	38	37	35	31	33	34	37	28	36	27	36	38	42	44	43	45	25	32	27	29	25	28	20	31	28	15	11	15	0	18	4	21	30	31	23
33 *Aphanothece halophitica*	35	32	32	32	36	35	31	29	34	35	39	34	39	36	43	43	44	44	30	33	29	31	24	31	24	27	23	24	16	16	18	0	18	22	32	34	26
34 *Spirulina maxima*	39	34	37	35	33	30	31	33	36	28	35	27	34	37	40	44	41	45	25	32	27	28	25	28	18	30	27	15	12	16	4	18	0	21	32	31	23
35 *Nostoc muscorum*	32	34	34	32	33	30	27	27	36	32	38	31	38	36	38	40	39	41	28	34	33	36	29	31	25	26	27	17	17	13	21	22	21	0	29	34	20
36 *Nostoc* strain MAC II	43	41	42	41	43	37	38	37	40	34	46	32	46	38	46	50	47	51	32	39	36	43	32	42	35	37	35	31	25	25	30	32	32	29	0	38	34
37 *Aphanothece sacrum* II	47	40	45	44	45	39	40	38	46	42	48	41	49	42	50	56	51	55	40	41	41	45	38	39	39	38	41	35	30	29	31	34	31	34	38	0	41
38 *Nostoc* strain MAC I	38	37	36	36	37	33	33	30	35	30	37	29	35	37	38	38	39	39	28	35	33	34	30	33	25	28	21	16	18	20	23	26	23	20	34	41	0

Fig. 2. Amino acid difference matrix of [2Fe-2S] ferredoxin sequences. One deletion is counted as one difference

		Angi.	Fern	Hors.	Chlo.	Rhod.	Chry.	Cyan.
Angiosperms	(13)	—						
Fern	(1)	34.8 ± 3.3	—					
Horsetails	(4)	40.5 ± 4.0	40.5 ± 5.8	—				
Chlorophyta	(3)	30.8 ± 4.7	31.3 ± 3.1	38.0 ± 5.3	—			
Rhodophyta	(2)	37.3 ± 1.9	39.0 ± 7.1	44.8 ± 4.7	32.2 ± 3.8	—		
Chrysophyta	(1)	42.1 ± 2.8	40	50.0 ± 2.9	35.3 ± 3.0	33.0 ± 2.8	—	
Cyanophyta	(13)	33.8 ± 5.2	35.5 ± 3.2	41.9 ± 4.6	30.3 ± 5.2	29.9 ± 8.2	33.4 ± 4.6	—

Fig. 3. Average amino acid difference matrix of [2Fe-2S] ferredoxin sequences for seven classes of plants. The numbers of ferredoxins belonging to each group are given *in parentheses*

Rhodophyta, and 4–41 for Cyanophyta. The fewest differences were found between ferredoxins of organisms belonging to the same genus. One difference was found between ferredoxin I and II, respectively, of two species of *Equisetum*, 2 and 6 were found between ferredoxin I and II, respectively, of two species of *Phytolacca*, and 4 were found between ferredoxins of two *Spirulina* species. However, presently there is no explanation of why there are 27 differences between ferredoxin I of *Aphanothece sacrum* and *A. halophitica*, and 34 between ferredoxin II of *A. sacrum* and ferredoxin of *A. halophitica*, since they both belong to the genus *Aphanothece*, order Chroococcales of blue-green algae (Rogers LJ, personal communication). As discussed below, we may need to reexamine not only the molecular phylogenetic system, but also the identification of the species *A. halophitica*.

Ferredoxins of several species show more amino acid differences from those of other species of the same group than from those of members of the other groups. This is particularly true for species of angiosperms and Cyanophyta. For example, there are 25–37 differences between ferredoxins of two species of *Phytolacca* and those of other angiosperms, while there are 22–34 between *Phytolacca* ferredoxins and those of *Synechocystis* 6714, and *Aphanizomenon flos-aquae*, members of the Cyanophyta. Therefore, if judged only from numbers of different amino acids, misleading relationships may be reported between remotely related organisms. One reason for this phenomenon is the rapid rate of evolution of ferredoxins. Mutation numbers seem to have been saturated in a relatively short period for a small protein and differences in numbers of different amino acids in ferredoxins of remotely related organisms do not reflect real phylogenetic distances. If the amino acid differences between ferredoxins of other angiosperms and *Phytolacca* are expressed as PAM (accepted point mutations per 100 residues; Dayhoff 1978), the values are 30 to 50. Assuming the time of divergence of angiosperms to be 135 million years, the evolutionary rate of ferredoxins is 11.1–18.5 PAMs per 100 million years. These values are much larger than that calculated by Dayhoff of 1.9 PAMs per 100 million years. Therefore, in the following discussion of the phylogenetic relationship among ferredoxins we evaluated higher plant ferredoxins separately from algal ferredoxins to avoid misleading results obtained by the comparison of amino acid differences between the two groups of plants.

4 Phylogenetic Tree of Algal Ferredoxins

We constructed a phylogenetic tree of algal ferredoxins using the amino acid difference matrix given in Fig. 2. The method was based on that described by Fitch and Margoliash (1967). Lengths of branches of the tree should correspond to the difference in number of amino acids among ferredoxins. Of several possible trees we selected the most probable one on the basis of the following two criteria: (a) the matrix reconstructed from the branch lengths of the tree should be close to the one constructed by the original amino acid differences, and (b) the calculated branch lengths should not be negative. Figure 4 illustrates one possible tree. The amino acid difference numbers among ferredoxins calculated reversely from the branch lengths

Fig. 4. Phylogenetic tree of algal ferredoxins. One gap is counted as one difference

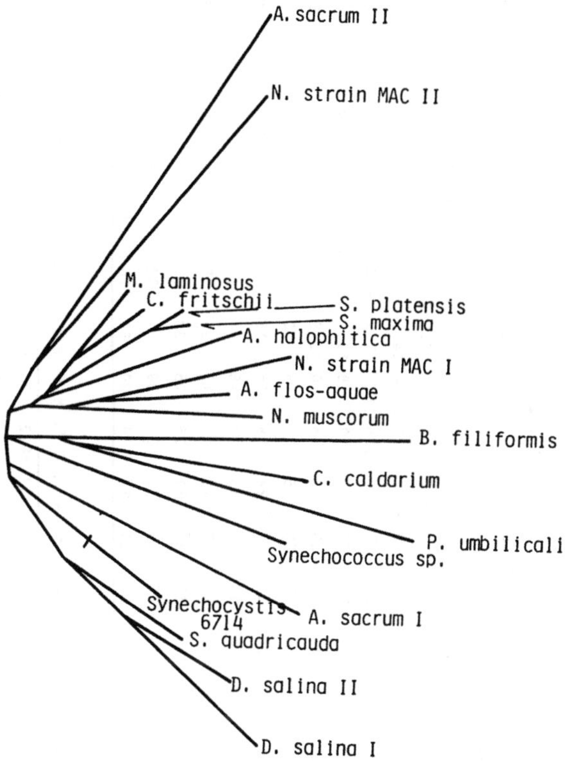

of this tree are given in the upper half of Fig. 5 together with the original matrix obtained by comparison of amino acid sequence data on ferredoxins. The average deviation between the two matrices was 8.37%.

One characteristic of this tree is that the ferredoxins of Cyanophyta do not constitute one group, but are separated into three groups. *Group* 1 includes the ferredoxins of *M. laminosus, C. fritschii,* two *Spirulina* species, *A. halophitica, N.* strain MAC (I), *A. flos-aquae,* and *N. muscorum,* and they are in positions close to each other. *Group* 2 includes the ferredoxins of *A. sacrum* (II) and *N.* strain MAC (II), and they are quite far from the others. These two were supposed to be derived by gene duplication and might have evolved at rather faster evolutional rates. *Group* 3 includes the ferredoxins of *Synechococcus* sp., *A. sacrum* (I), and *Synechocystis* 6714. The blue-green algal ferredoxins in group I are all from filamentous algae except that of *A. halophitica* and group 3 are all unicellular blue-green algal ferredoxins.

Eukaryotic algal ferredoxins were divided into two groups: (1) Those of *C. caldarium, P. umbilicalis* (Rhodophyta), and *B. filiformis* (Chrysophyta), and (2) those of *S. quadricauda* and *D. salina* (I and II) (Chlorophyta). Chlorophyta ferredoxins are located fairly close to group 3 of blue-green algal ferredoxins.

Another tree could be constructed to meet the criteria mentioned above, that is *Synechocystis* 6714 of group 3 in the three could be relocated in group 1. However, detailed inspection of the sequences of ferredoxins favors the tree presented in Fig. 4. The ferredoxin in groups 1 and 2 of blue-green algae, Rhodophyta, and Chrysophyta

	S.q.	D.s.	D.s.	P.u.	C.c.	B.f.	Syn.	A.s.	Syn.	A.f.	C.f.	M.l.	S.p.	A.h.	S.m.	N.m.	N.M.	A.s.	N.M.
Scenedesmus quadricauda	0	19.5	16.5	33.5	27.9	32.9	20.6	27.7	27.8	25	22.2	23	24.5	26.4	24.5	26.8	34.3	37.7	28.4
Dunaliella salina I	20	0	15	39.6	34	39	26.7	33.8	33.9	31.1	28.3	29.1	30.6	32.5	30.6	32.9	40.4	43.8	34.5
Dunaliella salina II	16	15	0	36.6	31	36	23.7	30.8	30.9	28.1	25.3	26.1	27.6	29.5	27.6	29.9	37.4	40.8	31.5
Porphyra umbilicalis	31	37	35	0	28	35.8	31.3	36.6	33.9	31.1	28.3	29.1	30.6	32.5	30.6	32.9	40.4	43.8	34.5
Cyanidium caldarium	27	34	29	28	0	30.2	25.7	31	28.3	25.5	22.7	23.5	25	26.9	25	27.3	34.8	38.2	28.9
Bumilleriopsis filiformis	32	38	36	35	31	0	30.7	36	33.3	30.5	27.7	28.5	30	31.9	30	32.3	39.8	43.2	33.9
Synechocystis 6714	20	26	25	32	27	35	0	25.5	25.6	22.8	20	20.8	22.3	24.2	22.3	24.6	32.1	35.5	26.2
Aphanothece sacrum I	29	34	31	41	33	38	24	0	30.9	28.1	25.3	26.1	27.6	29.5	27.6	29.9	37.4	40.8	31.5
Synechococcus sp.	29	33	30	32	25	35	28	29	0	25.4	22.6	23.4	24.9	26.8	24.9	27.2	34.7	38.1	28.8
Aphanizominon flos-aquae	22	34	28	31	25	31	20	28	27	0	15.6	16.4	17.9	19.8	17.9	16.8	29.7	33.1	16
Chlorogloeopsis fritschii	21	28	25	30	26	31	17	27	26	14	0	8	13.1	15.6	13.1	17.4	26.9	30.3	19
Mastigocladus laminosus	24	30	28	31	25	27	21	25	25	16	8	0	13.9	16.4	13.9	18.2	27.7	31.1	19.8
Spirulina platensis	25	32	27	29	25	28	20	31	28	15	11	15	0	17.9	4	19.7	29.2	32.6	21.3
Aphanothece halophitica	30	33	29	31	24	31	24	27	23	24	16	16	18	0	17.9	21.6	31.1	34.5	23.2
Spirulina maxima	25	32	27	28	25	28	18	30	27	15	12	16	4	18	0	19.7	29.2	32.6	21.3
Nostoc muscorum	28	34	33	36	29	31	25	26	27	17	17	13	21	22	21	0	31.5	34.9	20.2
Nostoc strain MAC II	32	39	36	43	32	42	35	37	35	31	25	25	30	32	32	29	0	38	33.1
Aphanothece sacrum II	40	41	41	45	38	39	39	38	41	35	30	29	31	34	31	34	38	0	36.5
Nostoc strain MAC I	28	35	33	34	30	33	25	28	21	16	18	20	23	26	23	20	34	41	0

	S.q.	D.s.	D.s.	P.u.	C.c.	B.f.	Syn.	A.s.	Syn.	A.f.	C.f.	M.l.	S.p.	A.h.	S.m.	N.m.	N.M.	A.s.	N.M.
Scenedesmus quadricauda	0	23.5	20.5	41.5	35.9	40.9	20.6	27.7	31.8	37	30.2	31	32.5	34.4	32.5	34.8	42.3	49.7	44.4
Dunaliella salina I	24	0	15	51.6	46	51	30.7	37.8	41.9	47.1	40.3	41.1	42.6	44.5	42.6	44.9	52.4	59.8	54.5
Dunaliella salina II	20	15	0	48.6	43	48	27.7	34.8	38.9	44.1	37.3	38.1	39.6	41.5	39.6	41.9	49.4	56.8	51.5
Porphyra umbilicalis	39	49	47	0	28	35.8	39.3	44.6	37.9	35.1	28.3	29.1	30.6	32.5	30.6	32.9	40.4	47.8	42.5
Cyanidium caldarium	35	46	41	28	0	30.2	33.7	39	32.3	29.5	22.7	23.5	25	26.9	25	27.3	34.8	42.2	36.9
Bumilleriopsis filiformis	40	50	48	35	31	0	38.7	44	37.3	34.5	27.7	28.5	30	31.9	30	32.3	39.8	47.2	41.9
Synechocystis 6714	20	30	29	40	35	43	0	25.5	29.6	34.8	28	28.8	30.3	32.2	30.3	32.6	40.1	47.5	42.2
Aphanothece sacrum I	29	38	35	49	41	46	24	0	34.9	40.1	33.3	34.1	35.6	37.5	35.6	37.9	45.4	52.8	47.5
Synechococcus sp.	33	41	38	36	29	39	32	33	0	33.4	26.6	27.4	28.9	30.8	28.9	31.2	38.7	46.1	40.8
Aphanizominon flos-aquae	34	50	44	35	29	35	32	40	35	0	19.6	20.4	21.9	23.8	21.9	20.8	33.7	41.1	20
Chlorogloeopsis fritschii	29	40	37	30	26	31	25	35	30	18	0	8	13.1	15.6	13.1	17.4	26.9	34.3	27
Mastigocladus laminosus	32	42	40	31	25	27	29	33	29	20	8	0	13.9	16.4	13.9	18.2	27.7	35.1	27.8
Spirulina platensis	33	44	39	29	25	28	28	39	32	19	11	15	0	17.9	4	19.7	29.2	36.6	29.3
Aphanothece halophitica	38	45	41	31	24	31	32	35	27	28	16	16	18	0	17.9	21.6	31.1	38.5	31.2
Spirulina maxima	33	44	39	28	25	28	26	38	31	19	12	16	4	18	0	19.7	29.2	36.6	29.3
Nostoc muscorum	36	46	45	36	29	31	33	34	31	21	17	13	21	22	21	0	31.5	38.9	28.2
Nostoc strain MAC II	40	51	48	43	32	42	43	45	39	35	25	25	30	32	32	29	0	42	41.1

Fig. 5. Amino acid difference matrix of algal ferredoxins. The *upper* and *lower half* matrix are derived by the calculation that one gap in the sequence comparison is counted as 1 and 5, respectively. The amino acid difference numbers calculated reversely from the tree branch lengths are given at the *upper right* region and those originally calculated at the *lower left* in each matrix

have amino acids at positions 12 and 16, while those of Chlorophyta, and *A. sacrum* (I) and *Synechocystis* 6714 in group 3 of blue-green algae have deletions at these positions. Further, the amino acid residues on either side of the deleted position 12 were Thr-11 and Pro-13 in all these ferredoxins. This characteristic was also found without exception in all higher plant ferredoxins and might be most reflecting the phylogenetic relationship among ferredoxins.

The phylogenetic tree in Fig. 4 is justified on the basis of results obtained in terms of applying special value to the deletion by the following manipulation. Since during evolution, a deletion is much more rare than a point mutation, in comparing sequences a deletion was scored as a large difference number of amino acids. For example, when one deletion was counted as five instead of one difference, the tree became as shown in Fig. 6, which is very similar to that in Fig. 4, but reflects the effect of gaps in constructing the tree of ferredoxins in group 3. The mean deviation of the tree in Fig. 6 is 7.05%, which was calculated reversely from the branch lengths of this tree (see the lower half of Fig. 5).

The positions of individual ferredoxins in Figs. 4 and 6 show several important taxonomic characteristics. The blue-green alga, *C. fritschii* was first classified in the order Chroococcales, then in the order Nostocales, and recently in the order Stigonematales (Rippka et al. 1979). On the basis of its ferredoxin, *C. fritschii* should be placed close to *Mastigocladus laminosus*, a member of the Stigonematales.

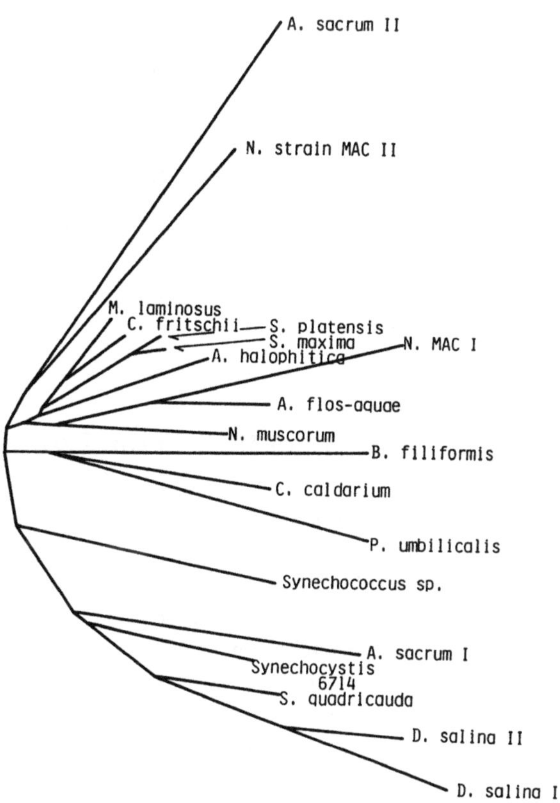

Fig. 6. Phylogenetic tree of algal ferredoxins. One gap is counted as 5 differences

Aphanothece halophitica and *A. sacrum* are placed in the same genus, but the ferredoxins have a large number of differences. However, the ferredoxins of *A. halophitica* are similar to the ferredoxins of the blue-green algae Nostocales and Stigonematales. As mentioned above, the reason for this discrepancy is unknown, but the taxonomic position of this alga should be carefully reexamined.

The rough division of blue-green algae into two types, unicellular and filamentous, is supported by protein structural characteristics including the presence of deletions in the ferredoxins of all these algae except that of *Synechococcus* sp., which has the intermediate features of a deletion at position 12, but none at position 16 and serine instead.

Rhodophyta are also roughly divided into Protoflorideophyceae (Bangiophyceae) and Florideophyceae. The ferredoxins of *Porphyra umbilicalis, Porphyridium aeruginеum*, and *Porphyridium cruentum* of the Protoflorideophyceae (only the 30 amino-terminal residues of the latter two are known; Andrew et al. 1981b) do not have the deletions at the positions mentioned above which are commonly found in filamentous blue-green algal ferredoxins. Further, *C. caldarium*, which was ambiguously classified before, belongs to this group as shown in Figs. 4 and 6.

Rhodymenia palmata ferredoxin was partially sequenced and found to have these deletions (Andrew et al. 1981a). Therefore, as in the case of blue-green algae, the two groups of red algae are remote from each other.

The ferredoxin of *B. filiformis* which belongs to the Xanthophyceae in the Chrysophyta is not clearly related to other algal ferredoxins. However, it has no deletions at the positions mentioned above, indicating its connection to ferredoxins of filamentous blue-green algae and Protoflorideophyceae. Structural studies of more ferredoxins of this class of algae are needed to help answer the classification questions.

The three Chlorophyceae ferredoxins are very similar. *Dunaliella salina* ferredoxins I and II occurred by gene duplication after divergence of *D. salina* from *Scenedesmus quadricauda*.

In summary, from the structures of their ferredoxins blue-green algae can be divided into two types, unicellular and filamentous. The former is closely related to Florideophyceae and Chlorophyceae, and the latter to Protoflorideophyceae and Xanthophyceae.

5 Phylogenetic Tree of Higher Plant Ferredoxins

Figure 7 shows a phylogenetic tree of higher plant ferredoxins. The positions of four ferredoxins of two *Equisetum* species, *Gleichenia japonica* (a fern) ferredoxin, and four of two *Phytolacca* species were readily determined, but those of other angiosperm ferredoxins were not, because the lengths of some branches showed negative values even when several branching possibilities were examined. Further the alternative trees gave nearly equal mean deviations of 9%. Therefore, these branching positions could not be definitely fixed, therefore they are indicated by dotted lines.

Figure 7 indicates that ferredoxins of angiosperms, a fern, and horsetails form three independent groups which reflect their systematic relationship. Only one amino acid

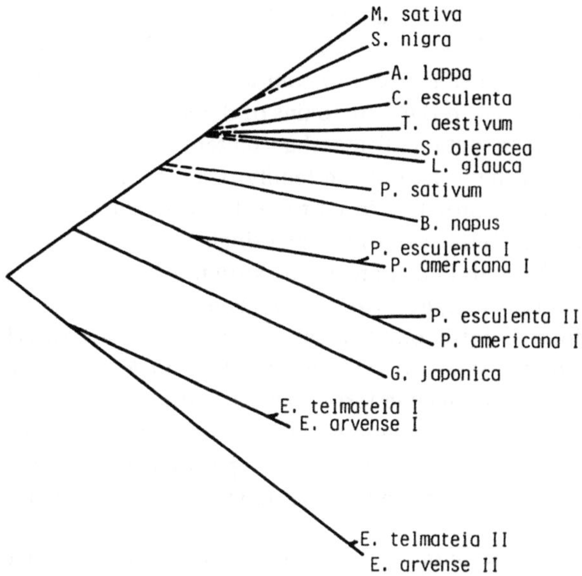

Fig. 7. Phylogenetic tree of higher plant ferredoxins. *Dotted lines* show ambiguous divergent points. See text for explanations

difference was detected in each case between ferredoxin I and II, respectively, of two species of *Equisetum*, suggesting the divergent point of the ferredoxins to be the point of speciation of *E. arvense* and *E. telmateia*. On the other hand, the differences between ferredoxins I and II of the two species were 29 and 30, respectively, which are obviously smaller than those found between ferredoxins of horsetails and other higher plants (33–48). This indicates that the divergent point of ferredoxins I and II corresponds to the time of gene duplication of ferredoxins of horsetails after divergence of a common ancestor of horsetails from other plant groups.

Among angiosperms, the species of *Phytolacca* were placed in the most remote position compared to other angiosperms. The relationships between their ferredoxins were similar to those of the horsetail ferredoxins mentioned above. There were 14 to 30 amino acid differences among angiosperm ferredoxins other than *Phytolacca* ferredoxins. The ferredoxins of *Petroselinum sativum* and *Brassica napus* showed differences of over 22 from those of all other angiosperms except for the difference of 19 between *P. sativum* and *Triticum aestivum*. Therefore, these two species must be placed relatively remote from other tested angiosperms. The least difference of 14 observed between ferredoxins of *Medicago sativa* and *Sambucus nigra* suggests their close relation. The differences between the ferredoxins of *Leucaena glauca* and *Sambucus nigra*, and between *Medicago sativa* and *Arctium lappa* or *Colocasia esculenta* were 15 and 16, respectively. However, comparisons of the ferredoxins of other pairs of angiosperms gave values of 17–27 and therefore, no specified angiosperms could be chosen to form a group giving minimal difference numbers. This difficulty is represented by the dotted lines in Fig. 7.

It was difficult from the present study to assign relative positions of various ferredoxins to those established hitherto from taxonomic data. For example, the amino acid differences of ferredoxins and the relative positions on the phylogenetic tree showed that *Leucaena glauca* and *Medicago sativa* both of which have been placed in the family Fabaceae, or *Spinacia oleracea* and two species of *Phytolacca*, order Caryophyllales, were not closely related. Ferredoxins of the monocotyledons *Colocasia esculenta* and *Triticum aestivum* did not show any closer similarity to each other than to the ferredoxins of dicotyledons.

These data indicate that presently it is not possible to deduce an angiosperm phylogenetic tree at the family or order level based upon their ferredoxins. This difficulty is probably in part due to the occurrence of divergence of angiosperms at the level of families or orders at a single time or within a short period of time long ago.

Only limited numbers of structures of ferredoxins of higher plants, particularly plants belonging to the same genus or species are known. The amino acid compositions of ferredoxins of several species of *Petunia* (Huisman et al. 1978) and several varieties of wheat (Shin et al. 1979) are known. These suggested that there is no difference among the ferredoxins of different varieties of *Triticum*, and there are only small differences among those of different species of *Petunia*. However, a comparison of ferredoxins among genera or species may give a good correlation between the ferredoxin structures and the taxonomic position of plant taxa.

Acknowledgements. The authors express their thanks to Prof. L.J. Rogers, University College of Wales, P. Böger, University of Konstanz, D.W. Krogmann, Purdue University, and S. Katoh, University of Tokyo, for permission to cite various sequences before their publications. They also thank Mr. K. Inoue, Mrs. N. Okutani, and Miss I.S. Lee for technical assistance. This work was supported in part by a Grant-in-Aid for a Special Project of Research (No. 57120005) from the Ministry of Education, Science, and Culture of Japan.

References

Andrew PW, Rogers LJ, Haslett BG, Boulter D (1981a) Phytochemistry 20:579–583

Andrew PW, Rogers LJ, Haslett BG, Boulter D (1981b) Phytochemistry 20:1293–1298

Benson AM, Yasunobu KT (1969) J Biol Chem 244:955–963; Proc Natl Acad Sci USA 63:1269 to 1273

Bryson V, Vogel H (eds) (1965) Evolving genes and proteins. Academic Press, London New York

Dayhoff MO (1978) Atlas of protein sequence and structure, vol 5,3. Nat Biomed Res Found. Georgetown Univ Med Center, Washington DC

Dickerson RE (1980) In: Sigman DS, Brazier MAB (eds) The evolution of protein structure and function. Academic Press, London New York, pp 174

Fitch WM, Margoliash E (1967) Science 155:279–284

Fukuyama K, Hase T, Matsumoto S, Tsukihara T, Katsube Y, Tanaka N, Kakudo M, Wada K, Matsubara H (1980) Nature (London) 286:522–524

Hase T, Wakabayashi S, Matsubara H, Rao KK, Hall DO, Widmer H, Gysi J, Zuber H (1978) Phytochemistry 17:1863–1867

Hase T, Matsubara H, Ben-Amotz A, Rao KK, Hall DO (1980) Phytochemistry 19:2065–2070

Hase T, Yamanashi H, Matsubara H (1982) J Biochem (Tokyo) 91:341–346

Huisman JG, Stapel S, Muijsers AO (1978) FEBS Lett 85:198–202

Lovenberg W (ed) (1973, 1977) Iron-sulfur proteins, vol I–III. Academic Press, London New York

Matsubara H, Hase T, Wakabayashi S, Wada K (1978) In: Matsubara H, Yamanaka T (eds) Evolution of protein molecules. Jpn Sci Soc Press Center Acad Publ Japan, Tokyo, pp 209

Matsubara H, Hase T, Wakabayashi S, Wada K (1980) In: Sigman DS, Brazier MAB (eds) The evolution of protein structure and function. Academic Press, London New York, pp 245

Nakano T, Hase T, Matsubara H (1981) J Biochem (Tokyo) 90:1725–1730
Rippka R, Derielles J, Waterburg JB, Herdman M, Stanier RY (1979) J Gen Microbiol 111:1–61
Shin M, Yokoyama Z, Abe A, Fukasawa H (1979) J Biochem (Tokyo) 85:1075–1081
Takahashi Y, Hase T, Matsubara H, Hutber GN, Rogers LJ (1981) Biochem Soc Trans 9:327; J Biochem (Tokyo) 92:1363–1368
Takruri I, Boulter D (1979a) Phytochemistry 18:1481–1484
Takruri I, Boulter D (1979b) Biochem J (Tokyo) 179:373–378
Takruri I, Boulter D (1980) Biochem J (Tokyo) 185:239–243
Takruri I, Gilroy J, Boulter D (1982) Phytochemistry 21:325–327
Tsukihara T, Fukuyama K, Nakamura M, Katsube Y, Tanaka N, Kakudo M, Wada K, Hase T, Matsubara H (1981) J Biochem (Tokyo) 90:1763–1773
Wakabayashi S, Hase T, Wada K, Matsubara H, Suzuki K (1980) J Biochem (Tokyo) 87:227–236

Polypeptide Composition of Rubisco[1] as an Aid in Studies of Plant Phylogeny

S.G. WILDMAN[2]

Abstract. Rubisco is an abundant protein found in photosynthetic organisms. It is composed of large subunits whose polypeptides are coded by chloroplast genes and small subunits with polypeptides coded by nuclear genes. Polypeptide resolution and analysis of composition is readily made by electrofocusing. Origin of amphiploids has been determined by rubisco electrofocusing analysis. Small subunit polypeptide compositions may be an indicator of the degree of ploidy. Both large and small subunit compositions, together or separately, could be used as markers in plant taxonomy and as indicators of the relative age of genera and species in plant phylogeny.

1 Introduction

Rubisco is the enzyme that catalyzes the initial act of carbon dioxide assimilation during photosynthesis. In higher plants, rubisco often amounts to about one-half of the total soluble protein. Because of its abundance, rubisco can usually be obtained by simple extraction techniques in either a pure or nearly pure state which facilitates analysis of the composition of this protein. Electrophoretical, electrofocusing, and serological properties of rubisco have been found useful in plant phylogenetic and evolutionary research. J.C. Gray (1980) has recently written a comprehensive review explaining how these rubisco properties have been used, and his review should be consulted for details on techniques as well as critical analysis and evaluation of the data. In the following paragraphs, I will concentrate on some of the electrofocusing properties of rubisco that had been extensively employed in my former laboratory.

2 Polypeptide Composition of Rubisco as Revealed by Electrofocusing

Rubisco has a molecular weight of about 0.5 million. The macromolecule is composed entirely of amino acids in polypeptide chains, the latter contained within two distinct types of subunits. Individual polypeptide chains in the "large" subunits (LS) have molecular weights of about 55,000 whereas the molecular weights of the individual

---

[1] Rubisco is a name proposed by Prof. David Eisenberg of UCLA as a more convenient designation for the protein in green plants previously called carboxydismutase, or Fraction 1 protein, or ribulose 1.5 bisphosphate carboxylase-oxygenase, or RuDPCase.

[2] Department of Biology, University of California, Los Angeles, Cal., USA

Proteins and Nucleic Acids in Plant Systematics
ed. by U. Jensen and D.E. Fairbrothers

polypeptides in the "small" subunits (SS) are approximately 13,000. The native macromolecule in plants is composed of 8 LS combined with 8 SS.

Exposure of native rubisco to urea causes the two kinds of subunits to dissociate from each other plus the further dissociation of their constituent polypeptides. Subjecting the dissociated polypeptides to the technique of electrofocusing results in separation of the individual polypeptides. The polypeptides differ from each other in electric charge which is the cause for the individual polypeptides coming to rest at their isoelectric points located at different positions in a pH gradient.

Figure 1 shows the actual results of having electrofocused different kinds of rubiscos obtained from nine species of *Nicotiana*. It is characteristic of the LS of rubisco to be composed of a cluster of three prominent polypeptides separated about equidistant from each other. The cluster of LS polypeptides usually locates in the pH gradient in regions more alkaline than the location of the polypeptides of the SS. One kind of rubisco can be distinguished from another kind if at least one out of the cluster of three polypeptides is different in isoelectric point, as for example, the difference between *N. africana* and *N. glauca* rubiscos displayed in Fig. 1. The isoelectric points of the three LS polypeptides are a reproducible phenotypic character for expression of chloroplast DNA genes inherited exclusively via the maternal line. Chloroplast DNA contains the code for the sequence of amino acids in the cluster of LS polypeptides, the sequence in turn determining the isoelectric points of the polypeptides. The SS of rubisco is composed of one to as many as four, different kinds of polypeptides, the latter often exhibiting wide differences in isoelectric points. The number and isoelectric points of the SS polypeptides are also reliable phenotypic characters that remain constant after alternation of generations of self-fertilized plants. In plant species hybrids each parent makes an equal contribution of genes governing the poly-

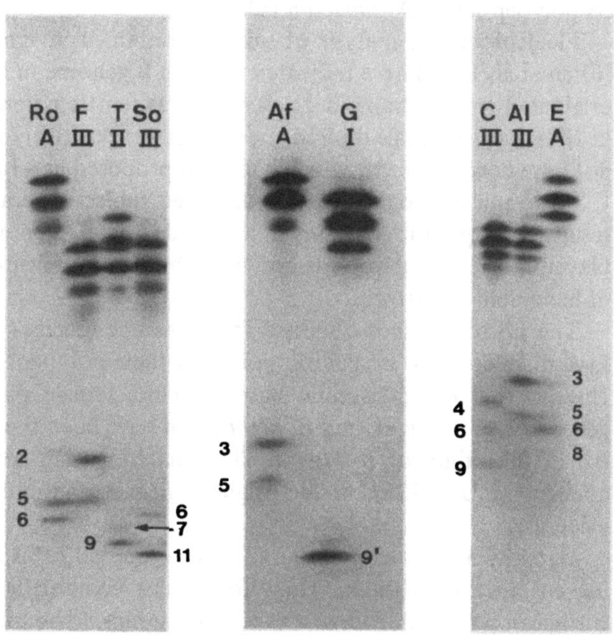

Fig. 1. Photographs of portions of three gels after electrofocusing rubiscos obtained from nine species of *Nicotiana. A* and *Roman numerals* designate positions of LS polypeptide clusters; *A* is the most alkaline cluster; *I* is less alkaline than *A*; *II* is less than *I*; and *III* is the least alkaline cluster with respect to isoelectric points. *Arabic numbers* refer to isoelectric points of SS polypeptides. **Ro** *N. rosulata*; **F** *N. forgetiana;* **T** *N. tabacum*; **So** *N. solanifolia;* **Af** *N. africana;* **G** *N. glauca;* **C** *N. clevelandii;* **Al** *N. alata*; **E** *N. excelsior.* (Chen and Wildman 1981)

peptide composition of the SS in offspring without dominance being a factor. The characteristic cluster of three polypeptides composing the LS, and variation in number of polypeptides constituting the SS has been found in rubiscos obtained from more than 150 plant species (Table 1).

3 Origin of Amphiploids as Determined by Rubisco Polypeptide Composition

Numerous attempts had been made to ascertain the origin of the tobacco plant, *N. tabacum*, which no longer exists in the wild. Whereas it seemed certain that *N. sylvestris* was one of the parents, a clear choice between *N. otophora* and *N. tomentosiformis* as being the other parent could not be made. Gray et al. (1974) analyzed the rubiscos obtained from the allopolyploid and the three putative parents. Electrofocusing showed that the phenotypic character of the cluster of LS polypeptides of *N. tabacum* rubisco was the same as the cluster in *N. sylvestris* rubisco, but different from the clusters in the rubiscos from the other suspected parents. Therefore, *N. sylvestris* was the female parent by this test. The SS of *N. tabacum* rubisco is composed of two kinds of polypeptides, one kind corresponding in isoelectric point to the single kind of polypeptide found in the SS of *N. sylvestris* rubisco. Thus, *N. sylvestris* as female parent also contributed the nuclear DNA coding for one of the two polypeptides found in the SS of *N. tabacum* rubisco. A difference in isoelectric points between the single kind of polypeptide composing the SS of rubiscos from *N. otophora* and *N. tomentosiformis* also made evident that only nuclear DNA from the latter species could have provided the code resulting in the isoelectric point of the second SS polypeptide of *N. tabacum* rubisco.

Electrofocusing analysis of rubisco contained in species of *Triticum* and *Aegilops* (Chen et al. 1975) have indicated that the B genome in hexaploid wheat, *T. aestivum*, originated from tetraploid *T. dicoccum* as female parent, the latter having originated in turn from a female diploid species not yet positively identified, but, at least, could not have been *T. urartu* which had been proposed as a likely possibility.

Steer and Kernoghan (1977) analyzed rubiscos obtained from *Avena* species with results that permitted them to conclude that C genome diploids could only have played a role as paternal parents in hybridization events leading to the development of hexaploid *A. sativa*.

The LS of rubiscos obtained from *Brassica* species (Uchimiya and Wildman 1978) have differences in isoelectric points of their polypeptides which permitted the conclusions that *B. nigra* could have been the female partner with *B. oleracea* in the genesis of *B. carinata*; and *B. nigra* could have been the male partner with *B. campestris* in the origin of *B. juncea*. The electrofocusing analyses were supportive of the concept of U (1935) on the origin of *Brassica* allopolyploids based on cytogenetic evidence.

Gatenby and Cocking (1978a) electrofocused rubiscos obtained from tuber-bearing species of *Solanum*. They found the composition of rubisco in the modern European potato, *S. tuberosum ssp. tuberosum*, to be inconsistent with the previously

Table 1. Rubisco polypeptide composition among plants representing several of the phyla of the Plant Kingdom. LS = large subunits; SS = small subunits. After Wildman (1982) with permission of Academic Press

Genus	Number of species analyzed	Kinds of *polypeptides* LS	SS	Range in number of kinds of small subunit polypeptides	References
Angiosperms					
Nicotiana	63	4	13	1–4	Chen and Wildman (1981)
Lycopersicon	8	1	3	1–3	Gatenby and Cocking (1978b), Uchimiya et al. (1979)
Solanum	7	2	3	1–3	Gatenby and Cocking (1978a)
Brassica, *Sinapis*, *Raphanus*	8	2	4	1–3	Uchimiya and Wildman (1978)
Beta, *Spinacia*	2	2	2	1–2	Chen et al. (1976)
Oenothera	12	1	1	1	Chen (unpublished)
Gossypium	19	4	8	2–4	Chen and Wildman (1981)
Zea	3	4	8	1–2	Uchimiya et al. (1978)
Sorghum	7	1	1	1	Chen et al. (1976)
Hordeum	4	1	1	1	Chen et al. (1976)
Triticum, *Aegilops*	8	2	1	1	Chen et al. (1975)
Avena	7	3	1	1	Steer and Kernoghan (1977)
Oryza	14	2	6	1–4	Uchimiya (unpublished)
Lemna, *Spirodela*, *Wolffiella*, *Wolffia*	11	4	8	1–4	Chen and Wildman (1981)
Other Tracheophytes					
Ginkgo	1	1	1	1	Chen et al. (1976)
Selaginella	1	1	1	1	Chen et al. (1976)
Equisetum	1	1	1	1	Chen et al. (1976)
Algae					
Chlamydomonas	1	1	1	1	Chen et al. (1976)

advanced idea that it might have arisen by chromosome doubling of *S. stenotonum*, or another idea that it had originated by hybridization between the latter species and *S. sparsipilum*. Also, they concluded that the form of the potato brought to Europe in the 16th century, *S. tuberosum* ssp. *andigena*, could not have been a female parent if hybridization with another species had been the route of origin of *S. tuberosum ssp. tuberosum*.

Both the LS and SS polypeptide compositions of rubiscos obtained from cultivated cotton, *Gossypium hirsutum* and *G. barbadense* containing AD genomes, are in accord with the view that the A genome donor was a female parent, with *G. herbaceum* being a likely candidate (Chen and Wildman 1981).

4 Multiple Small Subunit Polypeptides as an Indicator of Polyploidy

Rubiscos obtained from 63 species of *Nicotiana* display three different bands forming four different clusters of LS (represented as A, I, II, III in Fig. 1) and thirteen different bands of SS polypeptides with different isoelectric points (Chen and Wildman 1981). Various combinations of the LS and SS polypeptides have produced the 29 kinds of rubiscos that are now distributed among the different species of *Nicotiana*. Only four of these rubiscos have a SS composed of a single kind of polypeptide. The other twenty five rubiscos have a SS composed of two, three or four kinds of polypeptides with different isoelectric points. In several instances, it has been shown that rubiscos acquire more than one kind of SS polypeptide by polyploidization.

It seems clear that *N. tabacum* rubisco developed a SS containing two kinds of polypeptides by allopolyploidization. Since no evidence for segregation of the genetic information coding for the two kinds of polypeptides was obtained from analysis of more than 100 individual tobacco plants, the genes coding for the two kinds of polypeptides appear to be located on heterologous chromosomes contained within the two distinct genomes of *N. tabacum*. The origin of the four kinds of rubisco SS polypeptides in *N. digluta* is another result of polyploidy, two having come from *N. glutinosa* and the other two from *N. tabacum* (Kung et al. 1975).

Another clear example of where rubisco has acquired multiple SS polypeptides by polyploidy comes from parasexual hybrid plants. By fusion of *N. glauca* and *N. langsdorffii* protoplasts, Smith et al. (1976) were the first to create new species of self-fertile plants in a test tube. The SS of *N. glauca* rubisco contains only one kind of polypeptide that is different in isoelectric point from the two kinds of polypeptides in the SS of *N. langsdorffii* rubisco. The self-fertile hybrid has a rubisco SS with three kinds of polypeptides whose composition has remained constant through F_2 and F_3 generations. This provides further evidence that the coding information for each kind of SS polypeptide is located on heterologous chromosomes (Chen et al. 1977). It is of interest to note that parasexual hybridization created two distinct kinds of rubisco. The LS of *N. glauca* rubisco has a cluster of polypeptides of different isoelectric points than the LS cluster of *N. langsdorffii* rubisco. Among the various hybrids created by Smith et al. (1976), some of their rubiscos have a LS of the *N. glauca* type, others of the *N. langsdorffii* type without mixtures of both, but all of the rubiscos have a SS composed of the same complement of three kinds of polypeptides derived from the separate genomes brought together by fusion of protoplasts. In more recent times, other examples of the creation of new kinds of rubiscos by parasexual hybridization have been reported which display the same stability in polypeptide composition after alternation of generations (Iwai et al. 1980).

The polyploid origin of *N. rustica* (n = 24) from *N. undulata* (n = 12) x *N. paniculata* (n = 12), and polyploid *N. arentsii* (n = 24) from *N. wigandioides* (n = 12) x *N. undulata* have been revealed by the results of Gray (1978) using serological methods for analyzing the character of rubiscos contained in the different species. The results provide added examples of where the presence of multiple kinds of rubisco SS polypeptides is an indication of the polyploid origin of plants. The rubiscos of both *N. rustica* and *N. arentsii* contain a SS composed of two kinds of polypeptides whereas there is only a single kind of polypeptide in the SS of the rubiscos contained in the parental species.

Since the presence of multiple SS polypeptides in rubiscos from *Nicotiana* species is a strong indicator of polyploidy, the question arises as to whether a rubisco with multiple SS polypeptides obtained from plants outside the genus *Nicotiana* also suggests origin by polyploidy. A case in point is that of *Zea mays* whose rubisco SS is composed of two kinds of polypeptides as is also the case for teosinte, its close relative (Uchimiya et al. 1977). The suggestion of polyploid origin of the corn plant derived from rubisco polypeptide composition has been strengthened by the finding of numerous duplicated isozymes in maize by Gottlieb (1982). Since increase in number of isozymes by polyploidy is much greater than specific duplications of isozymes in diploids, Gottlieb views the increased level of isozymes in maize as cause for reexamining the conventional view of *Zea* being a diploid.

Having relatively few duplicated genes, the genome of *Lycopersicon esculentum* is considered to be at the diploid level. But the presence of three kinds of rubisco SS polypeptides in the rubisco (Gatenby and Cocking 1978b, Uchimiya et al. 1979) obtained from numerous accessions of this species provokes the question of whether the genome has not advanced beyond the diploid state, or whether the plant originated in antiquity by polyploidy followed by extensive elimination of the duplicated genes since its origin. The situation is somewhat analogous to that of the rubiscos in *N. langsdorffii* and three other species of *Nicotiana* with a haploid chromosome number of nine, the smallest number encountered among the Nicotianas, but all of whose rubiscos have a SS containing two kinds of polypeptides. However, the indication from rubisco analysis that these *Nicotiana* species originated by polyploidy may still be valid if Goodspeed's (1954) hypothesis of how the Nicotianas arose in the first place is correct. He proposed that they arose from allopolyploidization between diploid species of *Cestrum* (n = 6) so that some *Nicotiana* polyploids at the n = 12 level could have evolved n = 9 genomes by aneuploidy.

5 Examples where Rubisco Polypeptide Composition Could be Used as Markers in Plant Taxonomy

Creation of new, self-fertile hybrids by protoplast fusion has been used to overcome natural barriers to hybridization. As parasexual hybrids are created, there is the opportunity for plants closely similar in phenotypic appearance, but containing different cytoplasms to arise. If the rubisco LS polypeptides are different in the two species united by protoplast fusion, the new hybrids with different cytoplasms can be distinguished by electrofocusing their rubiscos. All experience so far indicates the composition of the rubisco LS to be an unusually stable property that is perpetuated without change as self-fertile parasexual hybrids are propagated by sexual reproduction, and could therefore serve as a taxonomic marker. The rubisco LS marker could be used to distinguish between two distinct self-fertile species of *N. glauca* + *N. langsdorffii* and *N. rustica* + *N. tabacum,* and no doubt numerous other opportunities will arise as work on parasexual hybridization intensifies.

There are cases where the composition of the rubisco SS could also be used as a marker in systematics. *N. suaveolens* displays considerable polymorphism in phenotypes of individual plants, and serves as an example of the potential of rubisco as a marker. Electrofocusing of rubisco from individual *N. suaveolens* plants (Chen and Wildman 1980) selected at random from a population displaying polymorphism revealed no differences in composition of the LS. However, differences in polypeptide composition of the SS among individuals were found. There were individuals whose rubisco SS contained three kinds of polypeptides. However, only two of the polypeptides were identical in isoelectric point in all plants whereas the third polypeptide had a more acidic isoelectric point in some plants than the third polypeptide in other plants. The difference allowed the *N. suaveolens* plants to be classified as containing either an "A" type or "B" type rubisco. Progeny derived from self-fertilized "A" type plants contained rubiscos exclusively of the "A" type. Similarly, the "B" type plants breed true as shown by their progeny containing only the "B" type rubisco. The rubisco compositions appeared to be of sufficient stability to distinguish between two kinds of *N. suaveolens* plants not otherwise conveniently distinguishable. In the natural population, there also existed individuals with a "C" type rubisco whose SS was composed of four kinds of polypeptides. However, progeny from self-fertilized plants with "C" type rubisco did not breed true. Instead, segregation of "A", "B", and "C" type rubiscos occurred in the ratio of 1A: 2C: 1B. Exactly the same ratio was produced by progeny derived from reciprocal crosses between *N. suaveolens* with "A" type rubisco x *N. suaveolens* with "B" type rubiscos. The last result suggests that the polypeptide composition of rubisco could serve as an indication of natural hybrids in other populations of self-fertile plants.

The difference in isoelectric point of the rubisco LS contained in *S. tuberosum* ssp. *tuberosum* compared to *S. tuberosum ssp. andigena* would seem to be a sufficiently stable marker to allow for the possible reclassification of the subspecies as distinct species.

Lycopersicon minutum, once considered to be a single species, has been designated by Rick et al. (1974) as two species: *L. parviflorum* and *L. chiemelewski*. The SS polypeptide composition of rubisco obtained from the two species is in accord with this decision since *L. parviflorum* rubisco has a single kind of SS polypeptide whereas *L. chiemelewski* rubisco has two kinds. The presence of multiple polypeptides may also indicate that the latter species originated by hybridization.

6 Polypeptide Composition of Rubisco as an Indicator of Comparative Age of Genera and Species

Twenty of the 65 species of *Nicotiana* and three of the 35 species of *Gossypium* are located in Australia. This unique geographical distribution vis-a-vis the distribution of the remainder of the species in South America and Africa led Goodspeed to postulate for the Nicotianas, and other authorities for the Gossypiums, that the two genera had evolved prior to the breakup of Gondwanaland. They proposed that the progenitors of modern day species of *Nicotiana* and *Gossypium* had arrived in Australia by crossing land bridges before the separation of Africa, Australia, Antarctica, and South America by continental drift. This view of early land immigration to Australia and consequent great antiquity of the two genera has been challenged on the basis of newer geophysical and phytogeographical evidence by Raven and Axelrod (1974). To them this indicated that the angiosperms were not much more ancient than *Nicotiana* and *Gossypium* would have to be if land migrations had been the mode of dispersal of *Gossypium* to Australia. Instead, they proposed that vicarious transport of seeds across oceans within the last 10 to 20 million years was the more likely cause for the presence of the two genera in Australia.

If Raven and Axelrod are correct, the LS composition indicates that the Australian Nicotianas must have originated from an old taxon probably with frequent occurrence in the Southern South America. *N. noctiflora* and *N. petunioides* are the only South American species with the same LS composition as Australian and African rubiscos. Therefore, they could be recent species left from the original populations. However, because of the very restricted recent geographical distribution of *N. noctiflora* and *N. petunioides* the dispersal probably did not happen in recent times, but in the Tertiary time, when the continents would have been more closely connected than they are presently.

References

Chen K, Wildman SG (1980) Biochem Genet 18: 1175–1184

Chen K, Wildman SG (1981) Plant Syst Evol 138:89–113

Chen K, Gray JC, Wildman SG (1975) Science 190:1304–1306

Chen K, Wildman SG, Smith HH (1977) Proc Natl Acad Sci USA 74:5109–5112

Chen K, Kung SD, Gray JC, Wildman SG (1976) Plant Sci Lett 7:429–434

Gatenby A, Cocking EC (1978a) Plant Sci Lett 12:117–181

Gatenby A, Cocking EC (1978b) Plant Sci Lett 13:171–176

Goodspeed TH (1954) The genus Nicotiana. Chron Bot Press, Waltham, MA

Gottlieb LD (1982) Science 216:373–380
Gray JC (1978) Plant Syst Evol 129:177–183
Gray JC (1980) In: Bisby FA, Vaughan JG, Wright CA (eds) Chemosystematics; Principles and practice. Syst Association, Spec Vol No 16. Academic Press, London New York, pp 167–193
Gray JC, Kung SD, Wildman SG, Sheen SJ (1974) Nature (London) 252:226–227
Iwai S, Nagao T, Nakata K, Kawashima N, Matsuyama S (1980) Planta 147:414–417
Kung SD, Sakano K, Gray JC, Wildman SG (1975) J Mol Evol 7:59–64
Raven P, Axelrod D (1974) Ann Mo Bot Gard 61:539–673
Rick C, Kesicki E, Fobes J, Holle M (1974) Theor Appl Genet 47:55–68
Smith HH, Kao KN, Combatti NC (1976) J Hered 67:123–128
Steer M, Kernoghan D (1977) Biochem Genet 15:273–286
U N (1935) Jpn J Bot 7:389–452
Uchimiya H, Wildman SG (1978) J Hered 69:299–303
Uchimiya H, Chen K, Wildman SG (1977) Stadler Genet Symp 9:83–100
Uchimiya H, Chen K, Wildman SG (1978) In: Redei GP (ed) Stadler Symp Vol 9, pp 83–100, Univ of Missouri Press
Uchimiya H, Chen K, Wildman SG (1979) Biochem Genet 17:333–341
Wildman SG (1982) In: Schiff J (ed) On the origins of chloroplasts. Elsevier/North Holland, New York, pp 231–242

Rubisco[1] in the Brassicaceae

M.P. ROBBINS and J.G. Vaughan[2]

Abstract. Leaf rubisco (Fraction 1 protein) was purified from a number of cruciferous species, and subunit polypeptide patterns were examined by isoelectric focusing in the presence of 8 M urea. Both cultivated and wild *Brassica* species were studied by this method and results suggest that the protein may be a useful taxonomic marker, especially where species may have a possible hybrid origin. One small subunit (designated SS 1) appeared from studies to have a restricted distribution within the Brassicaceae. Antiserum was raised to rubisco from *B. oleracea*. In double diffusion experiments, all cruciferous species examined gave reactions of complete identity when tested against *B. oleracea*, but outside the family solely reactions of partial and incomplete identity were noted.

1 Introduction

The high degree of morphological similarities have made the classification of the Brassicaceae into subfamily groupings such as tribes and subtribes difficult (Hedge 1976). The same type of difficulties especially exist within the genus *Brassica*, the most important crop genus within the mustard family, when comparing the morphological and anatomical characteristics. Although morphological characteristics have been useful for elucidating some relationships, major advances in understanding relationships have resulted from the interpretation of cytological, various types of chemical, and serological data. The fundamental cytological work of Morinaga and U in the 1930's (Fig. 1) helps to clarify the relationships of six basic *Brassica* species, all of which are agriculturally important. Cytological and breeding investigations (Harberd 1972, 1976) have also helped to enlarge our knowledge of the genus.

Serological (Vaughan and Waite 1967a,b) as well as biochemical (Vaughan and Gordon 1973, Phelan and Vaughan 1976) research have also helped to enhance our understanding of the relationships of taxa within the genus *Brassica*.

Investigations of both the large (LS) and small (SS) subunit polypeptide patterns of rubisco have provided valuable information which can be used to better understand the evolutionary histories. Large and small subunit patterns for six basic species of *Brassica* have been determined (Uchimiya and Wildman 1978) and the results obtained by isoelectrofocusing are included in Fig. 1. Large subunit data were used to derive the direction of two of the three crosses in the scheme presented in Fig. 1, as-

[1] Ribulose 1,5 biphosphate carboxylase-oxygenase or fraction 1 protein

[2] Biology Department, Queen Elizabeth College, Campden Hill Road, London, UK

Proteins and Nucleic Acids in Plant Systematics
ed. by U. Jensen and D.E. Fairbrothers

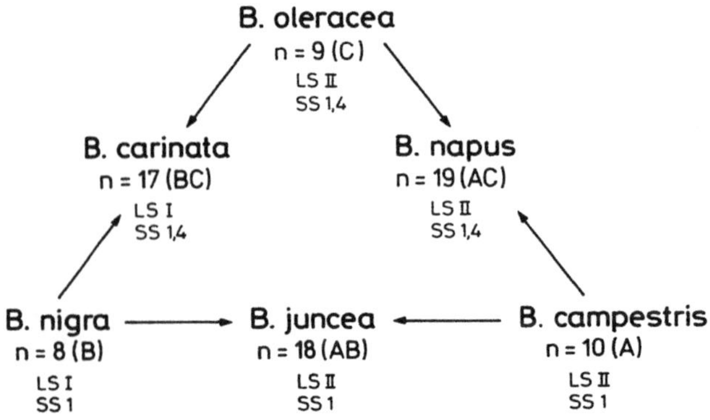

Fig. 1. Inter-relationships of six basic *Brassica* species. *n* haploid chromosome number; *A, B, C* genomic designations; *LS, SS* large and small subunit descriptions

suming maternal inheritance for the indicated large subunit types. The crosses were suggested as: *B. nigra* ♀ (LSI) x *B. oleracea* ♂ → *B. carinata* (LSI) and *B. campestris* ♀ (LSII) x *B. nigra* ♂ → *B. juncea* (LSII). The direction of the third cross could not be determined as *B. campestris* and *B. oleracea* have the same class of large subunit (LSII). With regard to small subunits, Gatenby and Cocking (1978b) noted subunit ratios for SS1 and SS4 (designated as polypeptides a and b) in the following species: *B. oleracea* 53.3:46.7, *B. carinata* 77.0:23.0, and *B. napus* 74.4:25.6. These results are consistent with a hybrid origin of *B. carinata* and *B. napus* with *B. oleracea* as one of the parental types.

In addition to *Brassica*, rubisco subunit patterns are reported for *Sinapis alba, S. arvensis, Raphanus sativus, Eruca sativa, Diplotaxis erucoides*, and *Hirschfeldia incana* (referred to as *B. adpressa* by Uchimiya and Wildman (1978) and Gatenby and Cocking (1978b)).

In this volume S.G. Wildman has included and discussed the properties and functions of the well analysed rubisco protein. He also reviewed the systematic and evolutionary research, and it clearly appears that rubisco research has made definite contributions at the infraspecific and generic taxonomic levels. The vast amount of rubisco research has made it possible to extract relatively large amounts of pure or almost pure research material. Gray (1980) has demonstrated that for comparative purposes even partially purified material can sometimes be used satisfactorily. Electrophoresis and isoelectrofocusing methods have both proven to be most satisfactory for such research. Dorner et al. (1958) performed the early serological screening of rubisco from diverse plant species. More recently Gray (1977, 1978) has again performed a few additional immunological experiments using rubisco for phylogenetic investigations.

The research in this present report includes isoelectrofocusing and serological techniques to investigate rubisco polypeptide subunit patterns from taxa of the Brassicaceae in our effort to expand the understanding of the relationships and evolutionary history of this family.

2 Methodology

2.1 Subunit Polypeptide Patterns

In these studies it was often necessary to purify rubisco from small quantities of leaf material. This was especially important for some wild accessions which exhibited erratic and poor germination and also comprised low amounts of leaf even when approaching maturity. Therefore a modification of the method of Chen et al. (1976b) was employed, which reduced leaf extraction from 50 g to 5 g. The leaves were homogenized in an MSE Atomix for 1 min with 40 ml 0.05 M Tris/HCl, pH 7.8, containing 0.2 M NaCl, 10 mM sodium metabisulphite, 1 mM KCN, 0.1% (w/v) bovine serum albumin, 20% (w/v) Amberlite IRA - 401 (Cl) and 1 mM phenylmethane sulphonyl fluoride; the latter component being present to combat endogenous protease activity. The homogenate was strained through muslin and then centrifuged at 26,000 *g* for 1 h. Protein was purified from the supernatant by salt precipitation 20%–50% $(NH_4)_2SO_4$, followed by gel filtration on an 8 x 2.5 cm column of Sephadex G-200 equilibrated with 0.05 M Tris/HCl, pH 7.8, 0.2 M NaCl. At this stage the protein was electrophoretically homogeneous with a yield of ca. 5 mg. Proteins at this point could not be crystallised either using low salt buffer or the vapour diffusion method of Johal and Borque (1979).

Protein preparations were S-carboxymethylated using a modification of the method of Kung et al. (1974); 0.2 ml of a solution of rubisco (ca. 10 mg/ml) was placed in a 2 x 8.5 cm Buchner tube. The contents were degassed and flushed with nitrogen for 15 min. Then 5 mg dithiothreitol was added to the tube in 1 ml of Tris-urea buffer; 0.5 M Tris/HCl, pH 8.5, 8 M ultra-pure urea, 2 mM EDTA. After a 2 h incubation the tube was covered with silver foil and 15 mg iodoacetic acid was introduced in 0.2 ml of the same buffer. 10 min incubation was followed by desalting on a 1.5 x 6 cm column of Sephadex G-25, also covered with silver foil and equilibrated with Tris-urea buffer. Void volume elution was followed by the addition of a few drops of Dextran Blue (10 mg/ml) in Tris-urea buffer. Then routinely 50 mg samples of protein were applied to 7 x 0.5 cm isoelectric-focusing tube gel containing 8 M urea and 2% (w/v) Ampholine pH 5–7, and focused for 7 h at 300 V. Bands on the gel were visualised using Kenacid Blue R. Five gels of each accession were focused and for cultivated types double replicates were employed. For scanning, a Unicam SP1809 scanning densitometry unit was used, operating at 540 nm.

It should be noted that this method is distinctly different from that employed by Uchimiya and Wildman (1978) in their *Brassica* experiments. Their procedure involved focusing of a carboxymethylated antibody-protein complex, and in *Nicotiana* this methodology results in a slight acidic shift of large subunit clusters (Uchimiya et al. 1979). Hence results obtained by Uchimiya and Wildman with cruciferous material may not be exactly comparable with results from our experiments. In view of the limit of accuracy when measuring polyacrylamide gels, this work has only distinguished classes of large subunit clusters when obviously different, i.e. when separated by ca. 0.10 pH units.

Material for cultivated Brassicas was grown under green house conditions using seed from commercial and research sources. Hybrid *Brassica* and x *Brassicoraphanus* accessions were received from S. Tokumasu, Ehime University, Japan. Most wild *Brassica* species were received from C. Gomez-Campo, Crucifer Germplasm Collection, Madrid. Other cruciferous and non-cruciferous seeds were obtained from commercial companies.

2.2 Serology

The basic method of Gray and Kekwick (1974) was followed, with antiserum being raised to rubisco in female New Zealand white rabbits. Injections were intramuscular employing 5 mg of antigen emulsified with Freund's complete adjuvant followed by a booster inoculation of 2.5 mg after 65 days. Eleven day post-inoculation sera were then used without further purification. Double-diffusion was carried out as Hudson and Hay (1976) using 2% Oxoid purified agar in 0.05 M Tris/HCl pH 7.8 buffer containing 0.2 M NaCl and 1% sodium azide. Diffusion was allowed to occur against 0.5 g/ml crude leaf extracts, prepared using 2.5 mM Tris/HCl 7.4 buffer containing 20 mM NaCl and 0.05 mM EDTA, at 37 °C overnight.

3 Results

3.1 Cultivated Brassica, Sinapis, and Raphanus Species

The polypeptide subunit patterns for the cultivated cruciferous vegetables were in general agreement with those of Uchimiya and Wildman (1978), not withstanding the technical differences mentioned in the previous section. Three or four bands made up the large subunit cluster while a variable number of small subunit bands was detected.

A brief investigation of the cultivated members of the *B. oleracea* (n = 9) cytodeme and the *B. campestris* (n = 10) cytodeme was conducted. Diagrammatic representation of the results is shown in Fig. 2 with the position of bands being indicated together with an average pH gradient. Included in the first group was *B. alboglabra*, which is a freely flowering Chinese Kale that crosses readily with *B. oleracea* yielding fully fertile hybrids (Morinaga 1933). *B. pekinensis, B. chinensis*, and *B. perviridis* were all included in the *B. campestris* group, these being Oriental vegetable and salad plants which appear from previous studies to be part of the *B. campestris* complex. The *B. oleracea* taxa had a general formula of LSII for the large subunits, and small subunits SS1 and SS4 were seen in a quantitative ratio of approximately 50: 50. For example in *B. alboglabra* the small subunit distribution was measured as 48.8: 51.2. The *B. campestris* accessions could be described using the nomenclature of Uchimiya and Wildman (1978) as LSII and SS1. Some faint bands were noted in the region of pH 5.1, but none of the components with this pH could be detected on scanning, and so are presumably at very low levels. These bands were found in both Oriental and non-Oriental vegetable types and so appear to have no geographical correlation. When reviewed together the two sets of results show good group identity for the two cytodemes examined, even though quite a large degree of morphological specialisation

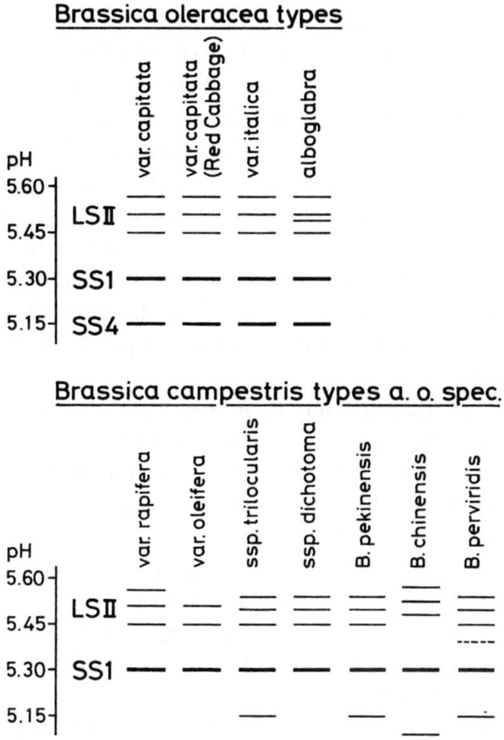

Fig. 2. Rubisco subunit patterns for *Brassica oleracea* and *Brassica campestris* types

exists in these crops. This constancy of pattern suggests that rubisco may be useful in helping to clarify difficult systematic problems.

Amphidiploidy has been suggested as a phenomenon found in the genus *Brassica. B. napus* was evaluated as a possible amphidiploid. The species exists in two crop forms, both as swede (var. *napobrassica*) which is grown as a root crop for animal and human consumption, and also as swede-rape (var. *oleifera*) which is an important source of vegetable oils. Figure 3 shows the subunit patterns for both types. Swede and swede-rape were noted to follow the description of LSII for the large subunits. SS1 and SS4 were also detected with a ratio approximating to 75:25. It has been

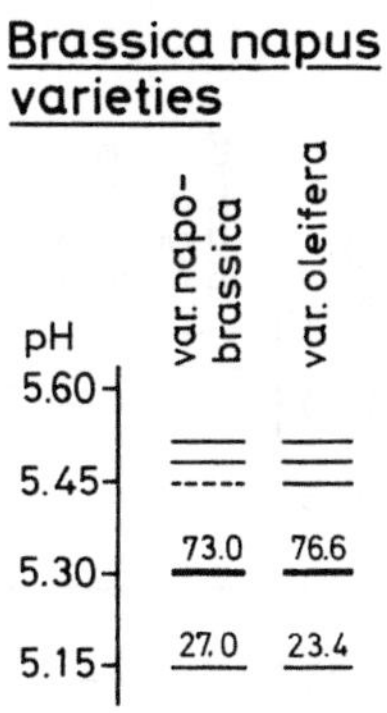

Fig. 3. Rubisco LS- and SS-subunit patterns for *Brassica napus* varieties. For the SS the relative amounts are indicated

suggested that this taxon exists as a result of interspecific hybridisation between *B. campestris* and *B. oleracea*, and the presence of LSII is consistent with this idea since it is found in each of the suggested parents. The small subunit ratio is as would be expected for such a cross, assuming equal contributions from both parents. In an attempt to verify this point an artificial *B. napus* was examined, which had been synthesised from *B. japonica* x *B. oleracea.* (*B. japonica* Sieb. is an oriental *B. campestris* type with a haploid chromosome number of 10.) When subjected to Chi^2 analysis with four degrees of freedom, *B. oleracea* could be considered as having a 50:50 ratio of SS1:SS4, with results being significant at $p = 0.05$. With similar treatment synthesised *B. napus* was successfully tested against a ratio of 75:25 for SS1:SS4. This is direct evidence that in *Brassica* equal biparental inheritance of small subunits may be noted.

B. carinata and *B. juncea* were also analysed. The following compositions were recorded: *B. carinata*, large subunits LSI with SS1 and SS4 present in a ratio of approximately 75:25; and for *B. juncea* large subunit type LSII together with SS1. These results agree with the formation and direction of the crosses suggested by Uchimiya and Wildman (1978). *B. juncea* is an interesting case in that there is evidence that two geographical races of the plant exist i.e. Indian and Oriental forms (Vaughan et al. 1963, Vaughan and Gordon 1973). The types may be distinguished by the presence or absence of a marked mucilaginous epidermis, and the relative levels of allyl and 3-butenyl isothiocyanate. In these characters Indian *B. juncea* taxa resemble *B. campestris* more than *B. nigra*, while for Oriental *B. juncea* taxa the opposite is true. The hypothesis was tested that the two races may reflect different directions of the formative hybridisation, but analysis of Indian and Oriental forms of *B. juncea* has suggested that the only cross to have occurred is *B. campestris* ♀ x *B. nigra* ♂.

Some research have indicated that *Sinapis* is closely related to *Brassica*. This idea is clear in the case of some American researchers who describe *S. alba* (white mustard) and *S. arvensis* (charlock) as *B. hirta* Moench, and *B. kaber* (DC.) Wheeler respectively. Figure 4 shows the subunit patterns of these two *Sinapis* species, together with that of *B. nigra*. Large subunit type LSI was detected in both *S. alba* and *S. arvensis*, the latter however may be distinguished by the presence of small subunit SS3. *B. nigra* was also noted to have LSI whereas LSII was detected in *B. campestris* and *B. oleracea*. This may be regarded as evidence that of the three elemental Brassicas, *B. nigra* is the closest to the genus *Sinapis*, which correlates with the seed electrophoretic results of Vaughan and Denford (1968).

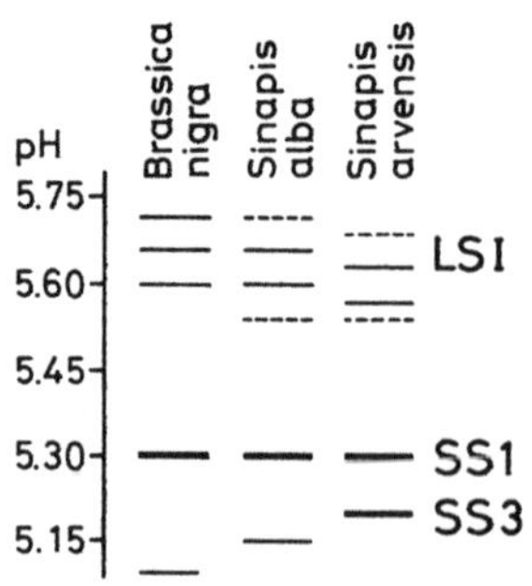

Fig. 4. Rubisco subunit patterns for *Sinapis* species and *Brassica nigra*

Raphanus is another closely related genus, and *Raphanus sativus* (radish) may be induced to hybridise with *Brassica* species to produce either x *Brassicoraphanus* or x *Raphanobrassica*, depending upon the direction of the cross. Extensive work has been carried out in order to produce these novel types, which may become agronomically important as forage or oil-seed crops (Mc Naughton 1972). An artificial x *Brassicoraphanus* (Tokumasu and Kato 1980) was analysed together with the parent material, *B. japonica* and *R. sativus*. The results are presented in Fig. 5 together with the rubisco subunit pattern determined by Gatenby and Cocking (1978b) for a x *Raphanobrassica* produced from a hybridisation between *R. sativus* and *B. oleracea*. When *B. oleracea* is used small subunit SS4 is noted in the hybrid, but this is not seen if *B. japonica* is employed as one of the parents. The observation that an AR genome may be distinguished from an RC genome, suggests that rubisco subunit considerations may be analytically useful for evolutionary hybrids produced by plant breeders.

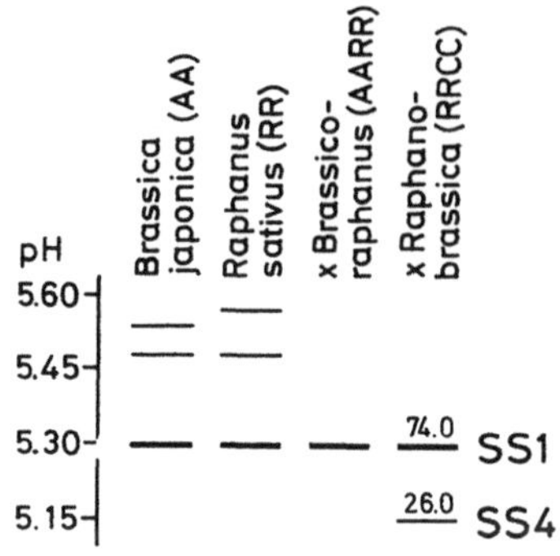

Fig. 5. Rubisco subunit patterns for *Brassica*, *Raphanus*, and their hybrids

3.2 Wild Brassica Species

A total of twenty wild *Brassica* species were investigated. Some of the species analysed are believed to be close relatives, and possible progenitors, of the cultivated Brassicas: *B. nigra*, *B. oleracea*, and *B. campestris*. The polypeptide subunit patterns of such wild taxa together with those of the three elemental *Brassica* species are shown in Fig. 6. It should be noted that especially with members of the *B. oleracea* cytodeme, large subunit visualisation was difficult, and this renders exact comparisons less easy. The wild and cultivated forms of *B. nigra* were both noted to conform to a formula of LSI, SS1. However, the two types were found to contain a faint small subunit polypeptide that focused at a different pH in the two accessions, emphasising some difference between the two types. Material detailed as wild *B. campestris* [syn. *B. rapa* ssp. *sylvestris* (L.) Janchen] was also studied and results from this were compared with a generalised pattern of the *B. campestris* cytodeme, and LSII and SS1 were common to both taxa. However, the sample was collected on Cambria beach, California, and this may be a crop "escape" or alternatively it may be a genuine wild turnip, in which case this data would be suggestive of a possible weed and crop relationship. *B. tournefortii* has been suggested as an ancestral turnip type though De Candolle (1821) was able to clearly distinguish this species from *B. campestris*. Only one small subunit, SS1, was detected in *B. tournefortii* as in *B. campestris* accessions. Therefore

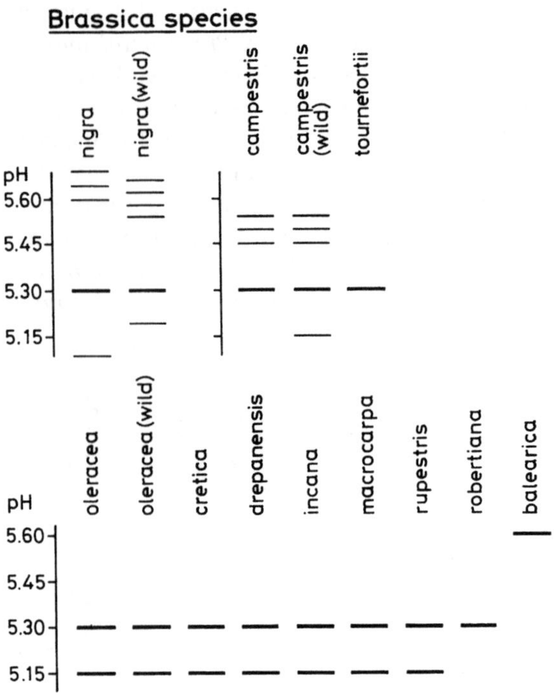

Fig. 6. Rubisco subunit patterns for cultivated *Brassica* species and possible crop progenitors

it is not possible to disprove this suggestion, but the individual occurrence of 3-methylsulphinylpropyl-glucosinolate in this species (Horn and Vaughan 1983) does imply some disjunction between the two types.

Harberd (1972) has commented that there are a number of wild species that comprise a group, the "sea cabbages", which are fleshy leaved perennials that have at various times been regarded as subspecies of *B. oleracea*. They exist in cliff habitats particularly along the northern side of the Mediterranean, usually in distinct geographical and morphological complexes (Snogerup 1980). Included in this group is *B. oleracea* L. s. str. which is a wild plant found in northern Spain, western France and on the south coast of Britain. This has close morphological similarity with certain cultivated forms. In his synopsis of the "sea cabbages", Harberd (1972) includes among others the following species: wild *B. oleracea, B. balearica, B. macrocarpa, B. montana* (syn. *B. robertiana*), *B. rupestris, B. insularis, B. atlantica, B. cretica, B. incana*, and *B. villosa*. Most of these taxa have been suggested as sources of cultivated *B. oleracea* forms even though some have characters quite atypical of *B. oleracea*. This is a situation that was investigated and the subunit patterns obtained from the "sea cabbages" that were examined are represented in Fig. 6 together with a generalised pattern for cultivated *B. oleracea*. Wild *B. oleracea, B. cretica, B. drepanensis, B. incana, B. macrocarpa*, and *B. rupestris* all contained small subunits SS1 and SS4 in visually similar quantities, in the manner of *B. oleracea*, and so may be crop progenitors. *B. robertiana* was only noted to have SS1 and appears to be an anomalous member of the group and an unlikely forerunner of cultivated forms. The subunit pattern of *B. balearica* is completely different from the others, small subunit SS1 is absent and only one small subunit is present which has a pI of 5.60. This may be evidence that *B. balearica* is not a relative of *B. oleracea* and this would be in accord with the work of Stork et al. (1980), who suggested that on account of seed morphology one would separate *B. balearica* from the *B. oleracea* cytodeme.

Its restricted distribution and the presence of a number of primitive characteristics seem to suggest that this is a Miocenic relict and it is difficult to classify in the genus *Brassica*. This may explain the presence of the detected anomalous subunit pattern.

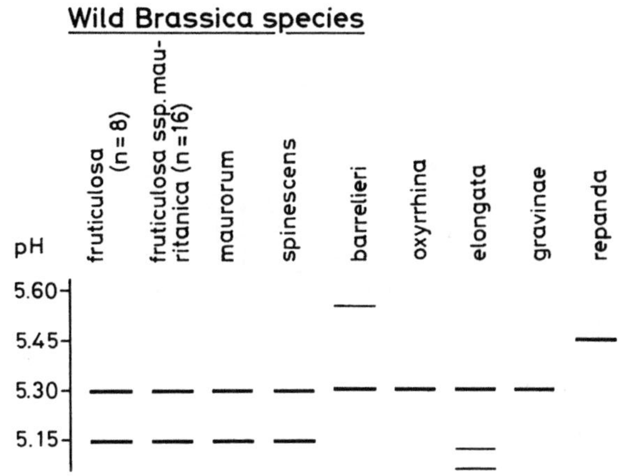

Fig. 7. Rubisco small subunit patterns for wild *Brassica* species

Figure 7 shows the results for a number of other wild species. It may be noted that the three species in the n = 8 *B. fruticulosa* cytodeme appear to form a group together with *B. fruticulosa* ssp. *mauritanica*. The latter has been recorded with a chromosome number of 16 (Takahata and Hinata 1978) and this may be a *B. fruticulosa* autotetraploid. Other groupings are perhaps difficult to discern, but it should be noted that *B. repanda* is another *Brassica* species that does not contain SS1; it has a small subunit polypeptide which focuses at pH 5.45. Gomez-Campo and Tortosa (1974) in their work on juvenile characters in the Brassiceae mentioned *B. repanda* as an exceptional member of the genus and rubisco data support this observation.

3.3 Cruciferous Tribes

In addition to studying *Brassica, Sinapis*, and *Raphanus* species, representative members of a number of other genera were investigated. Members of the tribe Brassiceae (as defined by Schulz 1919) were studied as well as individual taxa from other tribes. The data in Fig. 8 indicate the small subunit patterns. A variety of small subunit types were noted, but it is interesting that there appears to be a general distribution of SS1

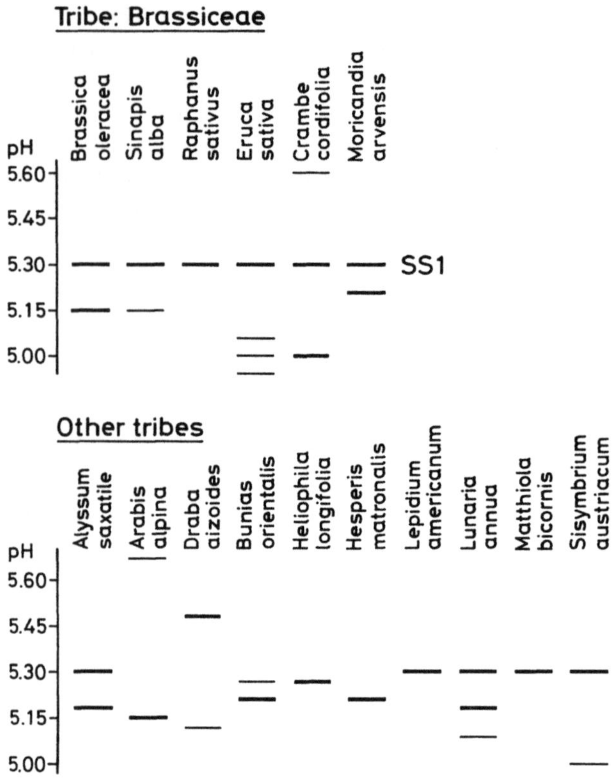

Fig. 8. Rubisco small subunit distribution in the Brassicaceae

within the Brassiceae, though as previously mentioned, *B. balearica* and *B. repanda* are exceptions. Uchimiya and Wildman (1978) in their rubisco research noted small subunit SS1 in *Diplotaxis erucoides* and *Hirschfeldia incana* which are also members of the same tribe. However, this small subunit does seem to show a restricted distribution within the family, and so such information can not really be regarded as supporting tribal disjunction within the Cruciferae. Perhaps this observation reflects Schulz's (1919) separated tribes based upon minor morphological distinctions, and thus many of these groups, with the possible exception of the Brassiceae, may be quite unnatural groupings (Hedge 1976).

3.4 Other Families

As mentioned in the introduction, protein serology may be used as a taxonomic tool as well as a phylogenetic method. However, when antiserum raised to *B. oleracea* rubisco was used in double diffusion experiments against crude leaf extracts of other *Brassica* species, it was not possible to detect any immunological differences in the protein within this group.

This appears to contrast with the observations of Gray (1977) in *Nicotiana*, who found that members of the genus when used as antigen sources in Ouchterlony plate experiments did exhibit spur formation when run against rubisco antisera from *N. gossei, N. glauca, N. tabacum,* and *N. glutinosa*. It is, however, difficult to compare rubisco serological data from the genera *Nicotiana* and *Brassica*, as titre levels for the experiments in the two systems may not be compared. Using the double diffusion method, all cruciferous species examined gave reactions of complete identity when compared with *B. oleracea* antiserum. However, reactions of partial identity and non-identity were noted in non-cruciferous species and these data are included in Table 1. As may be seen partial identity was noted in the Resedaceae and the Capparidaceae, which are often designated as the two nearest families to the Brassicaceae, suggesting that in this special case, and using a special antiserum as the reference system, reactions of complete identity delimit the Brassicaceae from other angiosperm families. Another observation of note is that generally the dicotyledon taxa examined gave reactions of partial identity, with the exception of *Kalanchoë* and *Papaver*. The latter is perhaps an interesting example in that some early angiosperm classifications placed the Papaveraceae close to the Brassicaceae due to some resemblance in floral structure; contemporary classifications (Dahlgren 1980, Takhtajan 1980), separate the families quite widely, and this result would support such an arrangement.

Also as regards the monocotyledon types examined; *Triticum, Hordeum, Tradescantia,* and *Limnobium* in contrast to the dicotyledon plants all gave reactions of nonidentity, with the exception of *Hordeum*, possibly suggesting a large evolutionary distance between such types and the Brassicaceae.

One serological reaction of particular interest is included in Fig. 9. As previously remarked, *Cleome spinosa* (Capparidaceae) and *Reseda lutea* (Resedaceae) gave reactions of partial identity, on comparison with *B. oleracea*. However, "secondary spurring" was noted in respect of these two species when compared with *Antirrhinum*

Table 1. Serological reactions of *Brassica oleracea* antiserum

Material	Family	
All Brassicas	Brassicaceae	Complete identity
Other genera within the Brassicaceae	Brassicaceae	Complete identity
Reseda (2 ssp.)	Resedaceae	Partial identity
Sesamoides sp.	Resedaceae	Partial identity
Cleome sp.	Capparidaceae	Partial identity
Antirrhinum sp.	Scrophulariaceae	Partial identity
Phaseolus sp.	Fabaceae	Partial identity
Begonia sp.	Begoniaceae	Partial identity
Lycopersicon sp.	Solanaceae	Partial identity
Beta sp.	Chenopodiaceae	Partial identity
Viola sp.	Violaceae	Partial identity
Hordeum sp.	Poaceae	Partial identity
Kalanchoe sp.	Crassulaceae	Non-identity
Papaver sp.	Papaveraceae	Non-identity
Triticum sp.	Poaceae	Non-identity
Tradescantia sp.	Commelinaceae	Non-identity
Limnobium sp.	Hydrocharitaceae	Non-identity

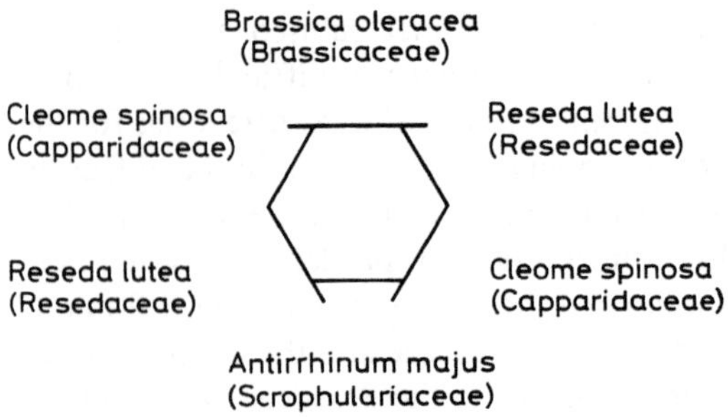

Fig. 9. Serological reactions of *Brassica oleracea* antiserum (*center*) and four antigenic systems

majus (Scrophulariaceae), suggesting that *Cleome* and *Reseda* are closer than *Antirrhinum* to *Brassica* in an evolutionary sense.

4 Discussion

For the Brassicaceae in analysing hybrid combination the mode of inheritance of rubisco, and the consistency of subunit patterns makes the enzyme useful in such studies. The determination of the directions of the formative crosses for *B. carinata* and *B. juncea* demonstrate this point. Such analysis also is valuable in monitoring artificially produced hybrids, for example *B. napus* and also *Brassica* x *Raphanus* hybridisation. In taxonomic research especially large and small subunit compositions are useful at and above the species level. Identity was noted in some cytodemic groups, such as *B. campestris, B. oleracea,* and *B. fruticulosa.*

The general occurence of small subunit SS1 in the Brassiceae and its presence in other cruciferous tribes is an indicator of the possible value of using rubisco in phylogenetic studies. This small subunit, however, was not found in *B. balearica* and *B. repanda.* This observation may correlate with the observation reported by Gomez-Campo and Tortosa (1974) that these two species may correspond to, or be related to, ancient types that are ancestral for contemporary cruciferous species.

The experimental determination of immunological identity within the Brassicaceae does support the morphological view that the family exists as a discrete and natural unit (Hedge 1976). Differences were only found on comparison with other angiosperm families, and 'secondary spurring' may permit ranking of such groups. Jensen and Penner (1980) have used similar methodology with single storage plant proteins and have found such studies to be of value in studying the Ranunculaceae. Rubisco is highly conserved entity and presumably has antigenic determinants that evolve at a slower rate (Chen et al. 1976a) than even those on storage proteins, hence it may have an importance in assisting the understanding of evolution of angiosperm plant families and their inter-relationships.

References

Baker TS, Eisenberg D, Eiserling FA, Weissman L (1975) J Mol Biol 91:391–399
Baker TS, Suh SW, Eisenberg D (1977) Proc Natl Acad Sci USA 74:1037–1041
Chen K, Sand SA (1979) Science 204:179–180
Chen K, Gray JC, Wildman SG (1975) Science 190:1304–1306
Chen K, Johal S, Wildman SG (1976a) In: Bücher T, Neupert W, Sebald W, Werner S (eds) Genetics and biogenesis of chloroplasts and mitochondria. North-Holland Biochem Press, Amsterdam, pp 3–11
Chen K, Kung SD, Gray JC, Wildman SG (1976b) Plant Sci Lett 7:429–434
Dahlgren RMT (1980) Bot J Linn Soc 80:91–124
De Candolle AP (1821) Regni vegetabilis systema naturale II, Paris, pp 582–616
Dorner RW, Kahn A, Wildman SG (1958) Biochem Biophys Acta 29:240–245
Gatenby AA, Cocking EC (1978a) Plant Sci Lett 12:177–181
Gatenby AA, Cocking EC (1978b) Plant Sci Lett 12:299–303
Gomez-Campo C, Tortosa ME (1974) Bot J Linn Soc 69:105–124
Gray JC (1977) Plant Syst Evol 128:53–69
Gray JC (1978) Plant Syst Evol 129:177–183
Gray JC (1980) In: Bisby FA, Vaughan JG, Wright CA (eds) Chemosystematics: Principles and practice. Academic Press, London New York, pp 167–193
Gray JC, Kekwick RGO (1974) Eur J Biochem 44:481–489
Gray JC, Kung SD, Wildman SG, Sheen SJ (1974) Nature (London) 252:226–227
Harberd DJ (1972) Bot J Linn Soc 65:1–23
Harberd DJ (1976) In: Vaughan JG, Macleod AJ, Jones BMJ (eds) The biology and chemistry of the Cruciferae. Academic Press, London New York, pp 47–68
Hedge IC (1976) In: Vaughan JG, Macleod AJ, Jones BMJ (eds) The biology and chemistry of the Cruciferae. Academic Press, London New York, pp 1–46
Horn PJ, Vaughan JG (1983) Phytochemistry (in press)
Hudson L, Hay FC (1976) Practical immunology. Blackwell Scientific Publ, Oxford, pp 110–115
Jensen U, Penner R (1980) Biochem Syst Ecol 8:161–170
Johal S, Borque DP (1979) Science 204:75–77
Kung SD, Sakano K, Wildman SG (1974) Biochim Biophys Acta 365:138–147
Link G, Bogorad L (1980) Proc Natl Acad Sci USA 77:1832–1836
Maire R (1965) In: Quezel P (ed) Flore de L'Afrique du Nord, vol XII. Lechevalier, Paris, pp 152–204
McIntosh L, Poulsen C, Bogorad L (1980) Nature (London) 288:556–560
McNaughton IH (1972) Euphytica 22:70–88
Morinaga T (1933) Jpn J Bot 6:467–475
Nishimura M, Akazawa T (1973) Biochem Biophys Res Commun 54:842–848
Olsson G (1954) Hereditas 40:398–418
Phelan JR, Vaughan JG (1976) Biochem Syst Ecol 4:173–178
Sakano K, Kung SD, Wildman SG (1974) Mol Gen Genet 130:91–97
Schulz OE (1919) In: Engler A (ed) Das Pflanzenreich, vol 70. Engelmann, Leipzig, pp 1–290
Snogerup S (1980) In: Tsunoda S, Hinata K, Gomez-Campo C (eds) *Brassica* crops and wild allies. Jpn Sci Soc Press, Tokyo, pp 121–132
Steer MW (1975) Can J Genet Cytol 17:337–344
Steer MW, Kernoghan D (1977) Biochem Genet 5:273–286
Stork AL, Snogerup S, West J (1980) Candollea 35:421–450
Sun VG (1946) Bull Torrey Bot Club 73:244–281, 370–377
Takahata Y, Hinata K (1978) Eucarpia Cruc Newsl 3:47–51
Takhtajan AL (1980) Bot Rev 46:225–359
Tokumasu S, Kato M (1980) Jpn J Breed 30:11–19
Tutin TG, Heywood VH, Burgess NA, Valentine DH, Walters SM, Webb DA (eds) (1964) In: Flora Europaea, vol I. Univ Press, Cambridge, pp 260–346

Uchimiya H, Wildman SG (1978) J Hered 69:299–303
Uchimiya H, Chen K, Wildman SG (1979) Plant Sci Lett 14:387–394
Vaughan JG (1968) In: Hawkes JG (ed) Chemotaxonomy and serotaxonomy. Syst Assoc, pp 93 to 102
Vaughan JG, Denford KE (1968) J Exp Bot 19:724–732
Vaughan JG, Gordon EI (1973) Ann Bot (London) 37:167–184
Vaughan JG, Waite A (1967a) J Exp Bot 18:100–109
Vaughan JG, Waite A (1967b) J Exp Bot 18:269–276
Vaughan JG, Whitehouse JM (1971) Bot J Linn Soc 64:383–409
Vaughan JG, Hemingway JS, Schofield H (1963) Bot J Linn Soc 58:435–447

Rubisco[1] as a Taxonomic Tool in the Genus Erysimum (Brassicaceae)

K. BOSBACH[2]

Abstract. A method is described whereby rubisco subunits can be separated into their polypeptides without preceeding separation steps, using polyacrylamide gel electrophoresis followed by iso-electrofocusing. Thus different rubisco polypeptide patterns are obtained for different *Erysimum* species. No intraspecific variation was observed. This, among other reasons, indicates that rubisco delivers a useful taxonomic tool for investigating the genus *Erysimum*.

1 Introduction

The genus *Erysimum* includes approximately 100 species which have a worldwide distribution with centers of morphological diversity reported in North America, Canary Islands, Eurasia, western and eastern Mediterranean Regions, and central Asia.

Many of the species of this genus are very difficult to distinguish morphologically. Several cytological studies have been reported for taxa of this genus (Favarger 1964, 1965, Jaretzky 1928, 1932, Manton 1932, Polatschek 1966). Rossbach (1958) published a taxonomic key for the North American species. Ball (1964) published a comprehensive study of the European species, and Polatschek (1966, 1973, 1974, 1976, 1979) has also pursued research involving the Européan species. However, even with these various studies, the integrating of anatomical, cytological, and morphological information remains to be done to produce a concise taxonomic treatment.

Hegnauer's (1964) review demonstrated that little chemotaxonomic data had been published. Latowski et al. (1979) evaluated data concerning the cardenolides found in European *Erysimum* species; and Rodman et al. (1983) discussed the value of the cardenolides as chemotaxonomic markers for the North American species of the genus.

The review by S.G. Wildman included in this volume demonstrates the value of rubisco in systematic research. Wildman's earlier publication (Wildman et al. 1975) and Uchimiya et al. (1977), also demonstrate how such research can be used in solving selected taxonomic problems.

A critical taxonomic subgroup within the genus *Erysimum* is the "sylvestre-group of species", which is geographically distributed in the mountains of central and southern Europe. Therefore, in this paper we are reporting preliminary information

[1] Rubisco = ribulose 1.5 bisphosphate carboxylase-oxygenase = fraction 1 protein

[2] Dept. of Botany, University of Osnabrück, 4500 Osnabrück, FRG

Proteins and Nucleic Acids in Plant Systematics
ed. by U. Jensen and D.E. Fairbrothers

about the polypeptide composition of rubisco of *E. sylvestre* (Crantz) Scop., *E. diffusum* Ehrh. = *E. canescens* Roth, *E. helveticum* (Jacq.) DC., and *E. decumbens* Dennst. = *E. ochroleucum* DC.

2 Materials and Methods

To resolve the polypeptide composition of rubisco we have developed both a more rapid and a more convenient method than that applied by Wildman and coworkers, and this method is described below.

2.1 Sample Preparation

Seed material was obtained from various Botanical Gardens, and plants were grown in greenhouses in the Botanical Garden of the University of Münster. Mature leaves were placed in a mortar containing a cold extraction buffer (0.05 M Tris/HCl, pH 6.8 containing 0.5% mercaptoethanol) in a ratio of 1:1 (ml/g leaf material) and ground. The crude extract was centrifuged for 20 min at 15,000 rpm, and the supernatant was directly used for polyacrylamide electrophoresis (PAGE).

2.2 Polyacrylamide Gel Electrophoresis

A discontinuous system was used similar to that described by Laemmli (1970). The stacking gel was 4.5% total polyacrylamide with 2.5% crosslinker concentration in a gel buffer of 0.125 M Tris/HCl pH 6.8, the separation gel contained 7.5% polyacrylamide with 2.5% crosslinker (N,N-methylenebisacrylamide) in a buffer of 0.375 M Tris/HCl pH 8.9. The gels were made by photopolymerization with riboflavine. The electrophoresis buffer contained 0.03 M Tris/HCl and 0.2 M glycine, pH 8.3, and electrophoresis was run in gel rods 9 cm long and 6 mm in diameter. A sample of 100 μl was applied on each gel-rod. Electrophoresis was carried out at 3–4 mA constant current per tube and was run for 2.5 h.

2.3 SDS-polyacrylamide Gel Electrophoresis

2.3.1 Sample Preparation

The supernatant obtained as described above (2.1) was precipitated by 3 M trichloric acid and centrifuged at 15,000 rpm for 15 min, the precipitate was redissolved in a buffer (of 0.06 M Tris/HCl, 2% SDS, 5% mercaptoethanol, 0.001 M EDTA and 5% sucrose). This mixture was subjected to SDS-PAGE.

2.3.2 Electrophoresis

SDS-PAGE was carried out essentially according to the method described by Laemmli (1970) with a separation gel of 12.0% acrylamide and 2.5% crosslinker pH 8.8, the stacking gel of 4.0% acrylamide and 2.5% crosslinker, pH 6.8 (crosslinker concentration expressed as a percentage of the total acrylamide). The electrophoresis buffer was 0.05 M Tris/HCl, 0.4 M glycine, pH 8.3, and 0.1% SDS. 100 μl of the samples were submitted on each gel rod and electrophoresis was carried out at 8 mA constant current per rod for about 5 h.

2.3.3 Staining

The gels were stained for total protein (max. 5 min) with Coomassie brilliant blue R-250 in 50% methanol and 10% acetic acid, and destained in 5% methanol and 10% acetic acid for only a few minutes (Cammaerts and Jacobs 1980). Following this procedure one rather intensively stained band could be observed which had migrated only a short distance in the gel, which was rubisco.

2.4 Isoelectrofocusing

The stained band was cut out and shaken in 1.0 ml of buffer (8 M urea, 5% sucrose and 2% pharmalyte pH 5–8). The gel-slices were directly used for isoelectrofocusing (IEF). IEF was carried out in acrylamide gels placed in a Pharmacia flatbed apparatus FBE 3000. The gels were composed of 5% acrylamide with 3% crosslinker (see above, Sect. 2.3.2) containing 2% pharmalyte pH 5–8 and made by photo-polymerization. The catholyte was 1 M NaOH, the anolyte 1 M phosphoric acid. Gel-slices cut out from the gel rods were directly laid onto the middle of the IEF-gel. Focusing was run at 30 W, constant voltage for a minimum of 2 h.

After the run was finished, the gels were shaken in perchloric acid for 2 h to remove urea- and pharmalyte, then stained with 0.05% Coomassie brilliant blue G-250 in 3.5% perchloric acid until the blue bands appeared (Reisner et al. 1975).

3 Results

The PAGE and SDS gel runs used for IEF yielded the same banding patterns. These results are in agreement with the findings of Cammaerts and Jacobs (1980, 1981) who investigated the polypeptide composition of rubisco from *Lycopersicon* species, using a technique very similar to the technique reported here.

For all four species a number of individuals were checked for the rubisco-pattern, and no intraspecific variation was detected. However, there were differences between the four taxa. The four species *E. helveticum, E. decumbens, E. diffusum,* and *E. sylvestre* revealed different large subunit clusters with three bands each (Fig. 1). In *E. diffusum*, some additional weak bands occurred within the range of the three bands

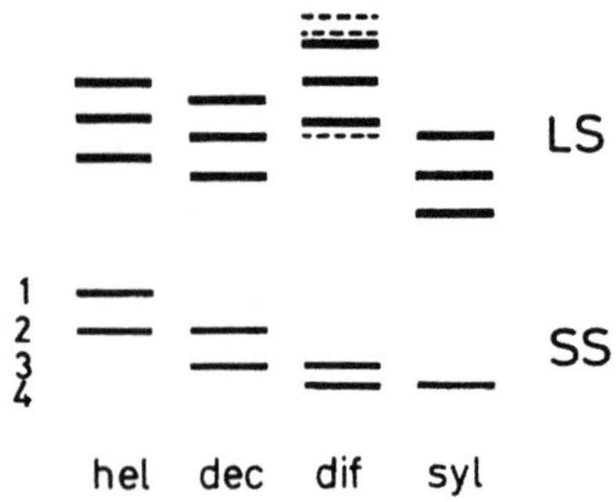

Fig. 1. Schematic representation of the rubisco pattern after isoelectrofocusing showing the small *SS* and large *LS* subunits of *Erysimum helveticum* (*hel*), *E. decumbens* (*dec*), *E. diffusum* (*dif*), and *E. sylvestre* (*syl*)

but we have no idea about their origin. They may be artifacts, resulting from some reaction with phenolic compounds (Wildman personal communication).

Also with regard to the small subunits differences were detected. *E. sylvestre* shows only one band (SS4 in Fig. 1), whereas in the other three species revealed two bands occur: SS1 and SS2 in *E. helveticum*, SS2 and SS3 in *E. decumbens*, SS3 and SS4 in *E. diffusum*.

References

Ball PW (1964) In: Tutin TG et al. (eds) Flora Eur 1:270–274
Cammaerts D, Jacobs M (1980) Anal Biochem 109:317–320
Cammaerts D, Jacobs M (1981) Plant Sci Lett 22:147–153
Favarger C (1964) Bull Soc Bot Suisse 74:5–40
Favarger C (1965) Ber Dtsch Bot Ges 77:73–83
Hegnauer R (1964) Chemotaxonomie der Pflanzen, vol III. Birkhäuser, Basel
Jaretzky R (1928) Jahrb Wiss Bot 68:1–45
Jaretzky R (1932) Jahrb Wiss Bot 76:458–527
Laemmli UK (1970) Nature (London) 227:680–685
Latowski K, Kortus M, Kowalewski Z (1979) Fragm Florist Geobot (Krakow) 25:261–268
Manton I (1932) Ann Bot (London) 46:509–556
Polatschek A (1966) Oesterr Bot Z 113:1–46
Polatschek A (1973) Ann Mus Goulandris 1:113–126
Polatschek A (1974) Ann Naturhist Mus Wien 78:171–182
Polatschek A (1976) Ann Naturhist Mus Wien 80:93–103
Polatschek A (1979) Ann Naturhist Mus Wien 82:325–362
Reisner AH, Nemes P, Bucholtz C (1975) Anal Biochem 64:509–516
Rodman JE, Brower L, Frey J (1983) Taxon (in press)
Rossbach GB (1958) Madrono 14:261–267
Uchimiya H, Chen K, Wildman SG (1977) Stadler Symp 9:83–99
Wildman SG (1983) In: Jensen U, Fairbrothers DE (eds) Proteins and nucleic acids in plant systematics. Springer, Berlin Heidelberg New York
Wildman SG, Chen K, Gray JC, Kung SD, Kwanyuen P, Sakano K (1975) In: Perlman PS, Birky CW, Byer TJ (eds) Genetics and biogenesis of chloroplasts and mitochondria. Ohio State Univ Press, Columbus, pp 309–329

Isozyme Number and Phylogeny

L.D. GOTTLIEB[1]

Abstract. The origin and establishment of certain gene duplications may reflect a unique and non-repeatable series of events in a given lineage. When this is the case, species which possess the same duplication inherited it from a common ancestor and constitute a monophyletic assemblage. Thus the duplication of the gene specifying the cytosolic isozyme of phosphoglucose isomerase, recently identified in *Clarkia* (Onagraceae), native in California, was used to realign the phylogenetic relationships of four sections of the genus previously thought to derive from different ancestral stocks. In another application, the ploidy level of certain species of Astereae was determined on the basis of the number of their isozymes, thereby providing new evidence regarding the basic chromosome number in the group. Isozyme number rather than allozyme variation may be applied in many phylogenetic studies above the species level.

1 Introduction

The pattern of enzymes visualized on a starch or polyacrylamide gel following electrophoresis contains substantial information, some of which has been applied to specific problems in plant systematics and evolution. For example, electrophoretic evidence has been used to assess the amount of genetic variation among individuals within populations and among populations within species (Brown 1979, Gottlieb 1981a). It has been used to estimate the amount of genetic divergence between closely related species, and to test progenitor/derivative relationships proposed on the basis of morphological and other evidence (Gottlieb 1973a,b, 1974, Gottlieb and Pilz 1976, Crawford and Smith 1982). It has also been used to show that homoeologous gene loci from different diploid genomes are expressed in allopolyploid derivatives, providing additional information to identify the diploid parents of polyploid species (Hart and Langston 1977, Hart 1979, Roose and Gottlieb 1976, 1980).

All of these applications have exploited the ability of electrophoresis to provide efficient data about the variation, heterozygosity, and divergence of alleles at many gene loci coding enzymes in large population samples. Most data have been applied to species or subspecific taxa rather than at higher taxonomic ranks. This is because as phylogenetic distance increases, electrophoretic identity of enzymes is less likely to reflect identity of coding genes since this is influenced, probably unevenly and to unknown extents, by historical factors such as fluctuations in population size (includ-

[1] Department of Genetics, University of California, Davis, Cal. 95616, USA

Proteins and Nucleic Acids in Plant Systematics
ed. by U. Jensen and D.E. Fairbrothers

ing founder effects), mode of speciation, and natural selection operating on species in different environments.

However, the electrophoretic pattern also includes information about the number of isozymes of particular enzymes and thereby the number of coding gene loci. In contrast to the often great variability of allozymes (allelic products), the number of isozymes of most enzymes, at least those assayed with natural (in vivo) substrates, appears to be highly conserved in plant evolution (Gottlieb 1982). However, occasional increases in isozyme number may occur as a result of the duplication of coding structural genes (this is the case in diploid species; in polyploid species, increased isozyme number is brought about by the addition of homoeologous genomes from parental diploid species). Increased isozyme number in diploids may have extraordinary value for phylogenetic analyses of taxa above the species level, when species which possess a particular duplication inherited it from the same common ancestor and belong to a single monophyletic assemblage (Gottlieb 1977, Gottlieb and Weeden 1979).

The validity of this proposal follows from the high probability that the origin and establishment of certain types of gene duplications represent a unique and non-repeatable series of events in a given lineage. The use of such a genetic character to detect phylogenetic relationships is substantially different than the more widely used assessments based on the relative degree of similarity or concordance of many varied characters derived from morphological, chromosomal, and biochemical evidence. Therefore, the appropriateness of this new source of evidence must be carefully evaluated.

2 Gene Duplication

Two types of gene duplication, depending on their mode of origin, must be distinguished. Those duplications that arise following unequal crossing-over during mispairing at meiotic synapsis appear to occur repeatedly and therefore are not relevant to the present discussion. However, a second class of duplications which arise by processes involving overlapping reciprocal translocation or insertional translocation is more likely to be unique because it requires the sequential occurrence of relatively rare and independent events. Assume that two partially overlapping reciprocal translocations have taken place in different plant lines. If they cross to each other, a chromosomally heterozygous type is produced (Fig. 1). Self-pollination or crosses between similarly heterozygous individuals will yield four progeny classes, including one which carries a duplicated chromosome segment as demonstrated by experimental manipulations in maize (Burnham 1962). Because of the translocation, the two duplicated segments are located on different nonhomologous chromosomes, and will assort independently in a linkage test. (In contrast, duplications that arise following unequal crossing-over are tightly linked). Certain crosses involving insertional translocations will also lead to duplications (Perkins 1972). Origin of duplications by translocation requires the occurrence of simultaneous multiple chromosome breaks, and the production of true-breeding (homozygous) duplicate progeny which must be viable and fertile. Conse-

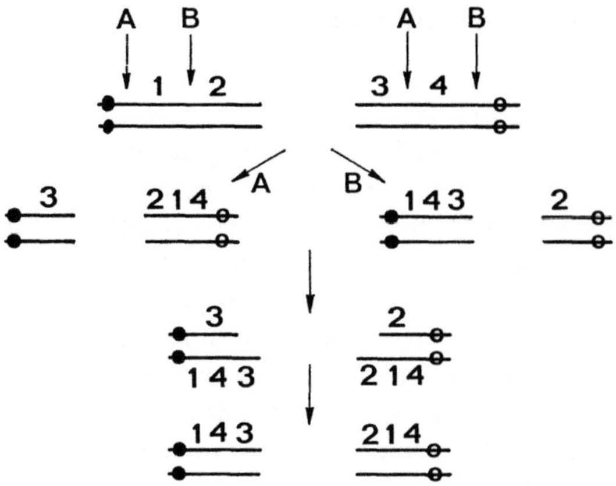

Fig. 1. Diagram (modified from Burnham 1962) illustrating how a cross between two overlapping reciprocal translocations can produce a progeny which is duplicated for the chromosome segments between the translocation break points. *A* and *B* represent two independent reciprocal translocations. Intercrossing them results in a chromosomally heterozygous individual. One of the four progeny classes is illustrated; it carries two duplicated segments. Self-pollination or crosses within this progeny class yield structurally homozygous individuals that have both duplications. Note that since the duplications are located on nonhomologous chromosomes, the duplicate genes are expected to assort independently

quently the establishment of a duplication by these processes has a very low probability of happening more than once in a given lineage. Based on this hypothesis, species which possess the same duplication must have inherited it from the same common ancestor.

Annual plants appear particularly likely to exhibit such cytological processes because the speciation process in annuals often involves complex chromosomal rearrangements so that even closely related species differ by multiple translocations. In addition, a high proportion of annual species are self-compatible so that the duplicated chromosome segments can rapidly be made homozygous by self-pollination.

More than a dozen sets of duplicated isozymes have been identified during surveys of electrophoretic variation in natural plant populations (Gottlieb 1982). Many of these were originally recognized because related diploid species had different numbers of isozymes of the particular enzyme. The duplicated enzymes include dehydrogenases and isomerases, all enzymes that catalyze essential steps in organized metabolic pathways, but are not responsible for regulating flux. The enzymes are dimeric molecules and their subunits associate to form catalytically active intra- and interlocus hybrid enzymes. The duplicated isozymes are always located in the same subcellular compartment; plastids, mitochondria, and the cytosol are represented. The enzymes include alcohol dehydrogenase, cytosolic phosphoglucose isomerase, plastid and cytosolic triose phosphate isomerases, and mitochondrial and cytosolic malate dehydrogenases. Only the isomerases have been utilized to test systematic models.

3 Isozyme Number

Before reviewing these studies, it is necessary to describe how many isozymes per enzyme are usually present in plant species. Evidence from cell fractionation studies and electrophoretic surveys of numerous species have shown that the number of isozymes reflects the number of subcellular compartments in which a particular catalytic reaction is required, at least for those enzymes assayed with natural substrates (Gottlieb 1982). For glycolysis and the oxidative pentose phosphate pathways, among the most completely studied, the number is two because the reactions are carried out both in the plastids and the cytosol. The number and subcellular location of isozymes of other enzymes, such as aminotransferases, are also highly conserved in plant evolution. Conservation of isozyme number extends both to conifers and many families of flowering plants.

Relatively little information is available about the number of isozymes of enzymes such as esterases, phosphatases, and peroxidases which are assayed with artificial substrates. Species differences in the apparent number of isozymes of these "non-specific" enzymes may reflect changes in substrate specificity, reaction requirements, developmental expression as well as duplication of coding genes. Consequently, changes in their numbers have not yet proven useful for systematics.

The origin of subcellular compartmentation in plant cells and the consequent presence of certain metabolic pathways in several locations within the cell may trace back to ancient symbiotic events. The evidence is strongest for the plastids and has been based on specific comparisons of components common to them and present-day cyanobacteria. Thus, it may be that the presence of a pair of isozymes, one in the plastids and the other in the cytosol, for enzymes of carbohydrate metabolism results from the phylogeny of the plant cell (Weeden 1981). A prokaryotic plastid-like symbiont may have provided the gene sequences for the plastid isozymes (its genes presumably transferred to the cell nucleus during the early evolution of the plant cell) whereas the "host" cell already possessed coding sequences for the cytosolic isozymes (Weeden 1981). Although the ancestral origins of many isozymes in plant cells can be explained by such a hypothesis, the increases in isozyme number that concern us result from nuclear gene duplications that lead to increases in the number of isozymes within single subcellular compartments.

4 Phosphoglucose Isomerase (PGI)

The most thoroughly studied duplicate isozymes in plants are the cytosolic isozymes of phosphoglucose isomerase (PGI; EC 5.3.1.9) which are present in about 12 diploid annual species of *Clarkia* (Onagraceae), native in California. Most diploid plants have two isozymes of PGI, one located in the plastids and the other in the cytosol, coded by independent nuclear genes (Weeden and Gottlieb 1980a).

PGI catalyzes the reversible isomerization of glucose-6-phosphate and fructose-6-phosphate, an obligatory step in glycolysis and gluconeogenesis. In the chloroplast, the PGI reaction is one of a sequence of steps that converts 3-phosphoglycerate to starch in the light. The reaction may also occur in the dark when starch is degraded to dihydroxyacetone phosphate for export to the cytosol. The cytosolic PGI isozyme functions in the pathway between the latter metabolite and sucrose. Both PGI isozymes are present in green and nongreen tissues.

In *Clarkia*, diploid species may have one or two cytosolic PGI isozymes. The 43 species of the genus have been divided into 10 sections, of which seven include diploid species (Lewis and Lewis 1955). Electrophoretic and genetic studies revealed that all of the diploid species in the three morphologically primitive sections *Rhodanthos* (formerly *Primigenia*), *Godetia*, and *Myxocarpa* have a single cytosolic PGI isozyme coded by a single gene locus, whereas all but one species in the phylogenetically advanced sections *Peripetasma, Phaeostoma, Fibula,* and *Eucharidium* have two gene loci specifying cytosolic PGIs (Gottlieb and Weeden 1979).

PGI is a dimeric molecule and the subunits coded by the two duplicate loci associate with each other in all possible combinations, indicating they possess a high degree of structural similarity. The combination of both intra- and inter-locus heterodimers, particularly in heterozygous plants, results in a marked increase in the number of cytosolic PGI isozymes. For example, if one of the duplicate loci is heterozygous, six distinct cytosolic PGIs are produced (2A2A, 2B2B, 2A2B, 2A3A, 2B3A, 3A3A), but if both are heterozygous, ten isozymes are present (2A2A, 2A2B, 2B2B, 3A3A, 3A3B, 3B3B, 2A3A, 2A3B, 2B3A, and 2B3B). In contrast, only three cytosolic PGIs are produced in heterozygous plants that lack the duplication (2A2A, 2A2B, 2B2B) (Gottlieb 1977).

The duplicated PGI genes assort independently in the four *Clarkia* species, representing two sections, that were tested. This is thought to mean that the duplication arose by a process involving overlapping reciprocal translocations or insertional translocations rather than by unequal crossing-over. The substantial chromosomal rearrangements characteristic of *Clarkia* species, and the self-compatibility of all the species which facilitates rapid achievement of structural homozygosity, provide a reasonable mechanism of origin, and permit the hypothesis that species with the duplication descended from a common ancestor.

Biochemical properties of the duplicated PGIs were studied in extracts from *Clarkia xantiana*, a species with the duplication for which true-breeding genetic stocks were available. The individual isozymes were partially purified and separated from one another (Fig. 2). The 18 homodimers and heterodimers examined, resembled each other closely in pH optimum, heat sensitivity, energy of activation, and apparent Michaelis constant (K_m) (F-6-P), although several homodimers extracted from two self-pollinated derivatives had unusually high K_m values (Gottlieb and Greve 1981). Overall the biochemical studies indicated relatively little catalytic divergence among the cytosolic PGIs.

Analysis of the molecular weight of the native enzymes by gel filtration and by electrophoresis in gels of different acrylamide concentrations suggested a molecular weight of approximately 130,000 (Gottlieb and Greve 1981). However, since these methods do not detect small differences, we have also examined the molecular weight of PGI subunits separated from purified enzymes on SDS-polyacrylamide gels. The more sensitive procedure showed that the subunits coded by the duplicate gene loci in *C. xantiana* differed by about M.W. 1,400, indicating a difference of about 12 amino

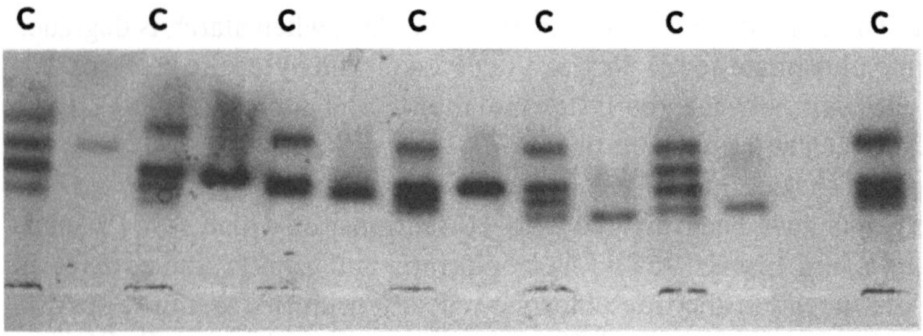

Fig. 2. Electrophoretic separation of PGI isozymes in crude extracts (*c*) and several partially purified single cytosolic PGI isozymes from *Clarkia xantiana* on a 6% acrylamide gel. The purified isozymes include both homo- and heterodimers. The single enzyme band at the *top* of each separation of the crude extracts is the plastid isozyme. The samples were loaded in the slots shown near the cathodal end of the gel (*bottom*)

acids (Higgins and Gottlieb unpublished). The subunits specified by the duplicate genes have a molecular weight of 59,000 and 60,400, respectively. The molecular weight identity of the smaller subunit in three additional species with the duplication supports the hypothesis that these species inherited the same duplicate locus. Since the tested species were in different sections of the genus, the mutation which affected final subunit size probably became established soon after the duplication arose or perhaps as a consequence of the duplication process, and was then passed to all descendent species.

The presence of the duplicate PGIs only in the four phylogenetically advanced sections and their absence from the three morphologically more primitive sections, identifies a definite and specific branching point in the phylogeny of the genus. The evidence suggests that the four sections with the duplication constitute a monophyletic assemblage (Gottlieb and Weeden 1979), although previously they had been thought to derive from different stocks of ancestral Clarkias (Lewis and Lewis 1955). This new evidence has been used by H. Lewis, the monographer of the genus, to modify his earlier phylogeny to reflect the presumed monophyletic relationship (Lewis 1980) (Fig. 3).

The concordance between possession of the duplication and membership in one of the four advanced sections and absence of the duplication in the primitive sections meant that it was not necessary to move species in or out of particular sections, but only to realign the sectional phylogeny. The single exception to a perfect concordance was *Clarkia rostrata*, in the section *Peripetasma*, which exhibited only a single cytosolic PGI, whereas all the other species in the same section have the duplicated isozymes (Gottlieb and Weeden 1979). It is not known if *C. rostrata* lacks an additional isozyme because the coding gene is not duplicated or because the duplicated gene is present, but silenced by mutation(s). Whichever is the case, the dichotomous phylogeny within the genus remains intact.

The PGI duplication provides new evidence which can be used to support the retaining of *Eucharidium* as a section within *Clarkia*, rather than assigning it generic status. *Eucharidium* had been treated as a distinct genus because the two species within it (*C. breweri* and *C. concinna*) have several unique morphological characters

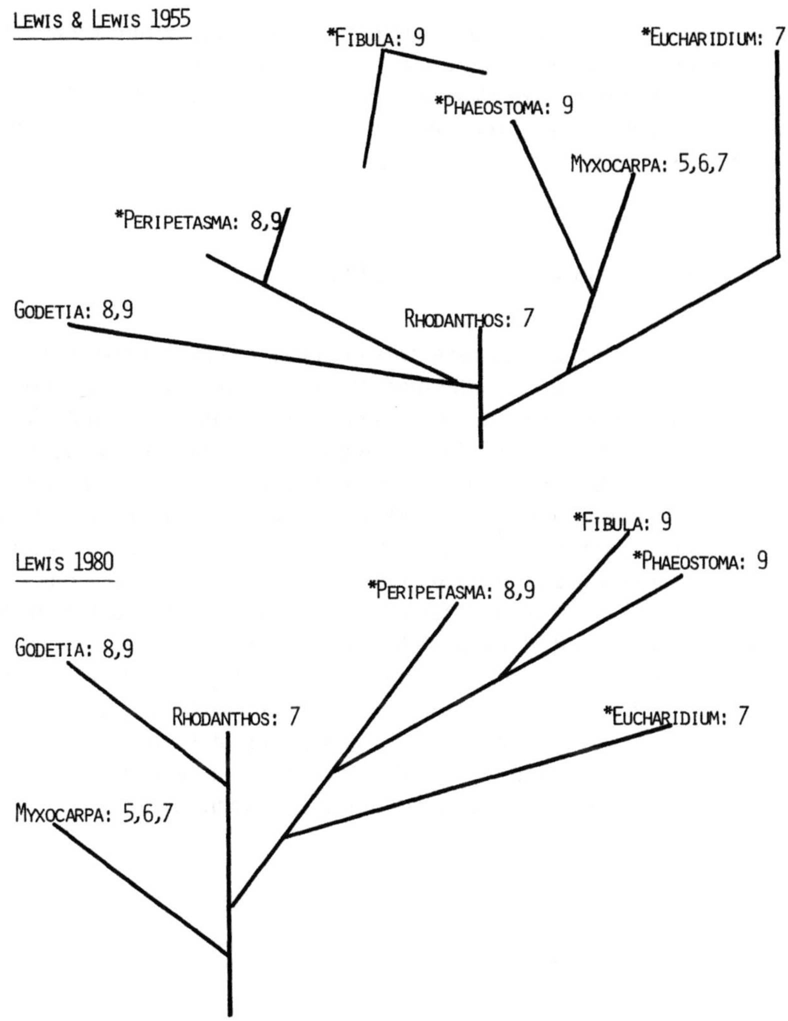

Fig. 3. Comparison of phylogenetic models of the seven sections of *Clarkia* that contain diploid species, modified from Lewis and Lewis (1955) and Lewis (1980). *Asterisks* indicate sections that have the duplicated cytosolic PGI isozymes. *Numbers* indicate the gametic chromosome numbers of species within each section

including four rather than eight stamens, and an extremely long floral tube. This unusual floral morphology appears to be associated with a distinctive pollinator system. Thus, unlike most outcrossed species of *Clarkia* which are pollinated by bees, *C. concinna* is pollinated by butterflies and long-tongued crytid flies, and *C. breweri* by moths as well (MacSwain et al. 1973). Lewis and Lewis (1955) retained *Eucharidium* within *Clarkia* because they believed it to be "almost certainly linked to section *Myxocarpa*" (p. 359) via the tetraploid *C. pulchella*. However, recent scanning electron microscopy of the surface of seeds of *Eucharidium* revealed that it does not resemble any other *Clarkia* species (Raven, personal communication), thus weakening the putative relationship of the section to other Clarkias. However, the biochemical evidence indicates that *Eucharidium* does belong within *Clarkia* because it possesses the PGI gene duplication (Gottlieb and Weeden 1979). Thus its morphological divergence is assumed to be a recent change.

5 Triose Phosphate Isomerase (TPI)

Most diploid plant species have two isozymes of triose phosphate isomerase (TPI; EC 5.3.1.1), one located in the plastids, and the other in the cytosol. TPI catalyzes the reversible isomerization of glyceraldehyde-3-phosphate and dihydroxyacetone phosphate, an essential reaction in glycolysis, gluconeogenesis, and photosynthetic carbon dioxide fixation. The only published genetic analysis to date was based on *Stephanomeria exigua* (Compositae) which revealed that the two isozymes are specified by independent nuclear genes (Gallez and Gottlieb 1982). Both isozymes are dimeric molecules, and heterozygotes at each locus exhibit three allozymes after electrophoresis. Hybrid enzymes are not produced between the monomers of the isozymes from the different subcellular compartments.

In contrast to this typical situation, all diploid species of *Clarkia* have true-breeding, multiple-banded electrophoretic patterns for both the plastid and the cytosolic isozymes (Fig. 4), that result from duplications of both structural genes (Gottlieb 1982, Pichersky and Gottlieb unpublished). Intra- and inter-locus hybrid enzymes are produced within each compartment, but not between them.

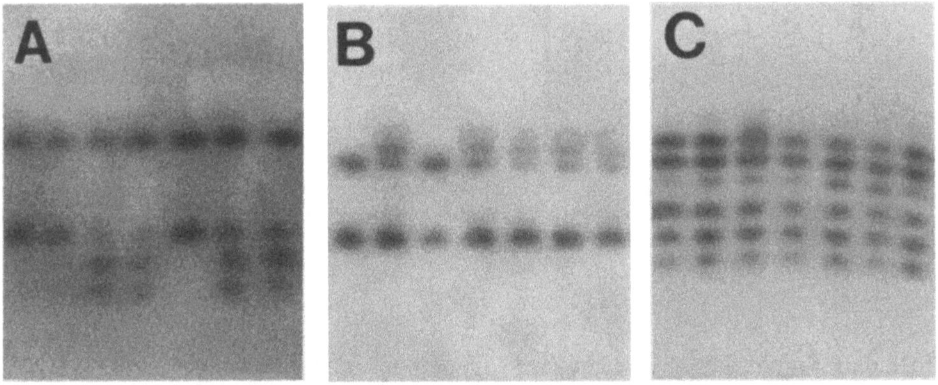

Fig. 4. Electrophoretic patterns of plastid and cytosolic triose phosphate isomerases in *Stephanomeria exigua* (**A** and **B**) and *Clarkia rostrata* (**C**) (revealed by activity stain). In each species, the more anodal isozyme(s) (*top*) are located in the plastids and the less anodal one(s) in the cytosol. *Stephanomeria exigua* has a single locus coding the plastid isozyme and a different single one specifying the cytosolic isozyme. (*Clarkia* species have two gene loci specifying isozymes in each compartment. **A** Segregation at the locus coding the cytosolic TPI. **B** Segregation at the locus coding the plastid TPI. **C** These multiple banded patterns are true-breeding because each individual has two pairs of duplicate gene loci coding them

In *Clarkia* species, the more anodal isozymes are found in the plastids and the less anodal ones in the cytosol. The subcellular locations were determined by cell fractionation tests as well as by a simple and rapid procedure we developed following our observation that only the cytosolic isozyme of many cytosolic-plastid isozyme pairs in plants is released from the pollen by soaking in an appropriate buffer for a short time period (Weeden and Gottlieb 1980b). The plastid TPIs are also present in pollen and can be demonstrated by crushing or other disruptive procedures. This identification test makes it possible to distinguish the plastid and cytosolic isozymes by simply comparing the electrophoretic patterns of extracts from leaf and soaked pollen, since the enzymes which are present in the pollen leachate are the cytosolic ones, whereas the enzymes which are absent are from the plastid.

The discovery of the two pairs of duplicated TPI gene loci in *Clarkia* will provide a second test of the value of gene duplications in plant systematics, this time at the generic level. The Onagraceae is a clearly demarcated family and a substantial amount of information is available from past studies of its systematics, morphology, and chromosomes (Raven 1979). The family includes 17 genera grouped into 7 tribes, 5 of which are monogeneric, one with two genera, and one, Onagreae, with ten genera. Jussiaeeae *(Ludwigia)* appears to be distinct from all the other tribes, and Fuchsieae *(Fuchsia)* seems the most generalized of the other members of the family (Raven 1979).

Discussion about phylogenetic relationships centers on the largest tribe Onagreae. Within the tribe, the presumed relationships between *Clarkia* (with the monotypic *Heterogaura*) and the other genera has recently been questioned on the basis of the discovery that certain unusual anatomical characters link *Clarkia* with the tribe Epilobieae (Raven 1979). The situation seems appropriate for the application of a biochemical character such as the TPI duplications. Consequently, we are presently surveying species representing each genus of the family to determine whether they possess either duplication.

Formal genetic studies have already been completed in at least one species of each of the diploid sections of *Clarkia* (Pichersky and Gottlieb 1983). The tests have revealed that the two plastid TPI coding gene loci in *C. breweri* and *C. rostrata* assort independently. Appropriate genetic variants are not yet available to test the linkage of the loci coding the cytosolic isozymes.

Since it will not be feasible to undertake genetic analyses in each genus of the family, the following criteria will be satisfied before claiming that one or the other duplication is present in a species of another genus. The plastid and cytosolic TPI isozymes will first be distinguished by the pollen/leaf electrophoretic comparison (Weeden and Gottlieb 1980b). Then the electrophoretic pattern of enzymes from each subcellular compartment will be examined for evidence of gene duplication.

In general two types of electrophoretic patterns will be observed for the plastid and the cytosolic isozymes, either a single enzyme band or three (or more) bands of activity. If the former is observed, a single coding locus is assumed. This may be incorrect if the duplicate genes code monomers that associate to form enzymes that overlap on the gels after electrophoresis, or if a "null" mutation has silenced one of the duplicated genes. If a three-banded pattern is observed, either it results from heterozygosity at a single gene locus or from a duplication, with each locus specifying monomers of a different homodimer, and the enzyme of intermediate mobility being an inter-locus heterodimer. The problem is to ascertain which of these possibilities is correct.

The analysis will be greatly simplified by another application of the pollen/leaf electrophoretic pattern comparison (Weeden and Gottlieb 1979). Both cytosolic and plastid TPIs are expressed in pollen, but since the pollen of diploid species is haploid, each pollen grain receives only one allele of each coding locus, and therefore can produce only a single monomer. Thus a plant which is heterozygous at a single locus will have pollen grains that contain one or the other monomer, but not both so that intra-locus hybrid enzymes cannot be formed in the pollen. However, in plants which possess duplicate loci, the pollen receives an allele of each locus, and therefore produces two different monomers which by pairwise association yield three isozymes: the two intra-locus homodimers and the interlocus heterodimer. Therefore, side-by-side comparison of zymograms of pollen and leaf extracts reveals if a three-banded pattern from the leaf tissue is retained in the pollen; if only the two homodimers are found, the pattern is specified by a heterozygous locus; if the pollen shows the three bands, then duplicate loci are present (Fig. 5). The test provides a rapid means of identifying the genetic basis of a multiple-banded (phenotypically heterozygous) isozyme pattern (Weeden and Gottlieb 1979).

The genetic basis for multiple isozymes can also be discerned by examination of selfed progeny of single individuals. If an individual has a three-banded pattern, but its progeny segregate genetically, then duplicate genes are not present, whereas if the three-banded pattern is true-breeding, then this is presumed to reflect the duplication.

Finally, observation of five or more enzyme bands for either subcellular compartment in an individual would constitute evidence for duplicate gene loci with at least

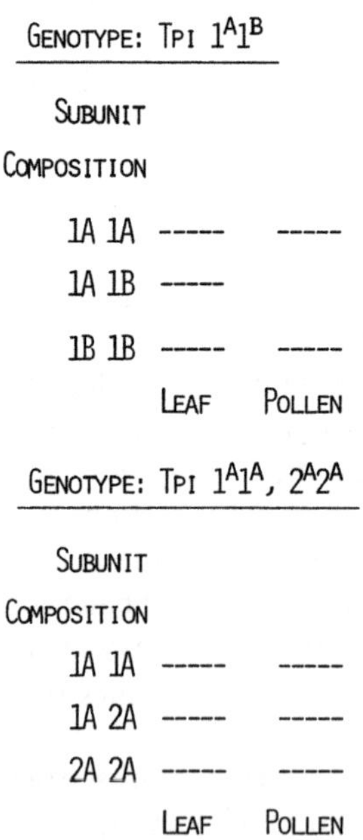

Fig. 5. Diagram to show how comparison of electrophoretic pattern of leaf and pollen extracts can determine if a three-banded TPI pattern results from genetic heterozygosity at a single gene locus (*top*) or from the presence of duplicate gene loci (*bottom*)

one of them heterozygous. Thus, even without extensive formal genetic breeding tests, the presence of duplicate genes can be determined.

Using these criteria, the duplication of the gene specifying the cytosolic TPIs has been identified in diploid species of four of the seven tribes in the family including: Jussiaeeae *(Ludwigia)*, Fuchsieae *(Fuchsia)*, Onagreae (*Clarkia, Heterogaura, Camissonia, Oenothera,* and *Gongylocarpus*), and Epilobieae *(Boisduvalia)*. Its presence in the phylogenetically primitive *Fuchsia* and in *Ludwigia*, considered a distinct branch of the family (Raven 1979), suggests an ancient origin for the cytosolic TPI duplication.

The duplication has not yet been observed in the monotypic Circaeeae *(Circaea)* or in the genus *Stenosiphon* of the Onagreae tribe. Assuming that further study fails to identify the duplication, its absence is best interpreted as a loss caused by mutation(s) either in the particular species examined or in their progenitors. This rationale follows from the fact that the duplication is known to be present in both *Fuchsia* and *Ludwigia* which represent the two most ancient lineages in the family.

Once a duplication has been eliminated by mutation, it has a very low probability of being "reinstated". This leads to the hypothesis that if a particular duplication has been established in a lineage, but lost in a derived taxon, it is unlikely to reappear in a descendent of the derived taxon. Thus certain conceivable phylogenetic lineages involving various genera can be ruled out, which may provide a valuable criterion to assess relationships in the family.

In contrast to the wide distribution of the cytosolic TPI duplication, the duplication of the gene coding plastid TPIs has not yet been demonstrated outside of *Clarkia*, although numerous genera have been examined. Its absence in other genera of the family suggests it arose more recently than the cytosolic TPI duplication.

6 Genetic Evidence for Ploidy Levels

The number of gene loci coding isozymes of particular enzymes is also relevant to other questions in plant systematics. One question is the ascertainment of ploidy level in species having gametic chromosome numbers between n = 9 and n = 12, which could be either diploid or tetraploid. Ploidy level has usually been determined primarily on the basis of chromosome number, but in many cases this can be ambiguous.

For example, sharp differences characterize the interpretation of chromosome numbers in the Astereae tribe of the Compositae. The most common numbers vary between n = 2 and n = 9, with many species having n = 4 or n = 5. Two hypotheses have been formulated: (1) n = 9 was the original base chromosome number of the group and that the lower numbers evolved by aneuploid reduction (Raven et al. 1960); (2) the ancestral base number was n = 4 or n = 5, therefore species with n = 9 were allotetraploids derived by hybridization between taxa with lower numbers (Turner et al. 1961, Turner and Horne 1964). Both proposals claimed support from various observations that were at best relevant correlations. Thus the association of n = 9 with the primitive woody habit, the high symmetry of the n = 9 karyotype, and the widespread phylogenetic occurrence of this gametic number were claimed as evidence favoring the first hypothesis. The rarity of species with n = 6 and n = 7, was considered to support the second proposal.

The essential attribute of polyploidy, however, is not relative chromosome number, or correlation with these other factors, but rather it is genome multiplication and its attendent increases in number of gene loci (Gottlieb 1981b). Therefore it was important to determine if species with n = 9 had more gene loci coding isozymes detectable by electrophoresis than those with n = 4 or n = 5. The test followed from previous findings that allopolyploid species invariably display more isozymes than diploids because they inherit homoeologous gene loci from their diploid parents that are frequently fixed or become so for alleles that specify enzymes having different electrophoretic mobilities (Hart 1979, Roose and Gottlieb 1976, Gottlieb 1981a). In addition, aneuploid decrease in chromosome number from an ancestral diploid condition does not change the number of structural genes coding isozymes (Roose and Gottlieb 1978, Crawford and Smith 1982).

The electrophoretic test was performed using 17 enzymes extracted from 5 species of *Machaeranthera,* with gametic chromosome numbers of n = 4, 5, and 9, and two species of *Aster* with n = 5 and n = 9. The number of isozymes of each of the enzymes was the same in all of the species and there was no evidence of isozyme multiplicity (Gottlieb 1981b). The isozymes appeared to be specified by a minimum of 26 gene loci. The result makes it most unlikely that the two species with n = 9 originated by allotetraploidy. The observed constancy of isozyme (gene) number is not unexpected if the species with lower chromosome numbers represent lineages that arose by aneuploid reduction. This process is generally conceived to involve translocation of essential euchromatin and loss of only heterochromatin and centromeres. Thus the electrophoretic test of isozyme number provided a direct and simple analysis, and one which seems appropriate to determine ploidy level in many other species when this *genetic* attribute is uncertain.

7 Conclusions

The phylogenetic studies reviewed here make use of increases in the number of isozymes in diploid plants brought about by duplications of coding gene loci, rather than genetic identity estimates based on presence/absence and frequencies of alleles. The latter evidence is subject to uncertainties associated with historical factors and the uneven accumulation of genetic divergence. In contrast, increases in the number of gene loci, although infrequent, appear to take place sufficiently often in plants that a priori arguments can be proposed to account for their origin and establishment. When caused by certain types of gene duplication, they have a high probability of being unique and can be used to group diverse taxa into a monophyletic assemblage descended from the single common ancestor in which the duplication arose. Thus attention to changes in isozyme number facilitates the application of electrophoretic evidence above the species level, whereas this is usually not appropriate with data derived from allelic divergence. Information on isozyme number is also useful in another context, such as the determination of ploidy independently of chromosome number, and has been valuable in solving a long-standing problem involving the basic number in the Astereae. Many other applications of isozyme number will probably be developed and the character deserves increasing attention in plant systematics.

References

Brown AHD (1979) Theor Popul Biol 15:1–42
Burnham C (1962) Discussions in cytogenetics. Burgess, Minneapolis
Crawford DJ, Smith EB (1982) Evolution 36:379–386
Gallez GP, Gottlieb LD (1982) Evolution 36:1158–1167
Gottlieb LD (1973a) Evolution 27:205–214
Gottlieb LD (1973b) Am J Bot 60:545–553
Gottlieb LD (1974) Evolution 28:244–250
Gottlieb LD (1977) Genetics 86:289–307
Gottlieb LD (1981a) Prog Phytochem 7:1–46
Gottlieb LD (1981b) Proc Natl Acad Sci USA 78:3726–3729
Gottlieb LD (1982) Science 216:373–380
Gottlieb LD, Greve CL (1981) Biochem Genet 19:155–172
Gottlieb LD, Pilz G (1976) Syst Bot 1:181–187
Gottlieb LD, Weeden NF (1979) Evolution 33:1024–1039
Hart GE (1979) Stadler Symp 11:9–30
Hart GE, Langston PJ (1977) Heredity 39:263–277
Lewis H (1980) The mode of evolution in *Clarkia* (Onagraceae). Paper presented at ICSEEB II, Vancouver, BC
Lewis H, Lewis ME (1955) Univ Calif Publ Bot 20:241–392
MacSwain JW, Raven PH, Thorp RW (1973) Univ Calif Publ Entomol 70:1–80
Perkins DD (1972) Genetics 71:25–51
Pichersky E, Gottlieb LD (1983) Genetics 105:in press
Raven PH (1979) N Z J Bot 17:575–593
Raven PH, Solbrig OT, Kyhos DW, Snow R (1960) Am J Bot 47:124–131
Roose ML, Gottlieb LD (1976) Evolution 30:818–830
Roose ML, Gottlieb LD (1978) Heredity 40:159–163
Roose ML, Gottlieb LD (1980) Biochem Genet 18:1065–1085
Turner BL, Horne D (1964) Brittonia 16:316–331
Turner BL, Ellison WL, King RM (1961) Am J Bot 48:216–223
Weeden NF (1981) J Mol Evol 17:133–139
Weeden NF, Gottlieb LD (1979) Biochem Genet 17:287–296
Weeden NF, Gottlieb LD (1980a) J Hered 71:392–396
Weeden NF, Gottlieb LD (1980b) Plant Physiol 66:400–403

Enzyme Profiles in the Genus *Capsella*

H. HURKA [1]

Abstract. Along a transect from Northern to Southern Europe GOT banding patterns within and between natural populations of the Brassicaceae genus *Capsella* have been investigated. Evidence is produced that some enzyme banding patterns display geographic variation while others vary randomly. Although species-specific banding patterns do exist, they cannot serve as discriminating characters because these patterns are also subject to variation within species. The banding patterns can be attributed to four GOT loci, which form a number of alleles and produce intra- and inter-locus enzyme bands. A gene duplication is likely to occur in populations from different parts of Europe.

1 Introduction

The genus *Capsella* is a member of the Brassicaceae and comprises several species. However, due to the enormous polymorphism detected within this genus, no clearcut species concept exists. The number of species recognized by taxonomists varies between three and ten. For our purposes we have adopted a broad concept and designated only three species. In general, these species are annuals and can easily be grown in glasshouses and experimental field plots. *Capsella*, therefore, has proven to be an ideal test plant for evolutionary biology research. This genus displays different ploidy levels, different breeding systems and different distribution patterns (Table 1). For several reasons, the well-known cosmopolitan annual weed *Capsella bursa-pastoris* is particularly interesting. As already stressed by Bradshaw (1972), critical examination of localized

Table 1. Ploidy levels, breeding systems and geographical distribution in the genus *Capsella*

Species	Ploidy level	Breeding system	Geographical distribution
C. rubella	2 n	Reported to be predominantly outcrossing	Mediterranean climates
C. grandiflora	2 n	Selfincompatibility	Only W-Greece and rarely in Northern Italy
C. bursa-pastoris	4 n	Predominantly selfing	World-wide except the hot tropics

[1] Fachbereich Biologie/Chemie der Universität Osnabrück, Postfach 4469, 4500 Osnabrück FRG

Proteins and Nucleic Acids in Plant Systematics
ed. by U. Jensen and D.E. Fairbrothers

differentiation for annual weeds is needed, and according to Nevo (1978) no cosmopolitan plant species has been analyzed satisfactorily.

An important problem for evolutionary biology, population ecology, and biosystematics is the origin, maintenance, and regulation of variation in natural populations. Research which includes the above phenomena eventually must determine the variability in natural populations, and must differentiate between phenotypic and genotypic components. In the last 15 years, electrophoresis of enzymes has been intensively used by population geneticists and has become equally important for systematists and evolutionary biologists. The main reasons for this rapid incorporation of isozyme data and their value as a taxonomic tool have been critically discussed by Lewontin (1973, 1974), Powell (1975), Ayala (1976), Nevo (1978), Hurka (1980), and Gottlieb (1982). However, it should be stressed that isozyme assays for systematic purposes are still being developed. From our *Capsella* studies and from a review of the literature, we have determined that it is unsound to draw far-reaching conclusions unless extensive efforts have been undertaken to detect the geographic variability in the enzyme profiles.

2 Enzyme Profiles in the Genus Capsella

2.1 Experimental Design

Along a transect from Northern Europe to Italy and Greece, we randomly collected individual seed samples from 25–30 specimens from each natural population of *Capsella*. Population size and habitat data were also recorded. The progenies of these seeds were grown under controlled climatic conditions in the glasshouse at Münster. The rosette leaves of equal age were analyzed for a variety of enzyme systems after their ontogenetic stability had been established.

For the purpose of estimating genetic variability within populations one offspring per collected plant was analyzed. In another set of experiments, a number of progeny per mother plant were grown thus providing well documented genetical profiles. Crossing experiments are still in progress.

Electrophoretic bands were characterized by Rf-values and scored by band numbers. All data are stored in a computer, and it is possible to obtain information retrieval about the frequency of each band, complete banding patterns or band combinations for each population or populations.

In the following discussions, I will include only examples from the GOT enzyme system. However, the following enzyme systems also have been analyzed: LAP, GDH, EST, LDH.

2.2 GOT Banding Patterns

A total of 22 GOT bands have been detected for *Capsella*, 17 for *Capsella bursa-pastoris,* also 17 for *C. grandiflora*, and 19 for *C. rubella*. Six of the twenty-two bands have a frequency lower than 1.0% and six higher than 10%. Although there are some interesting differences in the occurrence of single bands, the banding patterns should be considered first. The different bands combine to form different banding patterns. Each pattern represents a special genotype, and 214 GOT patterns have been manifested in *Capsella*. Of these 214, some patterns are observed in all three species; however, some are unique within each species. Most GOT genotypes occur with very low frequencies, and only 11 of the 214 patterns display an overall frequency higher than 1.0% (Sect. 2.5). In *Capsella bursa-pastoris*, the 80 GOT banding patterns which are unique for this species, were observed in 566 particular plants of the 2,172 analyzed. This corresponds to 26% (Tables 2 and 3). In the other two species, 40%–45% of the analyzed individuals displayed species-specific patterns (Tables 2 and 3). With regard to the number of GOT patterns, *Capsella grandiflora* is more variable than the other two species.

2.3 Variability within and between Populations

Monomorphic (only one pattern) and polymorphic (more than one pattern) samples were detected for GOT banding patterns per population. The index of heterogeneity in *Capsella bursa-pastoris* (patterns/number of individuals) varies up to 50%, but three quarters of the analyzed 87 populations fall into a range greater than 0%–25%, including the 14 (= 17%) monomorphic samples. These rather uniform samples come from Scandinavia and Switzerland. Homomorphic samples have also been detected in *Capsella rubella* (7 of 26 = 26%), but not in *Capsella grandiflora*.

The large number of banding patterns poses a question about recognizable geographic distribution patterns. One of the first ways to detect geographic variation is to plot the occurrence of single bands on a map. Plotting the GOT band 20 (Fig. 1), which has a frequency of 18% in *Capsella bursa-pastoris*, reveals that this band is not evenly distributed within the sampling area, but concentrated in the atlantic climatic region. However, the GOT band also occurs in the species *Capsella grandiflora* and *C. rubella*.

Table 2. GOT banding patterns in *Capsella*

Species	Population	Individuals	GOT-Patterns	Indiv./Patterns
C. bursa-pastoris	87	2,172	98 (80) [a]	22
C. grandiflora	16	715	80 (58) [a]	9
C. rubella	26	996	47 (28) [a]	21

[a] Patterns occurring only in this species

Table 3. Individuals with GOT patterns occurring only in the particular species

Capsella bursa-pastoris	566 out of 2,172 = 26%
Capsella grandiflora	313 out of 715 = 44%
Capsella rubella	432 out of 996 = 43%

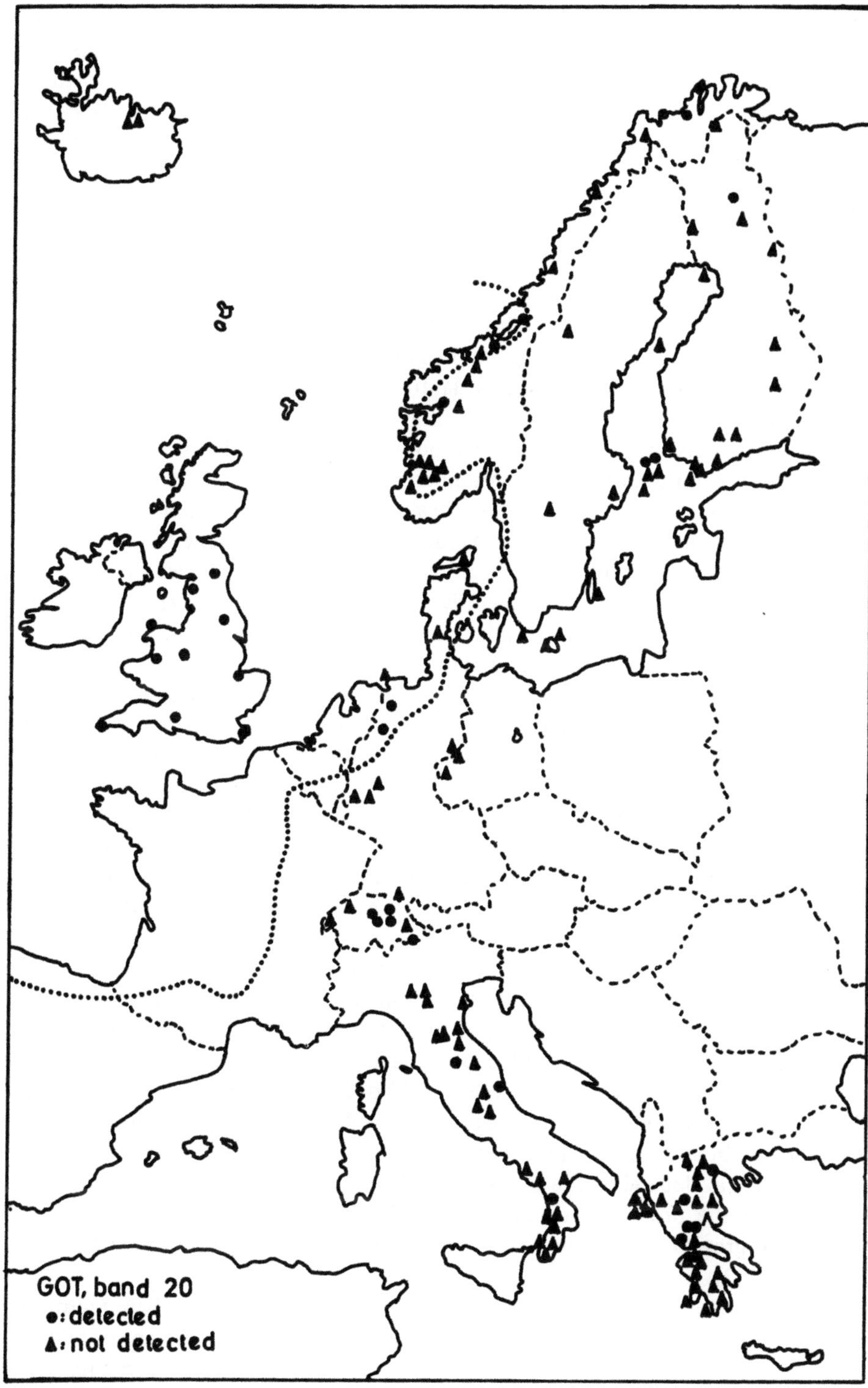

Fig. 1. Geographic distribution of the GOT band 20 within *Capsella* species

Another evaluation is possible if distinct banding patterns are plotted. This has been done for the genotype characterized by the bands 11 + 15. This pattern is mainly distributed in the southern parts of Scandinavia, i.e. south of the Limes Norrlandicus, but rarely outside this area (Fig. 2). A more complicated pattern is represented by the band combination 6 + 8 + 11 + 15. It is typical for Scandinavia, but often lacking outside this region (Fig. 3). This latter example clearly points to the existence of geographic variation. In contrast to this kind of distribution many other GOT bands and banding patterns, e.g. the joint occurrence of the bands 11 + 15 + 17, clearly display a patchy mosaic distribution (Fig. 4).

These data indicate that some characters in *Capsella* follow a clinal variation pattern, while others vary randomly. This holds true not only for isozyme patterns but also for the following characters: plant height, seed weight, leaf morphology, and indumentum, germination behavior, and flavenoids. These published and unpublished data (Hurka and Wöhrmann 1977, Hurka and Benneweg 1979), indicate the existence of a great variability within and between populations. In all these comparisons the interference of continuous with discontinuous mosaic variation was detected.

It is known that the analysis of gene flow events is of great importance in understanding the genetic structure and dynamics of populations. Gene flow within and between natural populations of *Capsella bursa-pastoris* is mainly restricted to seed dispersal because this plant is predominantly self-pollinated (Hurka et al. 1976). Therefore, it was essential to intensively study seed shedding, seed dispersal, and the formation of the soil seed bank (Hurka and Haase 1982). The establishment of the clinal variation patterns of *Capsella* can be understood by having knowledge of seed dispersal events. Most seeds fall to ground nearest to the parent plant, and are incorporated into the soil seed bank. Therefore, populations are mostly recruited from autochthonous seed material. Thus, selection can operate over years on the same gene pool, and a clinal variation pattern may eventually evolve.

Several reasons account for the existence of the discontinuous mosaic pattern: low gene flow between and within populations, properties of the soil seed bank and the founder principle. Long distance dispersal in *Capsella*, which is facilitated by the mucilaginous seed coat, may result in natural populations arising from the accidental transport of a small number of seeds to hitherto unoccupied sites. Therefore, such populations are often genetically rather uniform (e.g. the populations monomorphic for some characters). If different populations were established by small groups of founders drawn from a polymorphic parental population, then variation between these founder populations is to be expected. This interpretation of the origin of genetic divergence between populations can be applied to some populations of *Capsella bursa-pastoris*. In addition to these ideas, it has been shown that population variability in *Capsella* may depend on the alteration of the biotope (Bosbach and Hurka 1981). Knowledge of the properties of the soil seed bank, plus knowledge of germination behavior and growth rates, adds to the understanding of this variation pattern (Bosbach et al. 1982).

Apart from these arguments, historical factors are also involved. These factors are especially important for systematic evaluations. When genetic distances between species or populations are discussed, one has to consider ecological versus historical explanations. This implication points to the urgency of thorough biosystematic studies

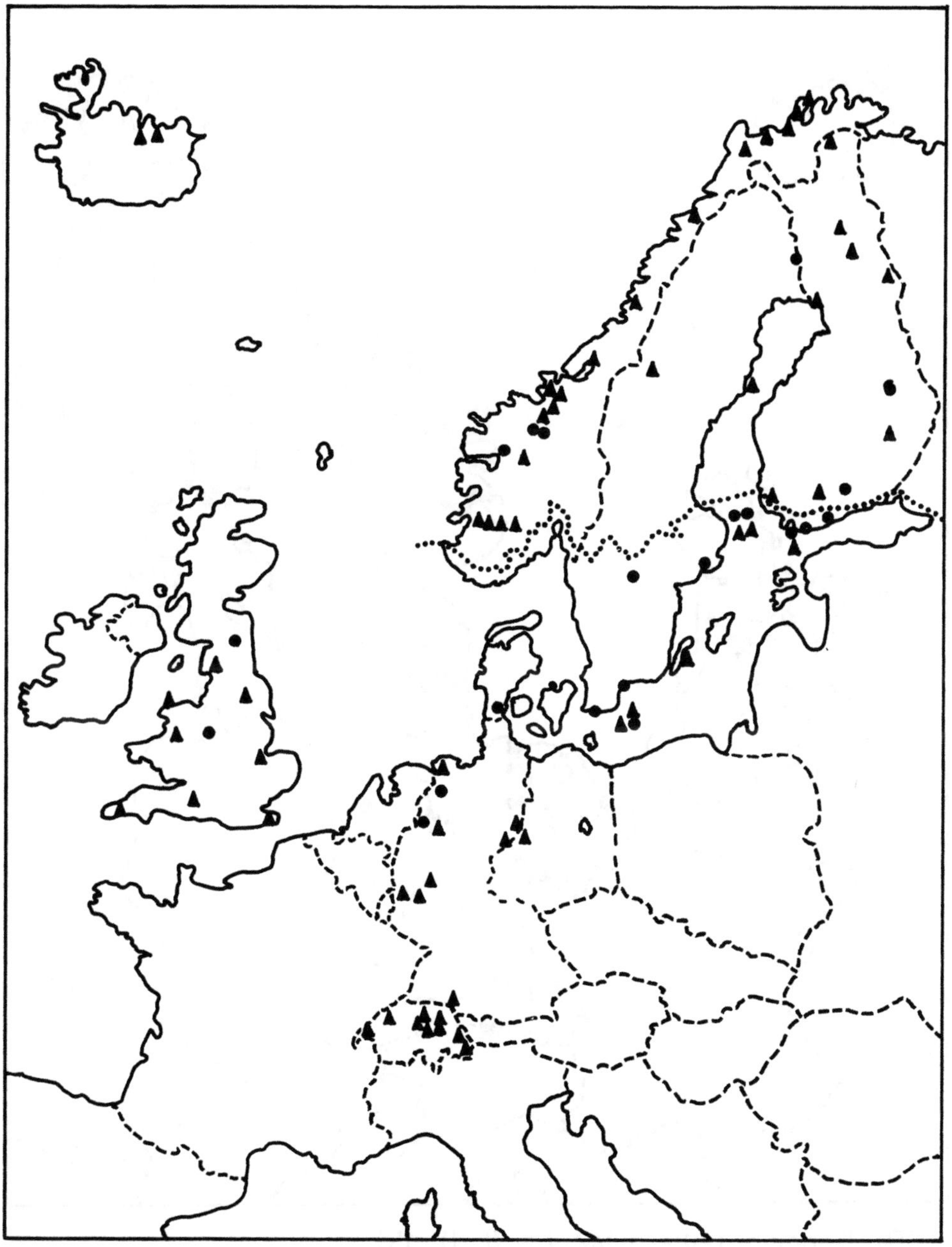

Fig. 2. Geographic distribution of the GOT banding pattern 42 within *Capsella* species, characterized by the bands 11 + 15. ... Limes Norrlandicus

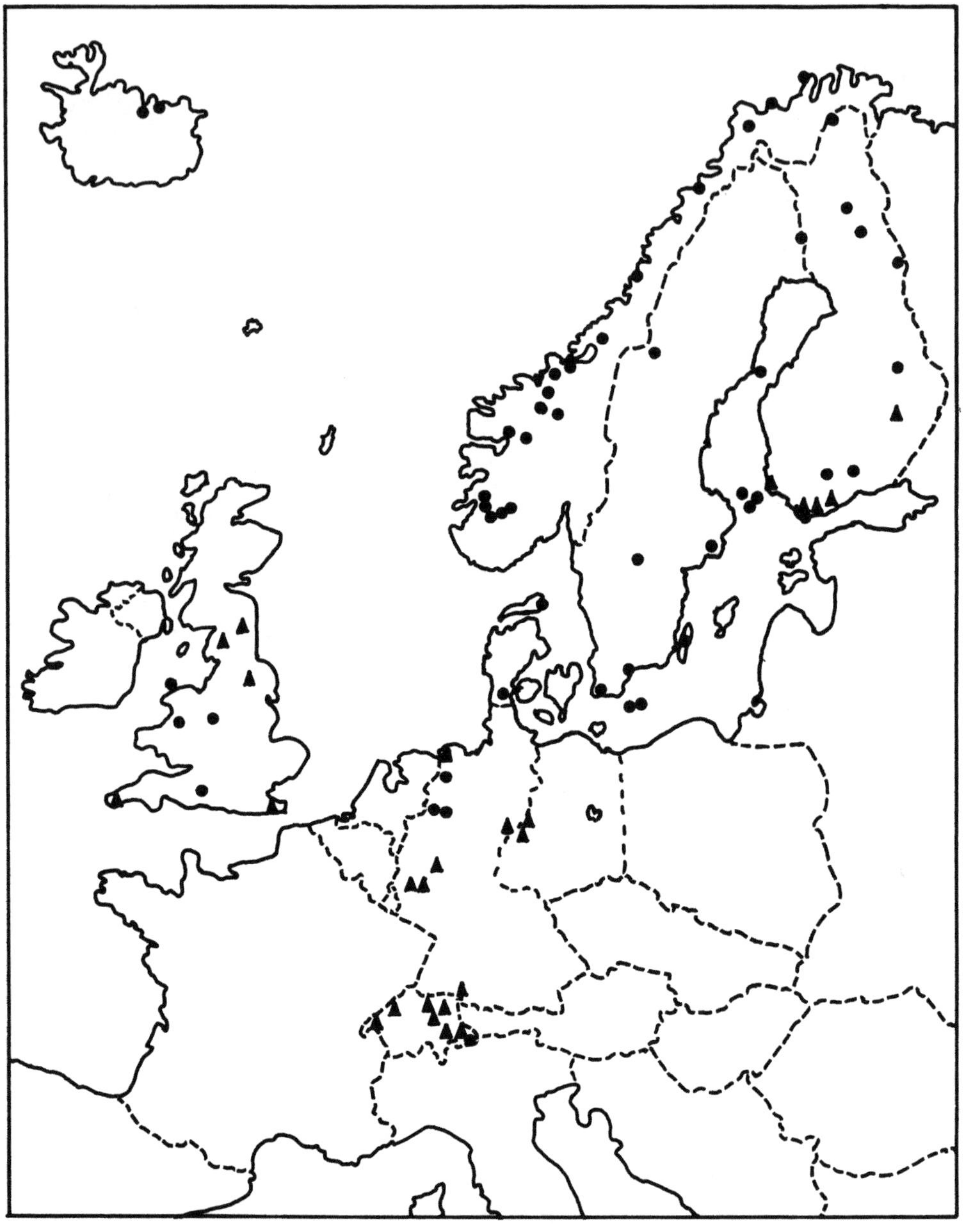

Fig. 3. Geographic distribution of the GOT banding pattern 142 within *Capsella* species, characterized by the bands 6 + 8 + 11 + 15

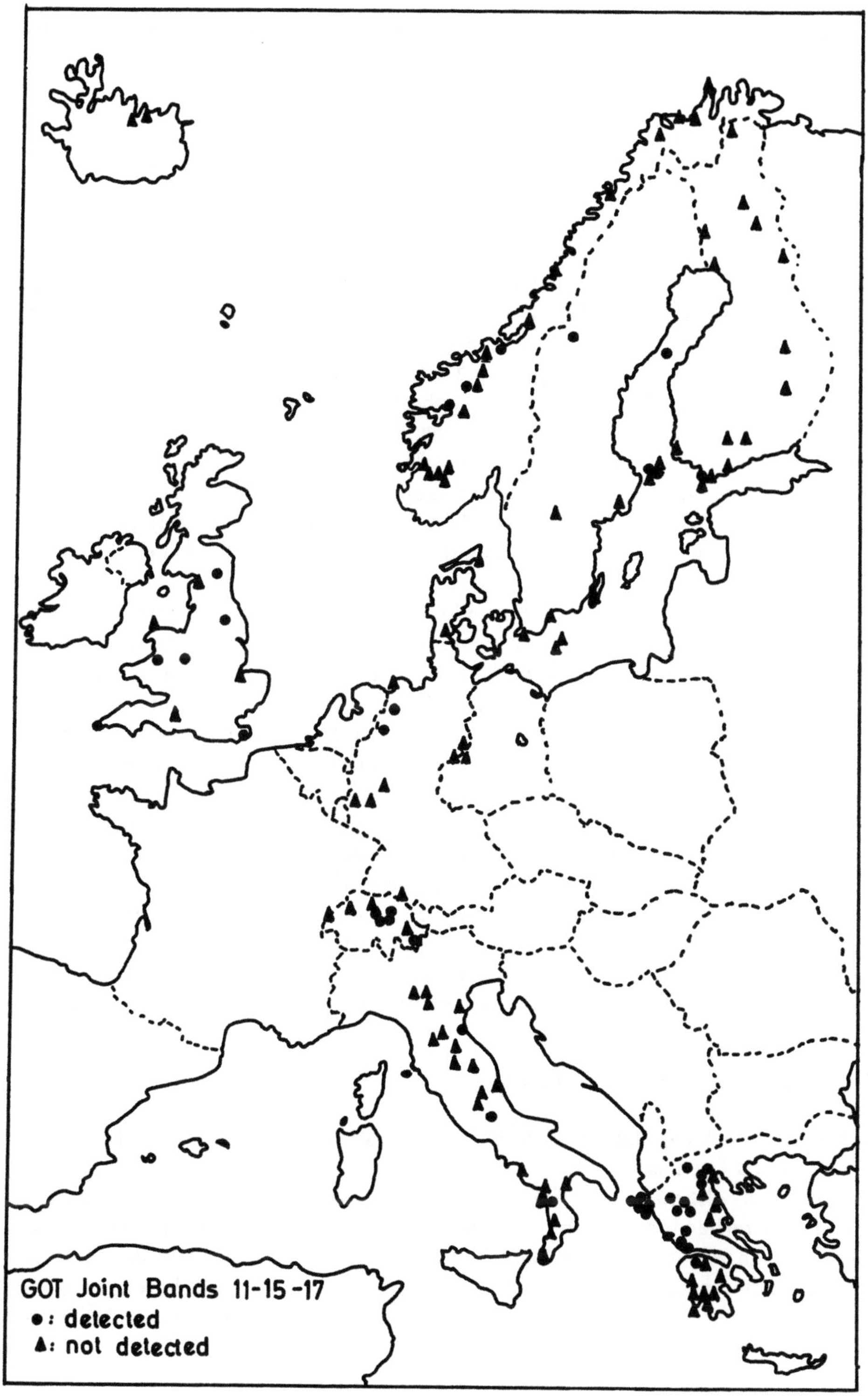

Fig. 4. Geographic distribution of the joint occurrence of GOT bands 11 + 15 + 17 in *Capsella* species

throughout the species area before statements concerning inter-specific differences can be presented.

2.4 Influence of the Breeding System

Average within-population variability in the self-incompatible *Capsella grandiflora* is higher than in *Capsella bursa-pastoris*, which is to be expected. However, this picture has to be modified for two reasons. (1) Different enzyme systems display different variation patterns. In LAP the differences between the three species are not so obvious as in GOT, despite the fact that the total number of different enzyme bands in LAP and GOT are nearly the same (25 and 22). (2) In *Capsella bursa-pastoris* and *C. rubella* there are both highly polymorphic and less polymorphic populations. These two species often display the same amount of polymorphism as detected in *Capsella grandiflora*. If, by chance or deliberation, less variable populations are selected, one can falsely believe that restricted gene flow (self pollination) will lead to a decrease in population variability. On the other hand, it also has been demonstrated (deliberately or unintentionally) that different breeding systems do not necessarily result in different population variability. Therefore, it is essential to analyze an array of natural populations.

Inbreeding does not necessarily decrease the genetic variability of natural populations. There are all transitions from homomorphic to highly heteromorphic populations up to that which is estimated for the strictly outcrossing *Capsella grandiflora*. However, it should be stressed that these differences are not correlated with actual population sizes. Nevertheless, with regard to the organization of the genetic variability, there are essential differences between the species. Individuals homozygous for allozyme loci are typical for *Capsella bursa-pastoris*, and heterozygous individuals for *C. grandiflora* (Figs. 5 and 6). At this juncture I want to stress that such statements are based on enzyme patterns. The comparative morphological characters are not in accordance with the enzyme data. An unexpectedly high degree of heterozygosity was revealed when measuring the segregation of morphological characters in progeny tests performed with *Capsella bursa-pastoris* (Hurka and Wöhrmann 1977). This indicates that qualitatively different characters can lead to different, even contradictory, statements about genetic variability within populations.

2.5 GOT Profiles in Capsella Species

Is the tetraploid *Capsella bursa-pastoris* more variable than the two diploid species *C. rubella* and *C. grandiflora*?

Concerning the number of GOT bands, there is no clear-cut difference in the number of enzyme bands coding for GOT (Sect. 2.2). The diploid *C. rubella* displays even more, though very rare, bands than the tetraploid *Capsella bursa-pastoris*. However, there are differences in the frequencies of major bands. Some bands have high to very high frequency values in one species, and low to very low frequency values in the others (Table 4).

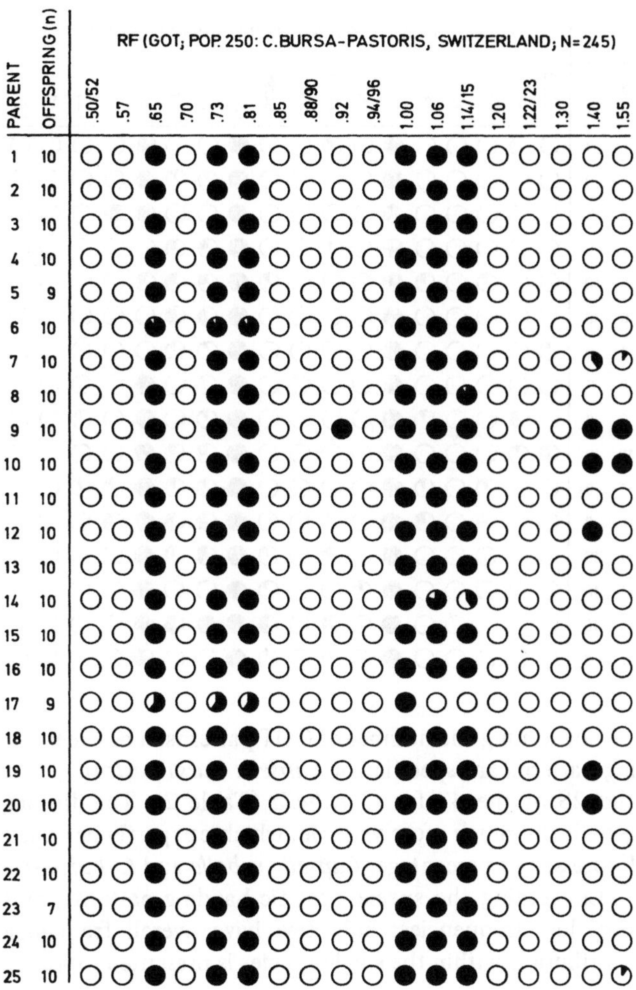

Fig. 5. *Capsella bursa-pastoris*, progeny tests for GOT banding patterns. *Circles* indicate band frequencies (○ = 0%; ● = 100%)

Table 4. Frequencies [%] of selected major GOT bands in *Capsella*

Band No.	C. bursa-pastoris n = 2,172	C. grandiflora n = 715	C. rubella n = 996
6	67	6	5
8	66	11	17
11	84	92	83
15	99	93	22
16	18	27	6
17	23	30	77
22	12	–	37

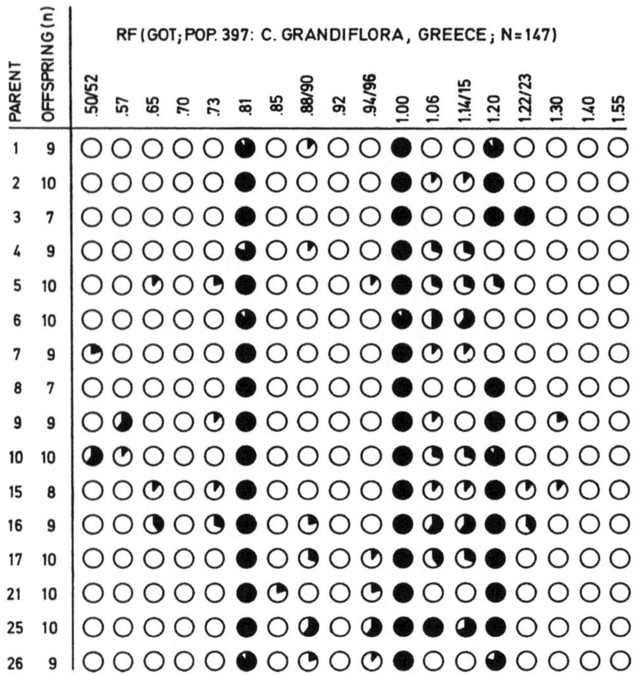

Fig. 6. *Capsella grandiflora*, progeny tests for GOT banding patterns. *Circles* indicate band frequencies (○ = 0%; ● = 100%)

Concerning the number of GOT patterns, it has been shown that 80 of the 98 banding patterns are specific for *Capsella bursa-pastoris*, 58 of the 80 for *C. grandiflora*, and 28 of the 47 for *C. rubella* (Table 2). The diploid *C. grandiflora* seems to be genetically much more variable, which can partly be explained by the breeding system. Expressed in percentages *Capsella rubella* and *C. bursa-pastoris* are equally variable.

Concerning the species-specific banding patterns, most GOT patterns occur with very low frequencies; only few have overall frequencies higher than 3.0%. Their distribution within the single species is presented in Table 5.

Table 5. GOT banding patterns in *Capsella* with frequencies higher than 1%

GOT banding pattern		Total		C. bursa-past.		C. grandifl.		C. rubella	
No.	Bands	n	%	n	%	n	%	n	%
35	(11–17)	395	9	–	–	22	3	373	37
36	(11–17–22)	272	6	–	–	–	–	272	27
42	(11–15)	505	11	325	15	175	25	5	< 1
50	(11–15–18)	129	3	–	–	129	18	–	–
61	(11–15–16–17)	103	2	2	< 1	96	13	5	< 1
110	(8–15)	128	3	–	–	24	3	104	10
142	(6–8–11–15)	721	16	685	32	12	2	24	2
143	(6–8–11–15–22)	87	2	60	3	–	–	17	2
145	(6–8–11–15–20)	64	1	62	3	2	< 1	–	–
163	(6–8–11–15–16–17)	222	5	217	10	3	< 1	2	< 1

It is interesting to note that band frequencies alone are not as valuable as comparing banding patterns for discriminating between species. The following figures provide some helpful examples. The GOT band 17 was mainly detected in southern Europe, that is in *Capsella rubella* and *C. grandiflora*, but it is also found in northern Europe, in *C. bursa-pastoris* (Fig. 7). However, the genotype which is characterized by the two bands 11 and 17 only, could not be detected in *Capsella bursa-pastoris*, but is often found in *C. rubella* (Fig. 8).

3 Conclusions

The study of allozymic variation has been a primary method by which population biologists have studied evolutionary processes during the last fifteen years. This has included the examination of both intra- and inter-specific variation. The results from these studies have had a tremendous impact on both evolutionary theory and thinking. However, as far as systematics is concerned, apparently the second step often has been performed before the first. The use of electrophoretic data to solve taxonomic problems has been mostly concerned with inter-specific variation, with only little being known about the intra-specific variation. This might be a serious pitfall, because one is faced with the problem of genetic differences between two species as compared to the genetic differences between two populations of the same species. Therefore, extensive efforts to uncover geographic variability in the enzyme profile is necessary, and there is strong demand for thorough biosystematic investigations throughout the species area before statements on inter-specific variations based on allozymes can be formulated.

There are also other problems in this connection. The so-called "genetic distance" estimated by isozyme banding patterns may represent only.a fraction of actual genetic differences because of limitations of the electrophoretic method itself (only those alterations of enzymes which influence their electrophoretic mobility are detectable). This has clear implications for the judgement of differences between species.

It is important to recall that the functional and evolutionary significance of electrophoretic variation, and the reasons for its maintenance in natural populations are still being debated and discussed (Le Cam et al. 1972, Kimura and Ohta 1974, Kimura 1977).

Thus it should be stressed that a critical interpretation of isozyme data is essential. In such evaluations the inclusion of enzyme data in systematic studies leads to a better understanding of the problems involved in the following: (1) extent of gene flow and the impact of the breeding system, (2) relationships between population variability and environment, (3) phylogeny and colonization history. In order to achieve these goals, careful analyses of isozyme banding patterns throughout the geographical area of a taxon and a more complete knowledge of the biology of that taxon are required.

Addenda. After the manuscript had been submitted, further investigations in our laboratory made it possible to separate allozymes and to detect single GOT loci. As far as the band numbers mentioned in the text are concerned, a brief account of these

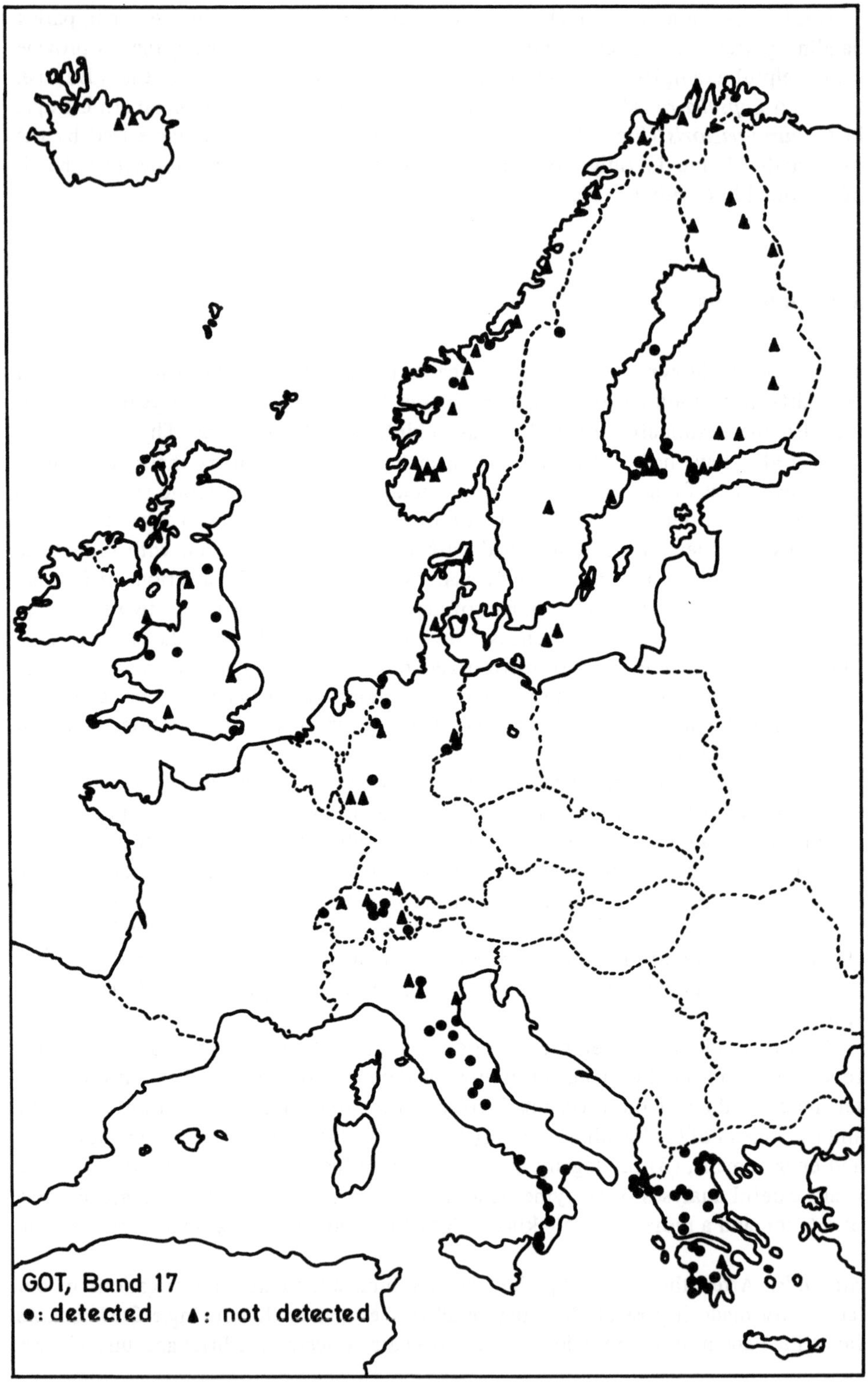

Fig. 7. Geographic distribution of the GOT band 17 within *Capsella* species

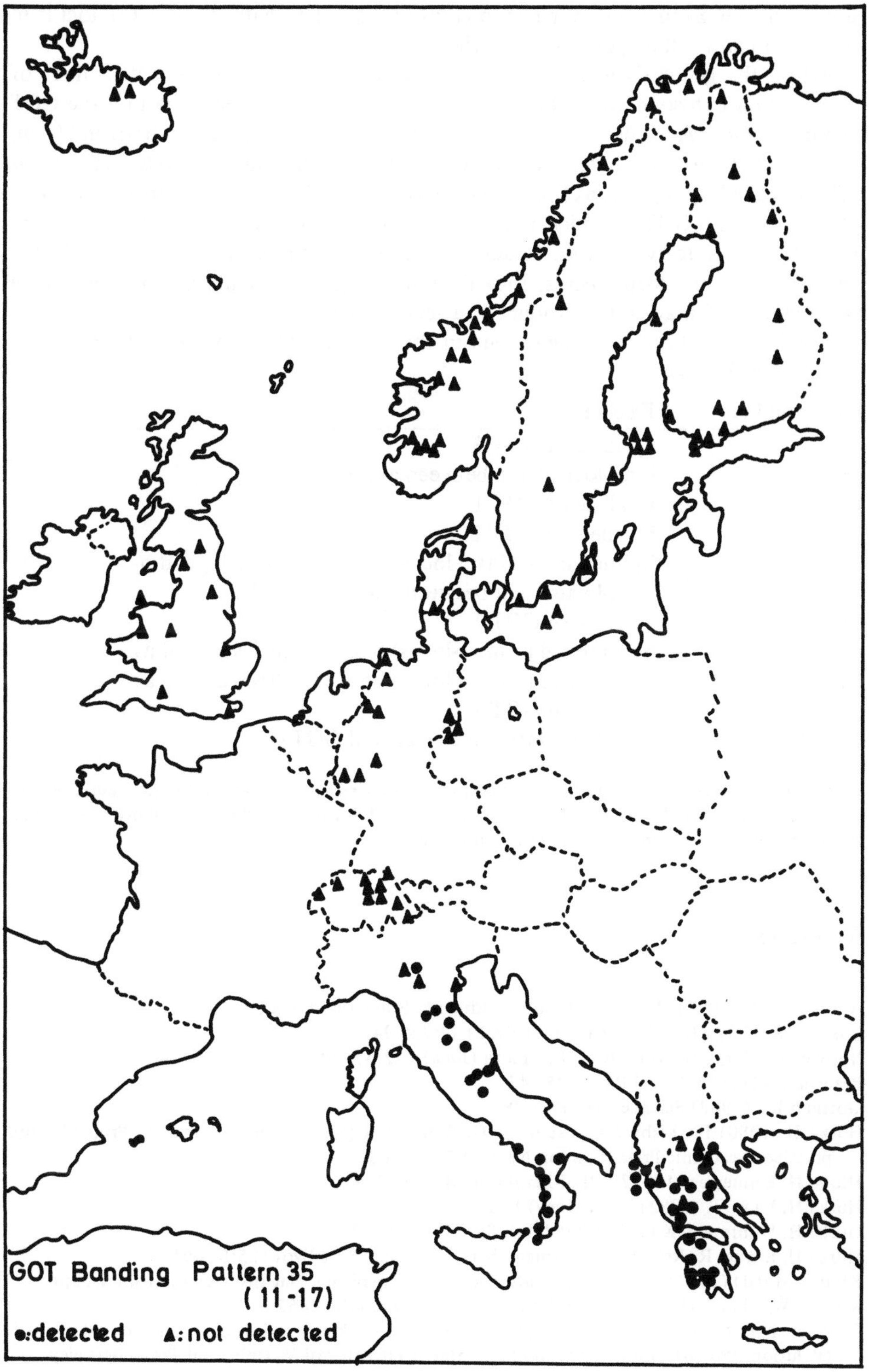

Fig. 8. Geographic distribution of the GOT banding pattern 35 within *Capsella* species, characterized by the bands 11 + 17

results will be given here. A paper presenting complete data for all GOT loci is in preparation and will be published elsewhere.

Within the genus *Capsella*, at least four GOT loci have been detected, three of them coding cytosolic GOT. In accordance with reports in the literature, the GOT enzyme in *Capsella* is also a homodimeric molecule, thus forming a hybridband (= intragenic) if the locus is heterozygous. So far, four alleles could be detected at locus GOT 1 (called a_1, a_2, a_3, a_4), three at locus GOT 2 (b_1, b_2, b_3), and two at locus GOT 3 (c_1, c_2). GOT 4, which probably codes a plastid GOT, has not yet been investigated thoroughly enough. In addition to intra-locus bands, GOT 2 and GOT 3 also produce inter-locus bands, thus demonstrating great similarity between these two loci. GOT 3 may be the product of a gene duplication.

The GOT band numbers mentioned in the text can be explained in terms of the GOT loci as follows:

Band number	Explanation
6	c_1 allele of GOT 3
8	inter-locus band between alleles c_1 and b_1
11	a_4 allele of GOT 1
	b_1 allele of GOT 2
	hybridband (= intra-locus) between alleles b_2 and b_3
	inter-locus band between alleles c_1 and b_2
15	a_1 allele of GOT 1
	hybridband (= intra-locus) between alleles a_2 and a_3
16	hybridband (= intra-locus) between alleles a_1 and a_2
17	a_2 allele of GOT 1
18, 20, 22	alleles and intra-locus bands of GOT 4.

Acknowledgements. Biosystematic investigations of *Capsella* are supported by the Deutsche Forschungsgemeinschaft (DFG), which is greatly appreciated. I would like to thank all my co-workers, especially Mrs. M. Wickenbrock and Dr. K. Bosbach.

References

Ayala FJ (ed) (1976) Molecular evolution. Sinauer, Sunderland, Mass
Bosbach K, Hurka H (1981) Plant Syst Evol 137:73–94
Bosbach K, Hurka H, Haase R (1982) Flora (Jena) 172:47–56
Bradshaw AD (1972) Evol Biol 5:25–47
Gottlieb LD (1982) Science 216:373–380
Hurka H (1980) In: Bisby FA, Vaughan JG, Wright CA (eds) Chemosystematics: Principles and practice. Academic Press, London New York, pp 103
Hurka H, Benneweg M (1979) Biol Zentralbl 98:699–709
Hurka H, Haase R (1982) Flora (Jena) 172:35–46
Hurka H, Wöhrmann K (1977) Bot Jahrb Syst 98:120–132
Hurka H, Krauss R, Reiner T, Wöhrmann K (1976) Plant Syst Evol 125:87–95
Kimura M (ed) (1977) Molecular evolution and polymorphism. Nat Inst Genet, Mishima, Jpn
Kimura M, Ohta T (1974) Proc Natl Acad Sci USA 71:2848–2852
Le Cam LM, Neyman J, Scott EL (eds) (1972) Darwinian, neodarwinian, and non-darwinian evolution. Proc 6th Berkeley Symp Math Statist Probab, vol V. Univ Calif Press, Berkeley

Lewontin RC (1973) Annu Rev Genet 7:1–17
Lewontin RC (1974) The genetic basis of evolutionary change. Columbia Univ Press, New York
Markert CL (ed) (1975) Isozymes, vol IV. Academic Press, London New York
Nevo E (1978) Theor Popul Biol 13:121–177
Powell JR (1975) Evol Biol 8:79–119

Seed Storage Proteins

U. JENSEN[1] and B. GRUMPE

Abstract. Besides the demonstrated value of serological properties of seed storage proteins in systematics, other characters of these proteins are also discussed. As a basis for a comparative study, the homology has been ascertained in several cases for which serological methods, using monospecific antisera, are helpful and easily performed. Amino acid sequence and serological data are very useful for phylogenetic studies if cautiously interpreted. Additional information is expected to arise from coding nucleic acids. For taxonomical descriptions of taxa, however, electrophoresis data as well as iso-electrofocusing, chromatography, and molecular weight data are valuable, at least for infrageneric taxa.

1 Introduction

Seed storage proteins have proved to be valuable macromolecules for the elucidation of phylogenetic relationships; this has been especially true for the majority of the comparative systematic serological investigations. In serosystematic research the whole pattern of the storage and non-storage seed proteins have usually been considered. However, since the main storage proteins are highly antigenic as well as quantitatively dominant among the seed proteins, *their* data have dominated the comparative data and, therefore, also the systematic considerations. There have been many practical arguments in support of using seed proteins, e.g. easy availability, good storage quality, often stable at room temperature, etc. Also their systematic usefulness has been proven repeatedly in numerous publications. In this volume Rolf Dahlgren presents a comprehensive discussion of the contributions of serological data in systematics and classification. Therefore, now is an appropriate time to discuss the different properties of seed storage proteins relative to other proteins and characters, and to evaluate their specific value in phylogenetic and taxonomic studies.

Our present knowledge of the molecular structure of seed storage proteins and their inherent properties is incomplete, and much important information is still lacking. However, some data are available which provide valuable insight into their structure and function, especially with proteins obtained from the two agriculturally important families, Fabaceae and Poaceae.

[1] University of Bayreuth, Department of Plant Ecology and Systematics, D-8580 Bayreuth, FRG

Proteins and Nucleic Acids in Plant Systematics
ed. by U. Jensen and D.E. Fairbrothers

2 Application of Seed Storage Proteins in Systematics

From earlier research the impression gained was that each group of higher plants is provided with unique storage proteins in the seeds: In the Fabaceae e.g., legumin and vicilin have been demonstrated for *Vicia faba*, glycinin and conglycinin for *Glycine max*, phaseolin for *Phaseolus vulgaris*. In the Poaceae, the quite distinct alcohol soluble prolamins and the base soluble glutelins have been demonstrated and investigated. If such proteins are to be useful in taxonomic and phylogenetic comparisons, the following criteria have to be true:

a) In order to assure accurate comparisons of taxonomic characters from different taxa, the *homology* of these characters has to be demonstrated.

b) The properties of these storage proteins have to be sufficiently variable to reveal differences between the compared taxa. If the *variability* is considered to be sufficient for a systematic study it has to be determined at which taxonomic level the degree of variability promises the most usefulness.

c) To demonstrate a posteriori phylogenetic *significance* for these character similarities they must be compatible with knowledge of relationships based upon other criteria and data obtained by numerical analyses or classical systematic investigations.

3 Homologous Seed Storage Proteins

Derbyshire et al. (1976) suggested similar physicochemical properties for each of the two major storage proteins present in many plant families, and that they are homologous with the legumin and vicilin of the Fabaceae. At least for the basic subunit of the legumin-like protein from *Vicia faba, Pisum sativum,* and *Glycine max,* Gilroy et al. (1979) reported high sequence homology of the N-terminal portion of the protein. Jensen and Büttner (1981) demonstrated serological cross-reactions for the two major storage proteins of the Ranunculaceae, i.e. nigellin and aquilegilin, with seed extracts from other angiosperm taxa. These reactions were interpreted as being due to reactions of homologous proteins present in almost all angiosperm families tested. In some cases direct indications of these homologies could be demonstrated, e.g. of the aquilegilin present in *Aquilegia vulgaris* (Ranunculaceae) and *Digitalis purpurea* (Scrophulariaceae) (Büttner and Jensen 1981). Unequivocal serological cross-reactions and the demonstration of similar physicochemical properties were the basis for the conclusions of homology. We now believe that the following statement is an appropriate working hypothesis: In angiosperms only a limited number of homologous storage proteins exist. These are:

a) A legumin-like protein, occurring as a major protein in *Pisum sativum* (legumin), *Vicia faba* (legumin), *Arachis hypogaea* (arachin; Tombs and Lowe 1967), *Glycine max* (Glycinin; Catsimpoolas and Wang 1971, Badley et al. 1975, Gilroy et al. 1979). The homology has been confirmed by a remarkable similarity in the tryptic peptide maps (Croy et al. 1979), identical or almost identical serological reactions (Dudman and Millerd 1975, Croy et al. 1979) , and a similar primary structure (Gilroy et al. 1979). These proteins are suspected to be homologous with a legumin-like protein from seeds of *Vigna radiata* (= *Phaseolus aureus*), *Phaseolus vulgaris* (Derbyshire and Boulter 1976) and a prominent Cucurbitaceae seed globulin (Hara et al. 1978, Gilroy et al. 1979). Nigellin, the primary storage protein of the Ranunculaceae as well as serologically cross-reacting proteins within the Magnoliaceae, Papaveraceae, Betulaceae, Scrophulariaceae, and other families (Jensen and Büttner 1981, Jensen unpublished manuscript) is suggested to be a homologe with legumin. If this homology is correct, all legumin- as well as nigellin-like proteins would be of the same genetic origin. To-date all reported data are in favor of the existence of such a homologous storage protein in the seeds of many dicotyledoneous families (Derbyshire et al. 1976, Jensen unpublished manuscript). In some taxa, however, 11/12 S proteins of this type are present in only small amounts, e.g. in *Vigna unguiculata* (Miège 1982), *Psophocarpus tetragonolobus* (Gillespie and Blagrove 1978), some *Delphinium* species (Jensen unpublished).

b) A *vicilin-like* protein is known to be another major storage protein found in *Vicia faba, Pisum sativum,* and other plant species within the Vicieae and other Fabaceae tribes (Bailey and Boulter 1972, Dudman and Millerd 1975). Following the data presented by Derbyshire et al. (1976) the existence of homologous storage proteins in other plant families has to be expected. Aquilegilin is generally the secondary storage protein of the Ranunculaceae. An aquilegilin homologous protein has been demonstrated for *Digitalis purpurea* (Scrophulariaceae) and suggested as being present in many angiosperm families by Büttner and Jensen (1981). The dominating or unique subunit of the aquilegilin from Ranunculaceae taxa is in the 50,000 molecular weight (M.W.) range and is similar in size to the dominating vicilin subunit (Boulter et al. unpublished). Aquilegilin is precipitated by concanavalin A, as is vicilin, because of its considerable glycosylation (Rosenkranz 1977), whereas nigellin and legumin are not (Fig. 1). Thus, the homology of vicilin and aquilegilin has to be assumed. For such proteins the term "vicilin-like proteins" should be used.

c) In Poaceae *prolamines* and *glutelins* are widespread and generally are the major storage proteins (Payne and Rhodes 1982). A partial sequence homology has been demonstrated (Shewry et al. 1980b) for a hordein prolamin from *Hordeum vulgare* and a gliadin prolamin from *Triticum monococcum*. Therefore, in the tribe Triticeae, phylogenetic comparable proteins have to be anticipated. Actually, prolamins have an even wider distribution, and occur in dicots as a minor component (Darmency et al. 1980).

d) In several families one, or a few, *minor storage proteins* might occur, e.g. γ-conglutin in *Lupinus angustifolius* (Elleman 1977), convicilin in *Pisum sativum* (Croy et al. 1980), narbonin in *Vicia narbonensis* (Schlesier et al. 1978). Nothing is known about their distribution in the plant system at the present time.

4 Protein Properties and their Systematic Value

4.1 Subunit Structure

Subunits and subunit molecular weights are known for some proteins from many plant species. However, different researchers have obtained different results. These

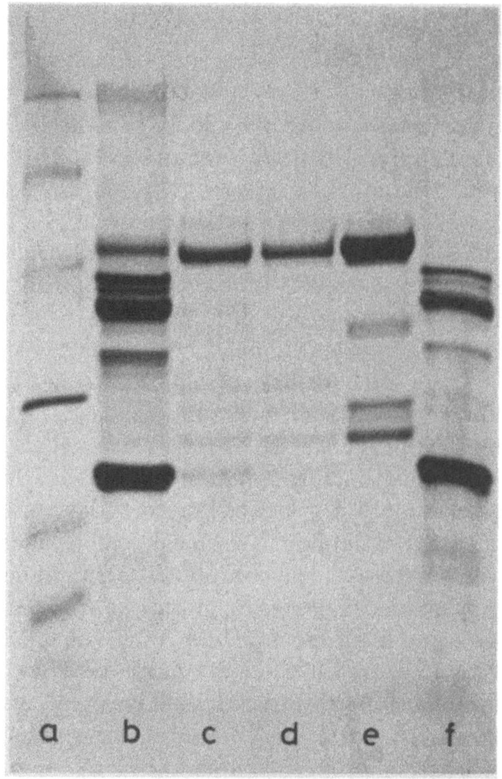

Fig. 1a–f. Gradient SDS-PAGE of *Nigella damascena* storage proteins
a marker proteins (see Fig. 3)
b mixture of both the major storage proteins prior to separation on a sepharose con-A gel column
c the vicilin-like protein eluated from the column using 10% α-D-glucose; the major subunit at 53,000 M.W. is visible
d the vicilin-like protein eluated from the column using 10% methyl-α-D-mannopyranoside
e a *Nigella damascena* vicilin-like protein standard
f a *Nigella damascena* legumin-like protein standard
Tris-borate buffer pH 7.2 was used for absorption

may be due to different extraction and purification techniques, and hence different and/or unknown contamination, or to alteration of the proteins during the elaborate processing. In particular, a different dissociation of the native protein resulting from different uses of detergents, pH, and ionic strength of the buffer systems has been reported. Additionally, significant heterogeneity, which is not influenced by the technical procedure, has been demonstrated. Nevertheless, there are dominating molecular weights which enable us to compare proteins from different taxa, and to present a plausible interpretation of their similarity in terms of phylogeny.

The currently accepted model for the structure of legumin is that each molecule consists of six pairs of subunits (approximately 60,000 M.W. for each pair), held together by non-covalent bonds, and each subunit pair consists of a larger acidic subunit ("40,000" M.W. type) and a smaller basic subunit ("20,000" M.W. type), linked by one or more disulphide bonds (Derbyshire et al. 1976, Matta et al. 1981a, b). This $\alpha_6\beta_6$-model was originally derived from extracts of *Vicia faba* and reported by Wright and Boulter (1974); but extended to *Glycine max* by Badley et al. (1975), and *Pisum sativum* and other taxa by Croy et al. (1979) and Matta et al. (1981b).

A problem in obtaining an exact comparison between taxa arises from the considerable heterogeneity of the acidic, as well as for the basic subunits of legumin, which is especially visible in isoelectrofocusing separations (Brown et al. 1980). For *Pisum sativum*, Matta et al. (1981b) revealed 5 basic subunits in the M.W. range between 20,700 and 22,700, whereas the 5 acidic subunits range between 35,000 and 43,000, apart from a subordinate subunit with a M.W. of 24,500. These data are similar to the findings of Przybylska et al. (1979); they demonstrated for *Pisum sativum* legumin a M.W. of 18,000–20,000 and 37,000–44,000 for the small and large subunit respectively, apart from a minor band at 27,000. For *Vicia faba* a greater heterogeneity was found for the large subunits (M.W. range 23,000–58,000), but 19,000–23,000 for the small subunit (Wright and Boulter 1974, Utsumi and Mori 1980). For the glycinin subunits of soybean seeds, molecular weights of 19,500 and 20,500 plus 33,500, 35,500, and 39,000 were reported (Spielmann et al. 1982, see also Kitamura et al. 1980). Heterogeneity has also been reported for vicilin (Manteuffel and Scholz 1975, Davey and Dudman 1979), nigellin (see below) and other seed proteins.

There are experimental indications that the heterogeneity is not only post-translationally produced, but also genetically encoded. At least in part heterogeneity is encoded in the genetic material (Matta et al. 1981a, Croy et al. 1982). Probably gene duplication and subsequent change produce this type of heterogeneity. The possession of such multiple genes coding for storage proteins may be useful in maximizing the rate of seed development (Bailey and Boulter 1972). It would be important to know whether these duplications and differentiations occurred only once in the early phylogeny of the angiosperms, giving rise to heterogeneous subunits which would be homologous in different families, orders or subclasses, or whether this process happened independently and therefore convergently in phylogeny.

Frequently post-translational changes have also been suggested (e.g. Beevers 1982, Croy et al. 1982), which result in different primary, secondary, or tertiary structure, or in different glycosylation (when glycoproteins are concerned, Beevers 1982). All these facts may account for the small differences detected in the heterogeneous subunit bands. Since intrinsic factors or a specific environment might influence these changes, subunit comparisons cannot automatically be considered to properly represent phylogenetic relationships.

Nevertheless, the question should be asked as to whether an electrophoretic comparison of subunits is expected to be a useful tool for systematic work and which limitations must be considered. To test this we investigated the infraspecific and infrageneric subunit composition of nigellin and aquilegilin within the genus *Nigella* (Rhanunculaceae) and other plant taxa. The use of single seeds i.e. genotypes (already reported for seeds from barley varieties by Shewry et al. 1978b) optimized the significance of the results and excluded possible effects of proteinases and other alterations during the analytical procedures. The possibility of obtaining information about the subunits of, e.g., the two major storage proteins from single seeds is quite easy, using SDS PAGE in gradient gels. One *Nigella* seed is large enough and its protein content great enough to produce a good visible banding pattern. The reason is that nigellin, which separates into subunits of two types, occurs as the primary protein which because of its high concentration allows even the subunits to be very visible. The secondary protein aquilegilin, which occurs in smaller quantities, is often visible because it is largely composed of only one main subunit type. All the other proteins usually occur in much smaller quantities and are invisible, or only slightly visible, when only one seed is used for electrophoresis.

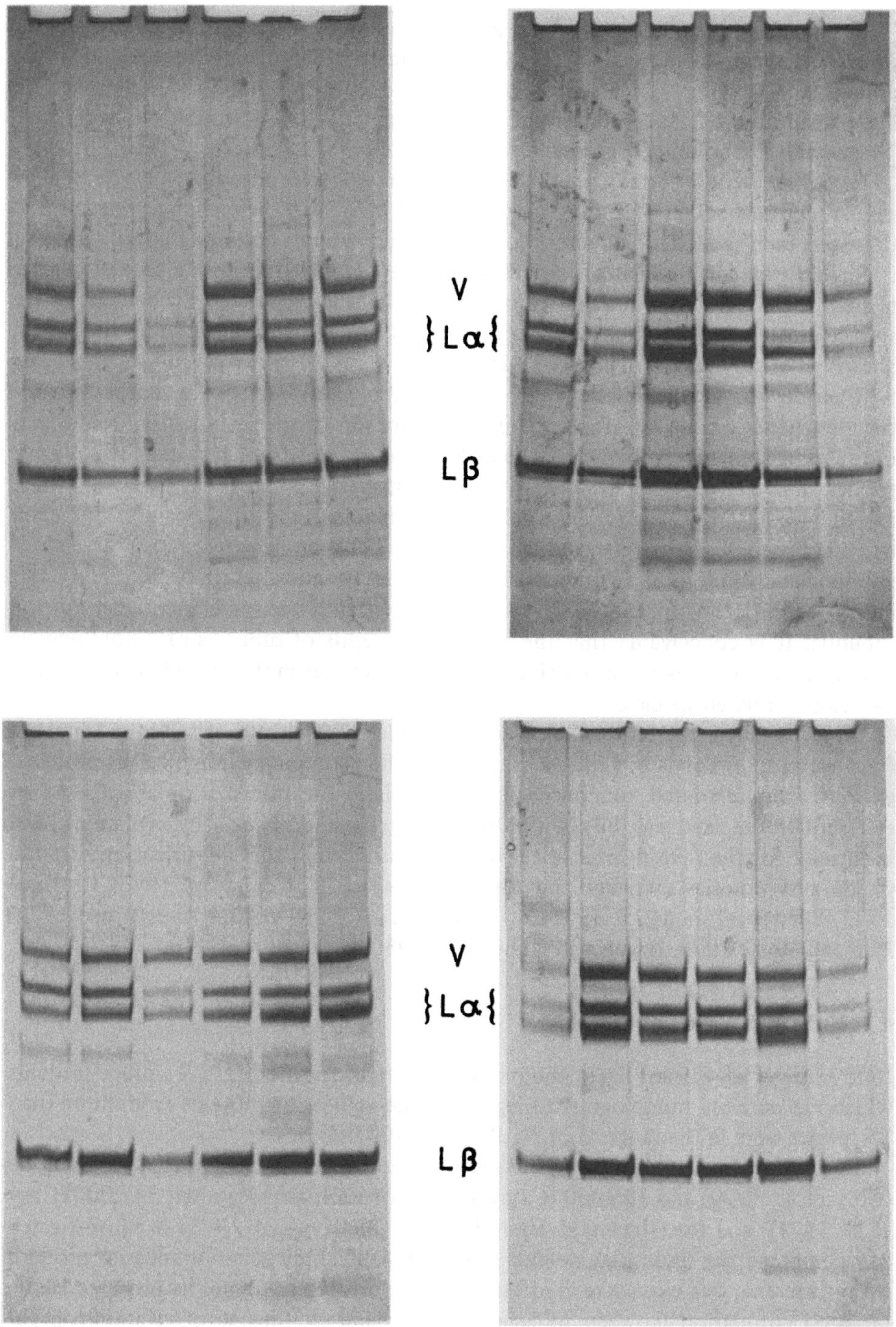

Fig. 2. Subunit separation of *Nigella damascena* storage proteins using single seeds from different origins. Gradient SDS PAGE. Tris-borate buffer, pH 7.9. *V* vicilin-like protein, main subunit; L_α legumin-like protein, 40,000-range subunits; L_β legumin-like protein, 20,000-range subunits

In any event, practically no infraspecific variability was detected within *Nigella damascena* seeds, of more than 50 different origins, which mainly revealed identical subunit patterns (Fig. 2). Only a few seeds revealed visible differences. To demonstrate these small divergences it is essential to use an appropriate electrophoretic technique (Stegemann this volume, Stegemann and Pietsch 1983).

For other *Nigella* species, the results, in the main, were the same, i.e. identical or almost identical subunit banding patterns of all individuals belonging to one species. However, each species is characterized by a distinct subunit pattern (Fig. 3).

Similarly a small range of variation has been reported from *Phaseolus vulgaris* phaseolin and the legumin and vicilin of *Pisum* lines in relation to the cereals, where storage protein variation is used to identify cultivars (Autran and Bourdet 1975, Shewry et al. 1978a,b, Villamil et al. 1982).

The question is whether the degree of subunit pattern similarity is useful evidence for phylogenetic considerations or for the separation of species. Indeed, from Fig. 3 the subunit similarity is higher between *Nigella damascena* and the other two *Nigella* species than between *Nigella damascena* and *Magnolia grandiflora*. However, when e.g. the 40,000 M.W. legumin-like subunit is concerned, the numerically higher similarity between *N. damascena* and *N. sativa* than between *N. sativa* and *N. arvensis* is unimportant. No logical argument can predict that genetic alterations, occurring during the evolution of species, always successively change the molecular weight of the subunits. It is conceivable that the molecular weights of subunits of a homologous protein in some cases are more similar between two distantly related than between two closely related species.

In systematics, therefore, the use of molecular weight data of subunits obtained from electrophoresis is of limited value, even when comparing homologous proteins, except when restricted to conspecific taxa, where the presence or absence of an electrophoretic band can be used to separate genotypes. Especially prolamins have been used for the investigation of close relationships and the identification of polyploids and varieties (Wrigley and Shepherd 1973, 1974, Johnson 1975, Dhaliwal 1977, Shewry et al. 1977, 1978a,b, Kim et al. 1978, 1979, Doll and Brown 1979, Kim and Mossé 1979, Hagen and Rubenstein 1980, Vitale et al. 1980).

4.2 Molecular Weight of Proteins

Derbyshire et al. (1976) listed the molecular weight for the 11/12 S storage proteins which are possibly homologous to legumin. The values for proteins from more than 25 species were in the range of M.W. 300,000–400,000.

Recent determinations for the *Pisum sativum* legumin were ca. M.W. 390,000 (Croy et al. 1979) and 395,000 (Casey 1979), for *Vicia faba* legumin, 347,000 (Croy et al. 1979), and for *Glycine* glycinin 320,000 (Badley et al. 1975). We investigated and compared the molecular weights of nigellin- and aquilegilin-homologous proteins from different, but closely related taxa. At least for the legumin-like protein nigellin no good correlation between molecular weight and the degree of relationship has been detected (Fig. 4.).

From all these data it can be concluded that the molecular weight of homologous proteins is expected to be different between closely related species. A slight shifting

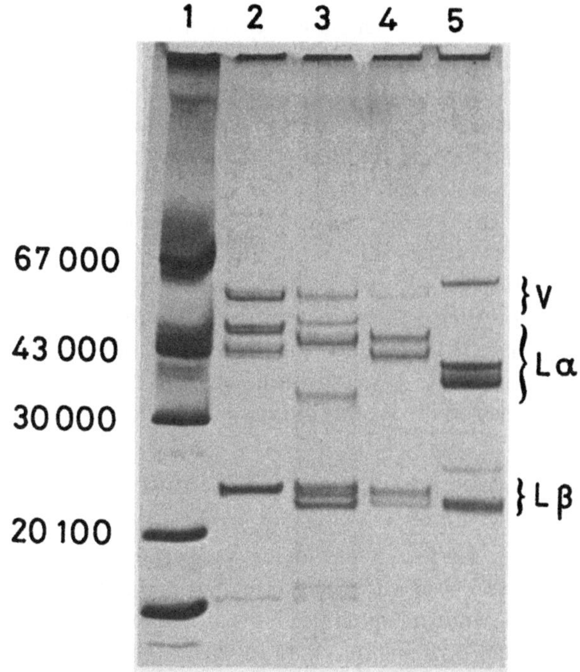

Fig. 3. Subunit separation of *Nigella* and *Magnolia* storage proteins using single seeds.
Gradient SDS PAGE; Tris-borate buffer pH 7.9
V vicilin-like protein, main subunit
L_α legumin-like protein, 40,000-range subunits
L_β legumin-like protein, 20,000-range subunits
1 marker proteins
2 *Nigella damascena* seed
3 *Nigella sativa* seed
4 *Nigella arvensis* seed
5 *Magnolia grandiflora* seed

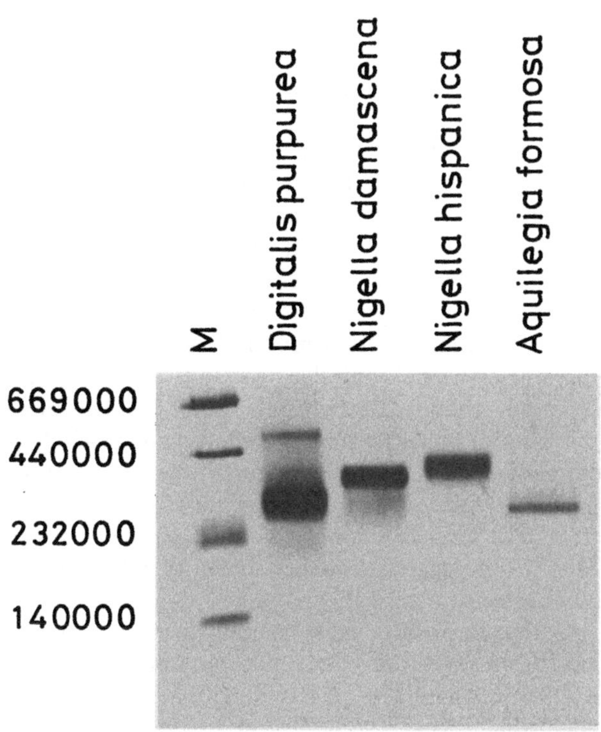

Fig. 4. Gradient PAGE of storage proteins from selected taxa of the Ranunculaceae indicating the molecular weight of the legumin-like proteins, i.e. the nigellins *(heavy stained band)*. *M* Marker proteins

of the molecular weight is expected even within infraspecific taxa, if the subunit heterogeneity tends to favor those with deviating molecular weights. However, a congruent electrophoretic behavior in PAGE does not necessarily indicate close relationships and might be interpreted in terms of accidental coincidence of the quarternary products, and homologous proteins from closely related taxa might have quite a different molecular weight. For the aquilegilin protein a different number of subunits forming the quarternary structure has been observed (Grumpe 1981) which additionally masks relationship similarities.

Therefore, the use of molecular weight data and electrophoretic properties of whole proteins can be recommended in the case of very closely related taxa, where a predominant amount of identical data contrasted with the amount of nonidentical data might reflect relationship. The use of these properties to establish relationships above the generic level is dubious.

4.3 Serological Properties

Although the use of serological properties of storage proteins in plant systematics is widely accepted and their contribution to the elucidation of plant relationships remarkable and indisputable, most of these contributions involved the total pattern of those seed proteins reflected in the antisera. However, the storage proteins and their determinants dominated such protein patterns, hence the data mainly represent these proteins. In this presentation we will restrict the discussion to isolated and purified storage proteins and their serological reactions.

4.3.1 Demonstration of Homology

The main benefit of serological reactions in plant systematics is the demonstration of homology. Of course, Boulter (1981) is correct when arguing that the homology of proteins can only be determined by amino acid sequence analysis, but in the absence of sequence data serological cross-reactivity can demonstrate the suspected protein homology (Metzger et al. 1968, Jensen 1981), because of the strict sequence specificity of the serological properties. Theoretically however, the occurrence of congruous determinants of different proteins cannot be excluded. This is especially possible in those cases where equivalent subunits are involved in the quarternary structure of different proteins (Derbyshire et al. 1976). However, until recently, no evidence of this possibility has been presented using polyclonal antisera. With regard to the major storage proteins, cross-reactions between legumin-like and vicilin-like proteins have not been reported, nor cross-reactions between nigellin-like and aquilegilin-like proteins. The cross-reaction between vicilin and convicilin from *Pisum sativum* (Croy et al. 1980) suggests sequence similarity, although this has not been proved by amino acid sequence analysis.

Although positive cross-reactions can be used to identify homologous proteins, lack of cross-reactions does not demonstrate non-homology. An example is provided by the distribution of phaseolin within the Fabaceae. From many investigations (e.g. Kloz and Turková 1963, Kloz et al. 1963, Kloz 1971, Kloz and Klozová 1974) it is shown that the primary storage protein of *Phaseolus vulgaris* cannot be serologically detected outside a limited group of highly related *Phaseolus* species, although this protein may be homologous with the widely distributed vicilin (Hirano et al. 1982). In similar cases three possibilities have to be considered: (a) If antideterminants with great taxonomic range reactivity are not represented in the particular antiserum used, then existing homology cannot be detected. (b) If by chance all external serological determinants are different between homologous proteins exhibiting a high number of corresponding amino acids, any serological cross-reactivity is suppressed. (c) No homologous protein occurs in the cross-reacting sample.

4.3.2 Gradation and Systematic Significance of the Serological Reactions

Our present knowledge discloses gradations of the cross-reactions in the following ways (Jensen and Penner 1980, Jensen and Büttner 1981):

a) Generally congeneric species reveal fully identical reactions. However, if one of the congeneric species is used as the reference system, occasionally congeneric species can be distinguished.

b) The highest resolving power is observed for confamiliar tribes and genera, provided that the reference system is derived from a member of the same family.

c) Although generally antigens from different families, orders, subclasses, or classes can cross-react with more or less decreasing binding capacities in the entire determinant-antideterminant-reaction, respectively, these data do not promise strong and useful taxonomic evaluations. Formed precipitations are weak and spurs in Ouchterlony immunodiffusion or reactions detected after pre-absorption are difficult to observe and to properly evaluate.

These statements indicating optimal evidence for infrafamiliar taxa are typical for the storage proteins but not necessarily correct for other proteins. The serological properties of many enzymes of the primary metabolism are of less evolutionary variability and therefore are of greater use within higher categories.

The evaluation of the usefulness of serological properties has to be done a posteriori in comparing serological results with those obtained from other disciplines. Published examples of this kind are few, but at least for the Ranunculaceae the possibilities are promising (Jensen and Penner 1980). Extending these principles to those multiprotein serological comparisons in which the properties of the storage proteins at least dominate, is even more promising for systematic application.

There have been attempts to use *monoclonal* antisera raised to storage proteins (Guttermann 1981). However, at this stage of development it is too early to draw conclusions about the value of such antisera. Because of the much higher specificity of monoclonal antisera, the occurrence of heteroclicity, and the influence of seniority and affinity (Fazekas de St.Groth this volume), the evaluation of such data has to be handled cautiously and differently.

4.4 Protein Sequence Analysis

The analysis of amino acid sequences of storage proteins and the incorporation of these results into phylogenetic evaluations is just starting. So far, essentially only

Poaceae prolamins and Fabaceae globulins have been considered. Valuable information is expected from additional analyses of the coding DNA molecules, and the first steps in this research have now been performed (Croy et al. 1982).

4.4.1 Cereal Prolamins

The prolamins of *Hordeum vulgare* are represented by the hordeins B and C. The high content (>50%) of glutamine and proline is reported to be typical for prolamin proteins. N-terminal amino acid sequences are known from C-hordein (Shewry et al. 1980a,b, Schmitt and Svendsen 1980). Although other prolamins from *Triticum aestivum* (Kasarda et al. 1974, Patey et al. 1975, Bietz et al. 1977, Autran et al. 1979), *Triticum monococcum* (Shewry et al. 1980b), *Secale cereale* (Autran et al. 1979), and *Zea mays* (Bietz et al. 1979) are used for N-terminal amino acid sequencing, only the C-hordein and the ω-gliadin of *Triticum monococcum* are similar in their sequence and therefore homologous (Shewry et al. 1980b). From the 27 N-terminal amino acids investigated, only a few positions are different (Fig. 5). On the other hand, all gliadins of wheat are completely different. This demonstrates that the preliminary names applied to proteins can prove to be misleading, when more data become available. The ω-gliadin apparently do not have close evolutionary connections with the

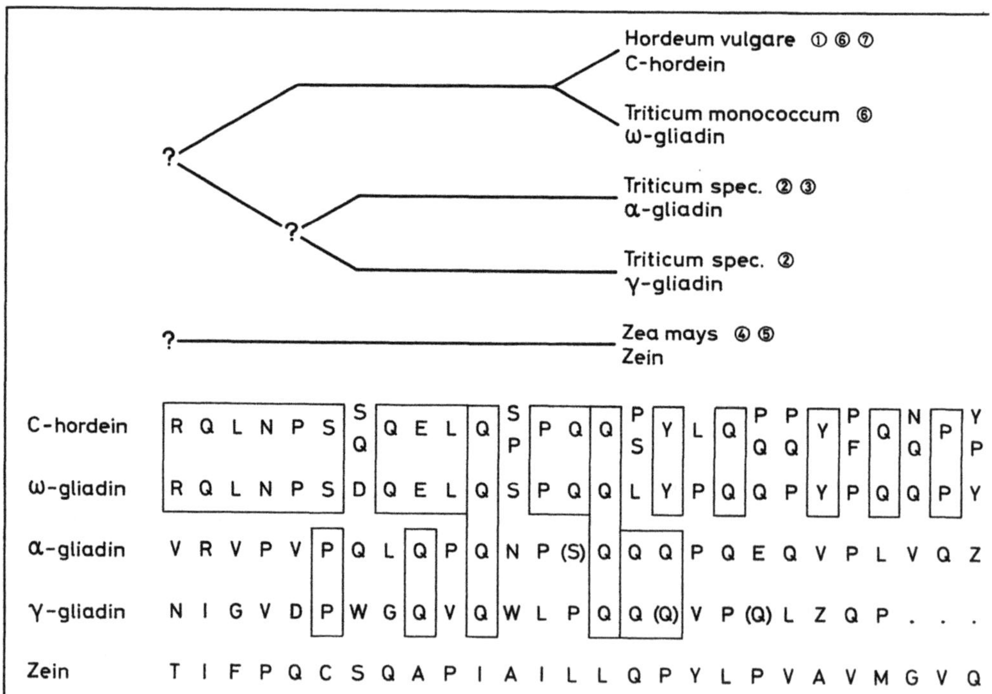

Fig. 5. N-terminal amino acid sequences of cereal prolamins. References: (1) Shewry et al. (1980a,b); (2) Bietz et al. (1977); (3) Kasarda et al. (1974); (4) Bietz et al. (1979); (5) Larkins et al. (1979); (6) Shewry et al. (1980b); (7) Schmitt and Svendsen (1980)

α-, β-, and γ-gliadins and should be associated with the C-hordein. Although Shewry et al. (1980b) concluded close relationships between *Hordeum vulgare* and *Triticum monococcum*, the degree of relationship should not only be based on the similarity of two homologous sequences. More data are required before phylogenetic conclusions can be considered valid.

The N-terminal amino-acid sequences from some other proteins have been elucidated, e.g. a lysine rich protein from barley (Svendsen et al. 1980) and a proline rich protein from *Zea mays* (Esen et al. 1982), both occurring in smaller quantities in the cereal grains, but no systematic evaluation is presently possible.

4.4.2 Fabaceae Globulins

Some sequence analyses have also been performed for the Fabaceae storage proteins. From these investigations the homology of the basic subunit from the legumin-like protein of *Vicia faba* and *Glycine max* has been confirmed by Gilroy et al. (1979), as previously suggested by Derbyshire et al. (1976). Croy et al. (1982) presented the whole amino acid sequence for the basic subunit of *Pisum sativum* legumin, partly by the analysis of tryptic peptides, and partly predicted from the DNA sequence coding for this polypeptide. Six N-terminal positions are known for *Cucurbita* sp. seed globulin δ_1- and δ_2-subunits (Hara et al. 1978, Ohmiya et al. 1980). They have demonstrated a good possibility of homology with the basic subunits of legumin (Fig. 6).

It has been proved that the three legumin-like proteins differ only slightly. There is no good reason for suggesting an isolated taxonomic position for *Glycine max* from the Vicieae, although the species has been placed in the separate tribe Phaseoleae. However, the serological comparison revealed a dissimilarity[2] of *Glycine max* legumin compared to *Pisum sativum* and *Vicia faba* legumin (Dudman and Millerd 1975). Either the immunological determinants do not reside at the N-termini of the basic subunits (Casey et al. 1981) or the exchange of specific amino acids at specific positions would be highly effective in the exchange of serological determinants. For example, for vertebrate cytochrome c it has been demonstrated that a determinant region becomes immunologically different or inert by the substitution of one amino acid (Arnon 1971, Berman and Harbury 1980). If such a substitution had effected leucine at position 2 within a separate phylogenetic branch (Phaseoleae), the greater sequence similarity between the Vicieae legumin and *Cucurbita* globulin than between the Vicieae legumin and the *Glycine* glycinin could be understood. (If a comparison of only 6 sequences is valid for making conclusions.)

A sequence of 185 amino acids has been presented from phaseolin, the 7S-storage protein of *Phaseolus vulgaris* (Sun et al. 1981). The analysis of the small subunit of γ-conglutin (154 amino acids), a minor storage protein from *Lupinus angustifolius*, should also be mentioned (Elleman 1977). However, these proteins do not reveal any sequence similarity with the legumin-like proteins.

---

[2] Recently a cross-reaction between legumin of *Pisum sativum* and *Glycine max* has been shown by Gatehouse and Boulter (unpublished)

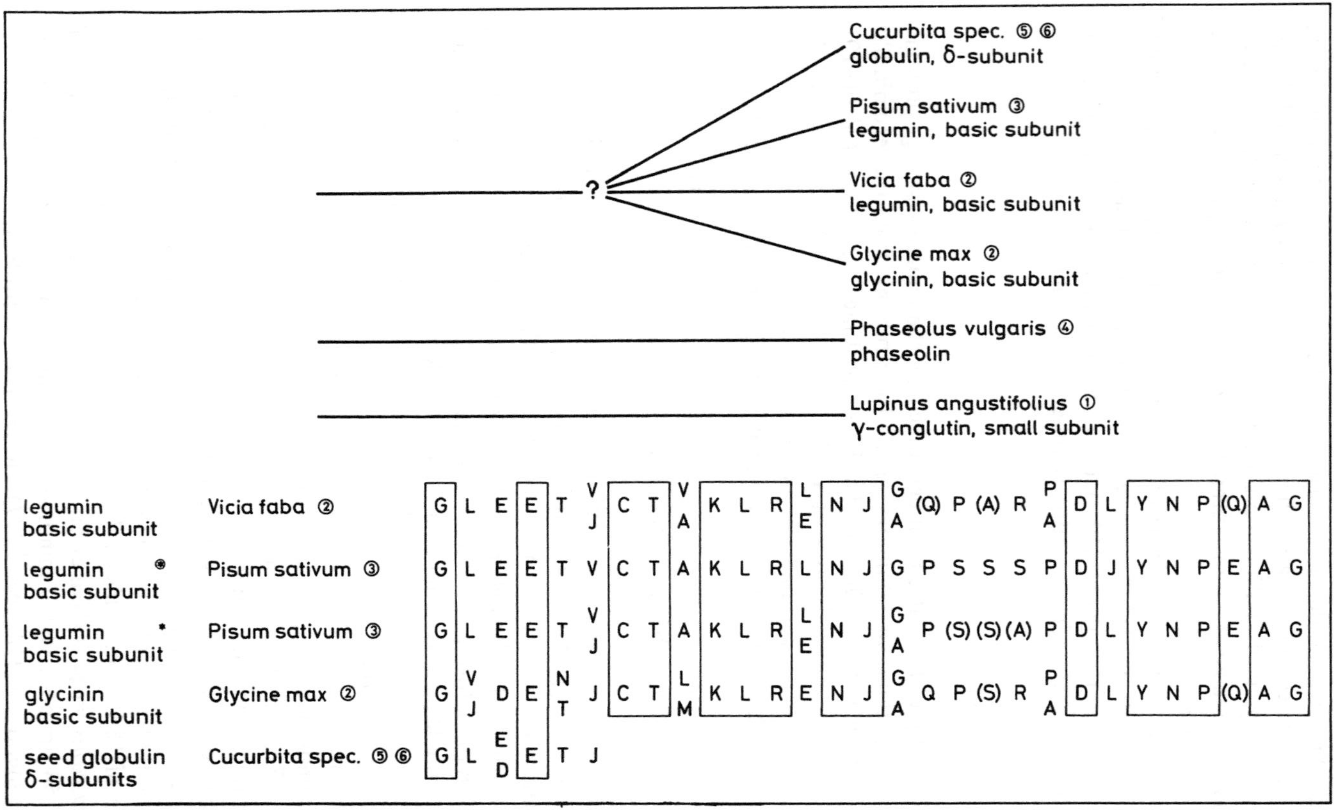

Fig. 6. N-terminal amino acid sequences of the basic subunit of legumin from different species.
References: (1) Elleman (1977); (2) Gilroy et al. (1979); (3) Croy et al. (1982); (4) Sun et al. (1981); (5) Ohmiya et al. (1980); (6) Hara et al. (1978)

⊛ from DNA sequence analysis

* from peptide sequence analysis

The presented examples illustrate the usefulness of sequence analyses for the demonstration of homology. For phylogenetic purposes, however, caution must be used for the evaluation and comparison of the data. From other comparative sequence data (e.g. cytochrome c, plastocyanin) it is known that the simple calculation of an overall sequence similarity does not reveal the most valuable information. Thus each amino acid, at each position, has to be considered as well as its significance for the protein structure and function; for it is at this level that the selection process has determined the stable arrangement of a changed protein sequence.

4.5 Coding Nucleic Acids

The significance of protein data in systematic investigations indicates that both cloning and analysis of cDNAs encoding plant storage proteins, or their precursors, should provide additional valuable information. The procedures for such comparative investigation can be deduced from the pioneering work of the Boulter-group, who isolated and analyzed the cDNAs coding for legumin and vicilin, the major storage proteins within the Vicieae.

For the *Pisum sativum* legumin Croy et al. (1982) demonstrated that the legumin mRNAs are of sufficient size to code for the 60,000 M.W. precursor of the acidic and basic legumin subunit. They also demonstrated the presence of four "single"-copy legumin genes per haploid genomic DNA, also suggested from the electrophoresis data as reported by Guldager (1978), Casey (1979), Spencer and Higgins (1980), and Matta and Gatehouse (1982). Additional heterogeneity observed, e.g. by Casey (1979), Croy et al. (1979), Krishna et al. (1979), Moreira et al. (1979), and Matta and Gatehouse (1982) may be accounted for by post-translational modifications (Croy et al. 1980, Gatehouse et al. 1980, Croy et al. 1982).

Similar 60,000 M.W. precursors have been demonstrated for *Vicia faba* (Croy et al. 1980) and *Glycine max* (Boulter-group, unpublished). Therefore, a common processing mechanism has been suggested (Croy et al. 1982).

For the storage protein vicilin the coding nucleic acids have been investigated for *Pisum sativum* (Croy et al. 1982). It has been demonstrated that the vicilin polypeptide precursors are synthesized only from 11S mRNAs, which are also derived from genes present in single or low copy number in the pea genome. The partial sequence of cDNA and genomic DNA has been elaborated also for phaseolin from *Phaseolus vulgaris* (Sun et al. 1981). If phaseolin really is homologous with vicilin the elucidation of structural similarities can be expected.

When more taxa have been included in such investigations it is anticipated that the structural differences might be interpreted in terms of relationships.

5 Conclusions

Proteins have been, and will continue to be, used to improve taxonomic classifications and phylogenetic evaluations. However, their value depends on the special circumstances and the protein properties which have been employed in the research.

According to the methods used, Boulter (1981) ranked them in decreasing power of resolution: amino acid sequence, serological properties, peptide finger prints, isoelectrofocusing patterns, gel electrophoresis patterns, chromatographic patterns, and molecular weight. Our research has confirmed his ranking in all cases where the interpretation of similarities in terms of *phylogenetic relationships* are concerned. The amino acid sequence data are closely related to the coding nucleic acids as well as to those properties of the proteins (e.g. structure, physiological properties) which are governed by selection pressure. Serological properties can be directly derived from the amino acid properties. On the other hand electrophoresis, chromatographic and molecular weight data are secondary properties, leading to deductions of relatively questionable value. However, when using different plant protein properties for a project, the balance between phylogenetic profit and excessive technical expense has to be considered carefully.

In *taxonomy*, which includes the complete description of a taxon, an important aspect is the precise identification of a taxon and the delineation from similar taxa. This aspect of taxonomy should ignore relationships, even though a complete description is used as the basis for phylogenetic considerations. In this taxonomic sense the leading position of amino acid sequence data is reduced because of their low practicability. Serological data are of even less use, because their contributions are only relative, when establishing the degree of similarity between two or more taxa. In this sense, electrophoresis and chromatographic data are of relatively high importance. This evidently demonstrates that the usefulness of each method, and of the properties compared, depends on the scientific question being asked, thus the procedures used must be weighed very carefully based upon the research objectives.

Why are *seed* proteins attributed with such great significance for systematic purposes? A priori such significance is not expected because of the low selection pressure operating on the "unspecific amino acid pool for the young seedling". Only vague suggestions can be offered for the physiological functions of the seed storage proteins (Ericson and Crispeels 1973, Cameron-Mills et al. 1978, Schmitt and Svendsen 1980, Ingversen 1975, v. Wettstein 1979). Probably the storage proteins are continuously under a moderately functional constraint from an evolutionary point of view (Schmitt and Svendsen 1980), not conceding an unlimited variability of amino acids at an entire position. However, their evolutionary rate of change is greater than for many other proteins, especially those which are involved in the basic metabolic processes. This fact seems to be the most important reason why such proteins are so valuable in systematic research.

Acknowledgements. The prudent technical assistance of Elke Albrecht and Renate Föckler is gratefully acknowledged. This work was partly supported by the Deutsche Forschungsgemeinschaft (DFG).

References

Arnon R (1971) Current Top Microbiol Immunol 54:47–93

Autran JC, Bourdet A (1975) Ann Amelior Plant 25:277–301

Autran JC, Lew EJ-L, Nimmo CC, Kasarda DD (1979) Nature (London) 282:527–529
Badley RA, Atkinson D, Hauser H, Oldani D, Green JP, Stubbs JM (1975) Biochim Biophys Acta 412:214–228
Bailey CJ, Boulter D (1972) Phytochemistry 11:59–64
Beevers L (1982) In: Boulter D, Parthier B (eds) Nucleic acids and proteins in plants, vol I. Structure, biochemistry, and physiology of proteins. Springer Berlin Heidelberg New York, pp 136 to 168
Berman PW, Harbury HA (1980) J Biol Chem 255:6133–6137
Bietz JA, Huebner FR, Sanderson JE, Wall JS (1977) Cereal Chem 54:1070–1083
Bietz JA, Paulis JW, Wall JS (1979) Cereal Chem 56:327–332
Boulter D (1981) In: Polhill RM, Raven PH (eds) Advances in legume systematics, vol II. Kew R Bot Gard Kew, pp 501–512
Brown JWS, Bliss FA, Hall TC (1980) Plant Physiol 66:838–840
Büttner C, Jensen U (1981) Biochem Syst Ecol 9:251–256
Cameron-Mills V, Ingversen J, Brandt A (1978) Carlsberg Res Commun 43:91–102
Casey R (1979) Biochem J 177:509–520
Casey R, March JF, Sharman JE, Short MN (1981) Biochim Biophys Acta 67:428–432
Catsimpoolas N, Wang J (1971) Anal Biochem 44:436–444
Croy RRD, Derbyshire E, Krishna TG, Boulter D (1979) New Phytol 83:29–35
Croy RRD, Gatehouse JA, Tyler M, Boulter D (1980) Biochem J 191:509–516
Croy RRD, Lycett GW, Gatehouse JA, Yarwood JN, Boulter D (1982) Nature (London) 295:76 to 79
Dahlgren R (1983) In: Jensen U, Fairbrothers DE (eds) Proteins and nucleic acids in plant systematics. Springer, Berlin Heidelberg New York
Darmency H, Gasquez J, Mosse J (1980) C R Acad Sci 290:435–438
Davey RA, Dudman WF (1979) Aust J Plant Physiol 6:435–447
Derbyshire E, Boulter D (1976) Phytochemistry 15:411–414
Derbyshire E, Wright DJ, Boulter D (1976) Phytochemistry 15:3–24
Dhaliwal HS (1977) Theor Appl Genet 51:71–79
Doll H, Brown AHD (1979) Can J Genet Cytol 21:391–404
Dudman WF, Millerd A (1975) Biochem Syst Ecol 3:25–33
Elleman TC (1977) Aust J Biol Sci 30:33–45
Ericson MC, Chrispeels MJ (1973) Plant Physiol 52:98–104
Esen A, Bietz JA, Paulis JW, Wall JS (1982) Nature (London) 296:678–679
Fazekas de St Groth S (1983) In: Jensen U, Fairbrothers DE (eds) Proteins and nucleic acids in plant systematics. Springer, Berlin Heidelberg New York
Gatehouse JA, Croy RRD, Boulter D (1980) Biochem J 185:497–503
Gillespie JM, Blagrove RJ (1978) Aust J Plant Physiol 5:357–369
Gilroy J, Wright DJ, Boulter D (1979) Phytochemistry 18:315–316
Grumpe B (1981) Manuscr, Univ Köln
Guldager P (1978) Theor Appl Genet 53:241–250
Guttermann D (1981) Manuscr, Univ Köln
Hagen G, Rubenstein I (1980) Plant Sci Lett 19:217–223
Hara I, Ohmiya M, Matsubara H (1978) Plant Cell Physiol 19:237–243
Hirano H, Gatehouse JA, Boulter D (1982) FEBS Lett 145:99–102
Ingversen J (1975) Hereditas 81:69–76
Jensen U (1981) In: Ellenberg H, Esser K, Kubitzki K, Schnepf E, Ziegler H (eds) Progress in botany, vol 43. Springer, Berlin Heidelberg New York, pp 344–369
Jensen U, Büttner C (1981) Taxon 30:404–419
Jensen U, Penner R (1980) Biochem Syst Ecol 8:161–170
Johnson BL (1975) Can J Genet Cytol 17:21–39
Kasarda DD, DaRoza DA, Ohms JI (1974) Biochim Biophys Acta 351:290–294
Kim SI, Mossé J (1979) Can J Genet Cytol 21:309–318
Kim SI, Charbonnier L, Mossé J (1978) Biochim Biophys Acta 537:22–30
Kim SI, Saur L, Mossé J (1979) Theor Appl Genet 54:49–54

Kitamura K, Toyokawa Y, Harada K (1980) Phytochemistry 19:1841–1843
Kloz J (1971) In: Harborne JB, Boulter D, Turner BL (eds) Chemotaxonomy of the Leguminosae. Academic Press, London New York, pp 309–365
Kloz J, Turková V (1963) Biol Plant 5:29–40
Kloz J, Klozová E (1974) Biol Plant 16:290–300
Kloz J, Turková V, Klozová E (1963) Serol Mus Bull 30:1–2
Krishna TG, Croy RRD, Boulter D (1979) Phytochemistry 18:1879–1880
Manteuffel R, Scholz G (1975) Biochem Physiol Pflanz 168:277–285
Matta NK, Gatehouse JA (1982) Heredity 48:383–392
Matta NK, Gatehouse JA, Boulter D (1981a) J Exp Bot 32:183–197
Matta NK, Gatehouse JA, Boulter D (1981b) J Exp Bot 32:1295–1307
Metzger H, Shapiro MB, Mosimann JE, Vinton JE (1968) Nature (London) 219:1166–1168
Miège M-N (1982) In: Boulter D, Parthier B (eds) Nucleic acids and proteins in plants, vol I. Structure, biochemistry and physiology of proteins. Springer, Berlin Heidelberg New York, pp 291–345
Moreira MA, Hermodson MA, Larkins BA, Nielsen N (1979) J Biol Chem 254:9921–9926
Ohmiya M, Hara I, Matsubara H (1980) Plant Cell Physiol 21:157–167
Patey AL, Evans DJ, Tiplady R, Byfield PGH, Matthews EW (1975) Lancet II:718
Payne PI, Rhodes AP (1982) In: Boulter D, Parthier B (eds) Nucleic acids and proteins in plants, vol I. Structure, biochemistry and physiology of proteins. Springer, Berlin Heidelberg New York, pp 346–369
Przybylska J, Hurich J, Zimniak-Przybylska Z (1979) Genet Pol 20:517–528
Rosenkranz J (1977) Manuscr, Univ Köln
Schlesier B, Manteuffel R, Rudolph A, Behlke J (1978) Biochem Physiol Pflanz 173:420–428
Schmitt JM, Svendsen I (1980) Carlsberg Res Commun 45:143–148
Shewry PR, Pratt HM, Charlton MJ, Miflin BJ (1977) J Exp Bot 28:597–606
Shewry PR, Faulks AJ, Pratt HM, Miflin BJ (1978a) J Sci Food Agric 29:847–849
Shewry PR, Pratt HM, Miflin BJ (1978b) J Sci Food Agric 29:587–596
Shewry PR, March JF, Miflin BJ (1980a) Phytochemistry 19:2113–2115
Shewry PR, Autran J-C, Nimmo CC, Lew EJ-L, Kasarda DD (1980b) Nature (London) 286:520 to 522
Spencer D, Higgins TJV (1980) Biochem Int 1:501–509
Spielmann A, Schürmann P, Stutz E (1982) Plant Sci Lett 24:137–145
Stegemann H (1983) In: Jensen U, Fairbrothers DE (eds) Proteins and nucleic acids in plant systematics. Springer, Berlin Heidelberg New York
Stegemann H, Pietsch G (1983) In: Gottschalk W, Müller HP (eds) Seed proteins – biochemistry, genetics, nutritional value. Nijhoff, Den Haag, in press
Sun SM, Slightom JL, Hall TC (1981) Nature (London) 289:37–41
Svendsen I, Martin B, Jonassen I (1980) Carlsberg Res Commun 45:79–85
Tombs MP, Lowe M (1967) Biochem J 105:181–187
Utsumi S, Mori T (1980) Biochim Biophys Acta 621:179–189
Villamil CB, Duell RW, Fairbrothers DE, Sadowski J (1982) Crop Science 22:786–793
Vitale A, Soave C, Galante E (1980) Plant Sci Lett 18:57–64
Wettstein v D (1979) In: Proc 17th Eur Brew Convention Congr, Berlin
Wright DJ, Boulter D (1974) Biochem J 141:413–418
Wrigley CW, Shepherd KW (1973) Ann NY Acad Sci 209:154–162
Wrigley CW, Shepherd KW (1974) Aust J Agric Res 14:796–804

Pollen Proteins

F. P. PETERSEN[1]

Abstract. If the problems involved with its collection, processing, and storage are understood, pollen can provide the researcher with a broad range of easily extractable and highly antigenic proteins that are readily amenable to both electrophoretic and immunological characterizations. Such characterizations, in addition to their usefulness in genetic and evolutionary studies, show the potential for being of value for the solving of systematic and classification problems at the species level as well as among higher categories. A review is provided of the research conducted on pollen proteins with specific reference to the studies done with *Amphipterygium* (Julianiaceae), *Leitneria* (Leitneriaceae), and *Quercus* (Fagaceae).

1 Introduction

With the knowledge that plant proteins possess properties useful for systematics has come their increasing appearance in the taxonomic literature (Fairbrothers et al. 1975). Owing to the relative ease of collection and storage, seed material has been in the forefront of these chemosystematic reports. This is in sharp contrast to the scarcity of taxonomic work oriented toward the use of pollen proteins.

Much of the systematic and taxonomic research, relating to pollen proteins reported in the literature, deals with either pollen/stigma recognition factors, or the clinical role of pollen proteins as allergens. Unfortunately, a large segment of the taxonomic speculation that is to be found in either of these two types of research are generally incomplete and hence of limited value. Such research, however, does give insight into the nature of pollen proteins which is, by itself, of vital interest for future systematic research employing pollen proteins.

2 Nature of Pollen Proteins

Since 1886, when Strasburger demonstrated that pollen releases a substance, amylase (capable of liquifying starch), it has been known that pollen possesses enzymes. Today it is realized that there are comparatively few enzymes of plant tissue which cannot

1 Department of Biological Sciences, Rutgers the State University, P.O. Box 1059, Piscataway, N.J. 08854, USA

Proteins and Nucleic Acids in Plant Systematics
ed. by U. Jensen and D.E. Fairbrothers

be localized in either the cytoplasm of the pollen grain or the pollen wall (Standley and Linskens 1965, Brewbaker 1971).

The pollen wall consists of an outer layer, the exine, composed of sporopollenin, and an inner pectocellulosic layer, the intine. Early information as to the nature of the proteins found in the pollen wall indicated that the proteins present in the exine cavities of pollen of the Malvaceae, were derived from the parental tapetal cells of the anther prior to pollen maturation (Heslop-Harrison et al. 1973). This sporophytic nature of the exine proteins is to be contrasted with the intine proteins, which are products of the gametophytic pollen grain. These intine proteins are concentrated in the germinal apertures (Knox et al. 1975, Heslop-Harrison 1975b, Clarke and Knox 1978).

Studies by other researchers have also dealt with the nature of the proteins located within the pollen grain. Weeden and Gottlieb (1980) reported that only cytoplasmic enzymes were being leached from pollen during the first 12 h of soaking in a buffer, whereas the organellar bound enzymes were released following either a prolonged soaking or the mechanical crushing of the pollen grain. They also noted that phosphoglucosisomerase (PGI), found in the pollen of *Clarkia* (Onagraceae), was synthesized by the pollen and not by the surrounding tapetal cells. Other enzymes that were leached from the pollen grain (e.g. acid phosphatase and esterase), were ascertained by them to be products of synthesis within the pollen grain, and not tapetal in origin. In an earlier study performed on pollen wall proteins of *Brassica oleracea* var. *acephala* the acid phosphatase of the intine was found to be the product of two periods of synthesis, the first associated with the deposition of intime polysaccharides, and the second with cytoplasmic synthesis during pollen grain maturation (Vithanage and Knox 1976). The same researchers noted that esterase activity, although initially found to be concentrated in the tapetal cells of the developing anthers, was transfered from the tapetum to the exine cavities during the maturation of the pollen grain.

Relating to the specificity of the proteins in pollen, Scandalios (1964), discovered that *Zea mays* pollen yielded three isozymes with differing mobilities from those of other tissues of the same plant. In a similar study, Roose and Gottlieb (1980) found that both seed and pollen of *Stephanomera exigua* (Asteraceae) possessed an anodally migrating alcohol dehydrogenase enzyme (ADH-2). Their previously described ADH-1 now appears to be unique to seed material.

When immunological tests were performed with rabbit antisera produced to pollen surface components of *Gladiolus gandavensis* (Clarke et al. 1977), it was shown that immunoprecipitation identity reactions could be produced among extracts from the petal, corm, pollen, and stigma tissues of the same plant. This cross-reactivity of plant tissue might well indicate that plant cells, like animal cells, are specified according to their family, origin, and tissue by surface determinants (Clarke et al. 1977).

As to the sources of variation inherent in any study of pollen proteins, it is known that pollen enzyme activities can vary depending on the species (Veidenberg and Safonov 1968), the age of the plant when pollen developed (Linskens 1966), methods of pollen protein extraction (King et al. 1964), and storage (Standley and Linskens 1974, Johnson and Fairbrothers 1975).

Isozymes can also arise as isolation artifacts resulting from partial proteolysis during extraction from the pollen grain, and thus may not always be related to their functional condition in the pollen (Dickinson and Davies 1971). In addition, detection of enzyme activities may be influenced by interfering compounds, such as phenolics. Bredemeijer (1970) found that only by careful extraction of the enzyme glutamine dehydrogenase (GDH) with polyvinylpyrrolidine (PVP), a phenolic binding agent, was the detection of high activities of GDH possible. Such modifying factors, as well as handling, bacterial contamination (Hitchcock 1956), and the developmental stage of the pollen (Paton 1921) may all contribute to detected variation in enzyme activities.

3 Pollen/Stigma Interactions

Much of the research involving the localization of pollen synthesis has been associated with questions relating to cell recognition factors, specifically pollen/stigma inter- and intra-specific compatibility control of breeding behavior. Research involving pollen/stigma interactions has tended to emphasize pollen interactions with "dry type" (e.g. *Gladiolus*) rather than "wet type" stigmas (e.g. *Petunia, Nicotiana, Prunus*). In the example of *Gladiolus gandavensis* its stigma was found to possess a sticky receptive layer composed of glycoproteins, proteins, and glycolipids secreted from the papillar cells (Lewis 1951, Linskens 1960, Knox et al. 1975 and 1976, Clarke et al. 1977, Heslop-Harrison 1977). Some of these secreted stigma substances appear to possess enzymatic activity (e.g. esterase), whereas in other instances antigenic activity has been noted (Knox et al. 1976, Heslop-Harrison 1977).

Pollination commences with the arrival of pollen on the stigma where the pollen, if compatible with the stigma, will take up fluid (hydrates). Upon completion of pollen hydration there is a release of tapetal derived exine borne fractions (proteins, glycoproteins, lipids) through the surface micropores of the pollen grain (Heslop-Harrison et al. 1973). Subsequent to this tapetal release, is the release of the intine proteins which in turn signals the start of pollen tube growth and germination (Linskens and Heinen 1962).

Genetic studies have also been used to explain the physiological barriers of the recognition events of the pollen/stigma interaction. Following East and Mangelsdorf's 1925 study which dealt with the role of the multiple alleles of the S gene in cellular recognition, Lewis (1960) presented evidence that closely linked cistrons determine the specificity of the S gene. In 1968, Heslop-Harrison combined the genetic and biochemical information relating to pollen/stigma compatibility, in an attempt to explain the different types of interspecific self-incompatibility that had been observed. It was hypothesized that the initial event in compatibility is determined by the S-genotype of the pollen producing plant's sporophytic system, with the second recognition determined by the S-genotype of the pollen (gametophytic system). In support of this hypothesis came the evidence obtained from ultrastructural and cytochemical experiments that indicated that pollen wall proteins were derived from two genetically separated sources: the parental tapetal tissue and the haploid pollen (Heslop-Harrison 1975a). Further, research reported by Vithanage and Knox (1976), found that *Brassica oleracea* var. *acephala*, had its self-incompatibility mechanism controlled by a single major locus S, with the behavior of the haploid pollen grains on the stigma determined by the genotype of its diploid parents via the tapetal derived exine proteins. Similar sporophytic self-incompatibility systems were found to be not uncommon within taxa of the Asteraceae and the Brassicaceae (Dickinson and Lewis 1973a,b).

Physiological experiments employing exine proteins of *Populus deltoides* and *P. alba* have provided additional evidence that tapetal derived exine proteins act in the initial recognition reactions determining interspecific incompatibility between the two species (Knox et al. 1972a,b). When inactivated self-compatible pollen or self-compatible pollen wall protein diffusates were applied to *P. deltoides* stigmas followed by incompatible *P. alba* pollen, hybrid seeds were produced. By employing electrophoretic techniques, Ashford and Knox (1980), were able to characterize these active *P. deltoides* diffusates and determined their location to be in the extracellular sites of the pollen grain.

The pollen diffusates of *P. deltoides* and *P. alba* were further shown to contain more than 20 protein bands by PAGE, four bands of which were determined by the use of Schiff's reagent to be glycoproteins (Ashford and Knox 1980). It is of interest that glycoproteins have been detected in the diffusates of pollen grains, since glycoproteins have been implicated in plant cell recognition (Clarke and Knox 1978).

4 Allergens and Antigens

Among the compounds considered as principle mediators of plant cell recognition are: lectins, allergens, and cell wall components, e.g. arabinogalactan proteins and arabinoxylans (Clarke and Knox 1978). Of these the allergens have been the subject of experimentation by epidemiologists. Indeed, the allergen research is also a primary source of information relating to proteins that has the potential of being important to taxonomists.

Allergens are agents that commonly cause the induction of immunoglobulin E antibody biosynthesis and the concomitant allergenic sensitivity in animals. Allergens from part of the family of substances, termed antigens, that are capable of illiciting cellular and humoral antibody responses in animals. Generally allergens include those proteins, glycoproteins, and carbohydrates with molecular weights of less than 40,000 and which are of foreign animal or plant origin. From a chemical standpoint there is little to differentiate allergens from antigens, thus it is reasonable to speculate whether all antigens might have a similar potential to be allergenic (Marsh 1975).

The early realization of the widespread occurrence of pollen grains (Bostock 1828, Wyman 1872, Blackley 1873) and their accumulation in quantities sufficient for study, are probable the principle reasons why pollen grains are the most widely analyzed of all the allergenic sources. Some of the earliest research relating to the isolation of allergenic components from pollen were reported by Kammann (1904, 1912), where it was deduced that the bulk of allergenic skin reactivity was to be found in the albumin fraction of pollen protein extracts precipitated using 50% to 100% saturated $(NH_4)_2SO_4$. This albumin fraction also revealed a marked insensitivity toward heating and extreme acid/alkaline pH. Investigations by Augustine (1959a,b) revealed the presence of this principle allergenic activity in the protein containing fraction of extracts from both ragweed and grass pollen.

The cross-reactivities of a large number of pollen extracts important in causing hayfever were investigated by immunodiffusion analysis using rabbit antisera prepared against crude pollen protein extracts (Wodehouse 1954, 1955, Augustine 1959a,b, Feinberg 1960). It was in these studies that the great antigenic, and potentially, allergenic complexity of pollen extracts were clearly revealed. Further research showed that pollen from plants of different families (e.g. Poaceae, Asteraceae), were antigenically unrelated, whereas pollen from closely related species within a family were antigenically similar, as revealed by cross-reacting antigens (e.g. *Betula, Alnus* (Betulaceae): Berstein et al. 1976, Löwenstein 1980; *Ambrosia, Artemisia, Chrysanthemum* (Asteraceae): Leiferman and Gleich 1976, Löwenstein 1980; *Avena, Dactylis, Lolium, Phleum* (Poaceae): Tangen and Nilssen 1975, Leiferman and Gleich 1976, Löwenstein 1980).

4.1 Purification of Major Allergens

Allergens have been purified from pollen of *Ambrosia elatior* and *A. trifida* (e.g. antigens E and K: King et al. 1964, antigens Ra_3 and Ra_5: Underdown and Goodfriend 1969, Griffiths and Brunet 1971), from grass pollen of *Phleum pratense, Dactylis glomerata, Lolium perenne* (e.g. Group I, II, III allergens: Johnson and Marsh 1965, Augustine and Haywood 1962), and birch pollen antigens of *Betula verrucosa* (Belin 1972).

Using gel filtration the major allergens of *Betula verrucosa* were separated from the non-allergenic and non-protein soluble components. These major allergens appeared homogenous as to molecular weight, and consisted of at least two, heat stable differentially charged but antigenically similar proteins. Further testing indicated that the major allergen of *Alnus* pollen was serologically related to one detected in *Betula* (Belin 1972, Apold et al. 1981, 1982). A chemical analysis of one of the major allergens (PI 5.18) of *Betula verrucosa* pollen showed it to be a glycoprotein that was rich in serine (Apold et al. 1981). Additional observations (Berstein et al. 1976, Löwenstein 1980, Apold et al. 1982) indicated that many of the tree pollen demonstrate a wide range of immunological cross-reactivity patterns. It is interesting to speculate on the possibility that the main allergens of pollen may eventually prove to be useful in the solving of taxonomic problems among genera of specific families, as have the main seed storage proteins (legumin and vicilin, respectively legumin-like and vicilin-like proteins) for e.g. Fabaceae and Ranunculaceae (Dudman and Millerd 1975, Jensen and Penner 1980, Jensen and Grumpe this volume). One such chemotaxonomic study already has been performed. Utilizing pollen antigens of *Ambrosia* and related genera of the Asteraceae, Lee and Dickinson (1979) ascertained the presence of substances closely related to antigen E, the main allergen of *Ambrosia artemisifolia*.

In a comparative study reported by Leiferman and Gleich (1976) on the cross-reactivity of IgE antibodies with pollen antigens of various grass species, it was found that *Poa pratensis, Dactylis glomerata, Festuca elatior,* and *Lolium perenne* (tribe: Festuceae (Gould 1968)) displayed similar reactivity in both inhibition and absorption experiments; *Anthoxanthum odoratum* (tribe: Aveneae) and *Cynodon dactylis* (tribe: Chlorideae) were considerably less reactive. *Phleum pratense* (tribe: Aveneae), on the other hand, appeared to possess unique determinants not displayed by the other pollen examined. Although the conclusion to be drawn here was that most grass pollen has closely reacting antigens, clinical experiments indicate that this may not always be true. In an investigation reported by Wright and Clifford (1965) the test responses to 12 different grass pollens were observed, despite the fact that similar patient responses were generally found between the pollen from grasses within the same taxonomic groups, the authors concluded that the skin test response to one species of pollen did not presuppose a positive response to other closely related species.

4.2 Storage Effects on Allergens/Antigens

Experiments which include *Phleum pratense* pollen (Anderson and Baer 1982) have revealed that when pollen was maintained at 4 °C the deterioration of antigen activity was slowed. If the storage temperature was gradually raised there was an increase in the degradation of some proteins, demonstrating that *Phleum pratense* extract was a mixture of heat-stable and heat-labile components, an observation that had been noted in pollen extracts of other species (Kammann 1912, Baer et al. 1980). Besides thermal denaturation, another major cause of protein destruction is due to enzymatic activity. Anderson and Baer (1982) found that the presence of 50% glycerol in the protein extract can serve as a competitive inhibitor of the sugar splitting enzymes found in pollen, and thus retard denaturation.

5 Botanical Serosystematics

The significance of taxonomic investigations employing pollen is not only of academic concern, but has immense practical value for the evaluation of allergenic tests. It is the serotaxonomic field, with its incorporation of the skills of immunological research, that has demonstrated the greatest potential for bridging the important gap between clinical and systematic research.

The field of botanical serosystematics deals with the use of serological techniques for the comparison of plant proteins (Fairbrothers 1977). These serological comparisons can be made with proteins extracted from seeds, spores (Petersen and Fairbrothers 1972), tubers, leaves, and pollen (Fairbrothers 1977). Pollen proteins have proven to be an excellent source of antigenic material for use in the serological investigations of the Rosaceae (Keszythüs 1957) and the Betulaceae (Brunner and Fairbrothers 1979). In the Betulaceae three serological groupings were detected: (1) *Alnus*, (2) *Betula*, and (3) *Carpinus, Corylus,* and *Ostrya*. The serological data supported the maintenance of all five genera in a single family, and the designation of these detected groupings as three distinct tribes within a single family. Additional pollen research was conducted with antisera produced from pollen of Typhaceae (Lee and Fairbrothers 1969), Anacardiaceae, Fagaceae, Juglandaceae, and Myricaceae (Petersen and Fairbrothers 1979). In a study performed using antisera raised to *Corylus avellana* (Betulaceae) and *Populus tremula* (Salicaceae) pollen, Chupov (1978) outlined the serological relationships among some amentiferous families. The phytoserological studies of Lee (1981) using pollen proteins extracted from members of the Asteraceae were among the first experiments to include members of this family. Pollen proteins from seven species of *Ambrosia* and one species from each of five other genera of Asteraceae were compared using the following serological techniques: Ouchterlony double diffusion, radial immunodiffusion, and rocket immunoelectrophoresis. The obtained data supported other taxonomic data which suggested the merging of the genus *Franseria* with *Ambrosia* (Lee 1981).

5.1 Quantitative Precipitation Response – Boyden Curve

A precipitation technique that has been employed in serotaxonomy is the quantitative precipitation reaction of Boyden and DeFalco (1943), otherwise designated the "Boyden curve". In this technique, precipitation occurs in a liquid medium, and which can be measured as changes in turbidity that can be quantified in a spectrophotometer. When different antigens are reacted against a single antiserum, the degrees of turbidity produced by the test samples reflect their serological correspondence. Figure 1 shows an example of a series of Boyden curves where rabbit antiserum against *Juglans regia* pollen protein material was tested against eight other taxa. The results of these and other tests were used to define the serological affinities among selected taxa of the Anacardiaceae, Fagaceae, Juglandaceae, and Myricaceae. There are principally three ways in which taxa from these four families have been classified: (1) all four families grouped in the Sapindales (Bessey 1915); (2) Anacardiaceae, Juglandaceae, and Myricaceae placed in the superorder Rutiflorae with the Fagaceae in the superorder, Hamamelidiflorae (Thorne 1981); (3) Juglandaceae, Myricaceae, and Fagaceae put in the subclass Hamamelidae, with the Anacardiaceae in the subclass Rosidae (Cronquist 1981). The results of the quantitative precipitation tests with pollen supported a classification which separates the four families as designated by Cronquist (1981).

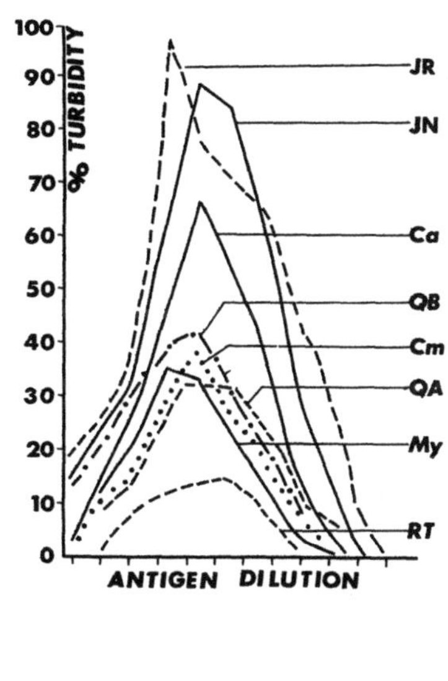

Fig. 1. Boyden precipitation curves showing the results of pollen proteins from eight taxa reacted with an antiserum produced to *Juglans regia* pollen proteins. *JR Juglans regia* L.; *JN J. nigra* L.; *Ca Carya cordiformis* Koch.; *QB Quercus borealis* Maxim.; *Cm Comptonia peregrina* Coult.; *QA Quercus alba* L.; *My Myrica pensylvanica* Loesel.; *RT Rhus typhina* L

5.2 Absorbed Antiserum

Using double diffusion, precipitation is performed in an agar medium and takes the form of visible immunoprecipitation lines. The immunoprecipitation lines produced by the protein extracts (antigen systems) to a particular antiserum can then be rated as to the number of immunoprecipitation lines and the degree of fusion of lines of the cross-reacting extracts with the reference reaction (Ouchterlony 1964).

A variant on the Ouchterlony double diffusion test is the absorbed antiserum (presaturation) technique (Moritz and Jensen 1961, Lester et al. this volume). In such experiments, proteins are tested as to their ability to deactivate an antiserum. Those extracts shown to have the greatest effect on the antiserum are deemed to possess a closer serological correspondence with the reference extract.

An example where absorption was employed using pollen antigens was a study to determine the affinities of *Leitneria floridana* (Leitneriaceae), a species endemic to the southeastern United States (Petersen and Fairbrothers 1983). In addition to antiserum against *Leitneria* pollen, 11 other antisera were used: anti-*Ailanthus*, anti-*Amphipterygium*, anti-*Betula*, anti-*Comptonia*, anti-*Corylus*, anti-*Fagus*, anti-*Juglans*, anti-*Ostrya*, anti-*Pterocarya*, anti-*Quercus*, and anti-*Rhus*. Using the mean absorption

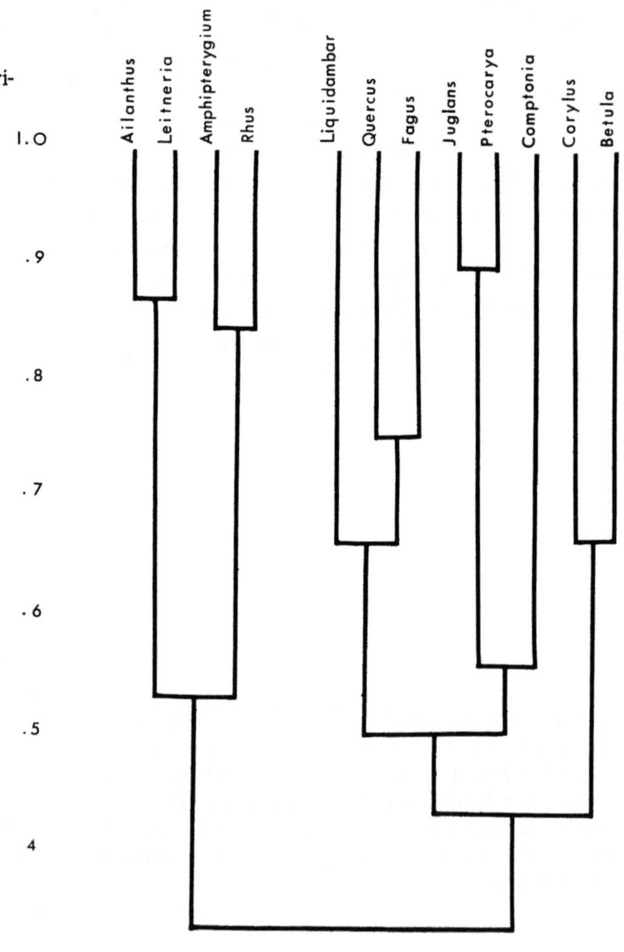

Fig. 2. Dendrogram showing the clustering of 12 taxa based on their weighted absorption similarity indices. (Petersen and Fairbrothers 1983)

similarity indices (Lester et al. this volume) for all 12 antisera, a UPGMA (unweighted pair group method analysis: Sneath and Sokal 1973) was performed, from which a dendrogram (Fig. 2) was generated. The results indicate two main groupings for the tested taxa: (1) Rosidae – *Leitneria* producing highest serological affinity (0.90) with *Ailanthus* of the Simaroubaceae; (2) Hamamelidae. The high affinity of *Amphipterygium* (Julianiaceae) with *Rhus* is noteworthy, as well as the grouping of *Juglans, Carya, Myrica, Comptonia* in close proximity to the Fagales *(Corylus, Ostrya, Quercus, Fagus)*. The latter results closely correspond with those obtained from the aforementioned turbidity analysis (Petersen and Fairbrothers 1979). The results from the various experiments with *Leitneria* were supportive of the contention of Abbe and Earle (1940) that the affinities of *Leitneria* are best found within the Sapindales.

5.3 Immunoelectrophoresis (IE)

Much of the serobotanical research that has employed immunoelectrophoresis, has used the technique of one-dimensional IE (Grabar 1959). The disadvantage of this technique lies in the inability to directly compare immunoprecipitation systems of one extract with that of another, there being no analog to the spur formation used in the analysis of Ouchterlony double diffusion.

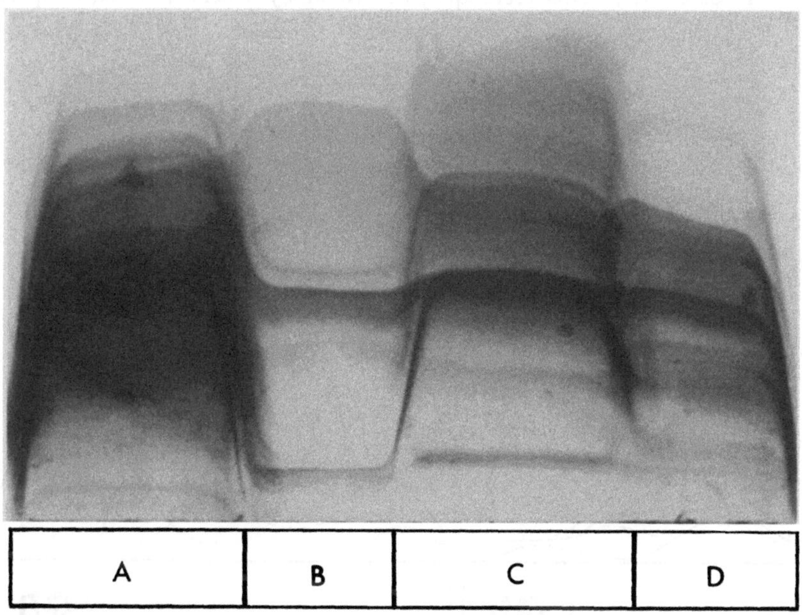

Fig. 3. Line immunoelectrophoresis of pollen proteins. Antigens: **A** *Amphipterygium glaucum* pollen protein, **B** *Ailanthus altissima* pollen protein, **C** *Rhus sylvestris* pollen protein, **D** *Rhus copallina* pollen protein. The content of pollen protein in the sample gels **A–D** was 20% (V/V). The sample gels were moulded at a temperature of 45 °C. Antibodies: Polyvalent rabbit antiserum against *Amphipterygium glaucum* pollen protein extract. The content of antiserum in the immunoelectrophoresis gel was 15% (V/V). Immunoelectrophoresis was carried out at 1.75 V/cm for 20 h *(anode at top)*. The bands were stained with Coomassie brillant blue R. The 0.75% agarose gel was prepared with barbitol buffer

A modification of one-dimensional IE is line immunoelectrophoresis (line IE) where the antigens from separate taxa are incorporated into small slabs of agar lying side by side on the gel. The antigens containing gels are then electrophoresed into a gel containing the test antibodies (Kröll 1975). Using this technique the immunoprecipitation patterns will take the form of straight lines moving across the gel. Where antigens between the adjacent taxa are identical, the immunoprecipitation patterns remain unbroken. Only the height of the line will vary, with a decrease in the elevation of the immunoprecipitation line reflecting a decrease in the amount of that particular antigen for the taxon investigated. Line IE affords both a qualitative evaluation of serological similarity (number and fusion of lines), and a quantitative evaluation (relative heights of the precipitation lines) for the tested taxa.

Figure 3 shows a line IE gel containing an antiserum produced against *Amphipterygium glaucum* pollen extract and tested against pollen extracts of *Amphipterygium, Rhus sylvestris, R. copallina,* and *Ailanthus altissima*. Results indicate the strong serological similarity of *Amphipterygium* with *Rhus* of the Anacardiaceae, as well as graphically demonstrating (i.e. presence of three shared immunoprecipitation systems) its association with the Simaroubaceae *(Ailanthus)*, another family within the Sapindales (Cronquist 1981).

Another useful immunoelectrophoretic technique is that of crossed immunoelectrophoresis (crossed IE). This technique, described by Weeke (1975), couples

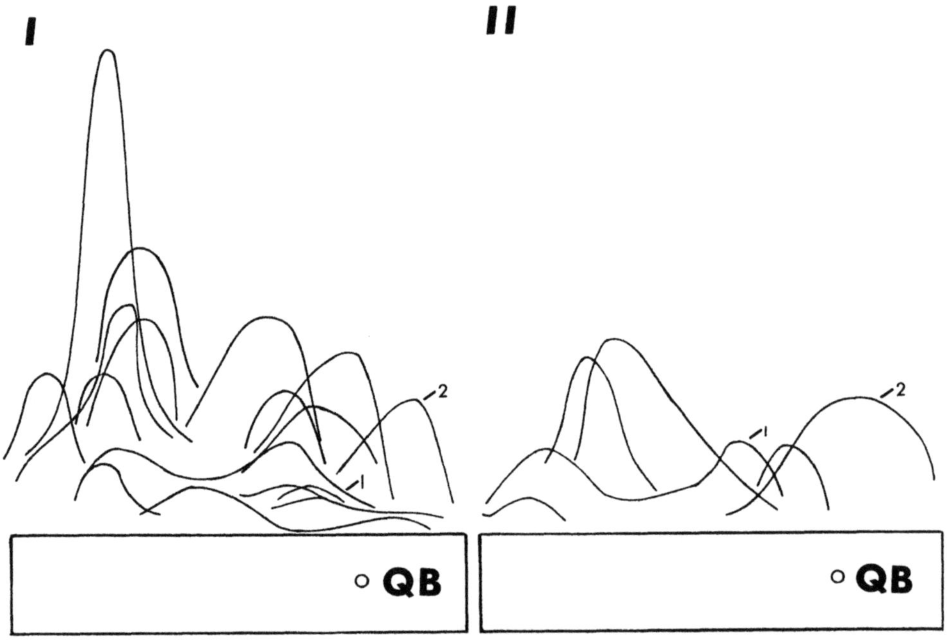

Fig. 4. Crossed immunoelectrophoresis of pollen proteins. Antigens: The application well was charged with 10 μl of *Quercus borealis (QB)* pollen proteins. Antiserum: Polyvalent rabbit antiserum against *Quercus borealis* pollen protein extract (**plate I**); same antiserum absorbed with *Fagus grandifolia* pollen (**plate II**). The content of antiserum in the immunoelectrophoresis gel was 15% (V/V). Immunoelectrophoresis was carried out at 6 V/cm for 1.5 h (first dimension); 1.5 V/cm for 16 h (second dimension).
Bands stained with Coomassie brillant blue R

one-dimensional IE with a second electrophoresis into a gel containing antiserum. This double electrophoresis produces a myriad of immunoprecipitation arcs when reacted with a general protein stain such as Coomassie brillant blue (Fig. 4, plate I; anti-*Quercus borealis* serum tested against the reference (QB) pollen antigens). When absorption is coupled with this technique the presence of only those immunoprecipitation systems unique to the reference extract can be ascertained. In plate II of Fig. 4, there are six major immunoprecipitation systems that are still remaining in the crossed IE gel of anti-*Quercus borealis* serum following absorption with pollen from *Fagus grandifolia*. The numbers (1,2) indicate the location of two of these immunoprecipitation systems in the unabsorbed gel.

The preceeding serological investigations, using several techniques, have proven to be useful for the elucidation of serological similarities at the generic and higher levels. The following section deals with the application of polyacrylamide gel electrophoresis of pollen proteins for the solving of taxonomic problems at the species level and below.

6 Polyacrylamide Gel Electrophoresis (PAGE)

6.1 Disc Electrophoresis

The zymogram, or pattern, produced by a substrate for a particular enzyme following the electrophoresis of a protein extract, is visualized as differences in the number, relative intensity, and spacing between protein bands. Such direct comparisons of zymograms by polyacrylamide gel electrophoresis when applied to analysis of pollen revealed up to 25 protein bands (Linskens 1966). Veidenberg and Safonov (1968) examined zymograms obtained from pollen of three species and a hybrid of *Malus*. Payne and Fairbrothers (1973) found that the positions of protein bands in disc electrophoresis gels were correlated with populational differences of *Betula populifolia*. Villamil and Fairbrothers (1973) reported that proteins from pollen of eight populations of the *Alnus-serrulata-rugosa* complex could not be consistently distinguished from each other by disc electrophoresis. Their protein data supported treatment of the complex as a single species.

6.2 Isoelectrofocusing

Isoelectrofocusing is an electrophoretic technique by which proteins are separated based on their isoelectric points (pH_I) in a stable pH gradient. Studies done with pollen from *Typha* (Lee and Fairbrothers 1973) were among the first systematic studies to employ isoelectrofocusing. Analyses of esterase, malate dehydrogenase, glutamate dehydrogenase, and alcohol dehydrogenase activity of pollen enzymes indicated that hybrid *Typha* x *glauca* populations were the products of hybridization between adjacent parent populations (*T. angustifolia* and *T. latifolia*). The isoelectrofocusing data were used in conjunction with standard polyacrylamide gel electrophoresis to verify the parents of the hybrid.

6.2.1 Isozyme Profiles of Oak Pollen

There are approximately 500–600 species of oak which are native to the temperate regions of the northern hemisphere. These species have been arranged into various infrageneric categories based on the relative importance of morphological and anatomical characteristics (Oersted 1871, Schwarz 1936, Lee this volume). The following is a preliminary account from our laboratory of the use of isoelectrofocusing of oak pollen for determining the infrageneric similarities among *Quercus* species native to eastern North America and Korea.

6.2.2 Materials and Methods

The method of gel preparation is that outlined by Righetti and Drysdale (1976). Gels were 7% acrylamide, catalyzed with riboflavin 5′-phosphate. Pharmalytes pH range 3–10 are the ampholytes used. Pollen proteins are extracted in a 0.1 M tris-glycine buffer pH 7.0 for 1 h at 4 °C. Additives to the buffer include polyvinylpyrrolidine, to bind phenols, and 0.5% phenylmethylsulfonyl-fluoride, to control for protein breakdown caused by protease activity. Gels are electrophoresed at 25 watts for approximately 1 h. At the termination of the run, the pH gradient is determined and the gels are stained with either a total protein stain, or reacted with substrates for specified enzyme activities.

6.2.3 Discussion

Table 1 provides a listing arranged according to Rehder (1940), along with abbreviations, of the 24 experimental species. Trial extractions of *Quercus acutissima* in tris-glycine buffers of varying molarity (0.01–0.1) with either PVP or protease inhibitor produced comparable results with bands of individual samples as demonstrated by the precise alignement of the bands. Using the general protein stain, Coomassie brillant blue, up to 40 bands in the pH range 3–10 were produced, thus making comparisons difficult. When the general substrates for either esterase or acid phosphatase activity were used, a more manageable number of bands for comparative purposes was obtained.

Figure 5 is a diagramatic representation of the esterase activity obtained for the tested taxa organized as to their respective subgenera (Rehder 1940). Bearing in mind the degree of variability to be found within an individual species, it is noteworthy to observe that many species within the subgenus Erythrobalanus reveal a moderate degree of pattern uniformity. This is especially true for the species *Quercus coccinea*, *Q. ilicifolia*, *Q. imbricaria*, and *Q. marilandica*. Within the Lepidobalanus grouping, a

Table 1. A listing of tested species of *Quercus* arranged to subgenera and sections as outlined by Rehder (1940). Also indicated are the abbreviations used for each taxon

Subgenus I:	Cyclobalanopsis Prantl.	
	Q. acuta Thunb.	AC
Subgenus II:	Ērythrobalanus Ŝpach.	
	Q. borealis Michx.	BO
	Q. coccinea Muench.	CO
	Q. ilicifolia Wangh	IL
	Q. imbricaria Michx.	IM
	Q. marilandica Muench.	MD
	Q. nigra L.	NG
	Q. palustris Muench.	PL
	Q. phellos L.	PH
	Q. velutina Lam.	VL
Subgenus III:	Lepidobalanus Endl.	
	Section 1: Cerris Lound.	
	Q. acutissima Car.	AT
	Q. cerris L.	CE
	Q. serrata Thunb.	SE
	Q. variabilis Bl.	VR
	Section 5: Robur Reichenb.	
	Q. robur L.	RO
	Section 6: Prinus Loud.	
	Q. alba L.	AB
	Q. aliena Bl.	AI
	Q. dentata Thunb.	DN
	Q. liaotungensis Koidz.	LI
	Q. lyrata Walt.	LY
	Q. macrocarpa Michx.	MA
	Q. mongolica Fisch ex Turcz.	MO
	Q. prinus L.	PR
	Q. saulii Schneid.	SL

much greater diversity in the pollen protein zymograms is observed, with some segregation corresponding to the designated taxonomic sections (e.g. Cerris, Robur, and Prinus). The one tested taxon of Cyclobalanopsis produced a zymogram that is somewhat similar to taxa included within the subgenus Lepidobalanus (e.g. *Quercus dentata; Q. liatungensis*). No such similarity was observed between *Q. acutissima* and tested taxa of Lepidobalanus. This distinctiveness from Erythrobalanus was also reflected in the serological data reported from *Quercus* pollen (Lee, this volume). Lee indicated that antisera produced to species belonging to Lepidobalanus gave high serological correspondence (>91%) with other species in that subgenus, but lower values (>51%) with species in Cyclobalanopsis.

A duplicate gel, stained for acid phosphatase activity, produced an even more distinctive separation of the subgenera then that obtained for esterase activity. Figure 6 is a diagramatic representation of that data, and it reveals a strong similarity among species within specific subgenera. The dominant bands for species of subgenus Erythro-

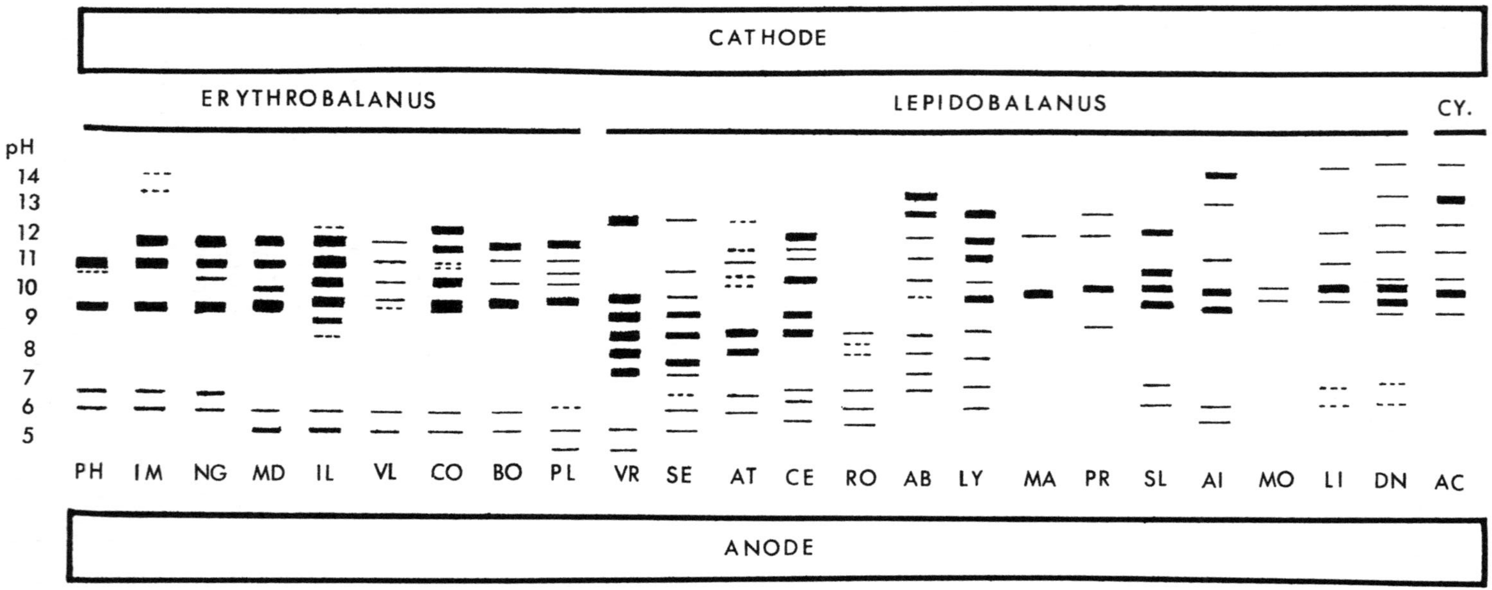

Fig. 5. Schematic representation of the zymograms of esterase activity produced by the isoelectrofocusing of pollen proteins from 24 taxa of *Quercus*. See Table 1 for the explanation of the abbreviations of the tested taxa. *CY* Cyclobalanopsis

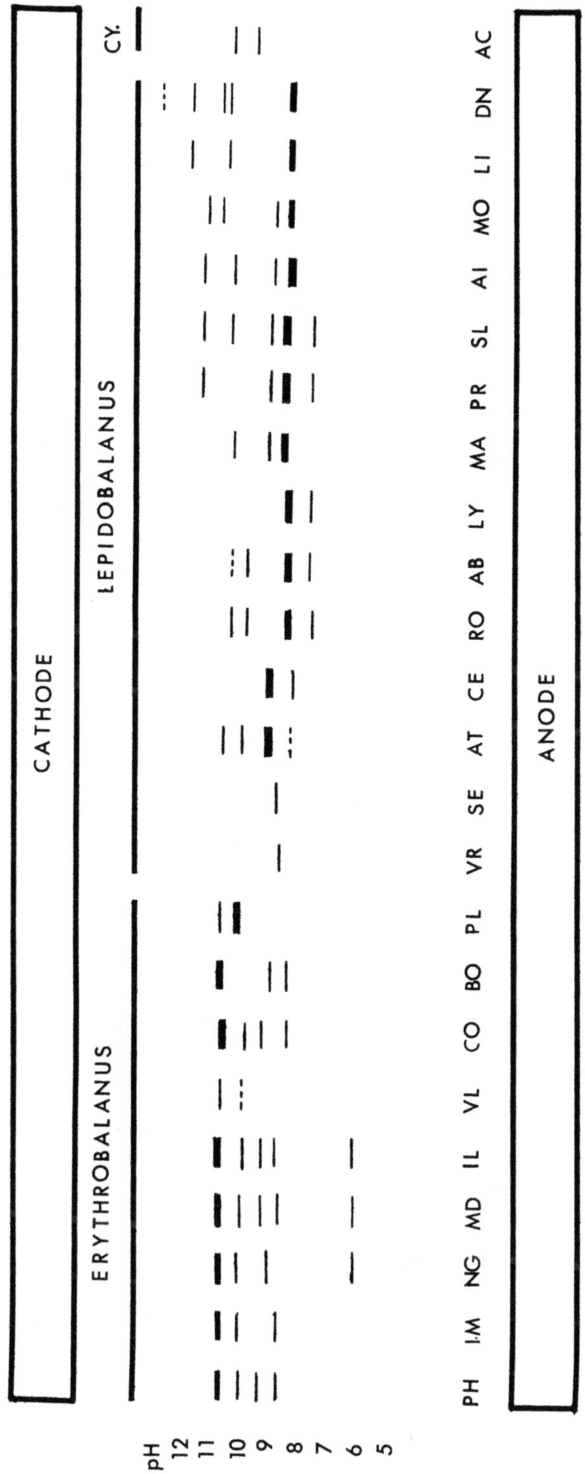

Fig. 6. Schematic representation of the zymograms of acid phosphatase activity produced by the isoelectrofocusing of pollen proteins from 24 taxa of *Quercus*. See Table 1 for the explanation of the abbreviations of the tested taxa. *CY* Cyclobalanopsis

balanus are found in the pH 10–11 range; while those for subgenus Lepidobalanus are in the 8–9 range. The acid phosphatase zymograms thus reveal a strong pattern uniformity within the two defined subgenera tested, with dramatic changes in electrophoretic mobility occurring when comparisons are made with species from different subgenera.

Substrate staining is an effective method for visualizing interspecific taxonomic relationships. This is especially true in pollen where the various enzymes are synthesized in either the gametophytic or sporophytic tissues. Indeed the example of *Populus* pollen (Ashford and Knox 1980) where esterase activity was found in both cytoplasmic and tapetal derived enzymes and acid phosphatase only in the cytoplasm, might have relevance to the great pattern diversity observed for *Quercus* pollen extracts reacted with similar substrates.

The diverse research findings included in this review clearly demonstrate that pollen contains an excellent source of proteins that can be readily extracted and utilized successfully for both serological and isozyme analyses, in evolutionary, genetic, systematic, and taxonomic investigations. The reviews also indicate the importance of understanding the significance of the origin of the proteins in different portions of the pollen grains. Adequate pollen data have been published to serve as an excellent guide for future chemosystematic and chemotaxonomic research.

References

Abbe EC, Earle TT (1940) Bull Torrey Bot Club 67:173–193

Anderson MC, Baer H (1982) J Allergy Clin Immunol 69:3–10

Apold J, Florvaag E, Elsayed S (1981) Int Arch Allergy Appl Immunol 64:439–477

Apold J, Florvaag E, Elsayed S (1982) Int Arch Allergy Appl Immunol 67:49–56

Ashford AE, Knox RB (1980) J Cell Sci 49:1–18

Augustine R (1959a) Immunology 2:148

Augustine R (1959b) Immunology 2:230

Augustine R, Haywood BJ (1962) Immunology 5:424–436

Baer H, Anderson MC, Hale R, Gleich GJ (1980) J Allergy Clin Immunol 66:281

Belin L (1972) Int Arch Allergy Appl Immunol 42:329–342

Berstein IC, Perera M, Gallagher J, Gabriel MJ, Johansson SGO (1976) J Allergy Clin Immunol 57:142–152

Bessey CE (1915) Ann Mo Bot Gard 2:109–164

Blackley CH (1873) In: Experimental researches on the cause and nature of catarrhus eastivus. Basillere, Tindall and Cox, London

Bostock J (1828) Med Chir Trans 14:437

Boyden A, DeFalco RJ (1943) Physiol Zool 16:229–241

Bredemeijer G (1970) Acta Bot Neerl 19:481

Brewbaker JL (1971) In: Heslop-Harrison J (ed) Pollen: Development and physiology. Butterworths, London, pp 156–170

Brunner F, Fairbrothers DE (1979) Bull Torrey Bot Club 106:97–103

Chupov VS (1978) Bot Zh (Leningrad) 63:1579–1584

Clarke AE, Knox RB (1978) Q Rev Biol 53:3–28

Clarke AE, Knox RB, Harrison S, Raff J, Marchalonis J (1977) Nature (London) 265:161

Cronquist A (1981) An integrated system of classification for flowering plants. Columbia Univ Press, New York

Dickinson DB, Davies MD (1971) In: Heslop-Harrison J (ed) Pollen: Development and physiology. Butterworths, London, pp 156–170
Dickinson HG, Lewis D (1973a) Proc R Soc London Ser B 183:21–38
Dickinson HG, Lewis D (1973b) Proc R Soc London Ser B 184:149–165
Dudman WF, Millerd A (1975) Biochem Syst Ecol 3:25–33
East EM, Mangelsdorf AJ (1925) Proc Natl Acad Sci USA 2:166–171
Fairbrothers DE (1977) Ann Mo Bot Gard 64:147–160
Fairbrothers DE, Mabry TJ, Scogin RL, Turner BL (1975) Ann Mo Bot Gard 62:765–800
Feinberg JG (1960) Int Arch Allergy Appl Immunol 16:1
Gould JW (1968) Grass systematics. McGraw-Hill, New York
Grabar P (1959) In: Glick D (ed) Methods of biochemical analysis, vol VII. Interscience, New York, pp 1–38
Griffiths GW, Brunet R (1971) Can J Biochem 49:396–400
Heslop-Harrison J (1968) Nature (London) 218:90–91
Heslop-Harrison J (1975a) Annu Rev Plant Physiol 26:403–420
Heslop-Harrison J (1975b) Proc R Soc London Ser B 190:275–299
Heslop-Harrison J, Heslop-Harrison Y, Knox R, Howlett B (1973) Ann Bot (London) 37:403–412
Heslop-Harrison Y (1977) Ann Bot (London) 41:913–922
Hitchcock JD (1956) Am Bee J 96:487
Jensen U, Penner R (1980) Biochem Syst Ecol 8:161–170
Jensen U, Grumpe B (1983) In: Jensen U, Fairbrothers DE (eds) Proteins and nucleic acids in plant systematics. Springer, Berlin Heidelberg New York
Johnson P, Marsh DG (1965) Eur Polymer J 1:63–68
Johnson RG, Fairbrothers DE (1975) Biochem Syst Ecol 3:205–208
Kammann O (1904) Beitr Chem Physiol Pathol 5:346
Kammann O (1912) Biochem Z 46:151
Keszythüs (1957) Acta Physiol Acad Sci Hung 11:399–407
King TP, Norman PS, Connel JT (1964) Biochemistry 3:458–467
Knox RB, Willing RR, Ashford AE (1972a) Nature (London) 237:381–383
Knox RB, Willing RR, Pryor LD (1972b) Silvae Genet 21:65–69
Knox RB, Heslop-Harrison J, Heslop-Harrison Y (1975) In: Duckett JG, Racey PA (eds) The biology of the male gamete. Academic Press, London New York, pp 177–187
Knox RB, Clarke AE, Harrison S, Smith P, Marchalonis J (1976) Proc Natl Acad Sci USA 73: 2788–2792
Kröll J (1975) In: Axelson NH, Kröll J, Weeke B (eds) A manual of quantitative immunoelectrophoresis. Universitetsforl, Oslo, pp 61–67
Lee DW, Fairbrothers DE (1969) Brittonia 21:227–243
Lee DW, Fairbrothers DE (1973) Bull Torrey Bot Club 100:3–11
Lee YS (1981) Syst Bot 6:113–125
Lee YS (1983) In: Jensen U, Fairbrothers DE (eds) Proteins and nucleic acids in plant systematics. Springer, Berlin Heidelberg New York
Lee YS, Dickinson DB (1979) Am J Bot 66:245–252
Leiferman KM, Gleich GJ (1976) J Allergy Clin Immunol 58:129–139
Lester RN (1979) In: Hawkes JG, Lester RN, Skelding AD (eds) The biology and taxonomy of the Solanaceae. Linn Soc Symp Ser 7:285–303
Lester RN, Roberts PA, Lester C (1983) In: Jensen U, Fairbrothers DE (eds) Proteins and nucleic acids in plant systematics. Springer, Berlin Heidelberg New York
Lewis D (1951) Proc R Soc London Ser B 140:127–135
Lewis D (1960) Proc R Soc London Ser B 151:468–477
Linskens HF (1960) Z Bot 48:126–135
Linskens HF (1966) Planta 69:79–91
Linskens HF, Heinen W (1962) Z Bot
Löwenstein H (1980) Allergy 35:198–200
Marsh DG (1975) In: Sela (ed) The antigens, vol III. Academic Press, London New York, pp 271 to 359

Moritz O, Jensen U (1961) Bull Serol Mus 25:1–5
Oersted AS (1871) K Dan Vidensk Selsk Skr 9:334–370
Ouchterlony Ö (1964) In: Achroyd JF (ed) Immunological methods. Blackwell Scientific Publ, Oxford, pp 55–78
Paton JB (1921) Am J Bot 8:471
Payne RC, Fairbrothers DE (1973) Am J Bot 60:182
Petersen FP, Fairbrothers DE (1979) Syst Bot 4:230–241
Petersen FP, Fairbrothers DE (1983) Syst Bot 8:134–148
Petersen RL, Fairbrothers DE (1972) Am Midl Nat 85:439–457
Rehder A (1940) In: Manual of cultivated trees and shrubs. MacMillan, New York
Righetti PG, Drysdale JW (1976) In: Isoelectricfocusing. North-Holland, Amsterdam
Roose ML, Gottlieb LD (1980) Genetics 95:171–186
Scandalios JG (1964) J Hered 85:281–285
Schwarz O (1936) Notizbl Bot Gart Mus Berlin-Dahlem 13:1–22
Sneath PH, Sokal RR (1973) Numerical taxonomy. Freeman, San Francisco, pp 230–234
Standley RG, Linskens HF (1965) Physiol Plant 18:37–43
Standley RG, Linskens HF (1974) Pollen: Biology biochemistry management. Springer, Berlin Heidelberg New York
Strasburger E (1886) Jahrb Wiss Bot 17:50–98
Tangen O, Nilssen BE (1975) Dev Biol Stand 29:175–187
Thorne RF (1981) In: Young DA, Seigler DS (eds) Phytochemistry and angiosperm phylogeny. Praeger, New York
Underdown BJ, Goodfriend L (1969) Biochemistry 8:980–989
Veidenberg AE, Safonov VI (1968) Dokl Akad Nauk SSSR 180:1242–1245
Villamil CB, Fairbrothers DE (1974) Biochem Syst Ecol 2:16–20
Vithanage HIMV, Knox RB (1976) J Cell Sci 21:423–435
Vithanage HIMV, Knox RB (1977) Phytomorphology 27:168–179
Weeden NF, Gottlieb LD (1980) Plant Physiol 66:400–403
Weeke B (1975) In: Axelson N, Kröll J, Weeke B (eds) Universitetsforl Oslo, pp 61–67
Wodehouse RI (1954) Int Arch Allergy Appl Immunol 5:337
Wodehouse RI (1955) Int Arch Allergy Appl Immunol 6:65
Wright GLT, Clifford HT (1965) Med J Aust:74–75
Wyman M (1872) In: Autumnal Catarrh (Hay Fever). Hurd and Houghton, Cambridge, Mass, pp 82

Serological Protein Properties Contributing to Phylogeny and Taxonomy

Analysis of Immunotaxonomic Data Obtained from Spur Identification and Absorption Techniques

R.N. LESTER[1], P.A. ROBERTS[2], and C. LESTER

Abstract. Immunotaxonomic techniques such as absorption or spur tests produce resolution of subsets of antibodies and hence of antigenic determinant sites. Researchers have used different experimental techniques and methods of analysis based on different theoretical considerations. It is explained that similarity of a pair of antigen systems is shown by their conjoint ability to produce precipitate with an absorbed antibody system, and also by their conjoint ability to absorb an antibody system and prevent precipitate production by another test antigen system. It is proposed that conjoint inabilities (i.e. conjoint absences of characters) should be omitted from numerical taxonomic treatments by using Jaccard's or similar coefficients, which are shown to be superior to Simple Matching Coefficient or Euclidean Distance Squared. Cladistic analysis was attempted with partial success. The data were obtained from antisera to *Solanum capsicoides, S. prinophyllum, S. quitoense, S. rostratum, S. sisymbriifolium,* and *S. torvum* absorbed by and reacted with a total of 20 taxa. The taxonomic relationships of the species compared are discussed.

1 Introduction

Immunotaxonomic comparisons of organisms are made possible by the immunological response of host animals to the injection of antigens from the investigated organisms. The antibody system (or antiserum) produced by the host animal can be reacted with cross-reacting (or heterologous) antigen systems extracted from other organisms and these cross-reactions are compared with the reference (or homologous) reactions produced by the reference antigen system used to induce the antibody system.

There are diverse techniques which enable the comparison of a series of antigen systems (Ags) with an antibody system (Abs) to be made. Theoretically the greatest resolution can be obtained by absorption of an antibody system by one or more antigen systems to leave a subset of antibodies which are more discriminating in their reactions with other antigen systems. Absorption (pre-absorption, pre-saturation, Vorabsättigung) may be achieved in a semi-solid medium by diffusion of the reagents through a gel: unwanted components of the antibody system are removed by precipitation with the absorbing antigen system, the remaining unreacted antibodies diffuse further until they encounter the test antigen system and produce further precipitate. Various techniques for absorption or spur tests in gel media have been reported and major attempts to understand and to describe precisely the results have been published (Moritz and Jensen 1961, Kirsch 1967, Moore and Goodman 1968).

[1] Department of Plant Biology, Birmingham University, UK
[2] Department of Nematology, University of California, Riverside, USA

Proteins and Nucleic Acids in Plant Systematics
ed. by U. Jensen and D.E. Fairbrothers

The present report extends the approach of Lester (1979), and owes much to earlier workers who have developed the description of the reactions of absorbed antibody systems in terms of set theory.

2 Terminology and Notations

The reference antigen system of an organism (A) consists of several proteins (a_1, a_2, a_3 etc.) each protein bearing several antigenic determinant sites (or epitopes) (a_1, a_2, a_3 etc.): the sum of these constitute the set of antigenic determinants of antigen system A. Immunisation produces an antibody system (A^{ab}) consisting of a set of antibodies (a_1^{ab}, a_2^{ab}, a_3^{ab} etc.), each kind of antibody being specific to one kind of antigenic determinant site. Unfortunately, chiefly due to inadequacies in the host animal, the set A^{ab} is not the complete image set, and does not map precisely and with full correspondence onto the original set A (Fig. 1). Furthermore in subsequent experiments the reference reaction only measures the extent to which the set of antigenic determinants (a_1, a_2, a_3 etc.) of the test antigen system A maps onto the set of antibodies (a_1^{ab}, a_2^{ab}, a_3^{ab} etc.) of A^{ab}. For these and other reasons the members of the set of antigenic determinant sites of A may not be detected totally or even exactly proportionally. Nevertheless the reference reaction of A with A^{ab} can be used as a standard and the cross-reactions of antigen systems B, C, D etc. may serve as measures of the extent to which their respective sets of antigenic determinants map onto the

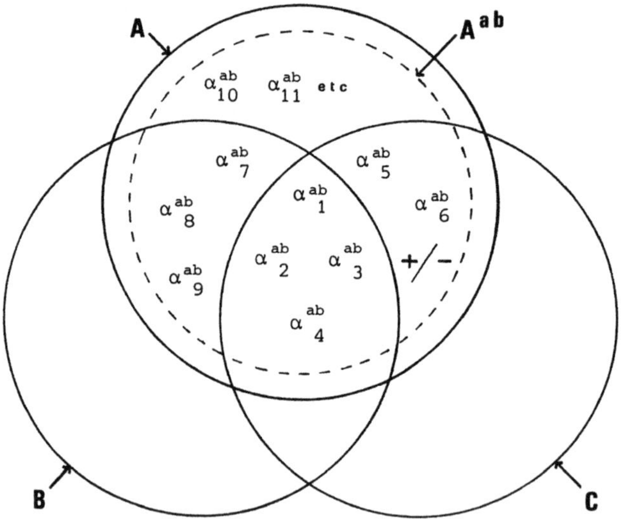

Fig. 1. Diagrammatic representation of the set of antibodies (a_1^{ab}, a_2^{ab}, etc.) of antibody system A^{ab}, produced by immunisation with antigen system A, showing that different members of this set may correspond to antigenic determinants which are variously members of sets A, B, and/or C.
In the intersect area $C \cap (A \cap B')$ a positive reaction (+) indicates similarity of C to A, whereas a negative reaction (−) indicates similarity of B to A

set of antibodies A^{ab} (Fig. 1). In the same way an antibody system B^{ab} may be produced by immunisation with antigen system B and may be reacted with the reference test antigen system B and other cross-reacting systems such as A, C, D etc. However since the antigenic determinant sites of both A and B may be detected neither completely nor exactly proportionally in their respective reference reactions, due to irregularities of host animal response, it is not surprising that the reaction of B with A^{ab} practically never is the same as the reaction of A with B^{ab}. This imperfect commutivity between $B \cap A$ and $A \cap B$ causes both theoretical and practical difficulties in the analyses of data.

When an antibody system A^{ab} is absorbed by a cross-reacting antigenic system B only those antibodies which are capable of reacting with antigenic determinant sites which are members of the set of A, but are the complement of the set of B (i.e. $A \cap B'$) remain. The absorbed antibody system may be described as $A^{ab}_{A \cap B'}$ or $A^{ab}_{B'}$ (Kirsch 1967, modified from Moritz and Jensen 1961). Likewise the antibody system B^{ab} absorbed by the antigen system A may be described as $B^{ab}_{A'}$. If the quantity (not quality) of the subset $A^{ab}_{B'}$ were the same as the quantity of the subset $B^{ab}_{A'}$ then these two values would be interchangeable, and the one measurement could be substituted for the other. This limitation was not considered by Goodman and Moore (1971) when explaining the "net spur size". In actual practice different results are often produced by two such absorbed antibody systems with their respective reference antigen systems, and some methods of analysis which are theoretically sound are not fully justified in practice. The problem is exacerbated when reactions such as $C + A^{ab}_{B'}$ are considered, i.e. the reaction of antigen system C with antibody system A^{ab} which has been absorbed by antigen system B, and when attempts are made to compare this with a reaction such as $A + C^{ab}_{B'}$.

The system of notation used here follows that of Kirsch (1967) and is derived from that of Moritz and Jensen (1961). Moore and Goodman (1968) and Goodman and Moore (1971) have used a different system of terminology and notation as partially explained below. [The equivalents used in the present paper are given in square brackets.] The antigen [antigen system] from species h [species H] consists of a set of antigenic sites [antigenic determinant sites] and can be called A_h [H]. Immunisation of an animal produces antibodies in the antiserum, $As(A_h)$ [antibody system, H^{ab}], which can be used for the homologous comparison [reference reaction] with the homologous antigen, A_h [reference antigen system, H] or in a heterologous comparison [cross-reaction] with an antigenic preparation from a heterologous species, A_r [cross-reacting antigen system, R]. When A_h and A_r are tested in a trefoil Ouchterlony plate against $As(A_h)$ a spur is formed: the net spur size between antigens A_h and A_r against antiserum $As(A_h)$ is called $S_{hr}{}^{h}$ or S^{h}_{hr}. If antigens A_j and A_k, from two heterologous species j and k, are tested against $As(A_h)$ the net spur size is S^{h}_{jk} or S_{jk} or $S_{j,k}$. This is a measure of the antigenic distance, H_{jk}, between A_j and A_k [and should not be confused with the usual numerical taxonomic notation where S_{jk} is a similarity coefficient between operational taxonomic units j and k].

The antigenic distance from species h to species r, H_{hr}, is measured by the quantity, N, of antigenic sites present in A_h but not A_r, i.e. $A_h \cap A_{r'}$ [$H \cap R'$]. Using antiserum $As(A_h)$ the net spur size between species h and r is described by $S^{h}_{hr} = N(A_h \cap A_{r'})$ or between species j and k by $S_{jk}{}^{h} = N(A_j \cap A_h \cap A_{k'}) - N(A_k \cap A_h \cap A_{j'})$

(i) Diagrams of three kinds of spurring and the coding systems used by Jensen and by Lester.

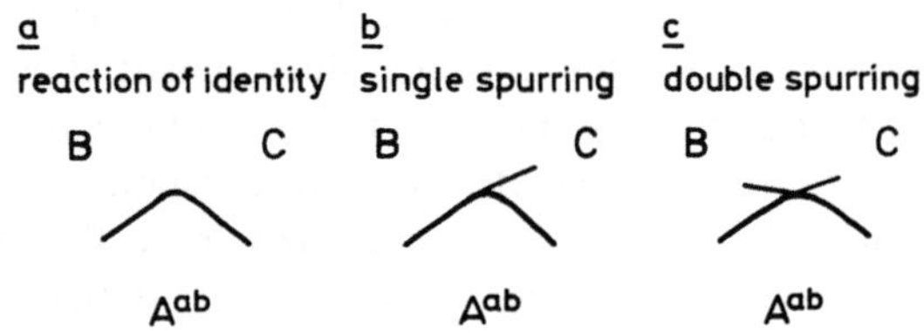

Code:	a		b		c	
Jensen	0	0	I	-	+	+
Lester	0	0	+	0	+	+

(ii) Description of spur reactions in set theory notation used in this paper

a. Reaction of identity

$A \cap B = A \cap C \qquad B \cap (A \cap C') = 0 \qquad C \cap (A \cap B') = 0$

b. Single spurring

$A \cap B \neq A \cap C \qquad B \cap (A \cap C') = + \qquad C \cap (A \cap B') = 0$

c. Double spurring

$A \cap B \neq A \cap C \qquad B \cap (A \cap C') = + \qquad C \cap (A \cap B') = +$

(iii) Description of spur reactions in notation used by Moore and Goodman to define net spur size, $S_{jk}{}^{h} \cdot A_h, A_j$, and A_k are A, B, and C respectively).

a. No spurs

$S_{jk}{}^{h} = 0$, i.e. $A_j \cap (A_h \cap A_{k'}) = 0$ and $A_k \cap (A_h \cap A_{j'}) = 0$

b. Net spur size

$S_{jk}{}^{h} = N(A_j \cap A_h \cap A_{k'}) - N(A_k \cap A_h \cap A_{j'})$

c. Two equal spurs

$S_{jk}{}^{h} = 0$, i.e. $A_j \cap (A_h \cap A_{k'}) = A_k \cap (A_h \cap A_{j'})$

Fig. 2. Interpretation of spur precipitates

(Fig. 2b). However in the comparison of A_j with A_k by As(A_h) the result will be evaluated as $S_{jk}{}^{h} = 0$ either if no spurs are formed (i.e. $A_j \cap (A_h \cap A_{k'}) = 0$ and $A_k \cap (A_h \cap A_{j'}) = 0$) or if two equal spurs are formed (i.e. $A_j \cap (A_h \cap A_{k'}) = A_k \cap (A_h \cap A_{j'})$) (Figs. 2a and c). [The formation of a spur by antigen system J and/or antigen system K when reacting with antibody system H^{ab} involves two reactions, namely $J + H^{ab}_{K'}$ —: sic/non and $K + H^{ab}_{J'}$ —: sic/non (Lester 1979, after Moritz and Jensen 1961, and after Kirsch 1967).]

Researchers have used set theory to varying degrees to discuss and analyse the results of absorption or spur tests. Moore and Goodman (1968) and Goodman and Moore (1971) have used a set theoretical approach to immunotaxonomy to develop a computer program for the analysis of spur data. Kirsch (1967) considered the theoretical principles, but did not follow them in his analysis of absorption data.

Jensen (1968a,b) avoided the problem by not making comparisons between antibody systems and by representing his absorption results graphically rather than arithmetically. Cristofolini and Chiapella (1977) used a relatively straightforward computer analysis of their spur tests, which was not derived from immunological principles. Lester (1979) used a manual method, based on immunological principles, which was relatively unsophisticated.

Jensen's method (Jensen 1968a, Jensen and Büttner 1981) utilises an analysis of spurs to determine, for each pair of test antigen systems, whether there is a reaction of identity with complete fusion of the two precipitation arcs (Fig. 2a), where there is single spurring indicating that the one antigen system has more determinants in common with the reference antigen system than has the other antigen system (Fig. 2b), or whether there is double spurring indicating that each test antigen system has some determinants in common with the reference antigen system, but lacking from the other antigen system (Fig. 2c). Careful analysis of all the spur reactions is used to deduce logically how many different subsets of antigenic determinants (or rather antibody determinants) have been resolved in the whole complex of experiments involving one antibody system. Each subset of determinants (or "determinant group") is then used as a character, the presence of each character in each taxon is tabulated and the taxa are then rearranged in rank order according to their similarity to the reference antigen system (Fig. 3). This method, which was developed before computer methods became available, is descriptive, but is approximately the same as the Mean Absorption Similarity Coefficient (Lester 1979) and produces a similar rank order. The visual impact of the bar diagrams approximates to using Jaccard's Coefficient.

The significance of absorption experiments should be considered carefully before devising a method of analysis. One antibody system can be absorbed separately by

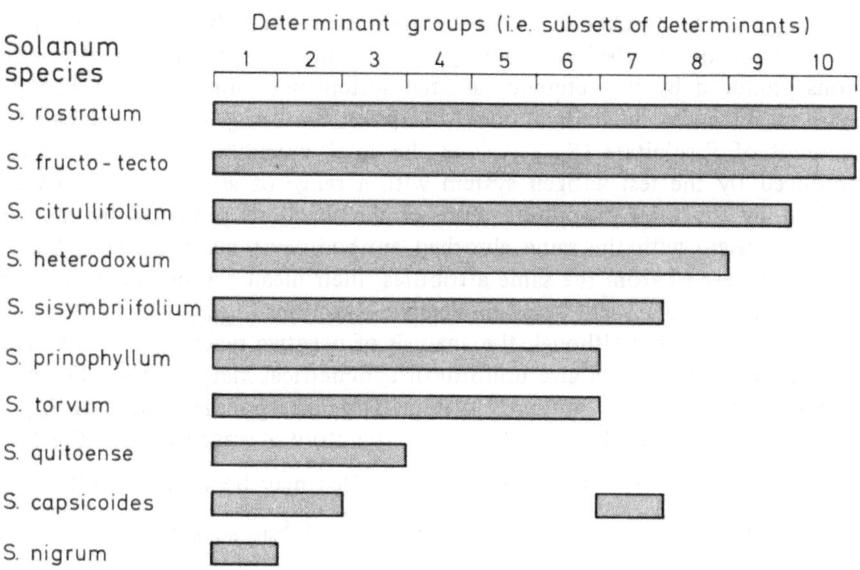

Fig. 3. Diagrammatic re-presentation of selected data from Table 2 to illustrate Jensen's method of serological determinant group analysis

each of a series of absorbing antigen systems and then tested separately with each of a series of test antigen systems. Ideally a complete square matrix of results can be produced (e.g. Tables 2–7) and can be analysed using the records of negative reactions or positive reactions in the appropriate columns or rows of the matrix.

Each negative reaction indicates that the absorbing antigen system is more like the reference system than is the test antigen system. Thus the affinity of B to A is indicated by the degree to which antigen system B absorbs antibody system A^{ab} and removes from it antibodies capable of reacting with the antigen systems of other taxa, e.g. C (indicated by – in $C \cap (A \cap B')$ in Fig. 1), D, E etc. Hence the number of negative reactions produced by the absorbed antibody system $A^{ab}_{B'}$ to a range of test antigen systems may be taken as an indication of the similarity of B to A. Lester (1979) defined the Absorption Similarity Coefficient based on Negative Reactions ($S_{no.neg.}$) as the number of negative reactions produced by the absorbed antibody system with a range of test antigen systems, divided by the number of negative reactions produced by the reference antibody system, absorbed by the reference antigen system, to the same test antigen systems.

An alternative analysis is based on the production of positive reactions. If a positive reaction is produced, it indicates that some antibodies of A^{ab}, which are capable of reacting with antigenic determinants of test antigen system C, have been left despite the previous availability for reaction of the antigenic determinants of B, which was used for absorption. A positive reaction indicates the existence, and to some extent the size of $C \cap (A \cap B')$ i.e. it is an indication of the similarity of C to A (see Fig. 1). The similarity of C to A may be assessed either by the number of absorbed antibody systems $A^{ab}_{B'}$, $A^{ab}_{D'}$, $A^{ab}_{E'}$ etc. which still react positively with C, or by the sum of the amounts of precipitate produced in all these reactions. Lester (1979) defined the Absorption Similarity Coefficient based on the Numbers of Positive Reactions ($S_{no.pos.}$) as the number of positive reactions produced by the test antigen system with a range of absorbed antibody systems, divided by the number of positive reactions produced by the reference antigen system with the same absorbed antibody systems. Likewise he defined the Absorption Similarity Coefficient based on the Amount of Precipitate ($S_{ppt.pos.}$) as the total precipitate score of the reactions produced by the test antigen system with a range of absorbed antibody systems, divided by the total precipitate score of the reactions produced by the reference antigen system with the same absorbed antibody systems. Since both these coefficients are derived from the same attributes, their mean can be taken as the Absorption Similarity Coefficient based on Positive Reactions ($S_{abs.pos.}$).

Lester found that although the analysis of negative reactions and the analysis of positive reactions did not give uniform or symmetrical matrices, they did give similar ranking orders for each antibody system. The corresponding values of $S_{no.neg.}$ and $S_{abs.pos.}$ for each antibody system were therefore averaged to give the Mean Absorption Similarity Coefficient ($S_{abs.mean}$). This may be restated as $S_{abs.mean} = \frac{1}{2}[S_{no.neg.} + \frac{1}{2}(S_{no.pos.} + S_{ppt.pos.})]$. The use of $S_{abs.mean}$ produced a more uniform and self-consistent matrix which was relatively symmetrical in that the ranking order of the various antigen systems with any one antibody system corresponded closely to the ranking order of the various antibody systems with the appropriate antigen systems.

A simpler coefficient, which gives almost identical results in square matrices such as those presented here, was suggested by Lester (1979) and may be called the Mean Absorption Similarity Coefficient based on the Numbers of Negative and Positive Reactions ($S_{abs.no.mean}$): the number of negative reactions caused by an absorbing antigen system is added to the number of positive reactions produced by it when it is used as a test antigen system with the whole range of absorbed antibody systems, and then divided by the corresponding total of negative and positive reactions of the reference antigen system. The „Korrelationszahlen" of Jensen (1968b, p. 271) appear to be the same.

Further analysis of a matrix of similarity coefficients, obtained by any of the foregoing methods, is made by averaging mutual values, such as those assessing $A \cap B$ and $B \cap A$, to produce a half matrix, and then clustering by a suitable sorting strategy to produce a phenogram.

Many computer programs for numerical taxonomy are readily available and several of these may be appropriate for analysis of immunotaxonomic data. The results of most kinds of spur analysis and absorption experiments can be presented as in Tables 2–7, where one antibody system is absorbed by each antigen system and is tested with each antigen system in all permutations. The results can be recorded as presence and amount, or else the absence, of precipitate in one or more arcs or spurs. Presence or absence can be treated as binary data and can be analysed by similarity coefficients such as the Simple Matching Coefficient or Euclidean Distance Squared, as used by Cristofolini and Chiapella (1977). Quantitative data can be treated in similar ways. These methods treat all conjoint absences of precipitate and all conjoint presences of precipitate as measures of similarity between any two antigen systems. However, Lester (1979) stated that these methods were inappropriate, and argued that when two antigen systems are used for absorption their conjoint ability to prevent a test antigen system from producing precipitate was a much better indication of their similarity to the reference antigen system, and hence to each other than was their conjoint failure to prevent such precipitate. Likewise the conjoint ability of two test antigen systems to produce precipitate with an antibody system absorbed by another antigen system was a much better indication of their similarity to the reference antigen system, and hence to each other, than was their conjoint failure to produce such precipitate. He recommended that data of negative and of positive reactions should be selected appropriately and should be analysed by Jaccard's or a comparable similarity coefficient.

Most of the methods discussed so far assess phenetic relationships and hence produce phenograms. Because proteins are such close reflections of the genetic code and also because immunology assesses so many antigenic determinant sites of the proteins, Goodman and Moore (1971) have asserted that the data from spur tests can be analysed by calculation of sum/cardinality ratios and hence cladograms which describe the phylogenetic relationships of the species being compared can be constructed. Mickevich (personal communication) has stated that cladistic analysis is appropriate for immunotaxonomic data obtained from absorption or spur analysis methods if the presence of a non-universal character can be taken as a departure from the condition of an unknown original ancestor. For a given group of organisms, any antigenic determinant sites shared by all members of the group probably were also present in their original common ancestor, and thus cannot serve as characters to differentiate between these organisms.

3 Materials and Methods

Seeds were obtained from 20 accessions of *Solanum* species, maintained in the Birmingham University Solanaceae Collection (Table 1). Proteins were extracted from defatted seed flour using phosphate buffered saline. Antisera were produced by immunising rabbits with protein extracts and Freund's adjuvant for about three months (Lester 1969). Immunodiffusion was carried out in borate-phosphate buffered 1.0% agarose on microscope slides (Sangar et al. 1972). Absorption was achieved by applying three aliquots of absorbing antigen system, followed by one aliquot of antibody system, to each central well, and one aliquot of appropriate reacting antigen systems to the four surrounding wells (Dray and Young 1959, Lester 1979). After 30 h incubation the gels were washed thoroughly (Lester 1969), dried, and stained with Crowle's Triple Stain (Crowle 1973). Reactions were recorded as very strong (4), strong (3), moderate (2), or weak (1) positives, uncertain positive-negatives (1/2) or definite negatives (0) (Tables 2–7). With two antisera, SIS^{ab} and CAP^{ab} (Tables 3 and 6), absorption was sometimes incomplete even after repeated absorption with more antigenic material. In these cases the score of the reaction of the reactant antigen system which had been used for absorption was reduced to zero and the same amount was deducted from all the other scores obtained using that absorbed antibody system, before further analysis. Full details of materials and methods were described by Roberts (1978).

4 Analysis of Data

The data were analysed by four main methods. For the present analyses all uncertain positive-negatives were changed to weak positives. Similar results were obtained when they were treated as definite negatives.

These numeric data were used to calculate the Mean Absorption Similarity Coefficients. The data were then converted into binary positive/negative form and used to calculate Simple Matching Coefficients, in which all conjoint positives and conjoint negatives were used to measure similarity, as described below.

The data in the columns were then reverse coded so that for Jaccard's Coefficient and Cladistic Analysis only conjoint positives of reactants and conjoint negatives of absorbants were used as measures of similarity. Thus the data from Table 2 for *Solanum fructo-tecto* and *S. torvum* became:

FTO oo+++o++++++++++++++ oo++++++++++++++++o
TOR oo++ooo+++o+++++++++ oooooooooo++oooo++oo
and S_J = a/a+b+c = 18/34 = 0.53
whereas S_{SM} = a+d/a+b+c+d = 24/40 = 0.60

Table 1. Taxonomic arrangement of *Solanum* species used

Subgenus and section	*Solanum* species	Accession Number	Abbreviation	Code No.
Subgenus *Leptostemonum* (Dun.) Bitt.				
Section *Acanthophora* Dun.	*S. capsicoides* All.	S. 0866	CAP	5
	S. chloropetalum Schl.	S. 0021	CHL	16
	S. viarum Dun.	S. 1418	VIA	15
Section *Androceras* (Nutt.) Bitt. ex M.				
Series *Androceras*	*S. rostratum* Dun.	S. 0097	ROS	1
	S. rostratum Dun.	S. 0399	ROS	7
	S. fructo-tecto Cav.	S. 0025	FTO	10
Series *Violaceiflorum*	*S. heterodoxum* Dun.	S. 0593	HET	9
	S. citrullifolium A. Br.	S. 0195	CIT	11
	S. citrullifolium A. Br.	S. 0127	CIT	12
Section *Cryptocarpum* Dun.	*S. sisymbriifolium* Lam.	S. 1099	SIS	2
	S. sisymbriifolium Lam.	S. 0136	SIS	8
Section *Lasiocarpa* (Dun.) D'Arcy	*S. hirtum* Vahl.	S. 1142	HIR	17
	S. quitoense Lam.	S. 0972	QUT	6
	S. tequilense A. Gray	S. 0973	TEQ	18
Section *Oliganthes* (Dun.) Bitt.	*S. indicum* L.	S. 1335	IND	19
Section *Torva* Nees.	*S. hispidum* Pers.	S. 0017	HIS	14
	S. torvum Swartz	S. 0839	TOR	4
Section (?)	*S. prinophyllum* Dun.	S. 0386	PRN	3
	S. prinophyllum Dun.	S. 1444	PRN	13
Subgenus *Solanum*				
Section *Solanum*	*S. nigrum* L.	S. 0498	NIG	20

Table 2. Scores with Abs R114/2 to S. 0097 *S. rostratum*

Reactant	Absorbant																			
	ROS 0097	ROS 0399	SIS 1099	SIS 0136	HET 0593	FTO 0025	CIT 0195	CIT 0127	PRN 0386	PRN 1444	TOR 0839	HIS 0017	CAP 0866	VIA 1418	CHL 0021	NIG 0498	QUT 0972	HIR 1142	TEQ 0973	IND 1335
ROS 0097	0	0	2	3	2	2	2	2	3	3	3	3	3	3	2	3	3	3	3	3
ROS 0399	0	0	2	3	2	2	2	2	3	3	3	3	3	3	2	3	3	3	3	3
SIS 1099	0	0	0	0	½	0	0	0	1	½	½	½	1	1	½	2	1	½	1	1
SIS 0136	0	0	0	0	0	0	0	0	1	0	½	½	1	1	0	2	½	½	1	1
HET 0593	0	0	1	0	0	0	0	½	1	1	1	½	1	2	½	2	2	½	1	2
FTO 0025	0	0	2	2	2	0	2	2	2	2	2	2	2	3	2	3	3	2	2	3
CIT 0195	0	0	1	1	½	0	0	0	2	1	1	1	1	2	1	2	2	½	1	2
CIT 0127	0	0	1	1	½	0	0	0	2	1	1	1	1	2	1	2	2	½	1	2
PRN 0386	0	0	½	½	½	0	½	0	0	0	2	1	2	1	1	2	½	½	½	½
PRN 1444	0	0	1	½	½	0	½	0	½	0	2	1	2	1	1	2	½	½	1	1
TOR 0839	0	0	½	½	0	0	0	½	1	½	0	½	1	1	½	1	1	½	1	1
HIS 0017	0	0	½	½	½	0	0	0	½	0	0	0	1	1	½	1	½	½	½	½
CAP 0866	0	0	1	0	0	0	0	0	½	0	½	½	0	1	0	1	½	0	½	½
VIA 1418	0	0	½	0	0	0	0	1	1	1	½	1	1	0	0	1	1	½	½	0
CHL 0021	0	0	1	0	0	0	0	½	1	½	½	½	½	1	0	1	1	0	1	½
NIG 0498	0	0	0	0	½	0	0	½	0	0	½	1	½	1	1	0	½	½	½	1
QUT 0972	0	0	½	0	0	0	0	0	0	0	0	½	1	1	0	1	0	0	½	½
HIR 1142	0	0	1	0	0	0	0	0	1	0	0	1	1	1	0	1	½	0	1	1
TEQ 0973	0	0	0	0	0	0	0	0	0	0	½	½	½	½	0	1	0	0	0	½
IND 1335	0	0	½	0	½	½	½	½	0	0	½	½	½	1	0	½	½	½	½	0
$S_{abs. mean}$	1.00	1.00	0.37	0.49	0.58	0.87	0.68	0.63	0.37	0.53	0.40	0.28	0.29	0.21	0.43	0.18	0.25	0.44	0.18	0.26

Table 3. Scores with Abs R119/2 to S. 1099 *S. sisymbriifolium*

Reactant	Absorbant																			
	ROS 0097	ROS 0399	SIS 1099	SIS 0136	HET 0593	FTO 0025	CIT 0195	CIT 0127	PRN 0386	PRN 1444	TOR 0839	HIS 0017	CAP 0866	VIA 1418	CHL 0021	NIG 0498	QUT 0972	HIR 1142	TEQ 0973	IND 1335
ROS 0097	½	½	0	½	½	½	½	½	½	1	1	1	1	3	2	1	1	1	1	½
ROS 0399	½	½	0	½	½	½	½	½	½	1	1	1	1	3	1	1	1	1	1	½
SIS 1099	2	2	0	½	1	2	1	2	3	3	3	3	2	4	3	2	3	2	2	2
SIS 0136	2	2	0	½	1	2	1	2	3	3	3	3	2	4	3	2	3	2	2	2
HET 0593	2	1	0	½	½	1	½	½	1	1	1	1	1	3	1	1	½	½	½	1
FTO 0025	1	½	0	0	½	½	½	½	½	1	½	½	1	2	2	1	1	½	1	½
CIT 0195	2	1	0	0	1	½	½	1	1	1	1	1	1	3	2	2	½	½	½	1
CIT 0127	2	2	0	0	2	2	1	1	2	2	2	2	2	3	2	2	2	1	2	2
PRN 0386	½	0	0	0	0	0	0	0	0	0	½	½	½	0	½	½	½	½	½	0
PRN 1444	½	½	0	0	0	0	0	0	½	½	½	½	½	0	½	½	½	½	½	0
TOR 0839	½	½	0	0	0	½	0	0	0	0	0	0	0	0	½	½	½	0	0	0
HIS 0017	0	0	0	0	0	0	0	0	0	0	0	0	0	0	½	½	½	0	0	0
CAP 0866	½	½	0	0	½	½	½	½	½	½	2	1	½	0	1	1	½	½	½	½
VIA 1418	½	½	0	0	½	½	½	½	½	½	1	1	½	0	0	0	0	0	0	0
CHL 0021	1	1	0	0	1	1	1	1	1	1	1	2	1	0	0	0	0	0	0	0
NIG 0498	1	1	0	½	1	1	1	1	1	2	2	3	1	0	2	1	1	1	1	1
QUT 0972	1	½	0	0	½	0	½	½	2	2	1	1	½	3	1	1	0	0	0	2
HIR 1142	2	1	0	½	1	½	½	1	2	3	2	2	1	3	2	2	1	½	½	2
TEQ 0973	1	1	0	0	½	½	1	1	2	3	2	2	1	2	1	1	½	½	0	2
IND 1335	2	2	0	½	1	1	1	2	1	1	1	1	2	3	2	2	2	½	1	1
$S_{abs. mean}$	0.46	0.52	1.00	1.00	0.61	0.55	0.66	0.86	0.21	0.31	0.12	0.12	0.40	0.26	0.25	0.62	0.37	0.72	0.48	0.70

Table 4. Scores with Abs R115/2 to S. 0386 *S. prinophyllum*

Reactant	Absorbant																			
	ROS 0097	ROS 0399	SIS 1099	SIS 0136	HET 0593	FTO 0025	CIT 0195	CIT 0127	PRN 0386	PRN 1444	TOR 0839	HIS 0017	CAP 0866	VIA 1418	CHL 0021	NIG 0498	QUT 0972	HIR 1142	TEQ 0973	IND 1335
ROS 0097	0	0	½	½	0	½	0	0	0	0	½	1	1	2	1	1	½	½	½	½
ROS 0399	½	0	0	½	½	½	½	0	0	0	½	1	1	2	1	1	½	½	½	½
SIS 1099	½	0	0	0	½	½	½	0	0	0	1	½	½	2	1	1	½	½	½	0
SIS 0136	½	0	0	0	0	½	½	0	0	0	1	½	½	2	1	1	½	½	½	0
HET 0593	0	0	0	0	0	0	0	0	0	0	0	0	1	2	2	1	0	0	½	½
FTO 0025	0	0	0	0	0	0	0	0	0	0	0	½	½	2	1	1	½	0	½	0
CIT 0195	0	0	0	½	0	½	0	0	0	0	0	½	1	2	2	1	½	½	½	½
CIT 0127	½	0	0	½	0	½	0	0	0	0	0	½	1	2	1	1	½	½	½	½
PRN 0386	2	1	2	1	2	2	2	2	0	0	1	2	2	2	2	2	2	2	2	1
PRN 1444	1	1	1	½	2	2	2	2	0	0	1	2	2	2	2	2	2	2	2	1
TOR 0839	½	½	½	0	1	0	½	½	0	0	0	0	1	1	1	2	2	1	1	0
HIS 0017	0	0	0	0	½	0	0	½	0	0	0	0	1	1	1	1	1	½	½	0
CAP 0866	0	0	0	0	0	½	0	½	0	0	0	½	0	1	0	1	1	½	½	0
VIA 1418	0	0	0	0	0	0	0	½	0	0	0	0	0	0	0	½	½	½	½	0
CHL 0021	0	0	0	0	0	0	0	½	0	0	0	½	½	1	0	1	1	1	1	0
NIG 0498	0	0	0	0	½	0	0	½	0	0	1	1	½	2	1	0	1	1	½	½
QUT 0972	0	0	0	0	½	0	½	½	0	0	0	0	½	1	0	½	0	0	0	0
HIR 1142	½	0	½	0	1	½	½	½	0	0	½	½	½	1	0	½	½	0	0	0
TEQ 0973	0	0	0	0	1	½	½	½	0	0	0	0	½	½	0	½	0	0	0	0
IND 1335	0	½	½	½	1	½	1	1	0	0	1	1	1	1	1	2	2	1	1	0
$S_{abs.\ mean}$	0.57	0.65	0.60	0.58	0.41	0.45	0.52	0.51	1.00	0.98	0.56	0.41	0.27	0.09	0.29	0.27	0.26	0.39	0.32	0.70

Table 5. Scores with Abs R116/2 to S. 0839 *S. torvum*

Reactant	Absorbant																			
	ROS 0097	ROS 0399	SIS 1099	SIS 0136	HET 0593	FTO 0025	CIT 0195	CIT 0127	PRN 0386	PRN 1444	TOR 0839	HIS 0017	CAP 0866	VIA 1418	CHL 0021	NIG 0498	QUT 0972	HIR 1142	TEQ 0973	IND 1335
ROS 0097	0	0	½	0	0	0	0	0	0	0	0	0	½	2	½	1	½	½	½	0
ROS 0399	0	0	0	0	0	0	0	0	0	0	0	0	½	2	½	1	½	½	½	½
SIS 1099	½	0	0	0	½	½	½	½	0	0	0	½	½	2	½	½	½	½	0	½
SIS 0136	0	0	0	0	½	½	½	½	0	0	0	½	½	2	½	½	½	½	½	0
HET 0593	0	0	0	0	0	0	0	0	0	0	0	0	0	2	½	1	½	½	½	0
FTO 0025	0	0	½	0	0	0	0	0	0	0	0	½	0	2	½	1	0	0	½	0
CIT 0195	0	0	½	0	0	0	0	0	0	0	0	½	½	2	½	1	½	½	½	0
CIT 0127	0	0	½	0	0	½	0	0	0	½	0	½	½	2	½	1	½	½	½	0
PRN 0386	1	0	½	½	1	1	½	½	0	0	0	0	0	1	½	1	½	1	½	½
PRN 1444	1	½	½	½	1	1	½	½	0	0	0	0	0	1	½	1	½	1	½	½
TOR 0839	2	1	½	½	1	2	1	½	½	½	0	0	½	2	1	2	1	1	1	1
HIS 0017	1	1	½	½	1	1	1	1	½	½	0	0	½	2	1	1	1	1	1	½
CAP 0866	½	0	½	½	½	0	½	½	0	0	0	0	0	1	0	½	0	½	½	0
VIA 1418	0	0	0	0	½	0	½	½	0	0	0	0	0	0	½	0	0	½	0	0
CHL 0021	0	0	0	0	0	½	½	½	0	0	0	0	0	1	0	0	0	½	½	0
NIG 0498	0	0	0	0	0	0	0	0	0	0	1	1	0	0	½	0	½	½	½	1
QUT 0972	0	0	0	0	½	½	0	0	½	½	0	0	½	1	0	1	0	0	0	½
HIR 1142	0	0	0	0	½	0	½	½	½	½	0	0	½	1	0	1	½	0	0	½
TEQ 0973	0	0	0	0	½	½	½	0	½	½	0	½	½	½	0	1	½	0	0	½
IND 1335	0	0	0	0	0	½	½	0	½	0	0	½	0	2	½	1	1	1	½	0
$S_{abs. mean}$	0.54	0.61	0.62	0.55	0.45	0.45	0.49	0.57	0.73	0.74	1.00	0.86	0.53	0.17	0.39	0.24	0.43	0.42	0.47	0.58

Table 6. Scores with Abs R117/2 to S. 0866 *S. capsicoides*

Reactant	Absorbant																			
	ROS 0097	ROS 0399	SIS 1099	SIS 0136	HET 0593	FTO 0025	CIT 0195	CIT 0127	PRN 0386	PRN 1444	TOR 0839	HIS 0017	CAP 0866	VIA 1418	CHL 0021	NIG 0498	QUT 0972	HIR 1142	TEQ 0973	IND 1335
ROS 0097	1	½	½	½	½	½	0	0	1	½	1	1	½	2	1	2	2	1	1	1
ROS 0399	1	½	0	½	½	½	0	0	1	½	1	1	½	2	1	2	2	1	1	1
SIS 1099	2	2	0	0	1	2	½	0	2	2	2	2	2	3	2	3	2	2	2	2
SIS 0136	2	2	0	0	1	2	½	0	2	2	2	2	2	3	2	3	2	2	2	2
HET 0593	2	2	0	0	1	1	½	0	1	1	2	2	1	2	2	3	2	0	½	1
FTO 0025	1	1	½	½	1	½	½	0	½	½	1	1	½	2	1	2	2	1	1	1
CIT 0195	2	1	½	½	1	1	½	0	1	1	1	1	1	2	1	3	2	½	½	2
CIT 0127	2	2	½	½	1	2	½	0	2	2	2	2	2	2	2	3	2	2	2	2
PRN 0386	½	0	0	1	½	½	0	0	0	0	0	½	0	1	½	2	½	½	½	½
PRN 1444	1	½	½	1	½	½	0	0	½	½	½	½	0	1	½	2	1	½	½	½
TOR 0839	½	½	0	1	½	½	½	0	0	0	0	0	0	2	½	2	0	0	0	0
HIS 0017	0	0	½	½	0	0	0	0	0	0	0	0	0	2	½	2	½	0	½	0
CAP 0866	3	2	1	½	2	1	2	1	1	2	1	2	½	2	1	3	1	½	½	½
VIA 1418	2	2	0	0	2	1	1	1	1	1	1	2	½	0	0	½	½	0	0	0
CHL 0021	3	2	0	0	2	2	2	1	2	2	2	2	1	2	0	3	1	0	0	0
NIG 0498	2	1	0	0	1	1	1	0	1	1	2	2	½	1	2	½	2	2	1	½
QUT 0972	1	½	0	½	½	0	0	0	½	½	½	½	0	2	0	2	0	0	0	½
HIR 1142	2	1	1	1	1	1	1	½	1	1	1	1	½	3	½	3	1	1	½	1
TEQ 0973	1	1	1	1	1	½	½	½	1	½	1	2	½	1	0	3	1	½	0	0
IND 1335	2	1	2	1	1	1	1	1	1	1	2	2	½	2	1	3	1	1	1	1
$S_{abs.\ mean}$	0.52	0.52	0.84	0.77	0.80	0.60	0.78	0.90	0.26	0.50	0.22	0.24	1.00	0.30	0.63	0.41	0.22	0.79	0.50	0.86

Table 7. Scores with Abs R118/2 to S. 0972 *S. quitoense*

Reactant	Absorbant																			
	ROS 0097	ROS 0399	SIS 1099	SIS 0136	HET 0593	FTO 0025	CIT 0195	CIT 0127	PRN 0386	PRN 1444	TOR 0839	HIS 0017	CAP 0866	VIA 1418	CHL 0021	NIG 0498	QUT 0972	HIR 1142	TEQ 0973	IND 1335
ROS 0097	0	0	0	0	0	0	0	0	1	1	1	1	0	1	0	½	0	0	0	½
ROS 0399	0	0	0	0	0	0	0	0	1	1	1	1	0	1	0	½	0	0	0	½
SIS 1099	0	0	0	0	0	0	0	0	1	1	1	1	0	1	0	½	0	0	0	½
SIS 0136	0	0	0	0	0	0	0	0	1	1	1	1	0	1	½	½	0	0	0	½
HET 0593	0	0	0	0	0	0	0	0	1	½	½	½	0	0	0	½	0	0	0	½
FTO 0025	0	0	0	0	0	0	0	0	1	1	½	½	0	0	0	½	0	0	0	½
CIT 0195	0	0	0	0	0	0	0	0	1	1	1	1	0	½	½	½	0	0	0	½
CIT 0127	0	0	0	0	0	0	0	0	1	1	1	1	0	½	½	½	0	0	0	½
PRN 0386	0	0	0	0	0	0	0	½	0	0	0	0	0	0	0	0	0	0	0	0
PRN 1444	0	0	0	0	0	0	0	½	0	0	0	0	0	0	0	0	0	0	0	0
TOR 0839	0	0	0	0	0	0	0	½	0	0	0	0	0	0	0	0	0	0	0	0
HIS 0017	0	0	0	0	0	0	0	½	0	0	0	0	0	0	0	0	0	0	0	0
CAP 0866	0	0	0	0	0	0	0	0	0	0	0	½	0	½	0	½	0	0	0	0
VIA 1418	0	0	0	0	0	0	0	0	0	0	0	½	0	0	0	0	0	0	0	0
CHL 0021	0	0	0	0	0	0	0	0	0	0	0	½	0	½	0	½	0	0	0	0
NIG 0498	0	½	0	0	0	0	0	0	0	½	1	1	½	1	1	0	½	½	½	1
QUT 0972	2	1	2	½	2	1	1	1	2	2	2	2	2	2	2	2	0	0	0	2
HIR 1142	1	1	1	0	1	1	1	1	2	2	2	2	2	2	2	2	0	0	0	2
TEQ 0973	1	1	1	0	1	1	1	1	2	2	2	2	2	1	2	2	0	0	0	2
IND 1335	0	0	0	0	0	0	0	0	0	0	0	0	0	0	0	0	0	0	0	0
$S_{abs. mean}$	0.57	0.56	0.57	0.65	0.51	0.52	0.57	0.52	0.24	0.24	0.24	0.21	0.45	0.52	0.40	0.48	1.00	0.96	0.95	0.30

4.1 Analysis by the Mean Absorption Similarity Coefficient ($S_{abs.mean}$)

This method compares each cross-reacting antigen system with the appropriate reference antigen system, but makes no comparisons between cross-reacting antigen systems.

The immunological principles and the procedure for this method of analysis have been described by Lester (1979) and have been re-iterated above. The rectangular matrix of results from each antibody system is analysed separately. Each antigen system is compared with the reference antigen system in its ability to absorb the antibody system being used and thus prevent other antigen systems from producing precipitates, and also in its ability to react with and produce amounts of precipitates with the antibody system after it has been severally absorbed by each of the other antigen systems. This may be summarised in the following formula (Lester 1979):

$$S_{abs.mean} = \frac{1}{2}\,[S_{no.neg.} + \frac{1}{2}\,(S_{no.pos.} + S_{ppt.pos.})]\,.$$

In effect $S_{abs.mean} = \frac{a}{a+b}$ where a is the total number and amount of significant positive reactions when that antigen system is reactant or of negative reactions when it is absorbant, and $a + b$ is the total number and amount of significant positive and negative reactions of the reference antigen system.

The values of $S_{abs.mean}$ for each antibody system (Tables 2–7) measure the serological correspondence of each cross-reacting antigen system to the reference antigen system, and are best interpreted by their rank order.

Where values are available for the reciprocal reactions between two antibody systems and their antigen systems (Table 8) these values can be averaged to form a half-matrix of similarity coefficients which can be analysed by Group Average, Centroid or other clustering methods to produce a phenogram (Fig. 4). Ideally more sophisticated methods (Gower 1977, 1980) should be used since the square matrix (Table 8) is asymmetric. Although the phenogram can only justifiably be constructed for those antigen systems against which antibody systems have been raised, other antigen systems can be positioned on appropriate arms of the phenogram (Fig. 4) according to their levels of similarity with any of the reference antigen systems (Tables 2–7).

Table 8. Mean Absorption Similarity Coefficient ($S_{abs.mean}$) values between the six antigen systems used for producing antibody systems

Ags	Abs	ROS^{ab}	SIS^{ab}	PRN^{ab}	TOR^{ab}	CAP^{ab}	QUT^{ab}
ROS 0097		1.00	0.46	0.57	0.54	0.52	0.57
SIS 1099		0.37	1.00	0.60	0.62	0.84	0.57
PRN 0386		0.37	0.21	1.00	0.73	0.26	0.24
TOR 0839		0.40	0.12	0.56	1.00	0.22	0.24
CAP 0866		0.29	0.40	0.27	0.53	1.00	0.45
QUT 0972		0.25	0.37	0.26	0.43	0.22	1.00

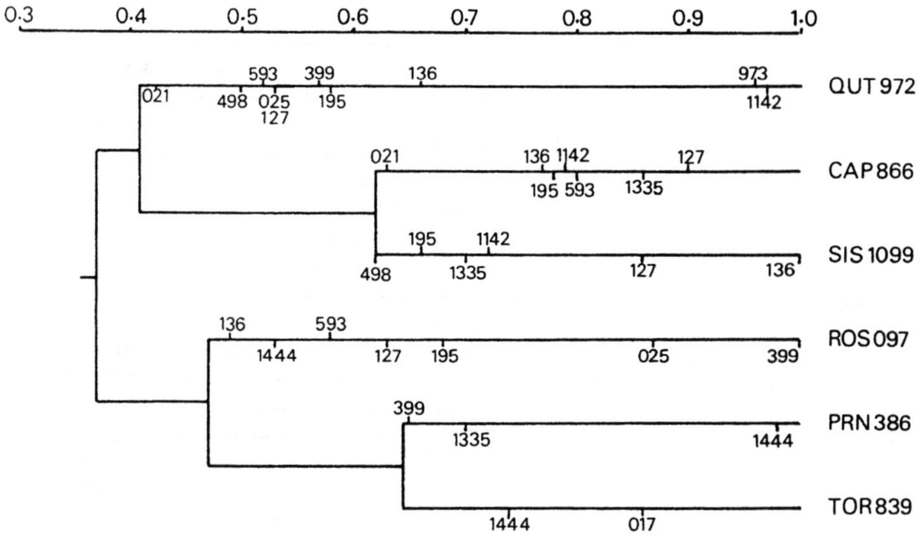

Fig. 4. Dendrogram of immunological relationships calculated by Mean Absorption Similarity Coefficient and Group Average clustering. (For explanation see text and Table 1)

4.2 Simple Matching Coefficient of Similarity (S_{SM})

The Simple Matching Coefficient compares all pairs of antigen systems or operational taxonomic units (OTUs) using all the data resulting from their employment both as reactants and as absorbants.

$$S_{SM} = \frac{a + d}{a + b + c + d} \quad \text{(Sneath and Sokal 1973)}$$

Where a is the number of characters present in both OTU A and OTU B (i.e. positive matches)
d is the number of characters absent from both OTU A and OTU B (i.e. negative matches)
and b and c are the numbers of characters present in only one or other OTU (i.e. mismatches)

Thus in each Table (2–7), for any one individual OTU, the production of a precipitate, when that OTU was employed as a reactant with each of the twenty reagents produced by absorption of the antibody system, was counted as a character being present, and likewise the production of a precipitate by any of the reactant antigen systems, when the same OTU had been used as the absorbant, was also counted as a character being present. Contrarily lack of precipitate in either of these situations was counted as a character being absent. (Counting the production of precipitate with an absorbed antibody system as a negative reaction and lack of precipitate as a positive reaction, as in the method using Jaccard's Coefficient, makes no difference to the final analysis by the Simple Matching Coefficient.)

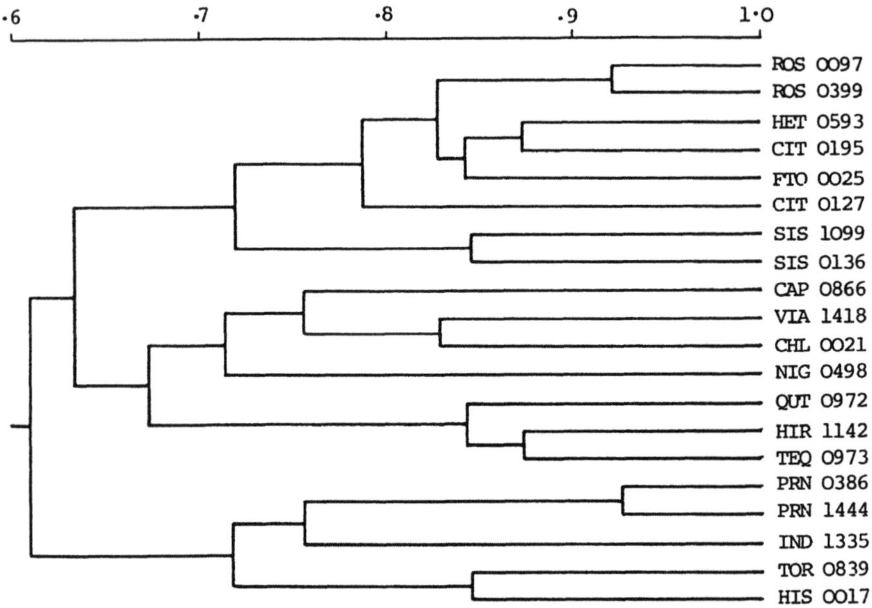

Fig. 5. Dendrogram of immunological relationships calculated by Simple Matching Coefficient and Group Average clustering. (For explanation see text and Table 1)

For each OTU there were 6x(20+20) = 240 characters available, and in comparing each pair of OTUs all these data were used, including not only conjoint presences (a) but also conjoint absences (d) of precipitate. Cristofolini and Chiapella (1977) used this or a very similar method to analyse spur test data.

All 20 OTUs were analysed by S_{SM} and clustered by UPGMA (unweighted pair group method using arithmetic averages or group average method) (Sneath and Sokal 1973) using the CLUSTAN 1C computer program (Wishart 1978) (Fig. 5). Analysis of the same data by Euclidean Distance Squared and clustering by the minimum incremental sum of squares (Ward's method) produced a topographically similar dendrogram.

4.3 Jaccard's Coefficient of Similarity (S_J)

Jaccard's Coefficient compares all pairs of operational taxonomic units (OTUs), but selectively disregards any characters which are not present in both of the OTUs being compared.

$$S_J = \frac{a}{a+b+c}$$

Where a is the number of characters present in both OTU A and OTU B (i.e. positive matches) and b and c are the numbers of characters present in OTU A but not OTU B and vice versa. The number of characters which is conjointly absent from both OTUs (*d*) is ignored (Sneath and Sokal 1973).

Before applying Jaccard's Coefficient some of the data were re-coded to follow the immunological principles stated earlier. The production of a precipitate by an antigen system when used as a reactant was counted as that character being present, but when that antigen system was used as absorbant the production of a precipitate was counted as that character being absent, and only the lack of precipitate was counted as that character being present. This is an essential prerequisite. The consequence of this was that when two OTUs were compared in this way, using Jaccard's Coefficient as suggested by Lester (1979), their mutual ability to react with antibody systems absorbed by other antigen systems, together with their mutual ability to absorb an antibody system and prevent further reaction with other antigen systems were taken as a measure of similarity, but, unlike the analysis using the Simple Matching Coefficient, their mutual lack of reaction with any absorbed antibody system or their mutual inability to absorb any antibody system and prevent further reaction were ignored.

This method does not suffer from the weakness of the Simple Matching Coefficient except in the assumption that precipitates with any one absorbed antibody system are produced by identical sets of antigenic determinant sites of the two antigen systems being compared.

All 20 OTUs were analysed in this way by Jaccard's Coefficient, and were clustered by UPGMA (group average) (Fig. 6).

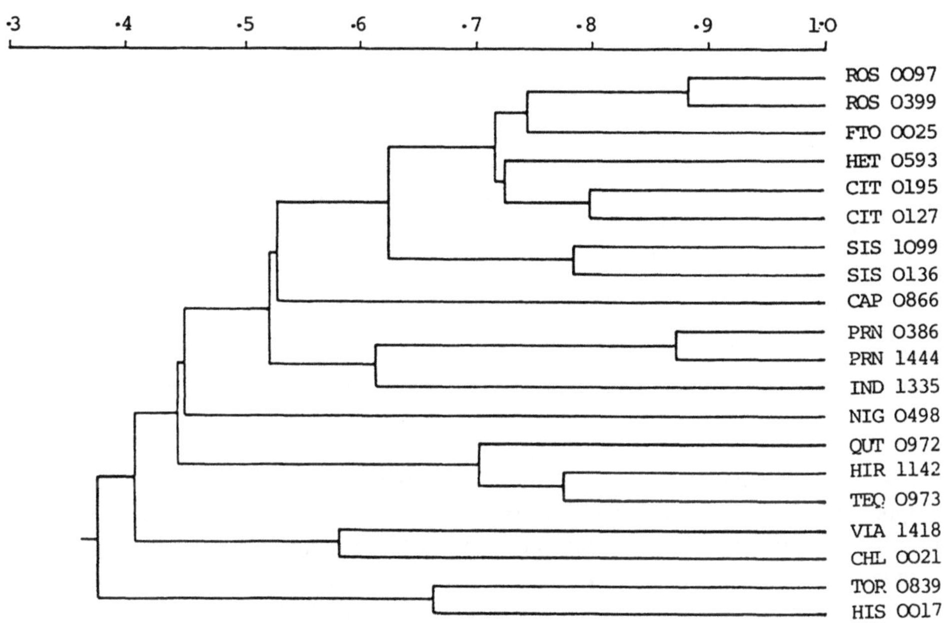

Fig. 6. Dendrogram of immunological relationships calculated by Jaccard's Coefficient and Group Average clustering. (For explanation see text and Table 1)

4.4 Cladistic Analysis

It can be argued that since the presence of a character (which may be either a positive or negative reaction depending on the circumstances already discussed for Jaccard's Coefficient) indicates the similarity of a particular species to the species used for the production of the antibody system, and that since the absence of any characters which discriminate between species can be taken as the condition of an hypothetical ancient ancestor of all the species, that therefore the number of discriminatory characters present in any species may be an estimate of its evolution or its patristic distance from that hypothetical ancient ancestor.

For any pair of organisms (Evolutionary Units, EU) the number of characters which is conjointly present in EU A and EU B, but not in all the other organisms (*a* in Fig. 7), is a measure of the amount of evolutionary history or patristic distance from the ancient common ancestor (Anc.) of the whole group to the immediate common ancestor (ICA) of EU A and EU B. The subsequent divergence or patristic distance of EU A and of EU B from their immediate common ancestor (ICA) is estimated by the number of characters specific to each of them (*b* and *c* respectively in Fig. 7). Any characters which are conjointly absent from EU A and EU B (*d*) are not involved in this analysis.

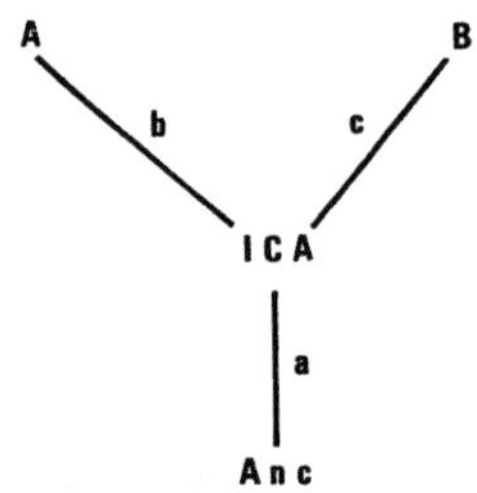

Fig. 7. Cladistic relationships of two evolutionary units (EU A and EU B) showing their patristic distances (b and c respectively) from their immediate common ancestor (ICA), and the further distance (a) from the ancient ancestor (Anc) of the whole group

This method of analysis follows the Patristic Distance (D) formula of Mickevich (personal communication) and Farris (1979, Farris et al. 1970). This method produces clustering which is an approximation to a parsimony method of the Camin and Sokal type (see review by Felsenstein 1982).

$$D\,[Anc,(A,B)] = \frac{1}{2}\,[D(A,Anc) + D(B,Anc) - D(A,B)]$$

Where D = patristic distance.

Using the terms as defined for Jaccard's Coefficient this can be rewritten as

$$a = \frac{1}{2}\,[(a + b) + (a + c) - (b + c)]$$

The estimates of patristic distance depend on the number of characters. It should be emphasised that in immunological absorption experiments the number of characters is not absolute, but depends on the design of the experiment. Furthermore one character which is in reality single may be assessed or treated as several characters by these techniques. Cladistic analysis of the *Solanum* data has been made by hand. Only the data from ROS^{ab} and SIS^{ab} were used for the cladogram in Fig. 8, but all relevant data were used to produce the cladogram of the six EU/OTUs used for production of antibody systems (Fig. 9).

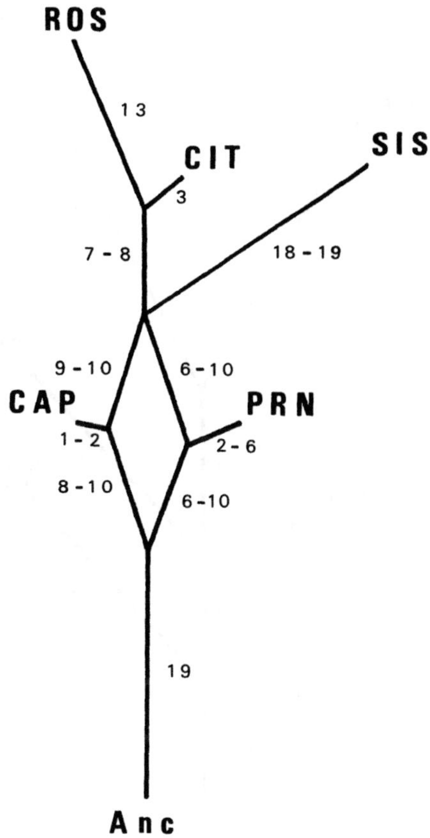

Fig. 8. Dendrogram of cladistic relationships of *Solanum capsicoides* (CAP), *S. citrullifolium* (CIT), *S. prinophyllum* (PRN), *S. rostratum* (ROS), and *S. sisymbrifolium* (SIS). Data were obtained from absorption experiments using antibody systems ROS^{ab} and SIS^{ab}

5 Discussion

5.1 Taxonomic Significance of the Results of Different Methods of Analysis

Each of the four methods of analysis resulted in the production of dendrograms. The cladograms are superficially dissimilar from the phenograms, but the topological differences between all the various dendrograms are much more important.

Analysis by the Mean Absorption Similarity Coefficient ($S_{abs.mean}$) and clustering by the Group Average method produced a six-pronged phenogram showing the relationships of the six reference antigen systems (Fig. 4). *Solanum sisymbriifolium* clustered with *S. capsicoides* and then with *S. quitoense; S. torvum* joined *S. prinophyllum,* and then *S. rostratum*; both clusters then fused. Onto this skeleton, produced by the six species used for antiserum production, were hung the other twenty species, according to their highest levels of similarity: several species showed high similarity to more than one of the six reference species. In all cases duplicate samples of species used for antiserum production were placed together with the relevant reference species (i.e. *S. sisymbriifolium* S.0136 with S.1099, *S. rostratum* S.0399 with S.0097, and *S. prinophyllum* S.1444 with S.0386). In several cases phenetic relationships are

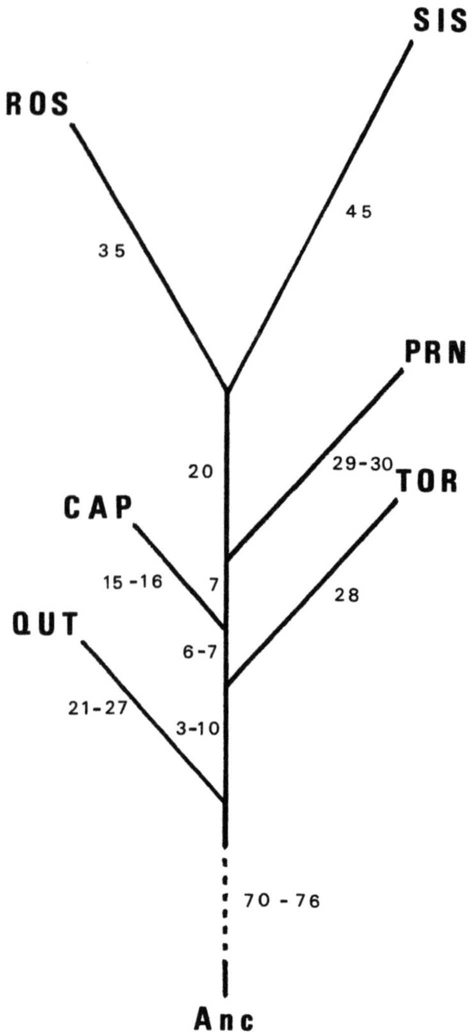

Fig. 9. Dendrogram of cladistic relationships of *Solanum capsicoides* (CAP), *S. prinophyllum* (PRN), *S. quitoense* (QUT), *S. rostratum* (ROS), *S. sisymbriifolium* (SIS), and *S. torvum* (TOR) derived from absorption experiments using antibody systems to all six species

supported, such as *S. hirtum* and *S. tequilense* with *S. quitoense* in section *Lasiocarpa*, *S. hispidum* with *S. torvum* in section *Torva* and most of the members of section *Androceras* with *S. rostratum*. However, there are several anomalies, noticeably with antiserum to *S. capsicoides* representatives of several other sections show unexpected kinship.

Analyses by Simple Matching Coefficient (S_{SM}) and by Jaccard's Coefficient (S_J) were both followed by Group Average clustering (Figs. 5 and 6). Both methods indicated close relationships for duplicate accessions of *S. rostratum*, of *S. sisymbriifolium*, and of *S. prinophyllum*, and general similarity of species within section *Androceras* and within section *Lasiocarpa*. However, the separation of the species into series *Androceras* (*S. rostratum* and *S. fructo-tecto*) and into series *Violaceiflorum* (*S. heterodoxum* and *S. citrullifolium*) was better by Jaccard's than by Simple Matching Coefficient. The placement by S_{SM} of *S. nigrum* (subgenus *Solanum*) amongst species of sections *Acanthophora* and *Lasiocarpa* (subgenus *Leptostemonum*) is unacceptable: by S_J *S. nigrum* showed little similarity to any of the other taxa, which is satisfactory.

Cladistic analysis was attempted on some of the more complementary and self-consistent parts of the data. Using only the data from ROS^{ab} it was fairly easy to make a cladistic analysis of the most similar taxa. The result was a single linear sequence (with patristic distances indicated in brackets): ancestral *Solanum* species (17) *S. sisymbriifolium* (6) *S. heterodoxum* (3) *S. citrullifolium* S.0127 (3) and S.0195 (5) *S. fructo-tecto* (4) *S. rostratum*. This arrangement is evolutionarily feasible; *S. sisymbriifolium* may be considered similar to the probable ancestor of section *Androceras* and is more primitive in being a perennial plant with lobed leaves, actinomorphic blue flowers and red berries, whereas *S. rostratum* is more advanced in being an annual plant with much divided leaves, zygomorphic yellow flowers and dry dehiscent capsules. The section *Androceras* is so distinct from other *Solanum* species that it was once separated as the genus *Androcera*. The evolutionary trends suggested herein are in agreement with the authoritative views of Whalen (1979).

An attempt was made to integrate selected data from ROS^{ab} with those from SIS^{ab} to show the cladistic relationships of *S. rostratum, S. citrullifolium, S. sisymbriifolium, S. capsicoides,* and *S. prinophyllum* (Fig. 8). Several discrepancies were found, the data do not meet the conditions specified by Friday (1980) for assigning distance, and the result is a split tree with two alternative routes from the ancestor. The position of *S. capsicoides*, which was supported by data from CAP^{ab}, is untenable since by many morphological criteria this is considered an advanced rather than a primitive species.

When data for the six species used for antibody production were subjected to cladistic analysis, considerable disagreement was found within this body of data. A compromise cladogram (Fig. 9) has been constructed using average distances, but the range of distance values for some branches is indicated. At this stage it is difficult to assess the taxonomic significance, except to note that the Australian *S. prinophyllum* appears amidst species which are all American.

To make a cladistic analysis of all twenty species using all the data would have been topographically impossible even if the time or computer programs had been available.

Comparison of the four methods of analysis is possible by comparing the topological relationships of the six reference species, *S. capsicoides, S. prinophyllum, S. quitoense, S. rostratum, S. sisymbriifolium,* and *S. torvum*, in the four dendrograms (Figs. 4, 5, 6, and 9).

Of the four methods of analysis there was greatest agreement between the procedure using Jaccard's Coefficient and the cladistic analysis. This might be expected since they both ignore conjoint absences of characters, unlike the Simple Matching Coefficient, and they both use all positive matches and all relevant mismatches between pairs of antigen systems with each antibody system. This is unlike the Mean Absorption Similarity Coefficient which only compares each cross-reacting antigen system with the reference antigen system for each antibody system.

Jaccard's Coefficient and the cladistic analysis both indicate that *S. rostratum* and *S. sisymbriifolium* are the most similar. This agrees with the morphological similarity between section *Androceras* and section *Cryptocarpum* noted by Whalen (1979). Lower levels of relationship are shown by *S. prinophyllum* and *S. capsicoides* to the above two taxa and even lower levels by *S. torvum* and *S. quitoense*. This corresponds fairly well with the levels of morphological similarities, especially of leaf shape, prickliness and hairiness, and of some inflorescence and calyx characters, as assessed subjectively by the senior author and objectively by principal components analysis of 57 morphological characters by Roberts (1978).

The analysis by Simple Matching Coefficient indicated not only the acceptable high similarity between *S. rostratum* and *S. sisymbriifolium*, but also suggested the similarity of *S. capsicoides* to *S. quitoense* and of *S. prinophyllum* to *S. torvum*, which seem unlikely when morphological characteristics are evaluated. Some of the relationships indicated by the Mean Absorption Similarity Coefficient are even less likely.

5.2 Discussion of the Numerical Taxonomic Methods

In this paper four different approaches to the numerical analysis of immunotaxonomic data which can be obtained from spur identification and absorption techniques have been described and criticised. An attempt has also been made to expound basic principles considered necessary for any method of analysis. The most fundamental principle is that the production of precipitate by any two test antigen systems with any one absorbed antibody system indicates similarity of these two test antigen systems, whereas the non-production of precipitate by any one test antigen system with any two absorbed antibody systems indicates similarity of the two antigen systems used for absorption. However, the converse situations of conjoint negative reactions of test antigen systems and of conjoint positive reactions of absorbed antibody systems do not indicate similarity of the reactants and absorbants respectively (Jensen 1968b, Lester 1979).

Cristofolini (1980) provided a discussion of the interpretation and analysis of serological data, including those from methods employing pre-absorbed antisera. However, although he described summing the scores of columns and rows and mentioned the use of Jaccard's Coefficient as suggested by Lester (1979), he regarded the use of an Euclidean distance or the correlation coefficient as equally valid, even though these two methods employ conjoint absences and therefore do not comply with the important fundamental principle stated above. Cristofolini pioneered the use of ordination techniques to present the results of multiple absorption experiments, and described their merits (Cristofolini 1980). On the other hand clustering methods have the merit of representing the highest levels of similarity most accurately, which is important since serological data are most reliable at the highest levels of similarity.

Although cladistic analysis of protein data appeals to researchers interested in evolution, it should be used with caution with immunological data. Cladistic analysis of the present data is inadequate on several counts. The data used for the calculation of patristic distances are not absolute and many characters (*viz* antigenic determinant sites) may be represented repeatedly due to the design of the absorption experiments. The geometry of the cladograms becomes impossible and even contradictory when several OTUs are incorporated. The computer analyses are rather complicated and not readily available. A perceptive critique of the status of immunological distance data in the construction of phylogenetic classifications has been provided by Friday (1980).

The Mean Absorption Similarity Coefficient ($S_{abs.mean}$) and similar coefficients (Lester 1979) are easy to use and do not require a computer. The matrices of similarity coefficient values can be manipulated to produce phenograms. The lack of congruity of reciprocal reactions results in discordant values for $S_{abs.neg.}$ and $S_{abs.pos.}$, but this is resolved pragmatically by using averages, a common immunotaxonomic expedient (Cristofolini 1980). This method produces estimates of the overall similarity of each cross-reacting antigen system to the appropriate reference antigen system: it does not compare cross-reacting antigen systems, and it does not make differential comparisons of the state of each character for each pair of OTUs.

The Simple Matching Coefficient and Euclidean Distance Squared, coupled with clustering strategies such as Group Average and Ward's Method respectively, are easily computed on most numerical taxonomy programs. However, they both suffer from the serious fault of treating conjoint absences of characters as equally important to conjoint presences of characters (described and discussed earlier). It is our opinion that this is not a valid procedure and these analyses should not be used.

Jaccard's Coefficient uses only those characters present in one or both OTUs being compared: it ignores conjoint absences of characters. Therefore it is very suitable for analyses when the columns of absorption data have been reverse-coded as described earlier in this paper. It is easily computed and is available in CLUSTAN and other packages. Clustering can be done by Group Average, Centroid or other appropriate methods.

For Jaccard's Coefficient the data must be coded in binary form as the presence or absence of each character (as treated in this paper). The absence of precipitin reaction in a data column is therefore weighted equally to the presence of precipitin reaction in a data row, and the amount of precipitate is not taken into account. If it is proper to consider the amount of precipitate this can be accommodated by using the numeric equivalent of Jaccard's Coefficient (CLUSTAN coefficient No. 28). Careful coding can avoid overweighting of large amounts of precipitate: standardisation of data is not appropriate. Principal Components Analysis is easy to use, but probably not appropriate.

Jaccard's Coefficient does not involve conjoint absences of characters in the numerator and therefore complies with the fundamental principles for analysis of absorption data discussed earlier. However, since it also omits conjoint absences from the denominator, it makes different comparisons on different scales: both 2/4 (two conjoint presences out of four total presences) and 50/100 give $S_J = 0.5$. This poor comparability can be ameliorated by using another binary coefficient such as Russell-Rao, where $S_{RR} = a/a + b + c + d$ (CLUSTAN coefficient No. 13), or its numeric equivalent the Dot Product Coefficient (CLUSTAN No. 26), or probably also Gower's Coefficient. Janowitz (1980) provides a good evaluation of similarity measures of binary data.

In our opinion it is essential to use a similarity coefficient which does not use conjoint absences in the numerator, and we are convinced that Jaccard's is superior to the Simple Matching Coefficient or Euclidean Distance Squared. However, we are not certain of the best compromise between binary and numeric coding of data, nor of the optimal weighting of conjoint absences in the denominator nor of conjoint presences in both numerator and denominator.

Acknowledgements. We are grateful to the Science Research Council for a CASE studentship (P.A.R.) and to Sarah Marsh for growing the plants.

References

Cristofolini G (1980) In: Bisby FA, Vaughan JG, Wright CA (eds) Chemosystematics: Principles and practice. Academic Press, London New York, pp 269–288
Cristofolini G, Chiapella LF (1977) Taxon 26:43–56
Crowle AJ (1973) Immunodiffusion, 2nd edn. Academic Press, London New York
Dray S, Young GO (1959) Science 129:1023–1025
Farris JS (1979) Syst Zool 28:200–214
Farris JS, Kluge AG, Eckardt MJ (1970) Syst Zool 19:172–189
Felsenstein J (1982) Q Rev Biol 57:379–404
Friday AE (1980) In: Bisby FA, Vaughan JG, Wright CA (eds) Chemosystematics: Principles and practice. Academic Press, London New York, pp 289–304
Goodman M, Moore GW (1971) Syst Zool 20:19–62
Gower JC (1977) In: Barra J (ed) Recent developments in statistics. North Holland, Amsterdam, pp 109–123
Gower JC (1980) In: Bisby FA, Vaughan JG, Wright CA (eds) Chemosystematics: Principles and practice. Academic Press, London New York, pp 399–409
Janowitz MF (1980) Syst Zool 29:342–359
Jensen U (1968a) Bot Jahrb 88:204–268
Jensen U (1968b) Bot Jahrb 88:269–310
Jensen U, Büttner C (1981) Taxon 30:404–419
Kirsch JAW (1967) PhD thesis, Univ West Aust, Nedl
Lester RN (1969) Arch Biochem Biophys 133:305–312
Lester RN (1979) In: Hawkes JG, Lester RN, Skelding AD (eds) The biology and taxonomy of the Solanaceae. Academic Press, London New York, pp 285–304
Moore GW, Goodman M (1968) Bull Math Biophys 30:279–289
Moritz O, Jensen U (1961) Bull Serol Mus 25:1–5
Roberts PA (1978) PhD thesis, Univ Birmingham
Sangar VK, Lichtwardt RW, Kirsch JAW, Lester RN (1972) Mycologia 64:342–358
Sneath PHA, Sokal RR (1973) Numerical taxonomy. Freeman, San Francisco
Whalen MD (1979) In: Hawkes JG, Lester RN, Skelding AD (eds) The biology and taxonomy of the Solanaceae. Academic Press, London New York, pp 581–596
Wishart D (1978) CLUSTAN user manual. Univ Program Library Unit, Edinburgh

Serological Investigation of the Annoniflorae (Magnoliiflorae, Magnoliidae)

D. E. FAIRBROTHERS and F. P. PETERSEN[1]

Abstract. Serological data obtained from seed meal extracts from 16 taxa belonging to the Annonaceae, Illiciaceae, Lauraceae, Magnoliaceae, and Schisandraceae, are analyzed by mean absorption similarity coefficients. Using unweighted pair group method analysis (UPGMA), dendrograms of serological similarity are generated. *Illicium* and *Schisandra* have 92% serological similarity. The five members of the Lauraceae reveal three serological groups. The descending order of serological similarity is as follows: Magnoliales → Illiciales → Annonales → Laurales. Both the Annonales and Laurales have approximately equal similarities with the Magnoliales.

1 Introduction

While a single measure of similarity is not an actual measure of taxonomic relatedness, it is believed that the compilation and evaluation of data obtained from several disciplines may reveal taxonomic affinities among taxa. Phytoserological data have proven to be valuable source of such information in systematic interpretations (Fairbrothers et al. 1975, Fairbrothers 1977, 1983, Jensen 1981, Dahlgren 1983). Data from systematic serological publications have been incorporated into Cronquist's (1981), Dahlgren's (Dahlgren et al. 1981), Takhtajan's (1980), and Thorne's (1981) revised Systems of Classification (Dahlgren 1983, Fairbrothers 1983).

The Annoniflorae (Magnoliiflorae or Magnoliidae) (Table 1), have been designated "living fossils" which have escaped extinction, and survived to the present-day. Thus they are valuable relics of the early stages of flowering plant evolution, and each member has retained features of a syndrome of characteristics making them pertinent to the basic understanding of evolutionary changes which have occurred.

The subclass Magnoliidae "consists principally of those dicotyledons that have retained one or more features of a syndrome of primitive characters" (Cronquist 1981). "Within the Magnoliidae it is clear that, the Magnoliales are the most archaic order" (Cronquist 1981). Dahlgren et al. (1981) also indicated that the superorder Magnoliiflorae contains the most primitive group of orders and families. Takhtajan (1980) stated, "Magnoliales retain many more archaic and primitive features in both vegetative and reproductive structures of its members than any order of flowering plants". Thorne (1981) stated, "the superorder Annoniflorae contains the greatest assemblage of taxa abounding in ancestral characteristics". Thus the four authors of

[1] Dept. of Biological Sciences and Bureau of Biological Research, Rutgers University, P.O. Box 1059, Piscataway, N.J. 08854, USA

Proteins and Nucleic Acids in Plant Systematics
ed. by U. Jensen and D.E. Fairbrothers

Table 1. Comparison of four recent classifications of the Annonaceae, Illiciaceae, Lauraceae, Magnoliaceae, and Schisandraceae

Cronquist 1981	Dahlgren et al. 1981	Takhtajan 1980	Thorne 1981
Magnoliidae[a]	Magnoliiflorae[b]	Magnoliidae[a]	Annoniflorae[b]
Magnoliales[c]	Annonales[c]	Magnoliales[c]	Annonales[c]
Magnoliaceae	Annonaceae	Magnoliaceae	Illiciineae[d]
Annonaceae	Magnoliales[c]	Annonaceae	Illiciaceae
Laurales[c]	Magnoliaceae	Illiciales[c]	Schisandraceae
Lauraceae	Illiciales[c]	Illiciaceae	Annonineae[d]
Illiciales[c]	Illiciaceae	Schisandraceae	Magnoliaceae
Illiciaceae	Schisandraceae	Laurales[c]	Annonaceae
Schisandraceae	Laurales[c]	Lauraceae	Laurineae[d]
	Lauraceae		Lauraceae

[a] Subclass [b] Superorder [c] Order [d] Suborder

the most contemporary classifications agree on the ancient status of this group of taxa (Table 1). Therefore, other taxa are often compared with and measured by how much they have differentiated from members of the Magnoliales (Annonales). It is this starting point position that causes botanists frequently to revisit this group when preparing classifications and evolutionary investigations.

The assemblage of characteristics which form the Annoniflorae (Magnoliiflorae) syndrome are the following: primitive tracheary elements, P-type sieve tube plastids, cellular endosperm, usually stipules, type of leaf epidermis, lack of differentiated perianth, free perianth parts, spiral arrangement of flower parts, apocarpy, type of placentation, seed coat structure, seeds with small embryo and copious endosperm, usually monocolpate or inaperturate pollen grains, essential oils present, benzylisoquinoline alkaloids, sesquiterpene lactones, rarely polyacetylenes, usually lack ellagic acid, lack iridoids (Smith 1947, Bailey and Nast 1948, Stern 1954, Periasamy and Swamy 1959, Tucker 1964, Johnson and Fairbrothers 1965, Baranova 1972, Gottlieb 1972, Walker 1972, Praglowski 1974, 1976, Takhtajan 1980, Cronquist 1981, Dahlgren et al. 1981, Thorne 1981).

Carlquist (1982) compared five anatomical characteristics of wood from *Illicium* (Illiciaceae) species. He indicated that Smith (1947) and Bailey and Nast (1948) reported that the wood characteristics of the Illiciaceae and Schisandraceae were nearly identical. Carlquist (1982) also indicated that the placement of the two families in the magnolioid alliance was justified. The Illiciaceae by having primitive wood features are similar to the Schisandraceae, and have more in common with the magnolioid family Magnoliaceae also possessing primitive wood features rather than the magnolioid family Annonaceae possessing more specialized wood features.

Young (1981) published a cladistic analysis of 33 taxa and indicated that the most parsimonious explanation of the absence of vessels in dicotyledons is that in all instances it is a derived feature. Thus the absence of vessels in the eleven extant genera of "primitive" woody dicotyledons is a secondary loss. His cladogram showed two major groups, each with two subgroups. Four families included in our serological research (Annonaceae, Lauraceae, Magnoliaceae, and Schisandraceae) were members of one major group (group A with 12 families), while the Illiciaceae was a member of another major group (group C with 13 families, e.g., Aristolochiaceae, Cabombaceae, Nymphaeaceae, and Piperaceae).

Recent published revision of the four Systems of Classification concerning the five families included in our investigation (Cronquist 1981, Dahlgren et al. 1981, Takhtajan 1980, Thorne 1981) are presented in Table 1. All four authors considered the Magnoliaceae and Annonaceae as being most similar. Cronquist, Dahlgren, and Takhtajan considered the Laurales most similar to the Magnoliales followed by the Illiciales. Thorne considered the Illiciales (Illiciineae) most similar to the Magnoliales (Annonineae) followed by the Laurales (Laurineae). All four authors considered the Schisandraceae most similar to the Illiciaceae and place both families in the same order or suborder.

The large family Lauraceae is divided into two subfamilies and several tribes. Table 2 shows Kostermanns' (1957) placement of *Cinnamomum, Laurus, Lindera,* and *Persea* into three tribes of the subfamily Lauroideae.

Taxa of the Illiciaceae and Schisandraceae were shown by preliminary experiments to have greater serological similarity with each other than either family had with tested genera of the Magnoliaceae (Johnson and Fairbrothers 1964).

2 Materials and Methods

2.1 Antigens and Antisera

Crushed and delipified seeds are designated seed meal to distinguish it from seed protein extracts. Proteins are extracted from seed meal in a 0.05 M sodium-potassium-phosphate buffer, in the ratio of 0.1 g seed meal to 1.0 ml of buffer. Following 12 h of extraction, the mixtures are centrifuged (14,000xg at 4 °C for 20 min). The protein concentration of the clarified supernatants are determined by means of the Bio-Rad protein assay (Bradford 1976). All protein extracts have concentrations ranging from 1.4–2.6 mg total protein per 1.0 ml of buffer. Both injections of the

Table 2. Kostermans' 1957 classification of the four genera of the family Lauraceae included as antigens and/or antisera

	Lauraceae
Subfamily	Lauroideae
1. Tribe	Perseae
Genus	Persea
2. Tribe	Cinnamomeae
Genus	Cinnamomum
3. Tribe	Laureae
Genus	Lindera
Genus	Laurus

experimental animals and processing of antisera follow procedures outlined by Petersen and Fairbrothers (1979). The 16 taxa belonging to five families from which protein extracts were obtained are presented in Table 3.

2.2 Ouchterlony Double Diffusion

The Ouchterlony technique of double diffusion in gels is performed on 9x10 cm glass plates, coated with 10 ml of 0.75% agarose. The stepwise procedure for double diffusion and presaturation (absorption) of antisera is described by Petersen and Fairbrothers (1983a,b); the only modification is seed meal used in place of pollen meal. Tests were made with rabbit antisera produced from injections of *Annona reticulata, Cinnamomum camphora, Illicium floridanum, Laurus nobilis, Magnolia kobus, Magnolia tripetala, Michelia champaca,* and *Persea americana* seed protein extracts.

2.3 Presaturation

Each antiserum is presaturated separately with seed meals of 16 taxa that are referred to as the absorbing antigen system. The complete series of presaturated (absorbed) antisera is then reacted against test antigens of 16 taxa. Results arė recorded on the basis of the type of precipitation patterns produced, and scored for statistical analysis. The method of analysis based on the indices of absorption similarity is used (Lester 1979).

2.4 Methods of Analysis

The similarity coefficient of the number of positive reactions $S_{no.pos.}$ records the occurrence of visible precipitation responses of the test antigen reacted against the complete series of absorbed antisera. The similarity coefficient of the intensity of precipitation of positive reactions $S_{ppt.pos.}$ records the differences in intensity of

Table 3. Sixteen taxa used in the serotaxonomic analysis of the Magnoliidae (Magnoliiflorae, Anoniflorae)

Annonaceae	Magnoliaceae
1. *Annona muricata* L.	1. *Magnolia kobus* DC
2. *Annona reticulata* L.	2. *Magnolia soulangeana* Soul.
3. *Annona squamosa* L.	3. *Magnolia splendens* Urb.
Illiciaceae	4. *Magnolia tripetala* L.
1. *Illicium floridanum* Ellis	5. *Magnolia virginiana* L.
Lauraceae	6. *Michelia champaca* L.
1. *Cinnamomum camphora* Nees & Eberm.	Schisandraceae
2. *Laurus nobilis* L.	1. *Schisandra chinensis* (Turc.) Baill
3. *Lindera benzoin* Blume	
4. *Persea americana* Mill.	
5. *Persea palustris* (Raf.) Sarg.	

positive reactions. The third index of positive response, the absorption similarity coefficient of positive reactions $S_{abs.pos.}$, represents the averaging of corresponding $S_{no.pos.}$ and $S_{ppt.pos.}$ values.

Absorption similarity coefficients of negative reactions $S_{abs.neg.}$ represent the effectiveness of individual seed meals as absorbing agents. Results of the cross-reacting absorbing seed meals are then compared to the control, the reference seed meal. Averaging of corresponding $S_{abs.pos.}$ and $S_{abs.neg.}$ readings for each taxon used, both as an absorbing seed reagent and a test antigen, produces the mean absorption similarity coefficient $S_{abs.mean}$. The final reading, coefficient of correlation of $S_{abs.neg.}$-values with $S_{no.pos.}$, $S_{ppt.pos.}$, and $S_{abs.pos.}$ records the degree of correspondence between the positive and negative similarity values. Using the Unweighted Pair Group Method Analysis (UPGMA) of Sneath and Sokal (1973) the dendrograms of serological similarity were generated (Roberts 1978). Details and applications concerning these analyses can be found in Lester (1979), Petersen and Fairbrothers (1983a,b), and Lester et al. (this volume).

3 Results and Discussions

3.1 Analysis

The following discussions are based on a compilation of data from eight antisera, each absorbed with 16 seed meals and reacted against 16 seed protein extracts. The analysis of absorption similarity coefficients is according to Lester (1979) and Petersen and Fairbrothers (1983a,b). The $S_{abs.neg.}$ and $S_{abs.pos.}$ readings are calculated for each set of data obtained from individual antiserum. The mean absorption similarity coefficients $S_{abs.mean}$ calculated from corresponding $S_{abs.pos.}$ and $S_{abs.neg.}$ readings are presented in Table 4. The calculated $S_{abs.means}$ values range from 0.065 as the least similarity (*Laurus* antiserum and *Schisandra* antigen), to 1.0 as the greatest similarities for each reference reaction and two *Magnolia kobus* cross-reactions with serologically very similar *Magnolia* species (Table 4).

The $S_{abs.mean}$-data are subjected to unweighted pair group method analysis (UPGMA), and the results from this analysis are recorded as a half matrix. From the calculated half matrix the two dendrograms are derived (Figs. 1 and 2). Figure 1 is a dendrogram that represents a cuboidal comparison of the eight tested antisera, and Fig. 2. is a dendrogram expanded to include eight additional taxa for which antibody systems were not raised. In order to add a new taxon, Table 4 is monitored for areas of similarity, and the highest of these values are used in the placement of the taxon on the approximate branches (arms) of the dendrogram (Roberts 1978). Thus the more comprehensive dendrogram of Fig. 2 is derived from monitoring $S_{abs.mean}$ values for all tested taxa included in the research. The addition of eight taxa presented no placement difficulties, since they are species placed in genera and/or families previously plotted on the dendrogram (Fig. 1).

Table 4. Values for the Mean Absorption Similarity Coefficients ($S_{abs.mean}$) based on negative and positive reactions of sixteen antigen systems with eight antibody systems within the Annoniflorae (Magnoliidae)

Antigens \ Antisera	anti-*Laurus nobilis*	anti-*Persea americana*	anti-*Cinnamomum*	anti-*Magnolia tripetala*	anti-*Magnolia kobus*	anti-*Michelia champaca*	anti-*Illicium*	anti-*Annona reticulata*
Laurus nobilis	1.000	0.761	0.513	0.250	0.156	0.219	0.250	0.391
Persea palustris	0.729	0.947	0.372	0.132	0.234	0.250	0.250	0.156
Persea americana	0.442	1.000	0.215	0.206	0.167	0.203	0.250	0.167
Cinnamomum camphora	0.692	0.763	1.000	0.206	0.264	0.172	0.250	0.316
Lindera benzoin	0.882	0.761	0.876	0.308	0.188	0.203	0.234	0.485
Magnolia tripetala	0.334	0.344	0.349	1.000	0.817	0.599	0.659	0.283
Magnolia soulangeana	0.416	0.328	0.519	0.913	1.000	0.930	0.659	0.431
Magnolia splendens	0.368	0.358	0.307	0.817	0.802	0.867	0.659	0.421
Magnolia virginiana	0.380	0.328	0.576	0.886	1.000	0.930	0.659	0.502
Magnolia kobus	0.369	0.328	0.591	0.884	1.000	0.836	0.659	0.422
Michelia champaca	0.280	0.375	0.246	0.729	0.784	1.000	0.659	0.389
Schisandra chinensis	0.065	0.344	0.272	0.692	0.634	0.604	0.935	0.536
Illicium floridanum	0.324	0.344	0.340	0.734	0.517	0.604	1.000	0.276
Annona reticulata	0.345	0.328	0.344	0.282	0.383	0.287	0.250	1.000
Annona muricata	0.292	0.328	0.138	0.314	0.343	0.371	0.261	0.805
Annona squamosa	0.458	0.442	0.320	0.268	0.304	0.259	0.250	0.859

3.2 Points to Remember

It is important to recall that all the comparisons are based on each antiserum being presaturated (absorbed) with each antigen (all 16 species), and testing 16 antigens with each of the eight antisera. The significance of such an experimental procedure is the generating of data from many comparisons of diverse taxa placed in the families Annonaceae, Illiciaceae, Lauraceae, Magnoliaceae, and Schisandraceae. It is also important to recall that the method of analysis includes similarities derived from several comparisons: (1) number of positive responses, (2) intensity of positive responses, and averaging of number and intensity of positive responses, (3) number of negative responses to absorbed antisera, and (4) the mean value of absorbed positive and negative responses. Thus all of the various kinds of responses are included in the evaluation and determination of final similarities, and the two dendrograms (Figs. 1 and 2) reflect these many comparison combinations.

Fig. 1. Dendrogram of serological similarity showing the clustering of eight Annoniflorae taxa derived by the Unweighted Pair-Group Method Analysis (UPGMA), and based upon the values for the Mean Absorption Similarity Coefficients in Table 4

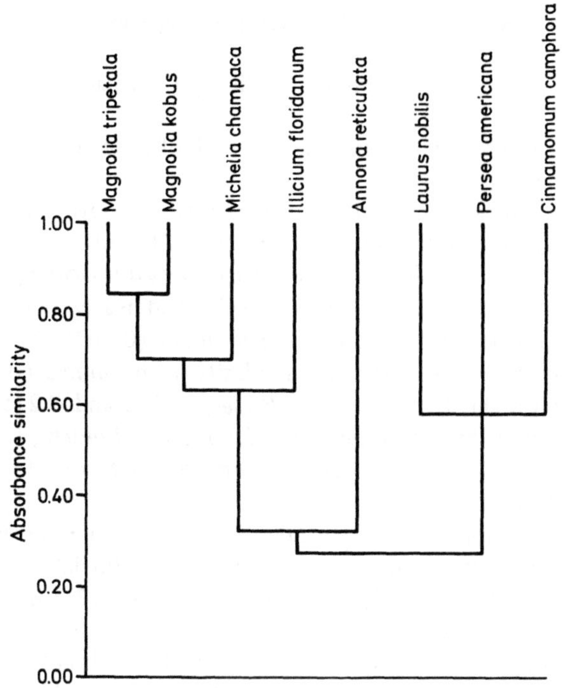

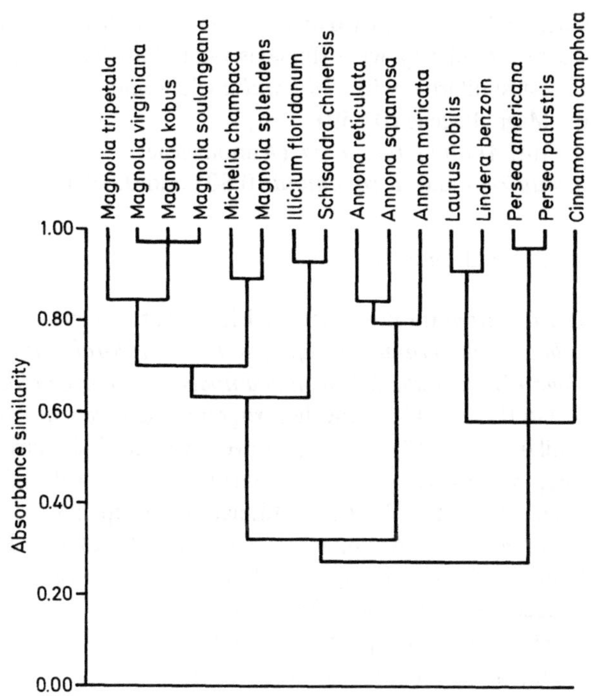

Fig. 2. Dendrogram of serological similarity showing the clustering of 16 Annoniflorae taxa derived by the Unweighted Pair-Group Method Analysis (UPGMA), and based upon the values for the Mean Absorption Similarity Coefficients in Table 4

3.3 Similarity and Taxonomic Groupings

3.3.1 Mean Absorption Similarity Coefficients

Table 4 reveals similarities for each of the eight antisera compared with each of sixteen antigens:

Anti-*Laurus nobilis* (Lauraceae) – reveals greatest similarity with *Lindera* (0.882), and less with *Cinnamomum* (0.692), and *Persea* (0.729 and 0.442) which also are members of the Lauraceae. Anti-*Persea americana* (Lauraceae) – reveals greatest similarity with *Persea palustris* (0.947), and nearly identical similarity with *Laurus* (0.761), *Lindera* (0.761), and *Cinnamomum* (0.763). Anti-*Cinnamomum camphora* (Lauraceae) – reveals greatest similarity with *Lindera* (0.876) and *Laurus* (0.513), and less with the two species of *Persea* (0.372 and 0.215). These three Lauraceae antisera reveal three groupings: (1) *Laurus* and *Lindera*, (2) *Persea* species, and (3) *Cinnamomum*. These serological groupings correspond to Kostermans' treatment (1957, see Table 2).

Anti-*Magnolia tripetala* (Magnoliaceae) – reveals greatest similarity with other members of the order Magnoliales and Illiciales. All species of *Magnolia* reveal close similarity (0.913, 0.886, 0.884, and 0.817). *Michelia* reveals a little less similarity (0.729). Anti-*Magnolia kobus* (Magnoliaceae) – reveals even greater similarity with other *Magnolia* species (1.000, 0.817, and 0.802), and with *Michelia* only a little less similarity (0.784). Anti-*Michelia champaca* (Magnoliaceae) – reveals close similarity with all *Magnolia* species, and next with *Illicium* and *Schisandra* (0.604). The three Magnoliales antisera (*Magnolia kobus, M. tripetala,* and *Michelia champaca*) reveal greatest similarity with Illiciales taxa. Anti-*Illicium floridanum* (Illiciaceae) – reveals closest similarity with *Schisandra* (Schisandraceae) (0.935) and next with six taxa of the Magnoliaceae (0.659).

Anti-*Annona reticulata* (Annonaceae) – reveals nearly the same similarity with *A. muricata* and *A. squamosa* (0.805 and 0.859).

3.3.2 Dendrograms

The dendrogram presented in Fig. 1 was generated from eight antisera: *Annona reticulata, Cinnamomum camphora, Illicium floridanum, Laurus nobilis, Magnolia kobus, Magnolia tripetala, Michelia champaca,* and *Persea americana*.

On the left side of the dendrogram *Magnolia tripetala* and *M. kobus* have serological similarity of 85%, and they have serological similarity of 70% with *Michelia*. These three Magnoliaceae taxa have serological similarity of 62% with *Illicium* of the Illiciaceae. This Magnoliaceae – Illiciaceae group has serological similarity of 30% with *Annona* of the Annonaceae. On the right of the dendrogram the Lauraceae group (*Laurus, Cinnamomum,* and *Persea*) have serological similarity of 25% with the Magnoliaceae – Illiciaceae – Annonaceae group.

Thus three groups were revealed by the dendrogram: (1) Magnoliaceae – Illiciaceae, (2) Annonaceae, and (3) Lauraceae.

The dendrogram presented in Fig. 2 was generated by adding eight taxa to appropriate branches belonging to genera and/or families included in the Fig. 1 dendrogram. On the left side of the dendrogram *Magnolia virginiana, M. kobus,* and *M. soulangeana* have serological similarity of 98%, and these three *Magnolia* species have serological similarity of 84% with *M. tripetala. Magnolia splendens* indigenous to Puerto Rico, and *Michelia champaca* have 89% serological similarity, and these two taxa have 68% serological similarity with the four mentioned species of *Magnolia.*

The genus *Illicium* (Illiciaceae) and *Schisandra* (Schisandraceae), members of the order Illiciales, have a serological similarity of 92%. These members of the Illiciales have a serological similarity of 63% with the Magnoliales.

Annona reticulata and *A. squamosa* have a serological similarity of 83%, and they have a serological similarity of 78% with *Annona muricata*. The three Annonaceae species have a similarity of 30% with the Magnoliales – Illiciales group.

The five members of the Lauraceae reveal three serological groups within the family. *Laurus* and *Lindera* have a serological similarity of 92%, and the two species of *Persea* have a 97% serological similarity. *Cinnamomum* has a serological similarity of 58% with the other Lauraceae taxa. The five Lauraceae taxa have a serological similarity of 25% with the Magnoliales, Illiciales, and Annoniales cluster. The three detected serological groups in the Lauraceae correspond to Kostermans' three tribes: Laureae (*Laurus* and *Lindera*), Perseae *(Persea)*, and Cinnamomeae *(Cinnamomum)* with the genus *Cinnamomum* being the least similar of the five members tested.

The data reveal that the genus *Magnolia* is divided into groups (subgenera and tribes). *Magnolia splendens* is the most distinct *Magnolia* species tested. The genus *Michelia* is relatively close to the genus *Magnolia*, and is as similar as some species of *Magnolia* are to each other.

Illicium and *Schisandra* are serologically very similar and the data do not support the separation of *Schisandra* as indicated by Young's (1981) cladistic data. The greatest similarity of the Illiciales is with the Magnoliales.

The Annonaceae has only slightly more similarity with the Magnoliales than does the Laurales. The serological data support the placement of the Annonaceae in the order Annonales as distinct from the Magnoliales. The descending order of serological similarity would be as follows: Magnoliales → Illiciales → Annonales → Laurales; both the Annonales and Laurales have approximately equal similarity with the Magnoliales.

Acknowledgements. Research was supported by NSF Grant DEB-78-24217, and a Rutgers Research Council Travel Grant.

References

Bailey IW, Nast CG (1948) J Arnold Arbor 29:77–89

Baranova MA (1972) Taxon 21:447–469

Bradford NM (1976) Anal Biochem 72:248–254

Carlquist S (1982) Am J Bot 69:1587–1598

Cronquist A (1981) An integrated system of classification for flowering plants. Columbia Univ Press, New York

Dahlgren RMT (1983) In: Jensen U, Fairbrothers DE (eds) Proteins and nucleic acids in plant systematics. Springer, Berlin Heidelberg New York

Dahlgren RMT, Rosendal-Jensen S, Nielsen BJ (1981) In: Young DA, Siegler DS (eds) Phytochemistry and angiosperm phylogeny. Praeger, New York, pp 149–204

Fairbrothers DE (1977) Ann Mo Bot Gard 64:147–160
Fairbrothers DE (1983) Nord J Bot 3:35–41
Fairbrothers DE, Mabry TJ, Scogin RL, Turner BL (1975); Ann Mo Bot Gard 62:765–800
Gottlieb OR (1972) Phytochemistry 11:1537–1570
Jensen U (1981) In: Ellenberg H, Esser K, Kubitzki K, Schnepf F, Ziegler H (eds) Progress in botany, vol 43. Springer, Berlin Heidelberg New York, pp 344–369
Johnson MA, Fairbrothers DE (1964) Proc 10th Int Bot Congr 1:145–146
Johnson MA, Fairbrothers DE (1965) Bot Gaz 126:260–269
Kostermans AJGH (1957) Reinwardtia 4:193–256
Lester RN (1979) In: Hawkes JG, Lester RN, Skelding AD (eds) The biology and taxonomy of the Solanaceae. Academic Press, London New York, pp 285–304
Periasamy K, Swamy BGL (1959) Phytomorphology 9:251–263
Petersen FP, Fairbrothers DE (1979) Syst Bot 4:230–241
Petersen FP, Fairbrothers DE (1983a) Syst Bot 8:134–148
Petersen FP, Fairbrothers DE (1983b) Bull Torrey Bot Club 110 (in press)
Praglowski J (1974) World Pollen Spore Flora 3:1–48
Praglowski J (1976) World Pollen Spore Flora 5:1–32
Roberts PA (1978) PhD thesis, Univ Birmingham, England, UK
Smith AC (1947) Sargentia 7:1–224
Sneath PHA, Sokal RR (1973) Numerical taxonomy. Freeman, San Francisco
Stern WL (1954) Trop Woods 100:1–73
Takhtajan AL (1980) Bot Rev 46:225–359
Thorne RF (1981) In: Young DA, Siegler DS (eds) Phytochemistry and angiosperm phylogeny. Praeger, New York, pp 233–295
Tucker SC (1964) Am J Bot 51:1051–1062
Walker JW (1972) Taxon 21:57–65
Young DA (1981) Syst Bot 6:313–330

Proteins, Mimicry and Microevolution in Grasses

P.M. SMITH[1]

Abstract. Serology is used to illuminate taxonomic problems of generic, specific and infra-specific levels. *Boissiera* is shown to be very similar to *Bromus* while *Agropyron, Elymus, Hordeum,* and *Festuca* differ from it. *Brachypodium* appears serologically very distinct from *Bromus*. Of cereal mimic species, *B. secalinus, B. grossus,* and *B. bromoideus* are serologically similar but *B. pseudosecalinus* is different. Within *B. hordeaceus* agg., seed proteins of roadside populations appear relatively uniform over wide areas, while plants from different habitats show protein variation, sometimes correlating with minor morphological differences. It is argued that the seed protein variation is selection dependent.

1 Introduction

"Critical" groups of plants are the bane of the taxonomist's life, yet also the main reason for his existence. Taxa which are difficult to define may be critical for several reasons, including variation in breeding system options, hybridisation, and developmental plasticity. These problems, plus others, conspire to make the genus *Bromus* a taxonomic nightmare (Smith 1970, 1980).

Taxonomists are concerned to know how their taxa originate and diverge, as well as how to interpret small-scale intraspecific differences in taxonomic terms. In *Bromus*, such small morphological differences are either overlooked or overweighted, according to the fashion of the day. Some extra line of evidence is necessary to judge fairly. In the past (Smith 1968, 1973) I have found basic biosystematic methods helpful, but they are time-consuming. How can serological techniques, which are quick and cheap, help assess critical patterns of variation? This question should be evaluated at generic, specific, and intraspecific levels, which are where most critical taxonomic problems are found.

At specific level, Smith (1972) suggested that the two cereal mimic species *Bromus secalinus* and *B. bromoideus* were closely related serologically. Genetic and chromosomal similarities had long been demonstrated, but there were small though distinct morphological differences. These grains are mimics in the sense that by natural selection they have come to resemble the cereal grains which they used to contaminate (mainly rye and spelt). Both have caryopses which, because of curious thickening and longitudinal rolling, approach the size, shape, and density of the cereals. *B. grossus* is another spelt mimic, thought to be close to *B. bromoideus*, but hitherto untested

[1] Department of Botany, University of Edinburgh, Mayfield Road, Edinburgh EH9 3JH, UK

Proteins and Nucleic Acids in Plant Systematics
ed. by U. Jensen and D.E. Fairbrothers;

serologically. *B. pseudosecalinus* is a contaminant of *Lolium* and *Festuca*, achieving its grain mimicry in the same way, but on a smaller scale. This is a critical group where morphological characters overlap between species, making circumscription and identification difficult.

Figure 1 shows the immunoelectrophoretic (I.E.P.) patterns for the four above mentioned species, *B. pseudosecalinus* is clearly different from the others, and is a distinct species. The serological evidence indicates that *B. bromoideus* and *B. grossus* are indistinguishable. The trivial morphological differences distinguishing them have probably been overvalued. *B. secalinus* shows great similarity to *B. bromoideus* and *B. grossus,* but also several clear electrophoretic differences. These electrophoretic differences, plus the morphological characters, probably indicate it is a "good" species.

Above the species level, *Bromus* itself is a critical genus. It may be in a tribe Bromeae, or it may be in Brachypodieae. Its affinities can be examined using serological methods.

A range of species in other genera, some supposedly close to, others distant from *Bromus*, were chosen for a generic comparison. *Boissiera squarrosa* (=*Bromus pumilio:* Smith 1969b) was found to be probably congeneric with *Bromus* using other antisera. *Festuca* was shown by chemical means to be distinct from *Bromus* in short chain fructosans (de Cugnac 1941, Smith 1969a). *Agropyron, Elymus,* and *Hordeum*, members of the tribe Triticeae, were shown (Smith 1969a) to be serologically related to *Bromus. Brachypodium* has tended more recently (e.g. Hubbard 1962, Tutin et al. 1980) to be placed in a tribe separate from *Bromus*. Serological evidence supporting such a separation was given by Smith (1969a), but the total lack of reactivity of the *Brachypodium* material was unexpected. This test needs to be repeated with other *Bromus* antisera.

At the intraspecific level, 30 accessions of *Bromus hordeaceus* were assembled. These populations (see appendix) were geographically widespread, from several habitats where the species is common, and puzzling variants are recorded. Between

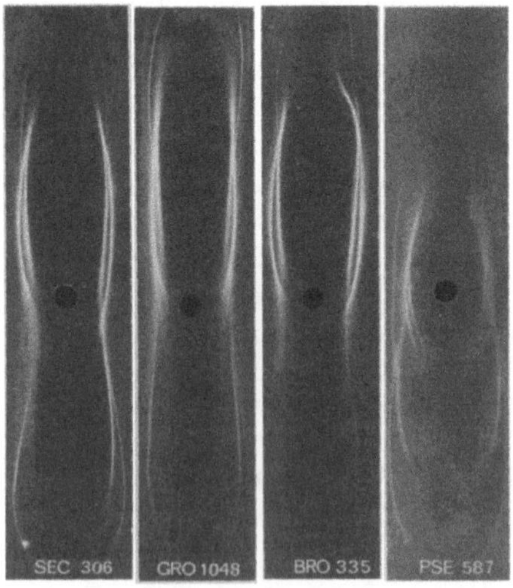

Fig. 1. Immunoelectrophoretic patterns of seed antigens of *Bromus secalinus, B. grossus, B. bromoideus,* and *B. pseudosecalinus*

most or all of these populations there will have been little if any gene flow, partly because the species is largely inbreeding, and partly because many of the populations have been spatially isolated from each other for a long period. *B. hordeaceus* represents very well the taxonomic problem of evaluating minor variants and, as an autogamous, opportunist, colonising species, might be expected to contain "accidentally" fixed, non-functional seed protein differences if, indeed, these occur.

If the morphological variants in the sample correlate with serological differences, it is unlikely that accidental character fixation is taking place, and much more likely that the differences, small though they may be, are selection based, hence functional and taxonomically more helpful. If the serological differences are correlated with habitat types, rather than just with population separation, the case for regarding them as indicators of "real", i.e. selection-dependent units, is further supported.

2 Materials and Methods

2.1 Technical

The general methods used have been described by Smith (1972). Seed sources are included in an appendix. Antisera were raised in rabbits by intramuscular injection series using double-emulsions of mineral oil adjuvant and seed protein solutions (Herbert 1965), or with aluminium hydroxide adjuvants, followed by intravenous injections of aqueous antigen. Injections were continued until antisera were ready for use (usually 10–15 injections over three months).

Antigens and antisera were reacted in agar gels buffered at pH 8.6 (barbitone-HCl). Immunoelectrophoretic separations were run for 7 h. Precipitin reactions were recorded after 4 days.

2.2 Interpretation

Difficulties arise in the interpretation of differences between those serological spectra which are, by definition, very similar overall. Where similarities are considerable, some method of transformation is necessary to bring what differences there are into prominence. Counting precipitin lines alone is not always reliable. Especially in double-diffusion tests, certain precipitin lines do not always diffuse to the same relative position, though others seem more consistent. Sometimes the same number of lines is made up in different ways. Absorption techniques can get around this problem, but can be expensive in raw materials, and the results are sometimes equivocal when one is working near the limit of resolution. In I.E.P. the line number is generally easy to establish, but there is the problem of slight variations in the mobility of certain antigens, some biological in origin, some due to experimental error. These variations matter little when larger, for example interspecific, differences are studied, but tend to frustrate intraspecific studies.

The procedure employed for analyzing *B. hordeaceus* reactions was to establish the routinely demonstrable I.E.P. lines (Fig. 2) (12 of a maximum seen of 14), and

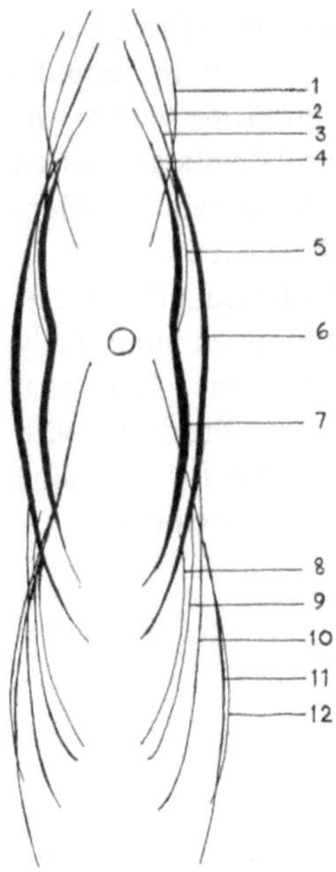

Fig. 2. A diagram of the 12 regularly occurring lines in the electrophoretic pattern of seed antigens of *B. hordeaceus*

to use only these as the basis for comparison. Five replicates of each accession test were made, and scored for all 30 accessions. The limit of experimental error in arc-peak position was 2.0 mm over a run length of 100 mm (2.0%). Arc-peak positions more than 2.0 mm from the mean position were taken to indicate an electrophoretic variant, i.e. a genuine taxonomic difference. Differences of this magnitude were always associated with particular precipitin arcs, and not a general feature of a particular separation.

From I.E P. results it thus becomes possible to record the serological variation in three parameters: (1) Number of lines produced (up to a maximum of 12). (2) Electrophoretic variants of the 12 lines. (3) Number of lines altogether missing. The homologous reaction can be recorded as 12/0/0, in these terms. A possible heterologous reaction would be 10/1/1, i.e. ten lines shared with the homologous reaction, one electrophoretic variant, and 1 missing line. The number of serological similarities (Ns) would here be 10, the differences (Nd) would be two. A coefficient (S) can be calculated, where $S\% = \frac{N_s}{N_s + N_d}$. The homologous reaction is 100%.

Finally, the spurring behaviour of adjacent antigen solutions in double diffusion reactions was used as a check on the groups of accessions defined by I.E.P. analysis.

3 Results

3.1 Bromus and Other Genera

Figure 3a shows double-diffusion reactions of the antigens tested, using a *B. scoparius* antiserum. *Boissiera squarrosa* here gives a reaction very similar to that of *Bromus scoparius*. Quite strong reactions are given by *Agropyron cristatum*, weaker ones by *Hordeum* and *Festuca*, and no detectable reaction is produced by *Brachypodium sylvaticum*.

3.2 Variation within Bromus hordeaceus

Figure 3b shows immunoelectrophoretic patterns of some of the *B. hordeaceus* accessions. Figure 4a–d shows double diffusion reactions of the 30 accessions. Table 1 summarises the I.E.P. differences observed between all thirty accessions, calculated as described above, using two *B. hordeaceus* antisera (HOR1 and HOR2). Table 2 shows how the accessions are distributed between 4 levels of similarity to the homologous reaction. Table 3 classifies the I.E.P. reactions of the 30 accessions into the seven observed categories of three-parameter serological response, to each antiserum.

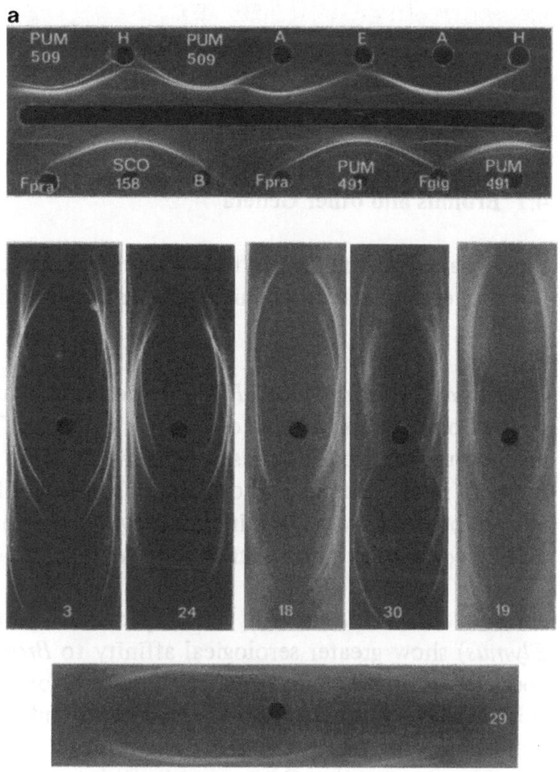

Fig. 3a,b. a Reactions of seed antigens of *Bromus pumilio (PUM)* (= *Boissiera squarrosa*) to a *Bromus scoparius* antiserum, with reactions also of *Bromus scoparius (SCO 158)*, *Agropyron cristatum (A)*, *Brachypodium sylvaticum (B)*, *Elymus canadensis (E)*, *Festuca gigantea (Fgig)*, *F. pratensis (Fpra)*, and *Hordeum distichum (H)*.
b Immunoelectrophoretic patterns of seed antigens of *B. hordeaceus* accessions *3* (roadside, UK), *24* (Cliff, UK), *18* (roadside, California), *30* (coast, Tripoli), *19* (roadside, California), *29* (coast, Palermo). Reactions of *3* and *24* are with HOR1 antiserum, the others with HOR2 antiserum

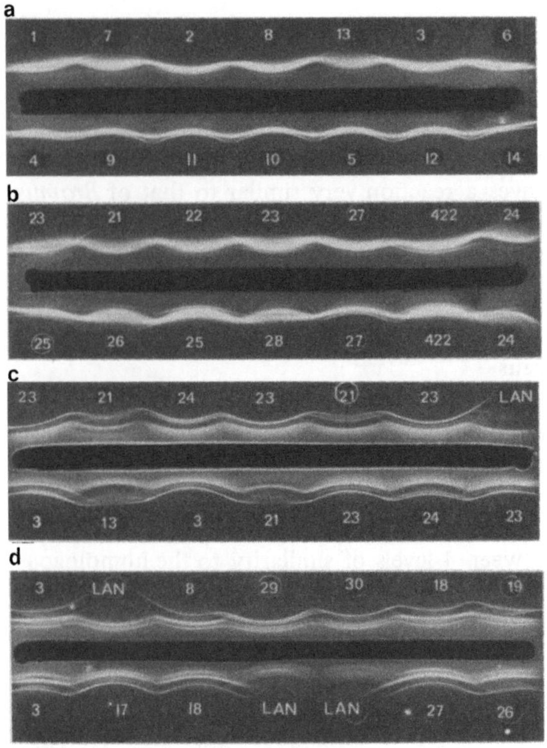

Fig. 4a–d. Double diffusion reactions of seed antigens of *B. hordeaceus* accessions. **a** European roadside populations (HOR2 antiserum); **b** Cliff and dune populations (HOR2 antiserum). 422 is the homologous reaction; **c** Dune populations, a Portuguese accession (13), with *B. lanceolatus* 154 (*LAN*) for comparison. (HOR1 antiserum). 3 is the homologous reaction; **d** Californian roadside populations (17, 18, 19), British cliff accessions (26, 27), Mediterranean coast accessions (29, 30), with *B. lanceolatus* 163 (*LAN*) for comparison. (HOR1 antiserum). 3 is the homologous reaction

4 Discussion and Conclusions

4.1 Bromus and other Genera

These data confirm and extend the earlier suggestions reported by Smith (1969a), but the recent data are technically superior and employ different antiserum of superior resolving power.

Bromus and Festuca. Both *Festuca* species (Festuceae) give a weak reaction, relative to the homologous reaction of *Bromus scoparius*. The difference detected here covers an undoubted tribal boundary. The work of de Cugnac (1941) demonstrated differences in the storage carbohydrate chemistry of *Bromus* and *Festuca*. Clearly, the storage proteins are also distinctive. *Bromus* species have simple starch grains, whereas all *Festuca* species so far examined have the compound type.

Bromus and Triticeae. All three species of Triticeae (*Agropyron, Hordeum*, and *Elymus*) show greater serological affinity to *Bromus* than do the two tested *Festuca* species. *Agropyron cristatum* gives quite strong reaction, as it did in previously reported tests (Smith 1969a). Brome-grasses and the Triticeae share the simple starch grain character. Given the genetic and morphological differences, however, it is not possible to argue that *Bromus* might fit into the Triticeae. Though this is another unequivocal tribal boundary, the serological evidence confirms that the Triticeae are nearer the bromegrasses than are the Festuceae.

Bromus and Brachypodium. This antiserum (*B. scoparius* 158) has failed to detect any serological similarity between the seed antigens of *Bromus scoparius* and *Brachypodium sylvaticum*. Earlier findings using antisera of *Bromus ciliatus* and *Bromus arvensis* (Smith 1969a), though a little surprising given morphological resemblances, are confirmed. Serological distinctness agrees with chromosomal evidence of a discontinuity between *Bromus* and *Brachypodium*. It would appear reasonable to regard *Bromus* as the principal genus in a tribe Bromeae, and to place *Brachypodium* and its annual relative *Trachynia,* in Brachypodieae.

Bromus and Boissiera. The strong reaction of *Boissiera squarrosa* antigens with *Bromus scoparius* antiserum agrees with earlier findings (Smith 1969b) and with the considerable morphological resemblances, supports the view, that *Boissiera* and *Bromus* are congeneric. *Bromus pumilio* is the correct name for *Boissiera squarrosa*, if it is to be recognized as a brome-grass. In this case the relationship of the two taxa should not be judged wholly on the very striking affinities now reported, partly because *Bromus* itself is a complex genus. Additional investigations into the serological affinities of *Boissiera* and *Bromus* are currently being conducted.

4.2 Serological diversity within Bromus hordeaceus

The thirty populations are in general ranked similarly by the two antisera (Tables 1, 2, and 3). There is total agreement between the two antisera on the level of similarity to the homologous reaction in over 70% of accessions, and in the remaining ones the difference is of only one level of similarity. HOR2 antiserum is perhaps a little more discriminating (Table 2).

Generally, roadside populations are rather similar to each other (and to the homologous species, both of which were roadside accessions). Cliff top and dune accessions are mostly somewhat different from the roadside types, and the two "mediterranean" collections are most different from the homologous reactions. All show a reasonably high degree of similarity – the method of analysis is aimed at exposing what differences there are most clearly. *Bromus interruptus*, a species only doubtfully distinct from *B. hordeaceus*, shows a resemblance of similar size to the least similar *B. hordeaceus* accession (30). *B. lanceolatus*, a distinct species, is considerably more different than any of these 30 accessions.

There is no evidence for suggesting that *B. hordeaceus* is sufficiently diverse to warrant its being split up into separate species, though some infraspecific taxa might be appropriate.

Considering similarities in the I.E.P. reactions (Table 3), there are seven categories of observed response. The homologous-type reaction (12/0/0) is shown by all roadside populations except 7, 13, 17, and 19. HOR2 antiserum also shows some difference in accessions 5, 10, 12, and 14. Additionally, both antisera elicit homologous reactions from dune population 23 and cliff population 25. Dune population 22 gives a homologous reaction with HOR1 antiserum, but a 92% reaction with HOR2 antiserum.

Both antisera show dune populations 21 and 24 similarly distinct from the homologous reactions. Cliff populations 26, 27, and 28 are also shown to be slightly, but consistently different from the homologous reaction, though HOR2 antiserum reveals that population 26 is more different than does HOR1 antiserum.

Both antisera show the two mediterranean coast accessions (29, 30) to be distinct from the bulk of the roadside populations. HOR2 antiserum indicates the Tripoli plants (accession 30) to be rather more different than does HOR1 antiserum.

With the minor exceptions mentioned above, the two antisera elicit similar responses from all the accessions.

Both from the 3-parameter analysis of I.E.P. resemblance (Tables 1, 2, and 3) and from the evidence of double diffusion (Fig. 4a–d) there can be little doubt that roadside populations are impressively uniform over a wide geographical area, at least based on these criteria.

Certain roadside populations (7, 13, 17, and 19) are nonetheless shown to be distinctly and consistently different from the usual roadside plants. It is intriguing to look at these in relation to the "mediterranean" populations (29, 30) which are also distinct. All six accessions have morphological attributes in common. They have stiffly erect, dense panicles, with many having spikelets with quite large (9–11 mm) lemmas, and flattened, patent, sometimes basally contorted awns. The mediterranean material would, in earlier days, have been quickly referred to the "portmanteau" species *B. molliformis* Lloyd (now *B. hordeaceus* ssp. *molliformis* Maire et Weiller; see Smith 1980, 1981). The four roadside accessions are all from habitats where soils and climate are not very different from those experienced by populations 29 and 30. Thus accession 7 is from a roadside near Rome, and is essentially therefore a "mediterranean" collection. Accession 13, from a coastal roadside near Lisbon, may similarly be placed in a "mediterranean – S. Europe" category, in terms of geographical origin. Both 17 and 18 are Californian, the former from dry, sandy soil in Yolo County and the other from a maritime situation near Santa Cruz. The coincidences of climate, well-drained soils, morphology and now serology are too great to be accidental.

Accessions 7 and 13 are simply *B. hordeaceus* ssp. *molliformis* plants growing in roadside habitats. The Californian plants show similar adaptations to arid conditions. They are most probably *molliformis*-types selected out from typical *hordeaceus* (like 18 and 20) introduced long ago to California, from Europe. It is interesting that the Californian *molliformis*-types show slight differences from each other, as well as from the European material (Fig. 4d).

Dune populations 21, 22, and 24 are serologically dissimilar from roadside populations. Similarly, most of the cliff populations (26, 27, 28) reveal serological differences from roadside plants. Even on the three-parameter I.E.P. basis, the reactions of two out of three dune populations (21 and 24) are identical to each other. Cliff populations 26, 27, and 28 are less uniformly different from the roadside type (see Fig. 4b and c).

There is a possibility that the British dune and cliff populations which display serological differences from roadside types but some similarities to supsp. *molliformis*, are of mediterranean or S. European origin. They may be recently imported aliens or may have evolved over an unknown length of time from earlier introductions of subsp. *molliformis*. The latter explanation is more likely in view of their morphological differences.

Table 1. Comparisons of I.E.P. spectra of 30 accessions of ***Bromus hordeaceus*** tested with two ***B. hordeaceus*** antisera (HOR 1 and HOR 2). HOR 2 data are italicised. Data for ***B. interruptus*** and ***B. lanceolatus*** shown for comparison. (* indicates no data available)

Accessions/Habitats	No. of shared lines (max 12)	Electro-phoretic variants of the 12	Missing lines	Serological similarities	Serological differences	Coefficient (%)
1. Roadside, UK	12/*12*	0/*0*	0/*0*	12/*12*	0/*0*	100/*100*
2. Roadside, UK	12/*12*	0/*0*	0/*0*	12/*12*	0/*0*	100/*100*
3. Roadside, UK	12/*12*	0/*0*	0/*0*	12/*12*	0/*0*	100/*100*
4. Roadside, UK	12/*12*	0/*0*	0/*0*	12/*12*	0/*0*	100/*100*
5. Roadside, UK	12/*11*	0/*0*	0/*0*	12/*11*	0/*1*	100/*92*
6. Roadside, Italy	12/*12*	0/*0*	0/*0*	12/*12*	0/*0*	100/*100*
7. Roadside, Italy	10/*10*	2/*2*	0/*0*	10/*10*	2/*2*	83/*83*
8. Roadside, France	12/*12*	0/*0*	0/*0*	12/*12*	0/*0*	100/*100*
9. Roadside, Yugoslavia	12/*12*	0/*0*	0/*0*	12/*12*	0/*0*	100/*100*
10. Roadside, Hungary	12/*11*	0/*0*	0/*1*	12/*11*	0/*1*	100/*92*
11. Roadside, Belgium	12/*12*	0/*0*	0/*0*	12/*12*	0/*0*	100/*100*
12. Roadside, Germany	12/*11*	0/*0*	0/*1*	12/*11*	0/*1*	100/*92*
13. Roadside, Portugal	10/*10*	1/*1*	1/*1*	10/*10*	2/*2*	83/*83*
14. Roadside, Spain	12/*11*	0/*0*	0/*1*	12/*11*	0/*1*	100/*92*
15. Roadside, Turkey	12/*12*	0/*0*	0/*0*	12/*12*	0/*0*	100/*100*
16. Roadside, Australia	12/*12*	0/*0*	0/*0*	12/*12*	0/*0*	100/*100*
17. Roadside, California	10/*10*	0/*0*	2/*2*	10/*10*	2/*2*	83/*83*
18. Roadside, California	12/*12*	0/*0*	0/*0*	12/*12*	0/*0*	100/*100*
19. Roadside, California	10/*10*	1/*1*	1/*1*	10/*10*	2/*2*	83/*83*
20. Roadside, W.Virginia	12/*12*	0/*0*	0/*0*	12/*12*	0/*0*	100/*100*
21. Dune, UK	10/*10*	2/*2*	0/*0*	10/*10*	2/*2*	83/*83*
22. Dune, UK	12/*11*	0/*0*	0/*1*	12/*11*	0/*1*	100/*92*
23. Dune, UK	12/*12*	0/*0*	0/*0*	12/*12*	0/*0*	100/*100*
24. Dune, UK	10/*10*	2/*2*	0/*0*	10/*10*	2/*2*	83/*83*
25. Cliff, UK	12/*12*	0/*0*	0/*0*	12/*12*	0/*0*	100/*100*
26. Cliff, UK	11/*10*	1/*2*	0/*0*	11/*10*	1/*2*	92/*83*
27. Cliff, UK	11/*11*	0/*0*	1/*1*	11/*11*	1/*1*	92/*92*
28. Cliff, UK	11/*11*	1/*1*	0/*0*	11/*11*	1/*1*	92/*92*
29. Coast, Palermo	10/*10*	1/*1*	1/*1*	10/*10*	2/*2*	83/*83*
30. Coast, Tripoli	10/9	2/*3*	0/*0*	10/9	2/*3*	83/*75*
B. interruptus						
482	9/*	2/*	1/*	9/*	3/*	75/*
B. lanceolatus						
163	*5/5*	1/*1*	6/*6*	*5/5*	*7/7*	42/*42*

Table 2. Distribution of 30 *Bromus hordeaceus* accessions between 4 levels of similarity to the homologous reactions of two *B. hordeaceus* antisera (HOR 1 and HOR 2)

	HOR 1		HOR 2	
	Accessions	Total	Accessions	Total
100%	1, 2, 3, 4, 5, 6, 8, 9, 10, 11, 12, 14, 15, 16, 18, 20, 22, 23, 25	19	1, 2, 3, 4, 6, 8, 9, 11, 15, 16, 18, 20, 23, 25	14
92%	26, 27, 28	3	5, 10, 12, 14, 22, 27, 28	7
83%	7, 13, 17, 19, 21, 24, 29, 30	8	7, 13, 17, 19, 21, 24, 26, 29	8
75%	–	–	30	1
		30		30

Table 3. Classification of reactions of 30 accessions of *Bromus hordeaceus* in terms of seven categories of serological response (the seven observed categories within the three parameters No. of Shared Lines/No. of Electrophoretic Variants/No. of Missing Lines) compared with two *B. hordeaceus* antisera (HOR 1 and HOR 2)

Categories	Accessions			
	HOR 1 reaction		HOR 2 reaction	
	Roadside	Dune, cliff, coast	Roadside	Dune, cliff, coast
12 / 0 / 0	1, 2, 3, 4, 5, 6, 8, 9, 10, 11, 12, 14, 15, 16, 18, 20	22, 23, 25	1, 2, 3, 4, 6, 8, 9, 11, 15, 16, 18, 20	23, 25
11 / 1 / 0		26, 28		28
11 / 0 / 1		27	5, 10, 12, 14	22, 27
10 / 1 / 1	13, 19	29	13, 19	29
10 / 2 / 0	7	21, 24, 30	7	21, 24, 26
10 / 0 / 2	17		17	
9 / 3 / 0				30

If morphological characters can be found to be correlated with these differences, as was suggested by Smith (1968), then there is a better case for recognizing subspecies. Additional samples are often the best way of strengthening the argument, but the serological evidence now reported was a factor in the recognition of ecotype-subspecies in *Bromus* (Smith 1981). Whether cliff or dune habitats are considered, there is the presence, in the immediate environs, of "roadside type" plants, of unmodified morphology. *"Molliformis"* plants grow admixed with normal roadside plants in some areas (e.g. Yolo County). Therefore, it is not possible to argue that the serological differences arise simply from spatial isolation. A dune plant, for example, is no more likely to be spatially isolated from a roadside type plant than two roadside plants are to be spatially isolated from each other. The differences are not mere inter-population differences, or pure-line "Jordanons", but are correlated with habitat-type.

Sufficient evidence does not exist to offer a convincing taxonomic explanation for the differences between the normal "roadside" homologous reaction and those of accessions 5, 10, 12, and 14. Neither antiserum could distinguish these four from each other, but HOR2 detected in each a one-antigen difference from the homologous reaction. The plants share no obvious morphological or geographical peculiarity, nor are the areas from which they come notable for any obvious climatic similarity.

Apart from taxonomic conclusions, outlined above, these results provide clear indications that intraspecific serological differences between grass seed proteins, are related to selection pressures and do not reflect accidental fixation of random variants. Generally speaking, it has been possible to illustrate that, on the one hand, differences which are found can be correlated with habitat differences. On the other hand, regardless of geographical isolation, where habitat remains a relative constant (i.e. in roadside populations in this example) so too do the serological patterns. Nowhere is this demonstration more needed or more welcome than in the serotaxonomy of the autogamous, annual brome-grasses. But it is one that could be extended with profit into many critical groups of annual and perennial species.

References

De Cugnac A (1941) Bull Soc Bot Fr 88:402–410

Herbert WJ (1965) Lancet II:771

Hubbard CE (1962) Grasses. Penguin Books, Harmondsworth

Smith PM (1968) Watsonia 6:327–344

Smith PM (1969a) Ann Bot (London) 33:591–613

Smith PM (1969b) Feddes Repert 79:337–345

Smith PM (1970) Not R Bot Gard Edinburgh 30:361–375

Smith PM (1972) Ann Bot (London) 36:1–30

Smith PM (1973) Watsonia 9:319–332

Smith PM (1980) In: Tutin TG et al. (eds) Flora Europaea, vol V. Cambridge Univ Press, Cambridge, pp 182–189

Smith PM (1981) Bot Jahrb Syst 102:497–509

Tutin TG et al. (1980) Flora Europaea, vol V. Cambridge Univ Press, Cambridge

Appendix: Provenance of Accessions Used

I. Antisera

For cereal mimicry: *B. hordeaceus* 21 (Roadside, Worcs., UK)
For generic comparisons: *B. scoparius* 158 (Origin Afghanistan, USDA PI no. 251, 976)
For comparisons within *B. hordeaceus* L.:
B. hordeaceus 21 (HOR 1) (Roadside, Worcs., UK)
B. hordeaceus 422 (HOR 2) (Roadside, Machynlleth, Cards., UK)

II. Antigens

a) Cereal Mimics

B. secalinus 240 (Origin Oregon, USDA PI no. 258, 466)
B. secalinus 306 (CSAZV, Brno, Czechoslovakia)
B. secalinus 117 (London, Ontario)
B. grossus 1048 and 1049 (Prof. de Cugnac, Rumigny, Ardennes)
B. bromoideus 132 (Brussels Botanic Garden)
B. bromoideus 335 (Dijon Botanic Garden)
B. pseudosecalinus 489 (Roadside, Worcestershire, UK)
B. pseudosecalinus 587 (Near haystacks, Tile Hill, Warks., UK)

b) Generic comparisons

Agropyron cristatum 908 (Agronomy Dept., UC Davis, Calif.)
Bromus pumilio 491 (Tabriz, Iran) } = *Boissiera squarrosa*
Bromus pumilio 509 (Kochka Valley, Afghanistan) }
Brachypodium sylvaticum 457 (Kidwelly, UK)
Elymus canadensis 934 (USDA PI no. 232, 249, origin Montana)
Festuca gigantea 1090 (Greta Bridge, N. Yorks, UK)
F. pratensis 552 (Trimsaran, UK)
Hordeum distichum 585 (CSIRO, CPI no. 13180)

c) 30 populations of *Bromus hordeaceus* L.

1. Roadside, Hartlebury, UK (Smith acc. 1)
2. Roadside, Spernall UK (16)
3. Roadside, Halesowen UK (21)
4. Roadside, Machylleth, UK (422)
5. Roadside, Burry Port, UK (144)
6. Roadside, Como, Italy (153)
7. Roadside, Rome, Italy (156)
8. Roadside, Paris, France (444)
9. Roadside, Sarajevo, Yugoslavia (449)
10. Roadside, Lake Balaton, Hungary (1055)
11. Roadside, Antwerp, Belgium (364)
12. Roadside, Bruchsal, Frankfurt, FRG (448)
13. Roadside, Estoril, Portugal (510a)
14. Roadside, Benasque, Spain (2)

15. Roadside, Adapazari – Geyne, Turkey (1020)
16. Roadside, Adelaide, Australia (357)
17. Roadside, Winters, California (949)
18. Roadside, Winters, California (947)
19. Roadside, Santa Cruz, California (950)
20. Roadside, West Virginia (493)
21. Dune, Ynys-Las, UK (429)
22. Dune, Culbin Forest, Moray, UK (753)
23. Dune, Sotherness, Dumfries., UK (1177)
24. Dune, Newborough, Anglesey, UK (464)
25. Cliff, Hampshire, UK (143)
26. Cliff, Morfa Bychan, UK (424)
27. Cliff, Holyhead, UK (459)
28. Cliff, Berry Head, Devon, UK (465)
29. Coastal scrub, Palermo, Italy (678)
30. Coastal rough ground, Tripoli, Libya (351)

B. interruptus: Little Abington, Cambs., UK (482)
B. lanceolatus: Turkey, USDA PI no. 170, 256 (163)
B. lanceolatus: Thompson and Morgan strain, Ipswich, UK (154)

Immunochemistry and Phylogeny of Selected Leguminosae Tribes

G. CRISTOFOLINI and P. PERI[1]

Abstract. The analyses of immunological reactions obtained from extracts of seed storage proteins have provided a substantial contribution to our understanding of the systematics and classifications of the Leguminosae. The Mimosoideae subfamily is rather isolated, while the two other subfamilies, Caesalpinioideae and Faboideae are more closely related to each other. Within the subfamily Faboideae, *Phaseolus* and *Vigna* are extremely isolated from all other genera, these two genera are also isolated from all members of the family Leguminosae tested. The tribes Robinieae, Abreae, Aeschynomeneae, and Amorpheae constitute a natural group with about equal similarity to each other, and as a cluster they are distant from all other groupings tested. Galegeae, Vicieae, and Trifolieae were determined to be a homogeneous group of tribes based upon immunological similarity. The correct position of the tribe Loteae in respect to the above three tribes was not possible to ascertain. The Loteae seem to be partially related to the Bossiaeeae and Crotalarieae. Sophoreae, Thermopsideae, and Genisteae form a strong coherent group with high similarity detected between the three tribes. The serological data verify most aspects of the recent systems of classification of the Leguminosae, and indicate various questionable taxa and groupings (genera and/or tribes) requiring additional data to clarify proper relationships.

1 Introduction

The Leguminosae have been intensively investigated using serological techniques; however, our present knowledge is very limited because only certain genera and tribes have been investigated, while many others have not been explored at all. Intensive serological research has been conducted in the past decade with members of the tribe Phaseoleae, with special emphasis on the genus *Phaseolus*, by several research groups, in particular the researchers in Prague (Kloz and Klozová 1974, Klozová and Turková 1978). Extensive research on Vicieae has been conducted by the group in Leningrad (Tarlakovskaya et al. 1976), on Trifolieae (Simon 1969), Abreae (Rougé and Lascombes 1977), Genisteae (Cristofolini and Chiapella-Feoli 1977, Feoli-Chiapella and Cristofolini 1981).

An extensive review was published by Kloz (1971) for research published before 1971, and by Cristofolini (1981) for publications appearing from 1971 until 1980.

Our present knowledge on legume sero-systematics can be summarized as follows: (1) seven tribes have been studied in detail, (2) three tribes are partially known (research restricted to selected genera), (3) eleven tribes are poorly known (one or very

[1] Institute of Botany, University of Bologna, Bologna, Italy

Proteins and Nucleic Acids in Plant Systematics
ed. by U. Jensen and D.E. Fairbrothers

few species have been investigated), (4) for ten tribes no serological research has been reported.

The above synthesis looks rather encouraging, since about two thirds of the tribes have been at least partially examined. However, a large portion of the research has been concerned with generic or tribal systematics. No intensive work has been conducted at the family level, nor has any major contribution come from serology to help explain inter-tribal relationships or to define the evolutionary patterns of the family.

The serological results are based on seed proteins and the physicochemical properties of some of these seed proteins are well known and have been subjected to intensive research. However, much of the physicochemical information has not proved to be of value for systematic comparisons. Of the proteins investigated the low molecular weight lectins seem to be specific for taxonomic categories at the level of genera and species (Rougé 1981), and less useful for establishing relationships of higher ranked taxa.

For the storage proteins the occurrence of legumin and vicilin in Faboideae tribes have been clearly demonstrated. The subunit structure and some immunochemical characteristics are already known from several species, and comparative data from electrophoresis and sequencing experiments are also known (Derbyshire et al. 1976, Jensen 1981, Miège 1982, Jensen and Grumpe this volume).

These storage proteins are surely the dominating proteins in the antigenic systems, derived from the whole pattern of seed proteins, which were used in our experiments. It was expected that these protein mixtures would yield a better differentiation between the tribes than those reported for the more or less uniform legumins and vicilins by Dudman and Millerd (1975) and Gilroy et al. (1979).

In our laboratory a stock of protein mixtures extracted from seeds of about 100 legume species has been prepared and stored. For the present preliminary survey 28 antigenic systems have been selected (Table 1) in such a way that the experimental taxa were evenly distributed throughout the family and represent the various tribes. They include the three subfamilies and 20 of the tribes within them, i.e. three of Mimosoideae, three of Caesalpinioideae, and fourteen of Faboideae. Ten of these extracts have been used to raise antibodies in rabbits. Also those taxa used for producing antisera are selected to evenly represent the different tribes. Thus the 10 reference systems include two tribes of the Mimosoideae, one tribe of the Caesalpinioideae, and seven tribes of the Faboideae.

For protein extraction and antisera production the methods described by Cristofolini and Chiapella-Feoli (1977) were followed. All antibody systems resulted from pooling antisera, obtained from two to three rabbits. All antigen systems were adjusted to a concentration of 10 mg/ml.

2 Analytical Results

A sampling of immunodiffusion tests is presented in the Figs. 1-5.

Table 1. Quantitative estimation of immuno-reactions between 28 species of Fabaceae (rows) and 10 antisera (columns). Reference reactions are encircled

	Antisera									
	Albizia 1	Acacia 2	Gleditsia 3	Robinia 4	Phaseolus 5	Vicia 6	Trifolium 7	Templetonia 8	Baptisia 9	Ulex 10
Mimosoideae										
Albizia lophanta Benth.	④	4	1	3	1	1	2	1	1	1
Acacia retinodes Schlecht.	2	④	1	1	1	1	1	1	1	1
Leucaena leucocephala (Lam.) DeW.	4	4	1	1	1	1	1	2	1	1
Caesalpinioideae										
Caesalpinia gilliesii Wall.	1	2	3	1	1	1	3	2	2	2
Gleditsia triacanthos L.	1	3	④	1	1	1	3	3	3	2
Ceratonia siliqua L.	1	1	2	1	1	1	1	1	1	1
Bauhinia acuminata L.	1	1	1	1	1	1	2	1	2	2
Faboideae										
Sophora microphylla Ait.	1	1	1	3	2	1	3	2	3	3
Abrus precatorius L.	1	1	2	3	1	1	1	2	1	2
Robinia pseudacacia L.	1	1	1	④	1	2	1	2	1	2
Phaseolus coccineus L.	1	1	2	1	④	1	2	2	1	2
Vigna catjang Walp.	1	1	1	1	4	1	2	1	1	2
Amorpha fruticosa L.	1	1	2	3	1	1	2	2	1	1
Arachis hypogaea L.	1	1	1	3	1	1	2	2	1	1
Astragalus cicer L.	1	2	1	1	1	2	2	2	1	2
Anthyllis vulneraria L.	1	2	1	1	1	2	1	2	1	1
Lotus corniculatus L.	1	2	1	1	1	1	1	2	2	1
Vicia angustifolia L.	1	2	1	1	1	④	1	1	1	2
Lathyrus heterophyllus L.	1	3	3	1	1	3	4	3	3	2
Lens culinaris Med.	1	3	2	1	1	3	4	2	3	2
Melilotus officinalis Lam.	1	3	3	1	1	3	4	2	2	2
Trifolium campestre Schreb.	1	3	3	1	1	3	④	2	3	2
Templetonia retusa R. Br.	1	3	2	1	1	2	2	④	1	2
Crotalaria ochroleuca G. Don	1	1	1	1	1	1	1	2	1	1
Thermopsis montana Nutt.	1	1	2	1	1	1	2	2	4	3
Baptisia australis R. Br.	1	2	2	1	1	1	3	2	④	3
Spartium junceum L.	1	1	3	1	1	1	2	3	4	4
Ulex europaeus L.	1	1	2	1	1	1	2	3	4	④

The antiserum raised to *Albizia lophanta* (Mimosoideae) reacted strongly with the proteins extracted from *Leucaena*, and somewhat less with those of *Acacia*. However, all reactions with proteins from taxa belonging to the Caesalpinioideae or Faboideae were very weak.

Stronger reactivity was obtained by the antiserum raised to *Acacia retinodes* (Fig. 1). Beside the reference-reaction and the very strong reactions of species belonging to the same subfamily, several species of other subfamilies produced a marked positive reaction. The strongest reactions were produced by the Caesalpinioideae *Caesalpinia*

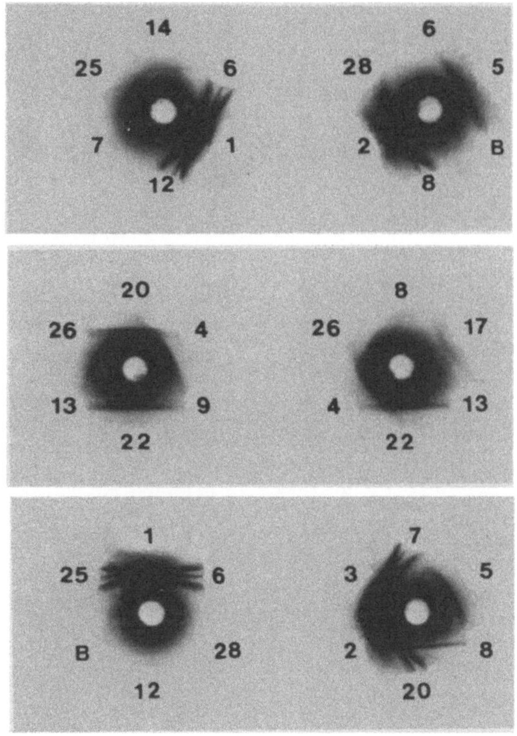

Fig. 1. Plates of immunodiffusion. *Central well* Antiserum to *Acacia retinodes. Outer wells* Antigen systems. (B) blank; (1) *Albizia lophanta* Benth.; (2) *Acacia retinodes* Schlecht.; (3) *Leucaena leucocephala* (Lam.) DeWitt.; (4) *Caesalpinia gilliesii* Wall.; (5) *Gleditsia triacanthos* L.; (6) *Bauhinia acuminata* L.; (7) *Sophora microphylla* Ait.; (8) *Abrus precatorius* L.; (9) *Robinia pseudacacia* L.; (10) *Phaseolus coccineus* L.; (11) *Vigna catjang* Walp.; (12) *Amorpha fruticosa* L.; (13) *Arachis hypogaea* L.; (14) *Astragalus cicer* L.; (15) *Anthyllis vulneraria* L.; (16) *Lotus corniculatus* L.; (17) *Vicia sativa* L. ssp. *nigra* (L.) Ehrh.; (18) *Lathyrus heterophyllus* L.; (19) *Lens culinaris* Med.; (20) *Melilotus officinalis* Lam.; (21) *Trifolium campestre* Schreber.; (22) *Templetonia retusa* R. Br.; (23) *Crotalaria ochroleuca* G. Don.; (24) *Thermopsis montana* Nutt.; (25) *Baptisia australis* R. Br.; (26) *Spartium junceum* L.; (27) *Ulex europaeus* L.; (28) *Ceratonia siliqua* L

and *Gleditsia*, and by several Faboideae tribes, especially Trifolieae, Vicieae, Galegeae, Thermopsideae, and Genisteae.

The cross-reactivity detected with taxa of the Mimosoideae is much higher than the cross-reactivity between taxa within other subfamilies, which is an indicator of high homogeneity in the Mimosoideae. However, there seems to be a remarkable difference in the reactivity of the two antisera toward species of other subfamilies. This may indicate either different qualitative titers of the antisera used, or a real higher affinity of *Acacia retinodes* toward Caesalpinioideae and Faboideae in comparison with *Albizia lophanta*. Obviously more species should be examined before extending the value of this result to the whole genus *Acacia* or even to the whole tribe Acacieae.

An antiserum specific to *Gleditsia triacanthos* (Fig. 2) was the only reference system for a member within the subfamily Caesalpinioideae. In addition to the refer-

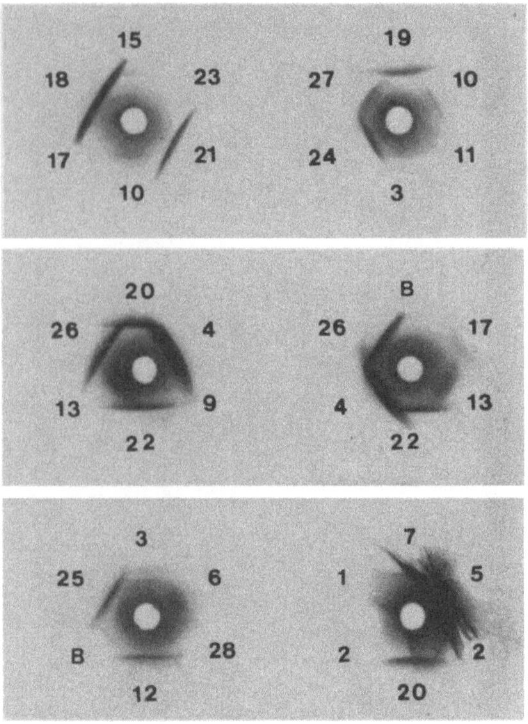

Fig. 2. Plates of immunodiffusion. *Central well* Antiserum to *Gleditsia triacanthos. Outer wells* Antigenic systems (for legend see Fig. 1)

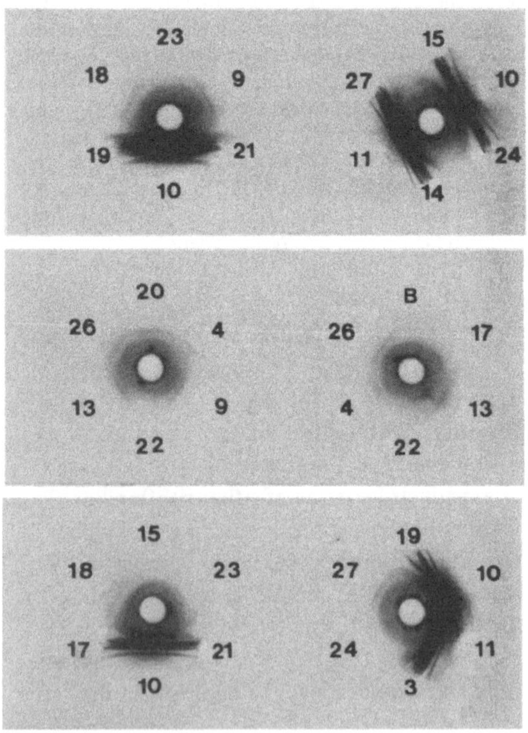

Fig. 3. Plates of immunodiffusion. *Central well* Antiserum to *Phaseolus coccineus. Outer wells* Antigenic systems (for legend see Fig. 1)

Fig. 4. Plates of immunodiffusion. *Central well* Antiserum to *Trifolium pratense. Outer wells* Antigenic systems (for legend see Fig. 1)

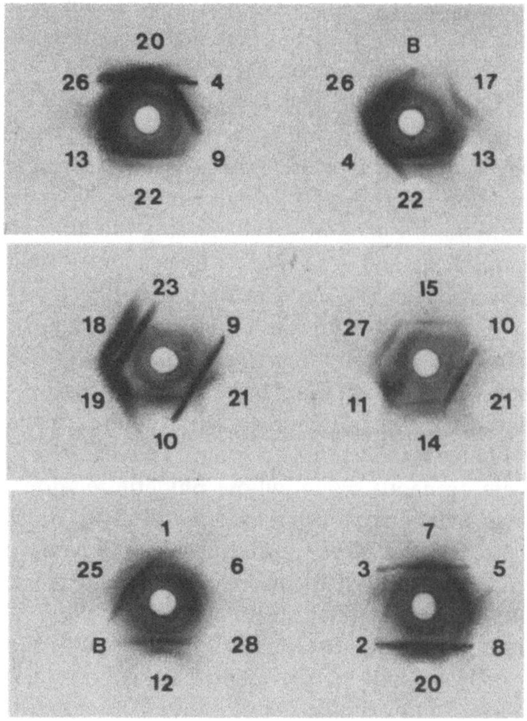

ence reaction and the strong reaction produced by *Caesalpinia*, good responses came from Trifolieae, Vicieae, Thermopsideae, Genisteae, as well as from *Amorpha* and *Templetonia*. The reactivity was quite diffuse among other Faboideae, while the reactions of Mimosoideae representatives were very weak or absent. Surprisingly *Ceratonia* and *Bauhinia* of the Caesalpinioideae only produced very weak reactions.

These limited data do not provide a definitive answer concerning the relationships among Mimosoideae, Caesalpinioideae, and Faboideae. However, they do not support the hypothesis that Mimosoideae and Caesalpinioideae are more closely related to each other than they are to Faboideae (El-Gazzar and El-Fiki 1977, El-Gazzar 1981). On the contrary, these data rather indicate a somewhat more isolated position for the Mimosoideae, and closer connections between Caesalpinioideae and Faboideae.

Among the Faboideae, *Robinia pseudacacia* proved to be isolated from most other species. A positive reaction against the *Robinia*-antiserum was only observed for *Arachis, Abrus,* and *Amorpha*, the main precipitating band (legumin?) of the two latter species being identical. All other species either reacted very weakly or not at all.

Vicia sativa presented a broader spectrum of affinities, mainly with other species of the Vicieae, with *Melilotus* and *Trifolium* (both Trifolieae) being identical, and with *Astragalus* and *Templetonia* as well. The reaction of *Abrus* was very weak, a result that endorses the recent opinion about the independence of *Abrus* from the Vicieae.

Phaseolus coccineus and *Vigna* presented a very high similarity in their reactions against the antiserum to *Phaseolus* (Fig. 3). All other species reacted very weakly;

the degree of intensity of the reactions was very uniform throughout the family. This marked isolation is a noteworthy feature of the Phaseoleae which also has been reported by other researchers.

Completely different was the behavior of *Trifolium pratense* antiserum, which gave a strong cross reaction with very many species (Fig. 4). *Melilotus*, another representative of the same tribe Trifolieae gave a reaction almost as strong as the reference-reaction. Moreover, the high similarity with the Vicieae is a valuable taxonomic information. Reactions of decreasing intensity were produced by the Sophoreae, Thermopsideae, and Amorpheae. Even Mimosoideae and Caesalpinioideae reacted positively with the *Trifolium* antiserum.

The reactions of the antiserum to *Templetonia* are fairly uniform: strong reactions are given by *Spartium, Sophora, Baptisia,* but also by *Abrus, Arachis, Astragalus,* and *Amorpha*. These reactions suggest that *Templetonia* might be related to the Genisteae, Sophoreae, and Thermopsideae. However, no decision is possible whether it is justified to group *Templetonia* with the Genisteae or the Bossiaeeae, a heterogeneous group, whose phyletic relations are reported to be toward the Sophoreae, and perhaps Tephrosieae (Polhill 1976, 1981).

Baptisia australis antiserum was chosen to represent the Thermopsideae. This antiserum proved to share many characters with *Thermopsis* (same tribe), with *Spartium* and *Ulex* (Genisteae) and with *Sophora*. Also the Vicieae and Trifolieae reacted rather strongly. These reactions agree with the idea of the placement of the Thermopsideae close to the Genisteae and the Sophoreae, with some links toward the Vicieae and Trifolieae.

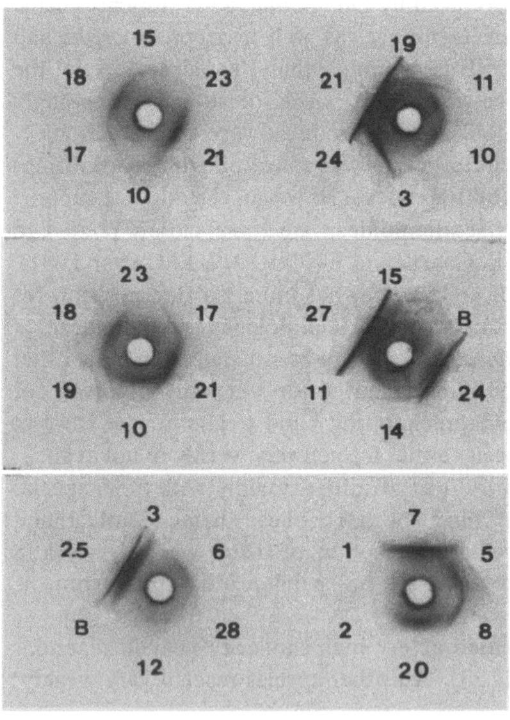

Fig. 5. Plates of immunodiffusion. *Central well* Antiserum to *Ulex europaeus*. *Outer wells* Antigenic systems (for legend see Fig. 1)

A serum directed against *Ulex europaeus* (Fig. 5) was used to represent the Genisteae. *Ulex* is one of the most derived genera within the Genisteae (Feoli-Chiapella and Cristofolini 1981); serologically it is strictly related to *Genista*. The homogeneity of the tribe is demonstrated by the very strong reaction of *Spartium*. Other strong reactions were detected for the Thermopsideae and, next, for the Sophoreae. Vicieae, Trifolieae, Crotalarieae, and Bossiaeeae reacted with a similar intensity, much lower than that for Thermopsideae and Sophoreae. If compared with the reactions of the *Baptisia* antisystem, *Ulex* seems to be placed more peripheral, connected through the Thermopsideae to the rest of the family.

3 Synthesis of Results

For the taxonomic evaluation of the precipitation data only quantitative data were used. The amount of immunoprecipitate detected on the immunodiffusion plates was estimated in the following four ways:

1. reaction not detectable, questionable or very weak
2. reaction weak, but distinctly observable
3. reaction strong, yet clearly weaker than the reference-reaction
4. reaction as strong as the reference-reaction or nearly so.

The reactions were tabulated (Table 1) and the correlation coefficient between rows was computed as a measure of interspecific similarity. The choice of the correlation coefficient was due to the following consideration. When two species react with the same intensity in the presence of the same antiserum, such reactions may be due to the same reacting systems or may not. However, when two species react in a similar way in the presence of several different antisera, such data do have a statistical meaning, in the sense that the two species are likely to have some common feature. Such a statistical correspondence is properly expressed by the correlation coefficient, a parameter that, unlike all similarity indexes, does not imply any interpretation about the real nature of the systems involved.

The matrix of correlation coefficients (Table 2) presents the species arranged according to a recent systematic scheme (Polhill 1981). Most of the highest correlation coefficients are located close to the diagonal of the matrix. This means that correlation coefficients roughly reflect the current systematics of the family.

This is especially true for:

a) The high similarity among the Mimosoideae taxa (average correlation coefficient: 0.80) and the low similarity between this subfamily and the two other subfamilies (average correlation coefficient Mimosoideae – Caesalpinioideae: −0.17; average correlation coefficient Mimosoideae – Faboideae: −0.14).

b) High similarity is detected among taxa of the Caesalpinioideae, except *Bauhinia* (average correlation coefficient: 0.67). The taxa of this subfamily are more closely connected to the Faboideae than to the Mimosoideae; their average correlation coefficient is about equal to the internal average correlation coefficient of the Faboideae. This is especially true for *Bauhinia*, which is more similar with the tribe Therm-

Table 2. Matrix of interspecific correlation coefficients. Correlation coefficients ⩾ 80 are italicized

Tribe	Genus	1	2	3	4	5	6	7	8	9	10	11	12	13	14	15	16	17	18	19	20	21	22	23	24	25	26	27	28
Ingeae	1–*Albizia*	–																											
Acacieae	2–*Acacia*	75	–																										
Mimoseae	3–*Leucaena*	*80*	*84*	–																									
Caesalpinieae	4–*Caesalpinia*	–24	–03	–18	–																								
	5–*Gleditsia*	–21	12	–03	*92*	–																							
Cassieae	6–*Ceratonia*	–25	–14	–20	53	56	–																						
Cercideae	7–*Bauhinia*	–30	–28	–38	47	–22	28	–																					
Sophoreae	8–*Sophora*	–27	–29	–56	15	–37	0	73	–																				
Abreae	9–*Abrus*	–06	–32	–31	0	25	0	–16	33	–																			
Robinieae	10–*Robinia*	05	–28	–29	–41	–22	–42	–19	36	*81*	–																		
Phaseoleae	11–*Phaseolus*	–48	–34	–36	06	11	–04	–02	12	–08	–27	–																	
	12–*Vigna*	–31	–24	–32	–14	–18	–30	11	24	–24	–24	*90*	–																
Amorpheae	13–*Amorpha*	06	–32	–31	19	25	14	–16	33	77	65	–08	–24	–															
Aeschynomeneae	14–*Arachis*	17	–26	–23	–04	–20	–11	–66	50	67	76	–13	–16	*90*	–														
Galegeae	15–*Astragalus*	–08	22	08	27	–33	18	21	0	–15	0	–11	–11	–15	0	–													
Loteae	16–*Anthyllis*	05	43	35	–11	–22	08	–43	–49	–16	05	–27	–35	–16	–06	65	–												
	17–*Lotus*	05	43	35	17	–22	49	05	0	–16	–19	–27	–35	–16	–06	22	52	–											
Vicieae	18–*Vicia*	–13	12	–04	–30	–18	–30	–12	–36	–24	12	–30	–18	–40	–33	54	59	–11	–										
	19–*Lathyrus*	–29	04	–15	76	20	75	38	0	29	–36	–19	–32	0	–09	59	38	38	21	–									
	20–*Lens*	–15	13	–12	60	–06	53	53	11	–46	–36	–26	–22	–15	–12	61	31	31	33	*92*	–								
Trifolieae	21–*Melilotus*	–15	13	–12	73	27	62	31	11	–30	–36	–16	–22	0	–12	61	31	09	33	*92*	*90*	–							
	22–*Trifolium*	–22	08	–17	74	23	68	46	0	–37	–41	–23	–27	–07	–18	50	24	24	27	*96*	*95*	*95*	–						
Bossiaeeae	23–*Templetonia*	–09	27	24	40	03	51	–16	–24	08	–04	–03	–29	08	06	74	76	53	17	56	35	45	35	–					
Crotalarieae	24–*Crotalaria*	–24	–14	08	09	–11	25	–22	0	25	15	11	–18	25	30	33	51	51	–18	20	–07	–07	–10	74	–				
Thermopsideae	25–*Thermopsis*	–52	–36	–39	49	06	51	*80*	57	0	–20	–07	–11	–15	–18	0	–31	36	–22	38	35	15	37	–02	07	–			
	26–*Baptisia*	–33	–11	–25	67	0	65	*87*	56	–15	–33	–11	–11	–15	–15	20	–22	44	–22	59	61	41	60	10	0	*92*	–		
Genisteae	27–*Spartium*	–60	–39	–39	57	24	59	66	46	18	–14	03	–13	–06	–17	08	–23	30	–22	37	23	15	30	18	25	*94*	*82*	–	
	28–*Ulex*	–55	–37	–35	45	0	47	73	57	13	–09	0	–09	–13	–13	17	–18	37	–18	33	26	08	25	18	28	*95*	*85*	*97*	–

opsideae, within the Faboideae, than with any other Caesalpinioideae taxon tested. For the other tested Caesalpinioideae species informative connections are observed with the Vicieae, Trifolieae, Bossiaeeae, Thermopsideae, and Genisteae.

c) Within the Faboideae the high species similarity within some tribes is obvious: *Phaseolus* and *Vigna* (Phaseoleae, correlation coefficient: 0.90); *Lathyrus, Lens* (Vicieae), and *Melilotus, Trifolium* (Trifolieae) (average correlation coefficient: 0.93); *Thermopsis, Baptisia* (Thermopsideae), and *Spartium, Ulex* (Genisteae) (average correlation coefficient: 0.91).

The matrix of correlation coefficients was the basis for a hierarchical classification (Fig. 6). The data available for Mimosoideae and Caesalpinioideae were inadequate for inclusion in this analysis. Therefore, from this point onwards only species of Faboideae will be discussed.

The first section (Fig. 6A) presents the result of clustering by average linkage analysis. It should be observed that low level branching may be disregarded, since low correlation coefficients have little or no significance. If only those clusters are considered, whose internal correlation exceeds 0.50, five groups of taxa are apparent.

Cluster 1 – Genisteae, Thermopsideae, Sophoreae, and *Lotus*. This seems to be a very natural group, with the noteworthy exception of *Lotus*. The highest affinity is between the Genisteae and Thermopsideae which cluster at decreasing levels; the more primitive *Sophora* joins this group with a lower coefficient. By complete linkage analysis (Fig. 6B) *Lotus* is slightly displaced, and clusters first with *Crotalaria*. By single linkage analysis (Fig. 6C) both *Lotus* and *Sophora* join separately, at lower levels. The grouping of Genisteae and Thermopsideae is consistent regardless of the method of analysis employed.

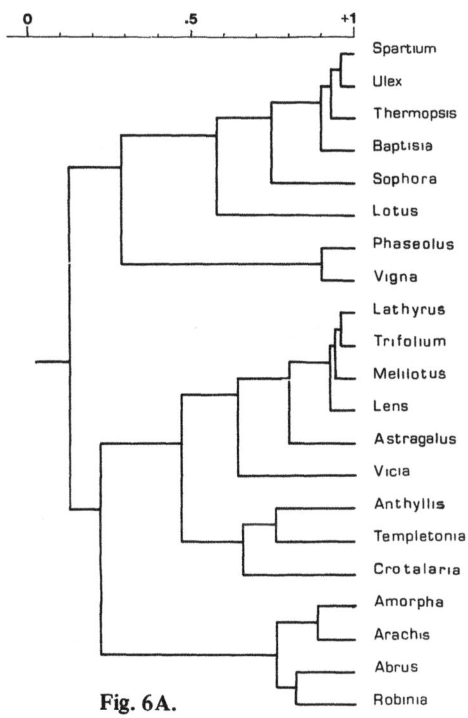

Fig. 6A.

Fig. 6A–C. Dendrogram of the phenetic similarity between species of Faboideae as computed from the correlation coefficients.
A clustering by average linkage = group average/UPGMA clustering sensu Lester.
B clustering by complete linkage = furthest neighbour clustering sensu Lester.
C clustering by single linkage = nearest neighbour clustering sensu Lester

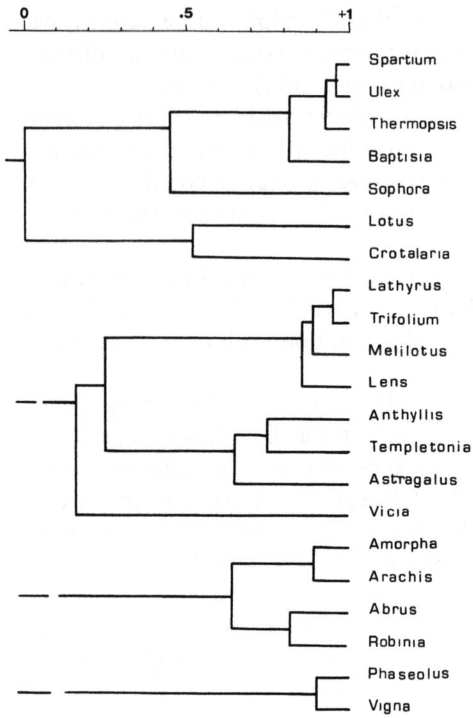

Fig. 6B. (Legend see p. 333)

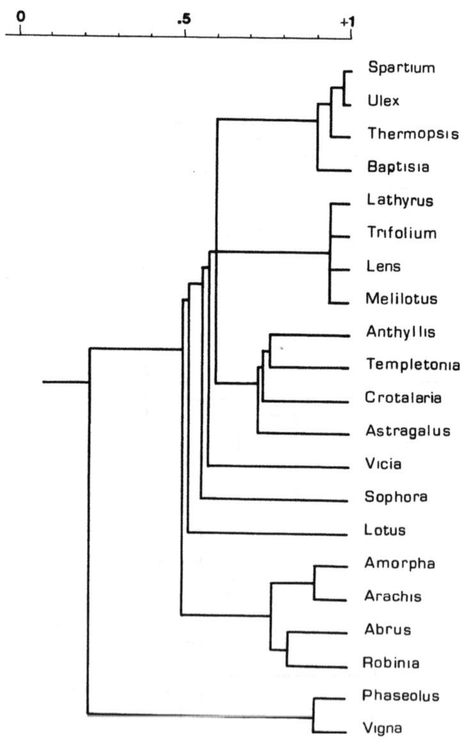

Fig. 6C. (Legend see p. 333)

Cluster 2 – Phaseoleae. The isolation of this tribe is fully confirmed. The similarity toward all other taxa is so low, that it is not possible to speculate about the phyletic relationship of this group. Its clustering with Genisteae and Thermopsideae by average linkage analysis takes place at a very low level. By single linkage analysis this low clustering is confirmed. By complete linkage analysis no clustering takes place, unless negative correlation coefficients are considered.

Cluster 3 – Vicieae, Trifolieae, Galegeae. Four of these taxa cluster at a very high level of correlation. *Astragalus* and *Vicia* join at lower levels to them. By single and complete linkage analyses the core of this cluster constantly forms in the same way, while the position of *Astragalus* and *Vicia* appears rather variable; in one instance *Astragalus* clusters with *Anthyllis* and *Templetonia*. This cluster corresponds to a natural group; however, the position of *Vicia* is anomalous and will require additional data.

Cluster 4 – *Crotalaria* unites with *Templetonia* and *Anthyllis*. The fusion of the tribes Crotalarieae and Bossiaeeae could be anticipated. However, less foreseen was the union of *Anthyllis* to them, and even less expected was the splitting of *Lotus* and *Anthyllis* into two different groups. This clustering is not very stable. By single linkage analysis *Astragalus* also aggregates with the group. The position of *Anthyllis* deserves some special comment. Indeed, the two representatives of the tribe Loteae, *Lotus* and *Anthyllis*, have a mutual affinity that is lower than their affinity toward species of other tribes. In addition, Loteae do not seem strictly connected to any other tribe tested. Their similarity tends towards the four tribes Vicieae, Trifolieae, Bossiaeeae, and Crotalarieae, and without a clear preference for any of them. In our opinion, the systematic relationship of Loteae will require further research before suggestions are presented.

Cluster 5 – Robinieae, Aeschynomeneae, Amorpheae, Abreae. Four species belonging to these four tribes cluster at a very high level (correlation coefficient higher than 0.67) and clustering indicates a low similarity toward all other groups. This grouping of the four tribes is very stable; it forms in the very same way, regardless of which analysis is employed. Some of the tribes of this group have been considered as related by recent systematic research (Polhill 1981, Polhill and Sousa 1981). The placement of *Abrus* in this context is particularly relevant. This genus has been placed within or close to the Vicieae in the past (Bentham 1865, Taubert 1894) and removed more recently to a monogeneric tribe, whose systematic relationship remained rather undefined (Kupicha 1977, Rougé and Lascombes 1977). The present serological data confirm the independence of *Abrus* from the Vicieae and indicate a possible natural alliance.

4 Systematic Inferences

A non-hierarchical ordination (Fig. 7) has been attempted in order to obtain a more flexible image of the network of intertribal connections. To this aim the species have first been grouped into their tribes according to the classical systematics. The intertribal correlation coefficients have then been computed as the mean of the coefficients

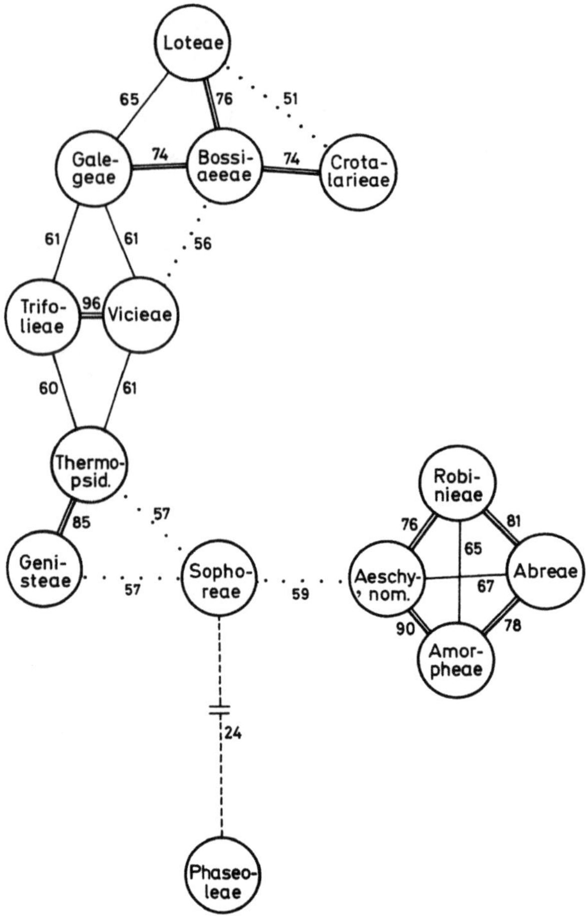

Fig. 7. A graph of intertribal serological relationships using the average correlation coefficients.
═══ correlation exceeding 0.70
——— correlation exceeding 0.60
...... correlation exceeding 0.50

of their members. Finally, the highest intertribal links have been represented in a graph.

At 0.70 correlation coefficient Trifolieae unite with the Vicieae, and Thermopsideae with Genisteae; Bossiaeeae have a multiple link, i.e. toward Crotalarieae, Galegeae, and Loteae. Robinieae, Abreae, Amorpheae, and Aeschynomeneae form a closed cluster. Only two tribes, the Sophoreae and Phaseoleae, do not join with any other tribe.

If the correlation level is lowered to 0.60, then the set of connections within Robinieae and allied tribes is completed. In addition, a "double bridge" connects the Thermopsideae to the Trifolieae and Vicieae, and these in turn to the Bossiaeeae alliance, via Galegeae. A further lowering of the level to 0.50 introduces a multiple connection of Sophoreae toward Aeschynomeneae on one side, and toward the Thermopsideae–Genisteae group on the other side.

The level must be lowered to 0.24 to detect the first connection of Phaseoleae, and this is toward Sophoreae.

In this network of tribes the Sophoreae are placed in the center, which is in agreement with the concept of this being a primitive group within the Faboideae (Fig. 8). Three groups of tribes are connected with the Sophoreae:

a) The Phaseoleae, which are serologically extremely distinct. This observation is mainly the result of the production of a different major storage protein. Since another protein, i.e. cytochrome c, has been detected to be quite different from the cytochrome c of the other Faboideae (Boulter 1976), more attention should be given to this indicated separation based on similarity of structure properties, rather than the morphology of the leguminous flower.

b) Four tribes mainly distributed in the new world: Robinieae, Aeschynomeneae, Abreae, and Amorpheae, which according to the classical taxonomic view form a separate evolutionary cluster (Fig. 8).

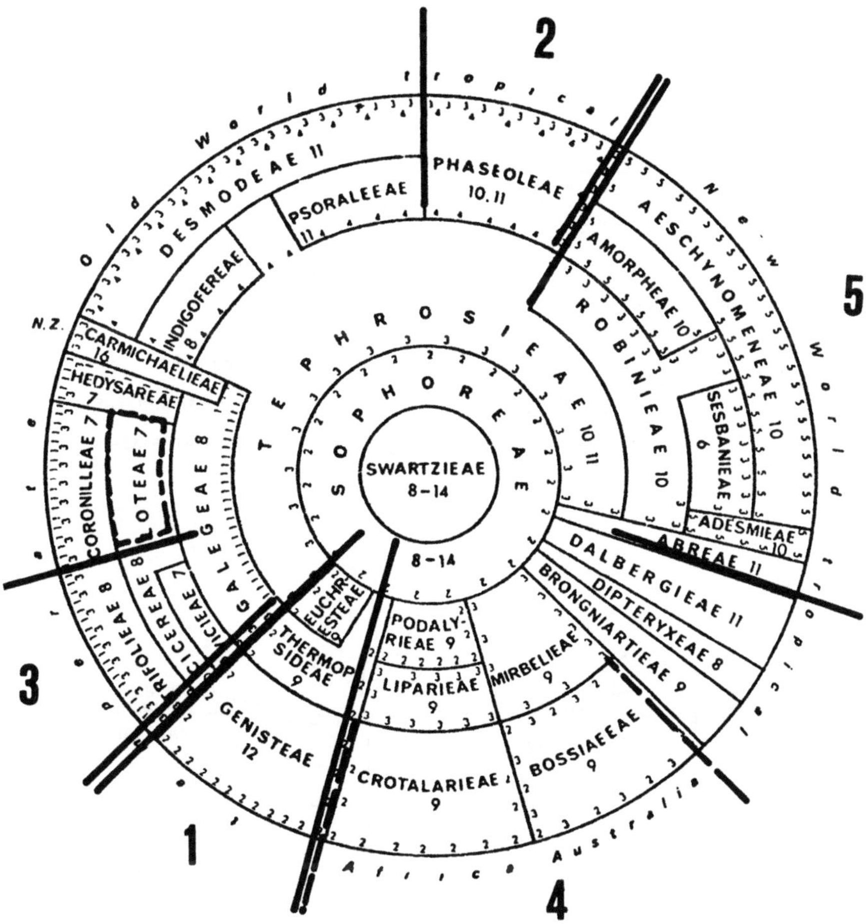

Fig. 8. A recent systematic ordination of Faboideae (Polhill 1981), and the cluster of tribes *(heavy lines)* as they resulted from serology (Fig. 6). The numbers *1* to *5* correspond to the clusters as they are discussed in the text.

Basic scheme published by the courtesy of Royal Botanic Gardens, Kew

c) For all the other tribes, their serological affinities are in good agreement with the classical data (Fig. 8). The only major discrepancy concernes the placement of Loteae; the mutual relationship between Crotalarieae and Bossiaeeae on one side and Vicieae and Trifolieae on the other side also requires additional consideration.

If it is assumed that "the number of discriminatory characters present in any species may be an estimate of its evolution or its patristic distance from the hypothetical ancient ancestor" (Lester et al. this volume), then those tribes which are connected by high correlation coefficients to several other tribes (that is: those tribes which have few discriminatory characters) can be considered as less derived. This condition is found in the Sophoreae, Thermopsideae, Bossiaeeae, Galegeae, Vicieae, and Trifolieae. On the contrary, those tribes or groups of tribes which present many discriminatory characters (low correlation to most other tribes) can be sup-

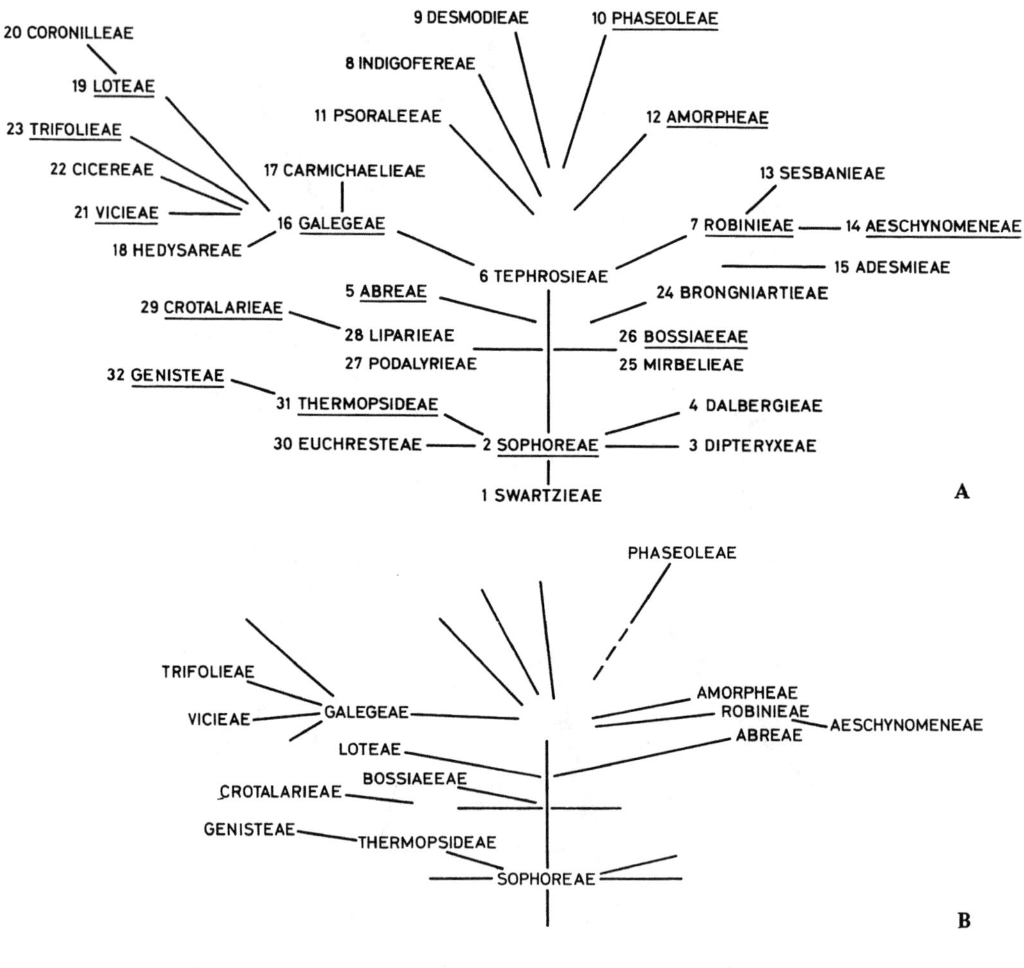

Fig. 9A,B. A Simplified representation of supposed relationship of tribes of the Faboideae after Polhill (1981). The serologically tested tribes are underlined. Courtesy of the Royal Botanic Gardens, Kew. **B** Some modifications to the above scheme to make it consistent with serological information. Only tribes considered in the present study are indicated

posed to be more distant from the hypothetical ancient ancestor. Such is the condition of Phaseoleae, Genisteae, Crotalarieae, Loteae, and the complex Robinieae-Aeschynomeneae-Abreae-Amorpheae as a whole.

A comparison of our phenetic graph (Fig. 7) with a phyletic tree of the family (Fig. 9A) also shows a remarkable agreement, in particular for the series Sophoreae - Thermopsideae - Genisteae, for the group Galegeae - Vicieae - Trifolieae, and for the isolated position of the Phaseoleae. On other points the following changes are suggested (Fig. 9B): (1) Abreae should be definitely removed from the Vicieae and located closer to Robinieae; we cannot suggest whether Abreae should lie on the same phyletic line or on a parallel line. (2) Bossiaeeae should be placed closer to Crotalarieae; if they are on the same phyletic line, then Bossiaeeae are probably a less derived group. (3) Loteae should be put on a line that is intermediate between Galegeae - Vicieae and Bossiaeeae. The scheme of Fig. 8B suggests how to modify a phyletic tree in order to make it consistent with the serological evidence and to point out which relationships in question are worthy of additional investigation.

Acknowledgements. The authors are indebted to Dr. P.H.A. Sneath, University of Leicester, U.K., for important suggestions in numerical taxonomic evaluations.

References

Bentham G (1865) In: Bentham G, Hooker JD (eds) Genera plantarum, vol I. Reeve, London

Boulter D (1976) Colloq Int CNRS, pp 371–377

Cristofolini G (1981) In: Polhill RM, Raven PH (eds) Advances in legume systematics, vol II. R Bot Gard, Kew, pp 513–531

Cristofolini G, Chiapella-Feoli L (1977) Taxon 26:43–56

Derbyshire E, Wright DJ, Boulter D (1976) Phytochemistry 15:3–24

Dudman WF, Millerd A (1975) Biochem Syst Ecol 3:291–298

El-Gazzar A (1981) In: Polhill RM, Raven PH (eds) Advances in legume systematics, vol II. R Bot Gard, Kew, pp 979–991

El-Gazzar A, El-Fiki MA (1977) Bot Not 129:371–375

Feoli-Chiapella L, Cristofolini G (1981) Nord J Bot 1:723–729

Gilroy J, Wright DJ, Boulter D (1979) Phytochemistry 18:315–316

Jensen U (1981) In: Ellenberg H, Esser K, Kubitzki K, Schnepf E, Ziegler H (eds) Progress in botany, vol 43. Springer, Berlin Heidelberg New York, pp 344–369

Jensen U, Grumpe B (1983) In: Jensen U, Fairbrothers DE (eds) Proteins and nucleic acids in plant systematics. Springer, Berlin Heidelberg New York

Kloz J (1971) In: Harborne JB, Boulter D, Turner BL (eds) Chemotaxonomy of the Leguminosae. Academic Press, London New York, pp 309–365

Kloz J, Klozová E (1974) Biol Plant 2:126–137

Klozová E, Turková V (1978) Biol Plant 20:129–134

Kupicha FK (1977) Bot J Linn Soc 74:131–162

Lester RN, Roberts PA, Lester C (1983) In: Jensen U, Fairbrothers DE (eds) Proteins and nucleic acids in plant systematics. Springer, Berlin Heidelberg New York

Miège M-N (1982) In: Boulter D, Parthier B (eds) Nucleic acids and proteins in plants, vol I. Springer, Berlin Heidelberg New York, pp 291–345

Polhill RM (1976) Bot Syst 1:143–368

Polhill RM (1981) In: Polhill RM, Raven PH (eds) Advances in legume systematics, vol I. R Bot Gard, Kew, pp 191–208

Polhill RM, Sousa M (1981) In: Polhill RM, Raven PH (eds) Advances in legume systematics, vol I. R Bot Gard, Kew, pp 283–288
Rougé P (1981) Biochem Syst Ecol 9:39–43
Rougé P, Lascombes S (1977) Bull Soc Bot Fr 124:84–91
Simon JP (1969) Bot Gaz 130:127–141
Tarlakovskaya AM, Leokene LV, Gavriljuk IP (1976) Bull Inst Plant Indust Vavilov 62:56–60
Taubert P (1894) In: Engler A, Prantl K (eds) Die natürlichen Pflanzenfamilien, vol III/3. Engelmann, Leipzig, pp 70–385

Seed Protein Characters in the Study of Inter- and Intraspecific Relationships within the Phaseoleae (Fabaceae)

E. KLOZOVÁ, V. TURKOVÁ, and J. ŠVACHULOVÁ[1]

Abstract. Seed proteins, especially the 7 S phaseolin, have been satisfactorily used to confirm the taxonomic re-arrangement reported by Verdcourt (1970), whereby *Phaseolus* is divided into two genera, *Phaseolus* and *Vigna*. Within the tribe Phaseoleae the genus *Phaseolus* s.str. has been shown to occupy an isolated position by possessing a serologically different major storage protein. For distinguishing intraspecific taxa the lectins, which are included in the albumin fraction, have proved to be most satisfactory. Thus, proteins have proved to be very valuable for the identification of certain taxa, aiding in establishing relationships, and helping to formulate a satisfactory classification.

1 Introduction

Originally the genus *Phaseolus* was a complex polymorphic genus containing more than 200 species native to both the New and the Old Worlds. However, for some time the heterogeneity of the genus has been recognized, and a new taxonomic arrangement of the taxa concerned was suggested. Thus it became evident that the Old World species differed quite distinctly from the New World species in significant morphological characters, e.g., the curvature of the keel and the insertion of the stipules.

Ohwi (1965) in his *Flora of Japan* separated five Asiatic species from *Phaseolus* and placed them in a new genus *Azukia*. Wilczek (1954) and Hepper (1956) transferred some Asiatic species of *Phaseolus* into the genus *Vigna*. Later Ohwi and Ohashi (1969) agreed that *Azukia* also should be fused with *Vigna*.

Verdcourt (1970) revised the proposed classifications, and based upon the comparison of many specimens, he proposed two distinct genera: *Phaseolus*, restricted to the New World, and *Vigna*, a complex genus including six subgenera from the New World representing the excluded *Phaseolus* taxa. This signified "a change from a system where we have a fairly well-defined *Vigna* and a 'rag-bag' *Phaseolus* to exactly vice versa" (Verdcourt 1970). According to him the different groups of *Vigna* appear to have closer phenetic affinities among themselves than with *Phaseolus*.

One of the significant characters valuable for the separation of *Vigna* and *Phaseolus* is the pollen sculpture which was reported in an extensive investigation by Maréchal et al. (1978). Mascherpa (1976) evaluated many different characters, including chemical properties. The various investigations all confirmed Verdcourt's view that

[1] Institute of Experimental Botany, Czechoslovakian Academy of Sciences, Prague, CSSR

Proteins and Nucleic Acids in Plant Systematics
ed. by U. Jensen and D.E. Fairbrothers

the restricted genus *Phaseolus* is well characterized, whereas the genus *Vigna* is more heterogeneous.

2 Seed Protein Characters in Interspecific Studies

2.1 Complex Protein Systems

Investigations which included proteins as taxonomic characters supported the new taxonomic concept. Thus in 1970 Konarev et al. characterized some species of the Phaseoleae by means of their seed protein spectrum and separated Asiatic species significantly from American taxa. Our early protein results were similar in supporting the reclassification (Kloz et al. 1966, Kloz 1971). By the comparison of seed proteins from 16 *Phaseolus* species and from species belonging to related genera, using the ring layering precipitin tests (Fig. 1) we discovered that the American endemics (*Phaseolus vulgaris* and its relatives) form a distinct group, sharply separated from the Old World species, the seed proteins of which proved to be more similar to former

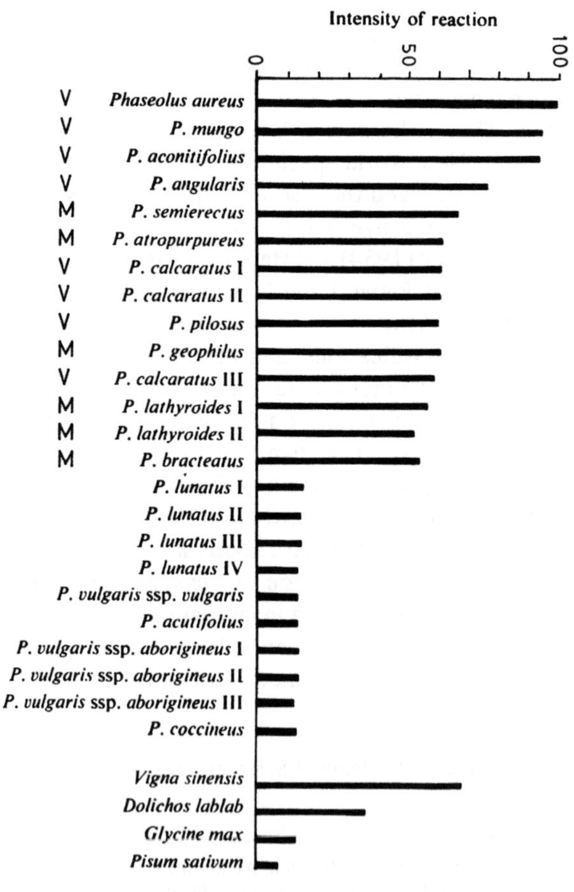

Fig. 1. The degree of similarity of seed proteins from different species of *Phaseolus*. The protein extracts were compared by means of quantitative ring layering precipitation. The antiserum used was raised against seed protein extracts from *Phaseolus aureus* (= *Vigna radiata*). (Kloz 1971); additionally the synonymous specifications according to Verdcourt (1970) and Maréchal et al. (1978) are indicated: *V Vigna; M Macroptilium*

Vigna species and even to the genus *Dolichos* than to the American *Phaseolus* species. Also the species of the South American section *Macroptilium* (according to Hutchinson 1964) are more similar to the Old World species of the section *Cochliasanthus* (according to Piper 1926) as far as their seed proteins are concerned (Kloz 1971). These data were confirmed by comparing seed proteins of *Vigna caracalla* with other 14 species of *Phaseolus* and *Vigna* immunoelectrophoretically and by double-diffusion experiments (Klozová et al. 1977). The seed protein spectrum of *Vigna caracalla* was similar to, but distinguishable from other South American *Vigna* and *Macroptilium* species (Fig. 2); *P. vulgaris* from *Phaseolus* s.str. revealed only a weak reaction.

On the basis of electrophoretic data obtained for seed globulins, Boulter et al. (1967) considered the separation of the "mung-group" (i.e., subgenus *Ceratotropis*) and *Macroptilium*. However, the comparison of esterase and leucine aminopeptidase isozyme patterns (West and Garber 1967) did not reveal additional data to help elucidate this suggested separation. The above data provide good examples which indicate the value of investigating macromolecules when attempting to solve taxonomic problems. For helping to solve the *Phaseolus-Vigna* taxonomic question the study of seed proteins provided valuable data for the establishment of more appropriate groupings at the generic level. It is important for future research to include more intensive investigations of the newly established "rag-bag" genus *Vigna*.

2.2 Defined Proteins

Jensen (1981) indicated that the phylogenetic differentiation and evolution of some well expressive seed proteins, i.e., the major storage proteins, may not have been as

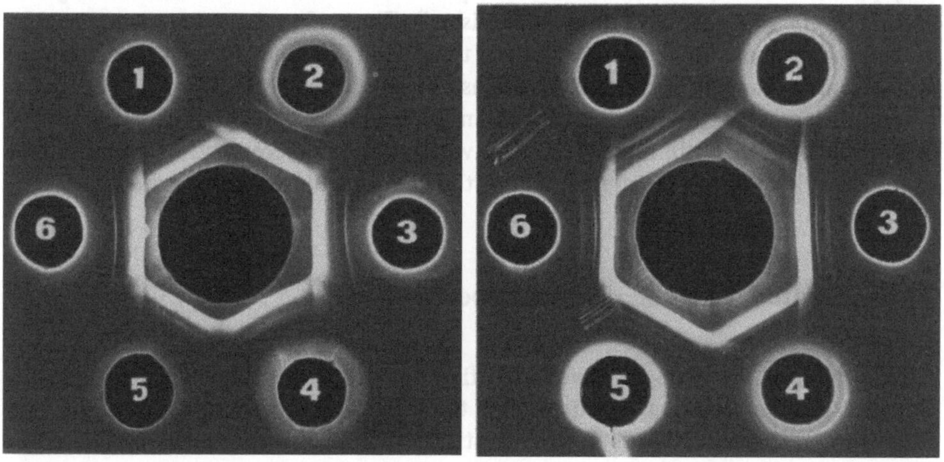

A B

Fig. 2A,B. Comparison of seed globulins of different species of *Phaseolus* and *Vigna* by means of double diffusion. The antiserum *(central well)* was raised against globulin of *Vigna caracalla* (= *Phaseolus caracalla*).

A *Antigen systems* from: *1 Phaseolus anisotrichus; 2 Macroptilium bracteatum* (= *Phaseolus bracteatus*); *3* and *6 Vigna caracalla* (= *Phaseolus caracalla*); *4* and *5 Phaseolus filiformis.*

B *1 Phaseolus anisotrichus; 2 P. vulgaris; 3* and *6 Vigna caracalla* (= *Phaseolus caracalla*); *4 Vigna umbellata* (= *Phaseolus calcaratus*); *5 Macroptilium bracteatum* (= *Phaseolus bracteatus*)

rapid a process as it was supposed formerly. Jensen and Büttner (1981) showed, based upon a rather wide species assortment of different genera and families within the angiosperms, that usually two major seed storage proteins occur, revealing partial serological similarity and cross-reactivity. One exception they reported were the Fabaceae. Especially the tribe Phaseoleae have a quite different major storage protein. This Phaseoleae protein revealed only questionable cross-reactivity with antisera against the major storage proteins from the Ranunculaceae, i.e., the legumin-like nigellin or the vicilin-like aquilegilin (see Jensen and Grumpe this volume). This is also true even for the other Fabaceae tribes (Dudman and Millerd 1975), which exhibit no cross-reactions at all with the phaseolin from members of the tribe Phaseoleae. The finding of a specific reacting Phaseoleae protein is in agreement with the discovery that the major seed storage protein of *Phaseolus* (i.e., phaseolin, G1 protein, 7S protein) has been demonstrated only for this genus (sensu Verdcourt 1970); whereas, in the genus *Vigna* (sensu Verdcourt 1970) a major storage protein of different serological specificity occurs (e.g., in the Asiatic species *Vigna radiata* and *V. aconitifolia* from the subgenus *Ceratotropis*). However, within *Phaseolus* the phaseolin protein is restricted. Using phaseolin extracted from *Phaseolus vulgaris* as the reference, serological identity was demonstrated with several New World species; a partial similarity was detected with the phaseolin from *Phaseolus acutifolius*. A lack of any serological cross-reactivity indicated a completely different serological determinant structure for those other *Phaseolus* species investigated which now are referred to the genus *Vigna*.

Recently we discovered a similar situation in the genus *Allium* (Alliaceae), where probably also the major seed storage protein (which is still not isolated and purified) seems to be serologically identical in the two subgenera *Rhizirideum* and *Allium*, but only partial similarity was observed with representatives of the more distant subgenus *Melanocrommyum* (Klozová et al. unpublished). These observations are in agreement with the hypothesis of Jensen and others that the major seed storage proteins (probably legumin-like and vicilin-like proteins) reveal serological cross-reactivity even among distant related taxa. But the examples of *Phaseolus* and apparently also of *Allium* indicate that still more detailed investigations are necessary. We propose that in some phyletic lines a rapid evolution of the mentioned proteins occurred.

3 Seed Protein Characters in Interspecific Studies

During the last few years we have intensified our investigations concerned with distinguishing intraspecific taxa to determine the variability of the protein patterns. The following was detected during our investigation of 1,200 specimens of *Phaseolus vulgaris*:

a) All cultivars investigated have the major 7S storage protein (phaseolin) of identical serological specificity, differing only very slightly in electrophoretic mobility (Kloz and Klozová 1974), indicating slight heterogeneity in the properties of the molecule (Brown et al. 1981, Pusztai et al. 1981).

b) In one cultivar (cv. *Krupnaya sacharnaya*) and one type of *Phaseolus vulgaris* ssp. *aborigineus* two expressed proteins of the water soluble fraction were absent

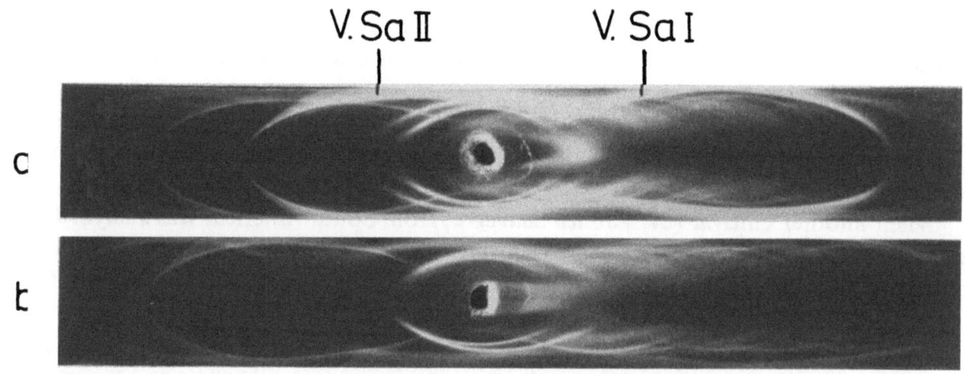

Fig. 3. Immunoelectrophoretic comparison of *Phaseolus vulgaris* cv. *Veltruská Saxa* (*a*) and cv. *Krupnaya sacharnaya* (*b*). Antiserum raised against albumins of *Phaseolus vulgaris* cv. *Veltruská Saxa*

(Fig. 3). In about 7% of the cultivars one of these expressed water soluble proteins (i.e. the cathodic one) was missing (e.g. Kloz 1971). This protein was identified as a lectin by Brücher (1968) and later also by us (Klozová and Turková 1978a). Brücher (1968) also stated that up to 10% of cultivars do not possess this lectin. It was demonstrated (Klozová et al. 1976), that cv. *Krupnaya sacharnaya* had, instead of the other mentioned proteins, two other proteins of similar mobility, but of quite different serological specificity (Fig. 4). This finding is correlated with a missing erythroagglutinating activity with human AB0 system and rabbit red cells. Later, 26 selected cultivars of the *Phaseolus vulgaris* specimen assortment were analyzed (Klozová and Turková 1978a) with different antisera (about 20 individual seeds from each cultivar) with regard to the presence or absence of the four mentioned proteins (i.e. two specific for cv. *Krupnaya sacharnaya* type and two specific for cv. *Veltruská Saxa* type). Different combinations of these four proteins were found to be expressed

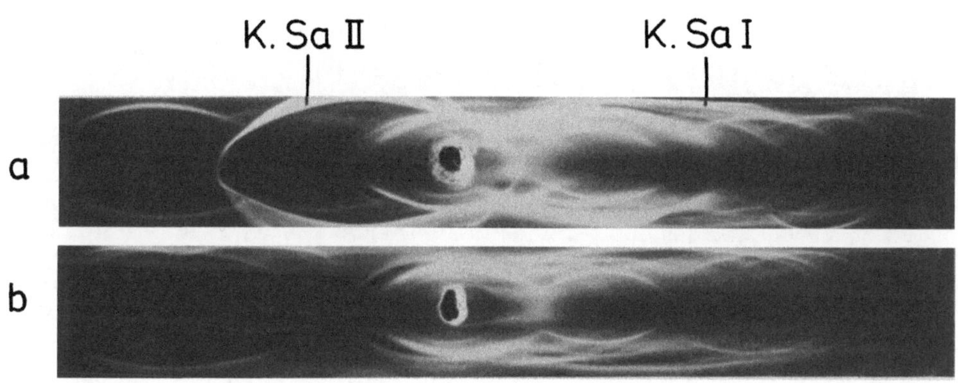

Fig. 4. Immunoelectrophoretic comparison of *Phaseolus vulgaris* cv. *Krupnaya sacharnaya* (*a*) and cv. *Veltruská Saxa* (*b*). Antiserum raised against albumins of *Phaseolus vulgaris* cv. *Krupnaya sacharnaya*

in seeds of 26 cultivars investigated (Fig. 5). The positive or negative erythroagglutinating activity was linked with the presence or absence of the cathodic lectin fraction (V.Sa.II in Table 1).

Thus in this seed assortment we were able to distinguish some cultivars on the basis of their protein pattern, including the variability within individual conspecific cultivars.

When another cultivar (cv. *Vainica Saavegra*, from South America) was analyzed a still more complicated situation was discovered. Single seeds of this cultivar differed more or less in the electrophoretic mobility of the cathodic lectin protein (Fig. 6). In

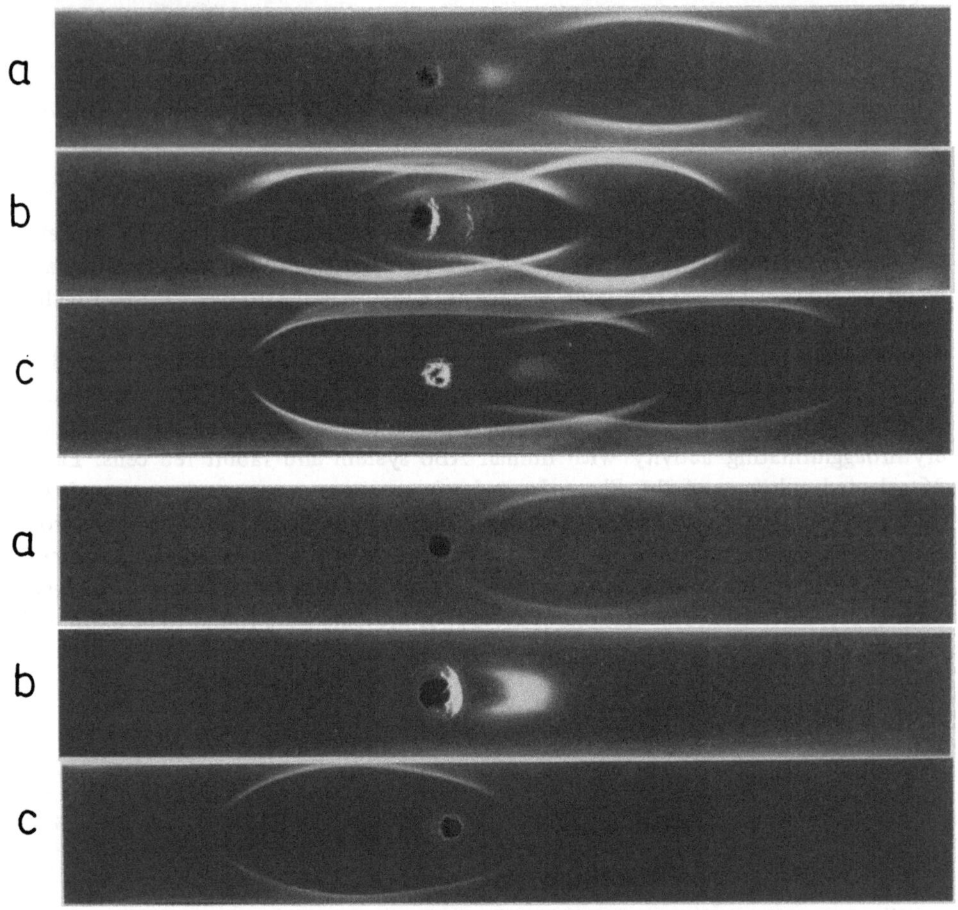

Fig. 5A,B. A Examples of occurrence of the two proteins from cv. *Veltruskd Saxa.* Detection of antigens by antiserum raised against albumins of cv. *Veltruskd Saxa* absorbed with antigens of cv. *Krupnaya sacharnaya. a Phaseolus vulgaris* cv. *Bonateana; b P. vulgaris* cv. *Bekeczy Libanai; c P. vulgaris* cv. *Dnepropetrovskaya.*
B Examples of the occurrence of the two proteins from cv. *Krupnaya sacharnaya.* Detection of antigens by antiserum raised against albumins of cv. *Krupnaya sacharnaya* absorbed with antigens of cv. *Veltruskd Saxa. a Phaseolus vulgaris* cv. *Chiao ton chiao; b P. vulgaris* cv. *Bekeszy Libanai; c P. vulgaris* cv. *Dnepropetrovskaya*

Table 1. Presence or absence of 4 proteins in single seeds of 26 cultivars of *Phaseolus vulgaris* and phythaemagglutinating activity (PHA)
V.Sa I – anodic protein typical for cv. *Veltruskd Saxa*
V.Sa II – cathodic protein typical for cv. *Veltruskd Saxa*
K.Sa I – anodic protein typical for cv. *Krupnaya sacharnaya*
K.Sa II – cathodic protein typical for cv. *Krupnaya sacharnaya*

Cultivar	Number of analysed seeds	Number of seeds containing				PHA activity
		K.Sa I	K.Sa II	V.Sa I	V.Sa II	
Ojo de Cabra	12	12	12	12	0	0
Pinto 5	15	15	15	15	9	9
Pinto VI III	17	17	17	17	0	0
Dobragea	40	40	40	40	1	1
Bonateana	20	19	19	20	0	0
Frühe weisse Brech	20	20	20	20	0	0
Azaryan	20	18	20	20	2	2
Olympia	10	10	10	10	0	0
Unselt's Meisterstück	20	20	20	20	0	0
Juwageld	20	20	20	20	1	1
Tocho Coquenes	10	9	9	10	1	1
Zeppelin	10	10	10	10	0	0
Chiao ton chiao	39	39	0	39	0	0
Allererste	20	19	0	20	1	1
Très Hatif de Nassy	10	10	0	10	0	0
Bekesy Libanaj	20	0	1	20	20	20
Early Giant	20	0	0	20	20	20
Bountiful	20	0	4	20	20	20
Delikat	20	0	0	20	20	20
Alpha	20	0	0	20	20	20
Ascherslebener Meisterwerk	20	0	16	20	20	19
Blue Coco	20	1	20	20	20	20
Dnepropetrovskaya	20	0	20	20	20	20
Flageoleo rote Parisien	20	0	20	20	20	20
Bombast	20	0	20	20	20	20
Banat	20	0	20	20	20	20

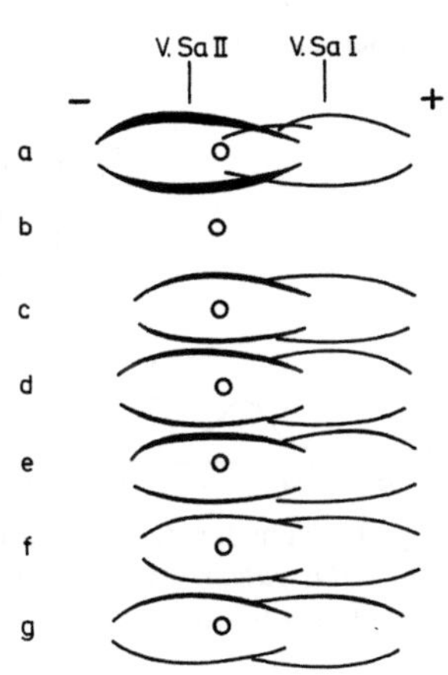

Fig. 6. Examples of immunoelectrophoretic analyses of individual seeds of *Phaseolus vulgaris* cv. *Vainica Saavegra* (*c–g*), using an antiserum raised against albumins of cv. *Veltruskd Saxa* absorbed by antigens of cv. *Krupnaya sacharnaya*. For comparison purposes also cv. *Veltruskd Saxa* (*a*) and cv. *Krupnaya sacharnaya* (*b*) were used as antigens

the seeds whose extracts had a remarkable changed mobility of the cathodic lectin (Fig. 6) no erythroagglutinating activity was detected, whereas in the seeds with normal or slightly changed mobility the erythroagglutinating activity against rabbit red cells was detected. In the double diffusion test with absorbed antisera this "lectin" with a remarkable changed mobility was only partial similar with normal lectin, i.e., both cross-reacted (Fig. 7). This indicates that structural variability in a seed protein exists even within a cultivar, and that a "uniform" cultivar can be formed by a complex "population" from the point of view of proteins (Klozová and Turková 1978b).

The albumin complexes of three different cultivars were analyzed in detail, i.e.: standard cultivar *Veltruská Saxa*, heterogeneous cv. *Vainica Saavegra* B (types with erythroagglutinating activity) and cv. *Krupnaya sacharnaya* (with a slightly different albumin complex and without erythroagglutinating activity). The elution profiles (on Sephadex G 100) were different (Fig. 8), especially that of cv. *Krupnaya sacharnaya*. Individual peaks revealed by SDS electrophoresis indicated different subunit composition (Švachulová et al. 1982).

The lectins of cultivars mentioned above, together with lectins of two other cultivars were isolated by affinity chromatography. Additionally affinity to the human ABO system red cells, to trypsinized and non-trypsinized rabbit red cells, as well as their mitogenic activity were proven. In these experiments seeds of cv. *Vainica*

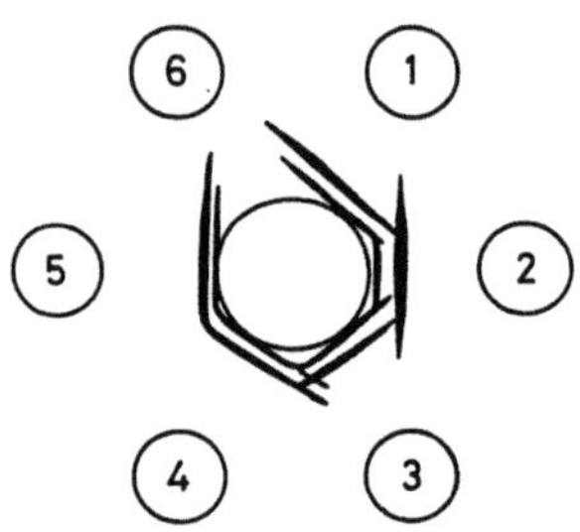

Fig. 7. Examples of antigen analyses of individual *Phaseolus vulgaris* seeds by means of double diffusion. Detection with an antiserum raised against albumins of cv. *Veltruská Saxa* absorbed by antigens of cv. *Krupnaya sacharnaya.*
Antigens: *1* and *3* cv. *Vainica Saavegra; 2* and *5* cv. *Veltruská Saxa; 4 Phaseolus vulgaris* ssp. *aborigineus; 6* cv. *Krupnaya sacharnaya*

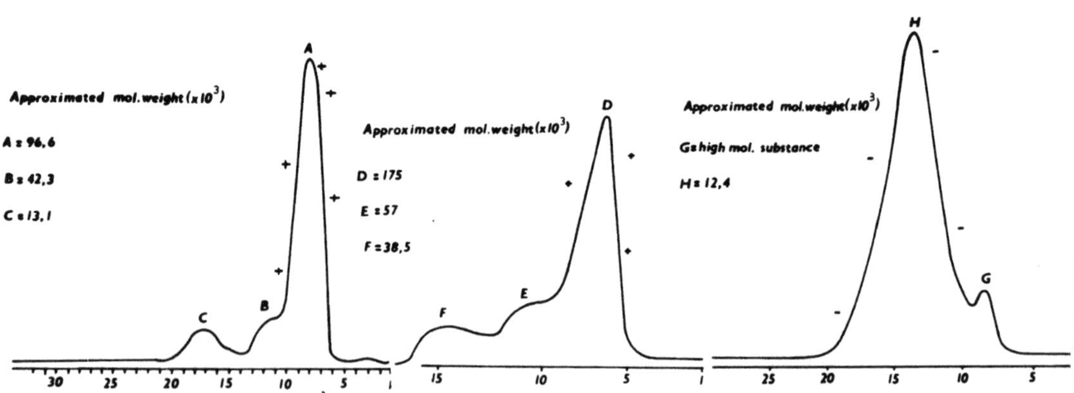

Fig. 8. Sephadex G 100 chromatography profiles of seed albumins of 3 different cultivars of *Phaseolus vulgaris*, i.e. cv. *Veltruská Saxa* (*left*) cv. *Vainica Saavegra* B (*center*) and cv. *Krupnaya sacharnaya* (*right*). + positive haemagglutination test; – negative haemagglutination test

Saavegra were divided into five groups according to the mobility of the cathodic lectin proteins.

It can be demonstrated (Table 2) that cultivars differ in their affinity to an AB0 system, and to trypsinized and non-trypsinized erythrocytes; they agglutinated all or none of the system tested. But in contrast the mitogenic activity was positive in all cases investigated. By isoelectrofocusing it was shown that most of the lectins isolated are composed of two or three components, except for that of cv. *Olympia* and cv. *Krupnaya sacharnaya*, where only one band could be detected. With regard to the pI values approximately three groups of isolectins could be distinguished: pI around 4, 5, and 7. However, no correlation could be found among the type of lectin activity and isolection type (Table 2). The isolectin with low pI seemed to be the lectin with mitogenic activity and formed by only L subunits (Felsted et al. 1981a). Immunoelectrophoretical analyses of purified lectins confirmed the existence of different isolectins in different cultivars (Fig. 9).

These findings do not disagree with Felsted's theory (Felsted et al. 1981a) regarding five isolectin types (composed of two subunits E and L). With regard to the relatively close pI values, more isolectins in our cultivars can probably be distinguished by using more precise pH measurements.

Recently Pusztai et al. (1981) published data on a new lectin type from cv. *Pinto III* previously regarded to be hemagglutinins-free (Table 1). This lectin had its pI value very close to our cv. *Krupnaya sacharnaya* lectin, with a very weak affinity to rabbit red cells, but very high activity to the pronase-treated rat erythrocytes. It also revealed a slight immunochemical cross-reactivity to common *Phaseolus vulgaris* lectins which was not observed for cv. *Krupnaya sacharnaya* lectin. By analyzing 62 cultivars of *Phaseolus vulgaris*, Felsted et al. (1981b) have found erythroagglutinating activity to rabbit and human red cells in all cultivars, whereas the mitogenic activity was absent in 13 cultivars. The authors established the existence of 2–3 groups of cultivars with regard to the lectin composition.

Thus the occurrence of specific lectin complexes in different cultivars is confirmed. More detailed investigation of the samples with missing erythroagglutinating activity

Table 2. Isoelectrofocusing, agglutinating, and mitogenic activity of lectins purified by affinity chromatography

Taxon: *Phaseolus vulgaris* cv.	pI			Agglutination of human ABO, rabbit red cells	Mitogenic activity
Veltruskd Saxa	4.4	4.9	7.3	+	+
Vainica Saavegra 1	4.5	5.2		–	+
Vainica Saavegra 2	4.5		6.6	–	+
Vainica Saavegra 3	4.4	5.3	7.4	±	+
Vainica Saavegra 4	3.7	5.4		+	+
Vainica Saavegra 5	4.1	5.2		+	+
Krupnaya sacharnaya	4.5			–	+
Olympia	3.8			–	+
Phaseolus vulgaris ssp. *aborigineus*	4.0		6.7	+	+

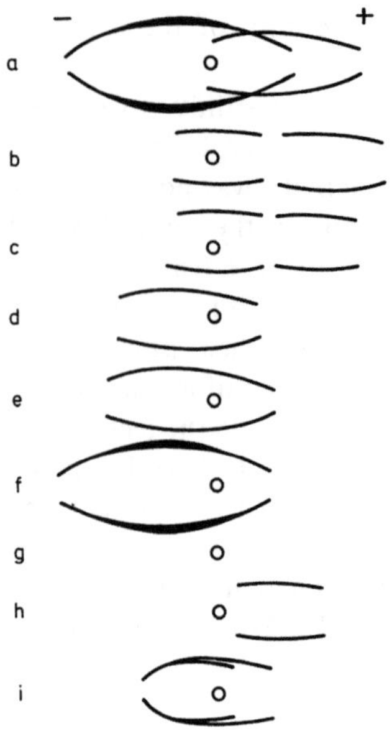

Fig. 9. Immunoelectrophoretic patterns of lectins from different *Phaseolus vulgaris* cultivars. Detection by an antiserum raised against seed albumin lectins of *Phaseolus vulgaris* cv. *Veltruská Saxa.*
a lectins of *Phaseolus vulgaris* cv. *Veltruská Saxa* – reference reaction; *b* lectins of *Phaseolus vulgaris* cv. *Vainica Saavegra* 1; *c* lectins of *P. vulg.* cv. *Vainica Saavegra* 2; *d* lectin of *P. vulg.* cv. *Vainica Saavegra* 3; *e* lectin of *P. vulg.* cv. *Vainica Saavegra* 4; *f* lectin of *P. vulg.* cv. *Vainica Saavegra* 5; *g* lectin of *P. vulg.* cv. *Krupnaya sacharnaya; h* lectin of *P. vulg.* cv. *Olympia; i* lectins of *P. vulg.* ssp. *aborigineus*

(e.g., *Krupnaya sacharnaya, Olympia, Vainica Saavegra 1* and *2*) is necessary, especially the testing of additional red cell types to determine the taxonomic significance.

4 Conclusions

When summarizing the information concerning the use of proteins as discriminating characters in *Phaseolus* we have concluded that seed proteins in *Phaseolus* are very valuable as taxonomic criteria for discerning intraspecific, interspecific, as well as higher taxonomic ranks. However, it is essential to recall that different classes of proteins have different taxonomic value. The major storage proteins (7S, 11S types), for example, are relatively non-specific, which has been reported by Jensen and Büttner (1981). According to Jensen and Büttner's investigations these proteins reveal serological cross-reactivity with almost all tested angiosperms. However, the proteins belonging to the "albumin" fraction (including most lectins) can be used for distinguishing taxa of lower ranks. With regard to the *Phaseolus* problem we are convinced that the investigation of the overall seed protein characters allows us to formulate a more natural classification for these taxa.

References

Boulter D, Thurman D, Derbyshire E (1967) New Phytol 66:27–36
Brown JWS, Ma Y, Bliss FA, Hall TC (1981) Theor Appl Genet 59:83–88
Brücher O (1968) Proc Trop Reg Am Soc Hortic Sci 12:68–72

Dudman WF, Millerd A (1975) Biochem Syst Ecol 3:25–33
Felsted RL, Leavitt RD, Chen CH, Bachur NR, Dale RMK (1981a) Biochim Biophys Acta 668:132–140
Felsted RL, Li J, Pokrywka G, Egorin MJ, Spiegel J, Dale RMK (1981b) Int J Biochem 13:549 to 557
Hepper FN (1956) Kew Bull 11:113–134
Hutchinson J (1964) The genera of flowering plants, vol I. Oxford Univ Press, Oxford
Jensen U (1981) In: Ellenberg H, Esser K, Kubitzki K, Schnepf E, Ziegler H (eds) Progress in botany. Springer, Berlin Heidelberg New York, pp 344–369
Jensen U, Büttner C (1981) Taxon 30:404–419
Jensen U, Grumpe B (1983) In: Jensen U, Fairbrothers DE (eds) Proteins and nucleic acids in plant systematics. Springer, Berlin Heidelberg New York, this volume
Kloz J (1971) In: Harborne JB, Boulter D, Turner BL (eds) Chemotaxonomy of the Leguminosae. Academic Press, London New York, pp 309–365
Kloz J, Klozová E (1974) Biol Plant 16:290–300
Kloz J, Klozová E, Turková V (1966) Preslia 38:229–236
Klozová E, Turková V (1978a) Biol Plant 20:129–134
Klozová E, Turková V (1978b) Biol Plant 20:373–376
Klozová E, Kloz J, Winfield PJ (1976) Biol Plant 18:200–205
Klozová E, Turková V, Platilová I (1977) Biol Plant 19:278–283
Konarev VG, Satbaldina SF, Gavrilyuk IP, Ivanov NR (1970) DAN USSR 190:975–978
Maréchal R, Mascherpa JM, Stainier F (1978) Abstr Int Legume Conf, Kew, UK
Mascherpa JM (1976) Thèse Fac Sci, Univ Genève
Ohwi J (1965) Flora of Japan. Smithsonian Inst, Washington DC
Ohwi J, Ohashi J (1969) J Jpn Bot 44:29
Piper CV (1926) Contrib US Nat Herb 22, part A
Pusztai A, Grant G, Stewart JC (1981) Biochim Biophys Acta 671:146–154
Švachulová J, Turková V, Klozová E (1982) Biol Plant 24:81–88
Verdcourt B (1970) Kew Bull 24:507–569
West NB, Garber ED (1967) Can J Genet Cytol 9:640–645
Wilczek R (1954) Papilionaceae – Phaseoleae – Phaseolineae. Flora Congo Belge Ruanda Urundi 6:260–409

Serological Investigation of a Hybrid Swarm Population of Pinus sylvestris L. x Pinus mugo Turra, and the Antigenic Differentiation of Pinus sylvestris L. in Sweden

W. PRUS-GŁOWACKI[1]

Abstract. By means of immunodiffusion, immunoabsorption, quantitative precipitation, and rocket-immunoelectrophoresis serosystematic relationships of *Pinus sylvestris* L., *P. mugo* Turra, *P. uliginosa* Neumann, and *P. nigra* Arnold were investigated. The high serological affinities between *P. mugo* and *P. uliginosa* demonstrate their close relationship. The antigenic properties of individuals from a hybrid swarm population of *Pinus sylvestris* x *P. mugo* allowed the establishment of characteristics, and the degree and direction of introgression in the hybrid population. Although the majority of the individuals usually resemble *Pinus uliginosa*, the genetic influence of both *Pinus mugo*, and, to a less degree, *Pinus sylvestris* has been detected. The Swedish *Pinus sylvestris* populations form two geographical groupings, northern and southern, with gene exchange detected between these two groups when they occur in the central part of Sweden.

1 Introduction

Hybrid swarm populations between Scots pine (*Pinus sylvestris* L.), dwarf mountain pine (*Pinus mugo* Turra), and probably peat-bog pine (*Pinus uliginosa* Neumann) occur in the peatbogs of the Tatra and Sudety mountains in Poland. A local population, growing in a peat-bog area near the village Zieleniec in the Góry Bystrzyckie mountains (Central Sudetes) at the elevation of ca. 720 m, has been characterized previously concerning their morphological and anatomical features (Szweykowski 1969, Szweykowski et al. 1976). Additionally isoenzymatic and serological studies were performed with this hybrid swarm (Prus-Głowacki and Szweykowski 1977, 1980, Prus-Głowacki et al. 1978, 1981).

The hybrid swarm population originated in the remote past (palynological data; Szweykowski, unpublished) by hybridization of a relic shrubby *P. mugo* population with the surrounding *Pinus sylvestris* forest. Since the population contains many trees resembling those described as *P. uliginosa*, probably *P. uliginosa* genes might also have been introduced into the recent hybrid swarm in question. However, morphologically and anatomically these hybrids are similar to *P. mugo* (Szweykowski 1969). These three pine species, being putative parents of the population in question, were compared serologically with individuals having intermediate morphological characters. Additionally *P. nigra* Arnold, which is a species native to southern Europe, was investigated since it is known to hybridize with *P. sylvestris* under natural conditions.

[1] Department of Genetics, A. Mickiewicz University, Poznań, Poland

Proteins and Nucleic Acids in Plant Systematics
ed. by U. Jensen and D.E. Fairbrothers

The main purpose of our research was to investigate the genetic structure of the hybrid swarm populations. As the first step it was necessary thoroughly to investigate serologically those species which were considered to be the putative parents of the hybrid populations. Then, particular individuals from the hybrid swarm populations were compared with the putative parental species to establish the degree, character and direction of introgression. Investigations of this kind are useful for understanding the process of speciation and to gain insight into population diversity and differentiation.

The amount of differentiation of antigenic proteins within Scandinavian Scots pine *(Pinus sylvestris)* was another aspect of this research. This pine species possesses considerable taxonomic variability throughout its geographical distribution, since it occurs in regions characterized by considerable climatic differences. The two subspecies *P. sylvestris* var. *lapponica* and *P. sylvestris* var. *septentrionalis* were investigated serologically to help estimate the amount of adaptation associated with the climate diversity.

2 Materials and Methods

2.1 Antigens and Antisera

The source of the antigens used in our investigations were 1-year-old fully developed needle leaves. They were collected during the winter to avoid unstability in protein spectra resulting from growth. Procedures of antigen extraction and antisera preparation are described in previous publications (Prus-Głowacki et al. 1978, Prus-Głowacki and Szweykowski 1979).

2.2 Serological Methods

In our investigation of antigenic similarity, double immunodiffusion, quantitative immunoprecipitation, immunoabsorption, and modified rocket-immunoelectrophoresis according to Kröll (1969) were used. Immunodiffusion plates were stained for detecting enzymatic activity following procedures described by Uriel (Clausen 1971) with minor modifications. Identification of the types of antigen-antibody reactions were performed as described by Ouchterlony (1967). Methodological details can be found in previous publications (Prus-Głowacki et al. 1978, Prus-Głowacki and Szweykowski 1979, Prus-Głowacki et al. 1981).

2.3 Numerical Analysis

On the basis of Ouchterlony immunodiffusion data, cluster analyses were performed (see: Lester et al. this volume). The individual immunoprecipitation lines were used as taxonomic characters. The appearance of individual precipitin arcs was coded depending on their intensity (0 = lack of an immunoprecipitin line; 1 = very feeble line; 2 = distinct line; 3 = strong line; 4 = very strong line). On the basis of the data matrix thus obtained, the Euclidean distances between the taxa were computed. In cases where the intensity of precipitate reactions was not considered, but only the presence or absence of lines, the simple matching similarity coefficient between taxa was calculated. The taxonomic distances D were then computed from the simple matching similarity coefficients S:

$$D_{j,k} = \sqrt{1 - S_{j,k}}$$

The matrices of similarity coefficients ($S_{j,k}$) were used for computing Anderson's hybrid indices (Anderson 1949) and also the principal components.

In the study of antigenic differentiation of *Pinus sylvestris* population from Sweden, frequencies of individual antigens were used for computing the serological similarity coefficients (I_N) according to the formula of Nei (1972). The serological distances (Dg_{I_N}) were computed from matrices of serological similarities (I_N) as:

$$Dg_{I_N} = \sqrt{1 - I_N}$$

Very useful information concerning antigenic differentiation of individual populations is obtainable from polymorphic indices calculated with reference to the formula by Marshall and Jain (1969):

$$P_i = \frac{1}{z} \cdot \left[\sum_{i=1}^{z} \right] q_i (1 - q_i)$$

where z denotes number of antigens distinguished in the investigated populations, and q denotes frequency of antigens in a single population.

3 Results and Discussion

3.1 Serological Correlations between Pinus mugo, P. uliginosa, P. sylvestris, and P. nigra

Previous serological tests, i.e. immunoabsorption, have already demonstrated specific antigens characteristic for each of the pine species investigated (Prus-Głowacki and Szweykowski 1979, Prus-Głowacki et al. 1981). Our serotaxonomic investigations concerning mutual affinities between the putative parental species of a hybrid swarm population clearly demonstrate antigenic differences between *P. mugo* and *P. sylvestris*. *P. uliginosa*, the peat-bog variety of *P. mugo*, is antigenically different from these two taxa, but shows a greater similarity to *P. mugo* than to *P. sylvestris*. Also, *P. mugo* displays a much closer serological affinity to *P. uliginosa* than to *P. sylvestris*. Proteins of *P. nigra* were found to be different from those of the other three pine species, but with closest connections with *P. sylvestris*. These findings are demonstrated by our data presented in Fig. 1, which presents a diagram of similarity among the investigated pines constructed on the basis of immunodiffusion analysis.

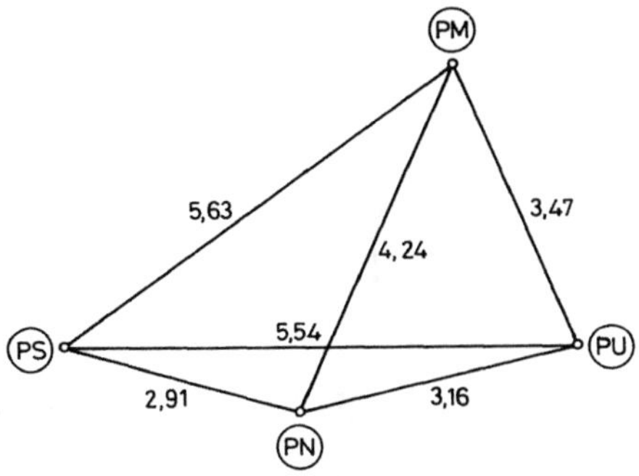

Fig. 1. Diagram showing the serological similarity of *Pinus sylvestris (PS)*, *P. mugo (PM)*, *P. uliginosa (PU)*, and *P. nigra (PN)* constructed on the basis of taxonomic distances $D_{j,k}$

3.2 Protein Affinities of a Pinus Hybrid Population

On the basis of serotaxonomic data it has been expected that the individuals investigated from a hybrid swarm population will have intermediate characteristics between *P. mugo* and *P. uliginosa*, and also between *P. uliginosa* and *P. sylvestris*. To obtain more data, common enzymes, i.e. MDH, esterases and peroxidases, are included in our comparative studies.

Figure 2 presents a three-dimensional diagram of the overall protein affinities of hybrid individuals using immunological and enzymatic data. Obviously the hybrid swarm individuals are heterogeneous. Some of them are similar to *P. mugo*, others, to *P. uliginosa*, or to *P. sylvestris*, or are intermediate. Most of hybrid individuals possess intermediate characters between *P. uliginosa* and *P. mugo*; and only a few are intermediate between *P. uliginosa* and *P. sylvestris*. Anderson's hybrid indices based on similarity coefficients (Table 1) indicated that one and the same individual

Table 1. Anderson's hybrid indices based on protein similarity coefficients. The values for reference taxa are –100 for *Pinus sylvestris* and +100 for *Pinus mugo*

Hybrids	Peroxidase	Esterase	Malate dehydrogenase	General proteins: precipitates	Mean Value
H1	+22.6	–12.9	–54.8	–27.6	–18.2
H2	+30.9	+21.8	+91.2	+27.5	+42.9
H3	+22.6	–74.3	–36.5	+15.8	–18.1
H4	+23.4	+64.6	+54.7	+49.3	+48.0
H5	+28.6	+14.4	+36.5	+ 9.8	+22.3
H6	+28.6	+39.5	+36.5	+ 8.6	+28.3
H7	+36.4	+16.9	+54.7	+59.9	+41.9
H8	+20.1	+42.0	+36.5	+36.1	+33.6
H9	+41.4	+12.1	+99.9	+62.5	+53.9
H10	+16.6	+60.5	+72.9	+27.7	+44.4
H11	+15.8	+59.7	+18.2	+13.2	+26.7
H12	+18.8	– 6.0	+57.7	+31.6	+25.5
H13	+41.4	–14.5	– 9.4	+ 3.1	+ 5.1
H14	– 6.7	+ 7.3	± 0	+43.4	+11.0
H15	+39.1	+20.0	+48.2	+19.1	+31.6
H16	+52.3	+54.0	+45.1	+30.1	+45.3

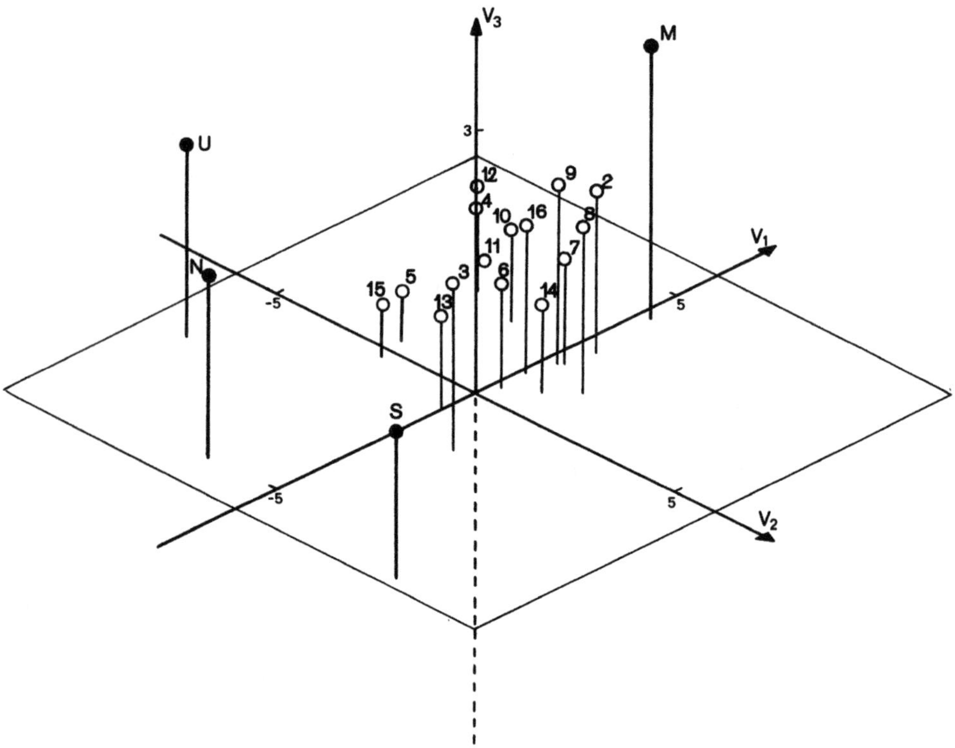

Fig. 2. Three-dimensional diagram showing the distribution of hybrid swarm individuals in the three first principal components space. Points marked with letters *S, M, U, N* represent pure species populations (*P. sylvestris, P. mugo, P. uliginosa*, and *P. nigra*, respectively)

can reveal different affinities to the pure species depending on the protein property used. For instance, the hybrid H1 has close affinities with *P. sylvestris* according to malate dehydrogenase and esterase patterns, whereas it is similar to *P. mugo* as far as the peroxidase pattern is concerned. Hybrid individual H9 is very close to *P. mugo* in its malate dehydrogenase pattern, whereas its esterase pattern is intermediate.

Figures 3 and 4 reveal similar independent behavior of particular protein properties in a group of intermediate individuals. These individuals are slightly more similar to *P. mugo* based upon peroxidase patterns and to *P. sylvestris* as far as esterase patterns are concerned (individuals H1, H3, H12, H13). In general, however, a predominant tendency towards *P. mugo* can be observed (11 hybrids as far as peroxidase, and 8 hybrids as far as esterase is concerned). Four hybrids each have a tendency towards *P. uliginosa*. The number of hybrids with only a faint tendency towards *P. sylvestris* is even less. Some of the *Pinus uliginosa* correlated hybrids show no similarity at all with *P. sylvestris* (e.g., hybrid H15). Thus the status of the hybrid swarm population is confirmed, as demonstrated by Fig. 2 (see also Prus-Głowacki and Szweykowski 1982).

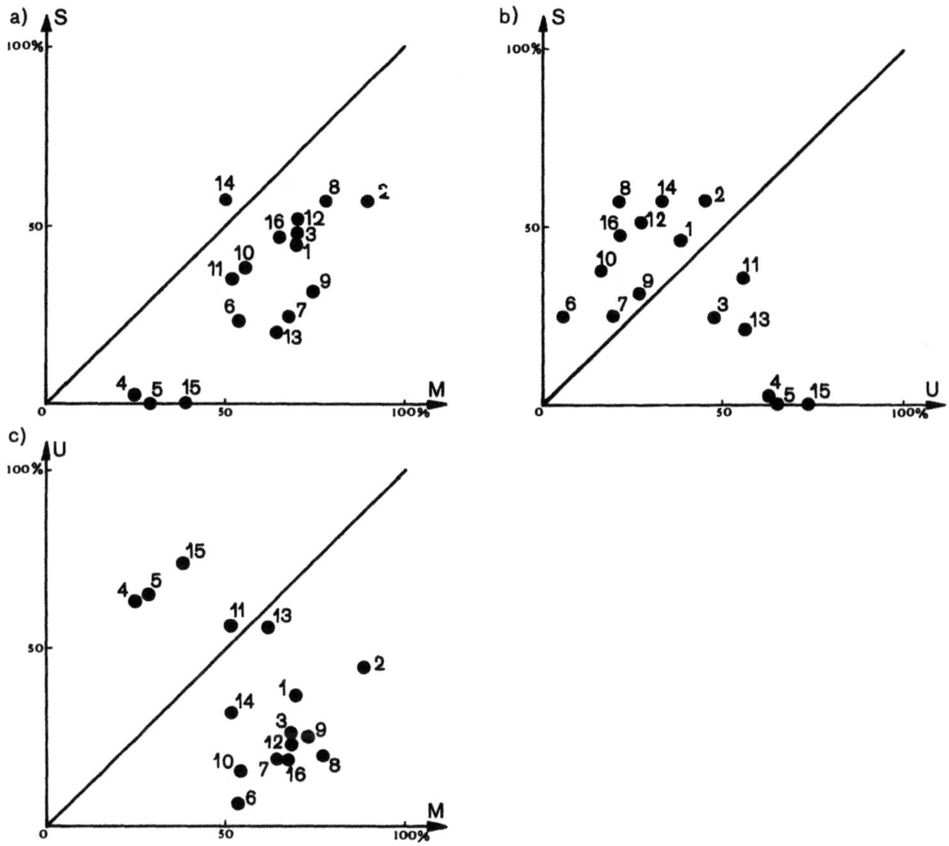

Fig. 3. Diagrams showing peroxidase activity affinities (in percent) of hybrid swarm individuals to: a *Pinus sylvestris* (*S*) and *P. mugo* (*M*); b *Pinus sylvestris* (*S*) and *P. uliginosa* (*U*); c *Pinus uliginosa* (*U*) and *P. mugo* (*M*), based on Euclidean distances (Sneath and Sokal 1973)

3.3 Antigenic Variability of Pinus sylvestris from Sweden

Pinus sylvestris is considered to have invaded Scandinavia, after glaciation, from both the north and south (Kjellander 1974). The migrating populations probably met in central Sweden between latidudes 60°–61°. It is in this geographical region where the greatest morphological variation has been observed. However, Scott (1907) and Langlet (1936) believe that in Sweden characteristics of *P. sylvestris* exhibit clinal variation from the south to the north without any prominent zone of gene exchange. Other researchers (Sylvén 1916, Kjellander 1974) are of the opinion that in Sweden two different varieties of *P. sylvestris* occur. The northern variety with narrow and fine branched crown is variety *lapponica*, while the southern broad-crowned is variety *septentrionalis*.

Proteins from needles of 20 to 22 trees from each of six Swedish populations were studied by means of Ouchterlony immunodiffusion and compared with the proteins

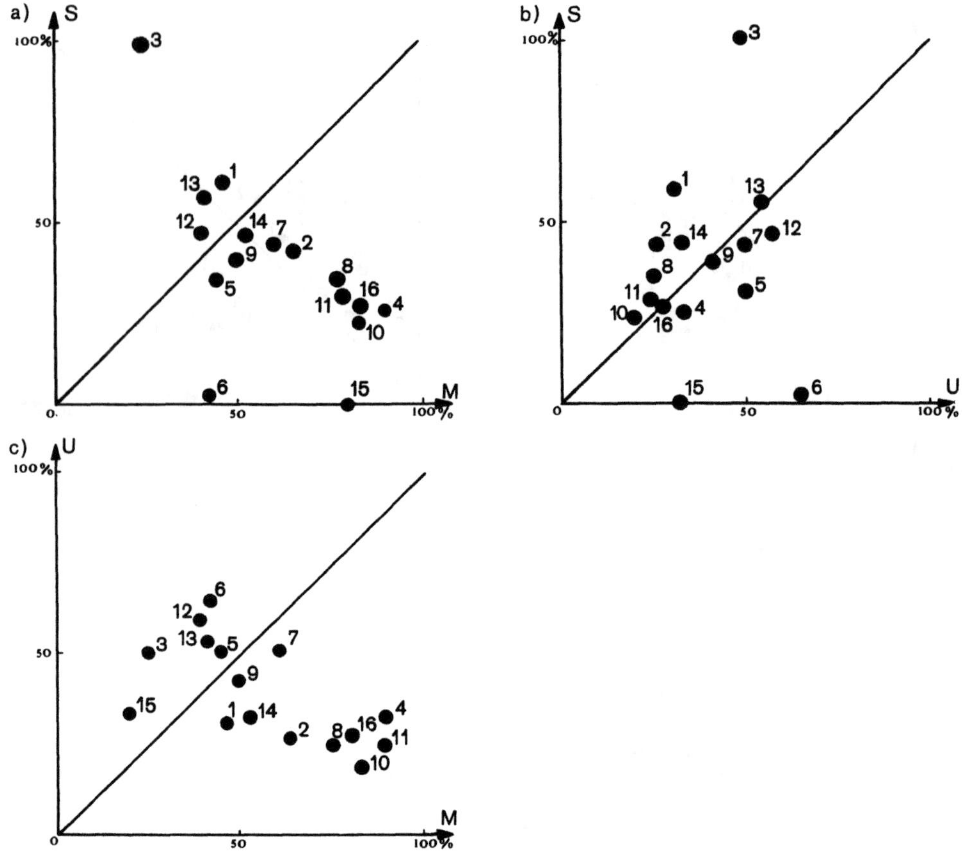

Fig. 4. Diagrams showing *a*-esterase activity affinities (in percent) of hybrid swarm individuals to reference species: **a** *Pinus sylvestris* (*S*) and *P. mugo* (*M*); **b** *Pinus sylvestris* (*S*) and *P. uliginosa* (*U*); **c** *Pinus uliginosa* (*U*) and *P. mugo* (*M*), based on Euclidean distances (Sneath and Sokal 1973)

of a highly heterozygous tree from the central part of Sweden as a reference. For each population investigated frequency of antigens, polymorphic indices (P_i), antigenic similarity coefficients (I_N) and serological distances (Dg_{I_N}) were calculated.

The frequency of each distinguished precipitin arc and the polymorphic indices are shown in Table 2. Two of the antigens (i.e. 3 and 6) show a tendency to increase and decrease respectively in populations from the south to the north of Sweden. This tendency is shown graphically in Fig. 5 for antigen 3. Two populations from the south (Nos. 1 and 6) have less than 45% of individuals possessing the antigen No. 3. In a population from the central part of Sweden (No. 17) the presence of that protein is observed in more than 70% of the trees of the given area. In populations originating from northern Sweden the antigen is noticed in approximately 90% of the individuals. At the same time antigen 6 shows the reverse tendency.

Serological distance values (Dg_{I_N}) between investigated populations distinguish two groups of populations, i.e. southern and northern ones, population No. 17 having

Table 2. Frequency of nine antigenic proteins, and polymorphic indices (P_i) for six investigated Swedish populations of *Pinus sylvestris*

Sample No.	Population	Latitude	Altitude (m)	No. of trees	Frequency of antigens									P_i
					1	2	3	4	5	6	7	8	9	
1	Glimåkra	56°20′	90	20	0.80	1.0	0.45	1.0	1.0	0.55	0.15	0.0	0.0	0.11
6	Floda	59°02′	80	22	0.77	1.0	0.41	0.91	0.95	0.59	0.23	0.09	0.05	0.12
17	Lillhärdal	61°45′	625	22	0.86	0.95	0.73	0.95	0.77	0.27	0.23	0.0	0.0	0.14
30	Anundsjö	63°31′	150	22	1.0	0.95	0.95	0.86	0.82	0.05	0.23	0.23	0.0	0.10
43	Älvsbyn	65°05′	150	22	0.91	1.0	0.86	0.91	1.0	0.14	0.09	0.05	0.0	0.07
58	Vittangi	67°38′	370	20	0.80	1.0	0.90	0.95	0.95	0.10	0.10	0.0	0.0	0.08
				ϕ	0.86	0.98	0.72	0.93	0.92	0.28	0.16	0.06	0.01	0.10

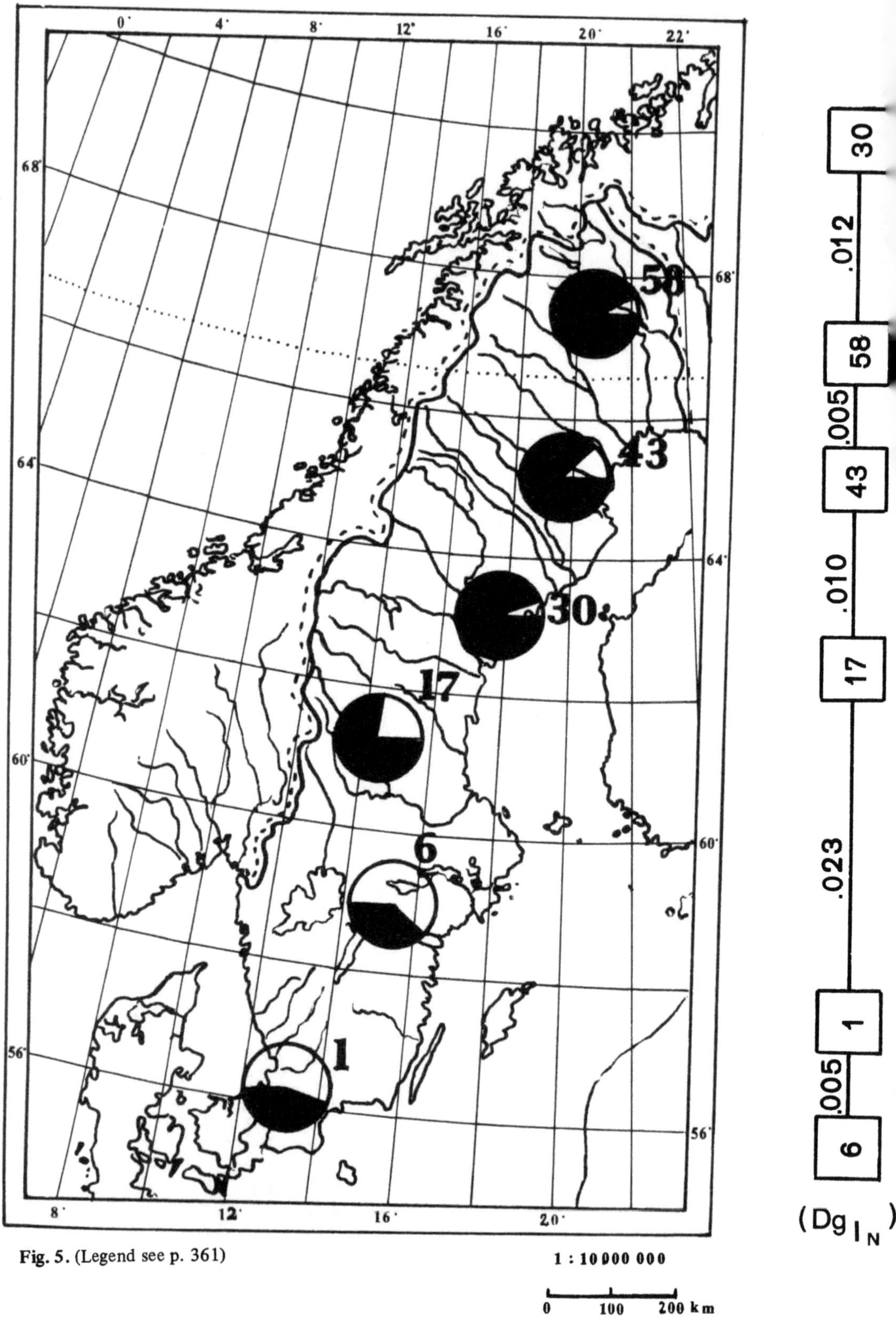

Fig. 5. (Legend see p. 361)

an intermediate position between these groups (Fig. 5). The polymorphic indices (P_i) computed on the basis of frequency of precipitin arcs (Table 2) show the highest value for a population from the central part of Sweden (No. 17). These results provide additional support to the hypothesis that two varieties of *P. sylvestris* are present in Sweden, and population No. 17 occupies an intermediate position having a mixture of characters of two varieties (Prus-Głowacki and Rudin 1981).

Our serological investigations allowed the characterization of systematic relationships of several pine species and also revealed the antigenic variability of *P. sylvestris* in Sweden. Data revealed the additional possibility of using serological methods, not only in chemotaxonomy, but also for helping to better understand hybridization and introgression in natural populations. Using the described techniques it was possible to serologically characterize each individual from a hybrid population as well as single individuals from populations of putative parents. This kind of approach also allowed the study of antigenic polymorphism in populations of *P. sylvestris* in Sweden. The detected antigenic differentiation of the populations is useful information when conducting both phytogeographic and ecologic research.

References

Anderson E (1949) Introgressive hybridization. Wiley and Sons, New York

Clausen J (1971) Immunochemical techniques for the identification and estimation of macromolecules. Work and Work, North-Holland

Kjellander CL (1974) In: Lecture notes in forest genetics. Swed Univ Agric Sci, Carpenberg, Sweden

Kröll J (1969) Scand J Clin Lab Invest 29:Suppl 124

Langlet O (1936) Medd Statens Skogsförsöksanst 29:219–406

Marshall DR, Jain SK (1969) Nature (London) 221:276–278

Nei M (1972) Am Nat 106:283–292

Ouchterlony Ö (1967) Handbook experimental immunology. Blackwell, Oxford

Prus-Głowacki W, Rudin D (1981) Silvae Genet 6:200–203

Prus-Głowacki W, Szweykowski J (1977) Bull Soc Amis Lettr Poznan 17:15–27

Prus-Głowacki W, Szweykowski J (1979) Acta Soc Bot Pol 48:217–238

Prus-Głowacki W, Szweykowski J (1980) Acta Soc Bot Pol 49:127–142

Prus-Głowacki W, Szweykowski J (1982) Bull Soc Amis Lett Poznan 22:107–122

Prus-Głowacki W, Szweykowski J, Sadowski J (1978) Gen Pol 19:327–338

Prus-Głowacki W, Sadowski J, Szweykowski J, Wiatroszak I (1981) Genet Pol 22:447–454

Sneath PHA, Sokal RR (1973) Numerical taxonomy. Freeman, San Francisco

Sokal RR, Michener CD (1967) Proc Linn Soc London 178:59–74

Szweykowski J (1969) Bull Soc Amis Poznan 10:39–54

Szweykowski J, Bobowicz MA, Kózlicka M (1976) Bull Soc Amis Lett Poznan 16:17–28

Scott PK (1907) Forstwiss Centralbl 29:199–218

Sylvén N (1916) Medd Statens Skogsförsöksanst 13–14:9–110

Fig. 5. Frequency of antigen No. 3 in six Swedish populations of *Pinus sylvestris* shown by the *filled circle sectors*, and, (*right*) serological distances among these populations according to Nei's (1972) formula (Dg_{I_N}). Antiserum raised against needle proteins of *Pinus sylvestris*

Preliminary Immuno-Electrophoretic Comparison of Selected Korean Quercus Species

Y.S. LEE[1]

Abstract. Protein analyses and multidisciplinary data have proved to be useful in the delimitation of *Quercus* taxa using pollen extracts from selected taxa. To calculate the degree of protein similarity, total rocket heights obtained from rocket immuno-electrophoresis (RIE), provided an index of serological correspondence (SC). Antisera raised to species, usually placed in the subgenus Lepidobalanus, gave very high SC values with other tested species in this subgenus, and lower values with species usually placed in the subgenus (or genus) Cyclobalanopsis. Obtained quantitative data are in agreement with the qualitative immunoprecipitin data.

1 Introduction

The taxonomic status of the genus *Quercus*, and the arrangement of subgenera and/or series, still remains in an unstable state after several intensive investigations conducted over a period of 100 years.

Cytologically *Quercus* species, as well as all other genera of the Fagaceae, have a relatively stable chromosome number of n = 12 and only slight differences in size and morphology. Fertile or partially fertile natural hybrids resulting from crosses between species are rather common. However, even with this frequent hybridization and resulting morphological recombinations, the classical species-concept has continued to be meaningful although in the species many populations of extraordinary complexity are found (Burger 1975).

A review of taxonomic treatments of *Quercus* taxa native in Korea reveals diverse classifications. In older treatments where Cyclobalanopsis was not considered, Oersted (1871) placed all taxa within three subgenera (Erythrobalanus, Lepidobalanus, and Macrobalanus) of *Quercus*. In Schwarz's (1936–1937) treatment, additionally Sclerophyllodrys was designated a subgenus of *Quercus*. In more recent years Cyclobalanopsis, which was considered a separated genus by Nakai (1952) and Chung (1957) was included into *Quercus* as a distinct subgenus (Lee 1961, Makino 1961). Melchior (1964) designated Cyclobalanopsis, Erythrobalanus, and Lepidobalanus subgenera within the genus *Quercus*, Lepidobalanus including Sclerophyllodrys and Macrobalanus of Schwarz (1936–1937). Thus, the intra- and intergeneric taxonomic problems, and proper relationships remain to be resolved.

Therefore, it has become necessary to investigate new characteristics and attempt to determine the significance of these characteristics in terms of *Quercus* taxonomic

[1] Department of Biology, Chungbuk National University. Chongju, Korea 310

Proteins and Nucleic Acids in Plant Systematics
ed. by U. Jensen and D.E. Fairbrothers

treatments. To obtain additional data as an aid in the interpretation of relationships, pollen proteins using different immuno-electrophoretic techniques were employed in this research. Recent publications have emphasized the value of the application of immuno-electrophoretic methods in contributing useful data in solving many systematic problems (Smith 1972, Lee 1977, Fairbrothers 1980, Lee 1981a,b). The characteristics that make rocket immuno-electrophoresis particularly attractive for taxonomic research, is the fact, that both quantitative and qualitative measurements can be obtained simultaneously for various antigens reacting with a specific antiserum. The detected amount of cross reactivity is valuable in comparing taxa, because it provides a measurement of protein similarity among them. Since amentiferous pollen proteins have been shown to provide an excellent source of extractable antigens for systematic serological research (Brunner and Fairbrothers 1979, Petersen and Fairbrothers 1979, Petersen this volume), pollen from species of Korean *Quercus* species were analyzed by immunoelectrophoresis.

2 Materials and Methods

2.1 Materials

Pollen of oaks were collected from native populations in Seoul, Gyonggi, and Busan in Korea. Table 1 presents the eight taxa included in this preliminary investigation, arranged according to Melchior's (1964) classification. All the pollen was desiccated and stored in containers at 1°–3 °C. Extraction of pollen antigens and preparation of antisera followed the procedures of Lee and Dickinson (1979).

2.2 Immuno-Electrophoretic Techniques

In this research the precipitin reaction was measured quantitatively and qualitatively in agarose gel media. Rocket immuno-electrophoresis, one of the immuno-electrophoretic techniques developed recently, was used to obtain data for analyses.

Table 1. *Quercus* species used in the various experiments

Taxa	Abbr.	Subgenus	Area collected
Quercus acuta Thunb.	AC	*Cyclobalanopsis* Prantl.	Busan
Q. acutissima Carr.	CT	*Lepidobalanus* Endl.	Busan
Q. acutissima x *variabilis*	VA	*Lepidobalanus* Endl.	Gyonggi
Q. aliena Bl.	AM	*Lepidobalanus* Endl.	Gyonggi
Q. dentata Thunb.	DN	*Lepidobalanus* Endl.	Seoul
Q. donarium Nakai	DO	*Lepidobalanus* Endl.	Seoul
Q. glauca Thunb.	GL	*Cyclobalanopsis* Prantl.	Gyonggi
Q. serrata Thunb.	SR	*Lepidobalanus* Endl.	Gyonggi

The basic outline of rocket immuno-electrophoresis of Axelsen et al. (1975) was followed. The 16 ml of heated 1.5% (w/v) agarose solution (55°–60 °C) was poured onto a horizontal glass plate (8.4 x 9.4 cm). A barbiture-glycine/tris buffer, pH 8.8 (Axelsen et al. 1975) and 0.03% (w/v) sodium azide were added to the agarose. After incubation overnight at 4 °C, each agarose gel strip was divided into four segments and transferred to the second glass plate (Fig. 1) to which 8 ml of antibody-containing agarose was poured. The wells for the samples were punched directly on the antibody-containing agarose gel. The wells were 7 mm apart, and the first and last wells were 14 mm from the edge of the gel slab. Approximately 7 µl of each sample was placed in each well. The electrophoresis was performed at 2 V/cm with the initial setting of 100 V and 75 mA and run for 2 h at 8 °C.

Rocket height analyses of the precipitin reactions revealed the different immunoprecipitin systems involved in the antigen-antibody reactions, and indicated the degree of serological similarity among the antigenic determinants of the taxa compared.

3 Results and Discussion

Pollen proteins extracted from the 8 species listed in Table 1 were used as antigens for the various experiments. Protein extracts from *Q. aliena, Q. donarium, Q. glauca,* and *Q. dentata* were also used to raise antisera to be used in various experiments. The observed rocket-shaped bands resulting from the multiplicity of a polyvalent antiserum, were generally well-separated and well-defined. This greatly facilitated the tracing of the precipitin bands (rockets) from an antibody to the corresponding antigenic peak.

The immunoplate preparation for the rocket immunoelectrophoresis was designed as illustrated in Fig. 1 to maximize the advantages of this technique, speed and specificity, and to minimize the amount of antiserum required. Following this procedure the rockets obtained for individual taxa were adequately formed (Fig. 2) to allow analysis.

To obtain the maximum development of the immunoprecipitin systems different periods of incubation were required for the diverse reacting systems. Although some

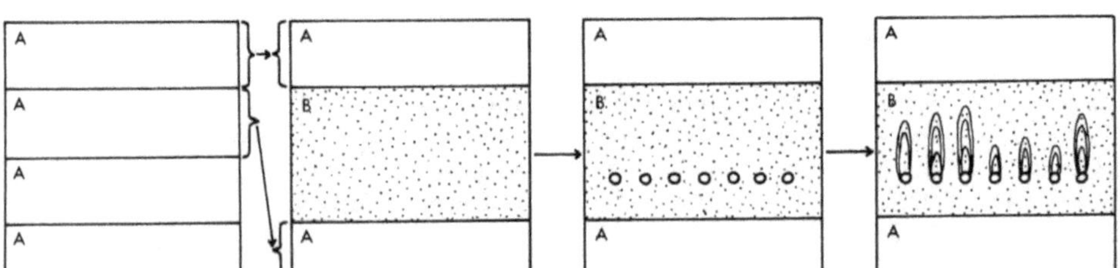

Fig. 1. Illustration of preparation for immunoplates used for rocket immunoelectrophoresis.
A agarose gel strip (plain agarose plus barbiture-glycine/tris buffer solution);
B antibody-containing agarose gel, 2 ml of antiserum was mixed with 6 ml of agarose solution for antibody-containing agarose gel. The wells (2.5 mm) were punched in the gel using an LKB template and suction gel puncher, and 7 µl samples were applied

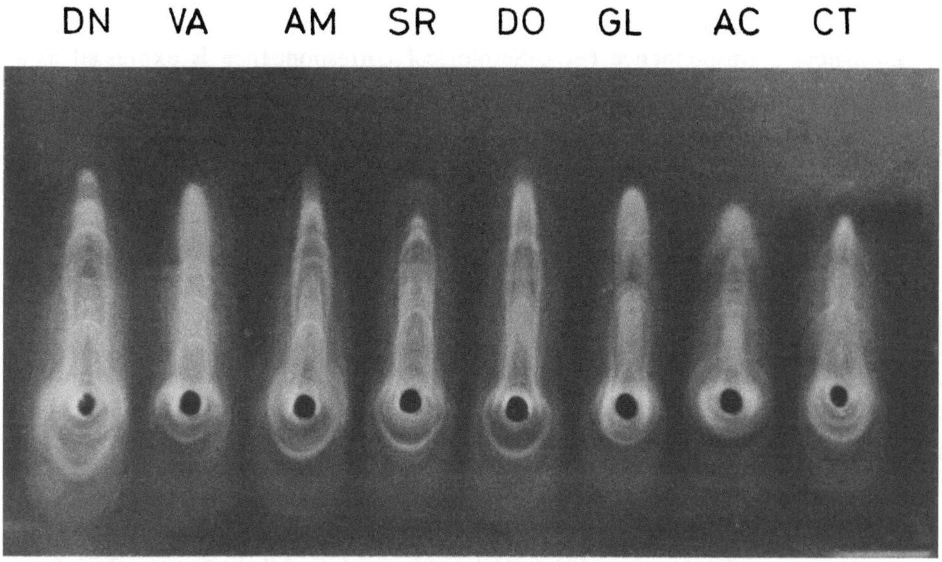

Fig. 2. Rocket immunoelectrophoretic patterns. The agarose gel was charged with *Q. dentata* antiserum. The plate was photographed after 2 days on a dark background with direct light. Antigens: *(DN) Q. dentata*; *(VA) Q. acutissima* x *variabilis*; *(AM) Q. aliena*; *(SR) Q. serrata*; *(DO) Q. donarium*; *(GL) Q. glauca; (AC) Q. acuta; (CT) Q. acutissima*

of the systems were faint and do not appear on the photographs, all of them were included in the total count. In these preliminary experiments quantification of all the rockets and all available rocket heights were evaluated, although it was difficult to identify each and every corresponding rocket among taxa tested with this technique. However, it is worthy to note that two major and sharp immunoprecipitin systems detected in the *Q. dentata* reference reaction were shared in many species. One of these systems remained near the well in which the sample was applied, and the other was highly charged and moved rapidly toward the anode. These are tentatively designated as the two major immunoprecipitin systems m_1 (near the well) and m_2. *Q. dentata* antiserum tested with extracts from *Q. aliena, Q. dentata, Q. donarium, Q. serrata, Q. acutissima,* and *Q. acutissima* x *variabilis* revealed the major immunoprecipitin system m_1 with the rocket heights being: 5, 5, 5.5, 6.5, 6, and 4 mm, respectively (Fig. 2).

Extracts from all the above species, except *Q. acutissima* and *Q. acutissima* x *variabilis*, produced the major immunoprecipitin system m_2 with their rocket heights being: 13, 12.5, 9.5, and 12.5 mm, respectively. Antigen extracts from *Q. acuta* and *Q. glauca* revealed neither the m_1 nor m_2 system in the agarose gel plate. These same results were also substantiated by data obtained from conducting experiments using conventional immunoelectrophoresis (data not presented). These data obtained with *Q. dentata* antiserum indicate that *Q. dentata* is serologically similar to *Q. aliena, Q. donarium,* and *Q. serrata;* less similar to *Q. acutissima* and *Q. acutissima* x *variabilis*; and least similar to *Q. acuta* and *Q. glauca*.

As revealed in Table 2, rocket height data were used as an index of protein similarity or serological correspondence (SC). Serological correspondence is expressed as a percentage value of total rocket heights in the reference reaction (cross reaction/reference reaction x 100). The sum of all rockets resulting from a reaction between an antiserum and antigenic material used to stimulate its formation (reference reaction) was designated as 100%. Serological comparisons among taxa obtained by this procedure are presented in Table 2 and Fig. 3. Antisera produced to *Q. aliena* and *Q. donarium* produced strong reactions resulting in high rockets with *Q. aliena, Q. dentata, Q. donarium, Q. serrata, Q. acutissima,* and produced relatively weak reactions with *Q. acuta, Q. glauca*, and *Q. acutissima* x *variabilis*. Theoretically the maximum quantification of immunoprecipitin systems, that is 100% in serological correspondence, would be expected in the reference reaction. The values slightly over 100% in the cross reaction in Table 2, Figs. 2 and 3 reflect experimental error. Therefore, serological correspondence should be used as a relative value for comparisons, but not as absolute ones.

Antisera produced to the species grouped in the Lepidobalanus (Table 1) gave very high serological correspondence values (⩾ 77%) with other species placed in the same subgenus, and low values (51%–73%) with species of the subgenus Cyclobalanopsis (Table 2, Fig. 3). These quantitative data agree with the qualitative data where the species of Lepidobalanus revealed more immunoprecipitin systems (10–12 rockets) than did species of Cyclobalanopsis (8–10 rockets), when *Q. aliena* antiserum was tested (Table 2). Such data support the placement of Cyclobalanopsis as a subgenus of the genus *Quercus* (Melchior 1964).

Overall quantitative and qualitative data indicate that the subgenus Cyclobalanopsis is distinct from Lepidobalanus. However, the elevation of the Cyclobalanopsis to the

Table 2. Quantification of rocket heights (mm) for eight species of *Quercus* obtained by rocket immunoelectrophoresis. The agarose gel was charged with *Q. aliena* antiserum. Serological correspondence (SC), is the percentage value of total rocket heights in the reference reaction (cross reaction/reference reaction × 100). Underlined numbers are the rocket heights of the two detected major immunoprecipitin systems

Taxa	Rocket heights	Number of rockets	Total rocket heights	SC (%)
Q. acuta	1, 2, 2.5, 3.5, 5, 6, 7, 8, 11, 15	10	61.0	66
Q. acutissima	3, 3.5, 5, 6, 6.5, 7, 8, 9.5, 12, 13.5, 14.5, 16	12	104.5	112
Q. acutissima x *variabilis*	0.5, 1.5, 4, 5, 7, 9.5, 10, 15, 16.5	10	72.0	77
Q. aliena	1, 1.5, 3, 5, 7, 9, 10, 11, 13, 15, 18	11	93.5	100
Q. dentata	1, 5, 5.5, 6.5, 8, 10.5, 12.5, 14.5, 18, 19	11	115.0	118
Q. donarium	0.5, 1, 1.5, 4, 5.5, 8.5, 9, 9.5, 13, 17, 19	11	88.5	95
Q. glauca	3, 4, 6, 7, 8, 9, 14, 17	8	68.0	73
Q. serrata	1.5, 3, 4.5, 6.5, 7, 8.5, 9.5, 11, 12.5, 14, 15, 16.5	11	109.5	115

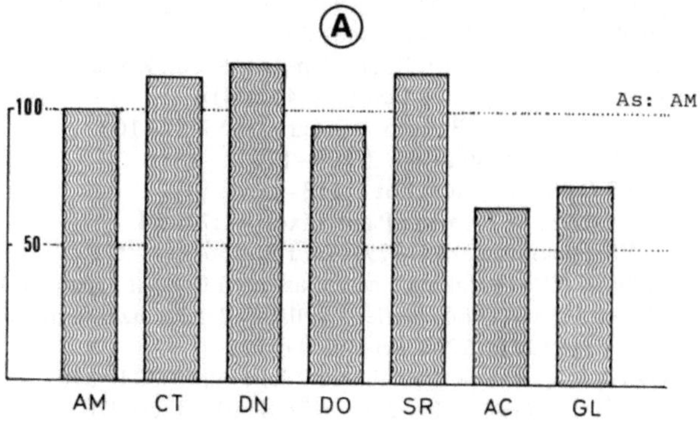

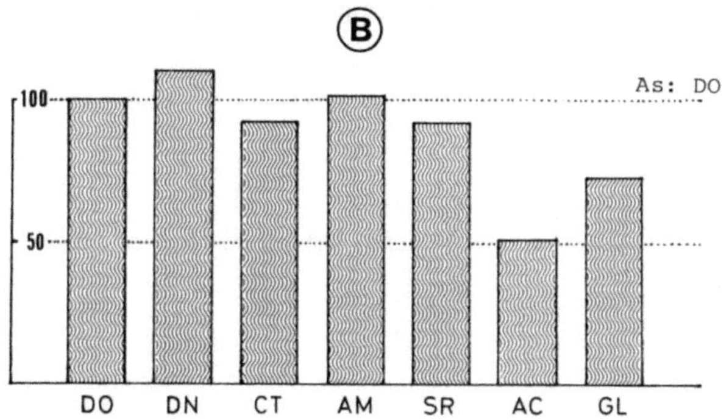

Fig. 3. Bar-graphs comparing the serological correspondence for *Quercus* species as measured by rocket heights. Antisera: (**A**) *Q. aliena;* (**B**) *Q. donarium*. Antigens: (*AM*) *Q. aliena*; (*CT*) *Q. acutissima*; (*DN*) *Q. dentata*; (*DO*) *Q. donarium*; (*SR*) *Q. serrata*; (*AC*) *Q. acuta*; (*GL*) *Q. glauca*

rank of a separate genus does not seem warranted. An understanding of the relationships of Lepidobalanus to Erythrobalanus requires additional research.

In this preliminary research rocket immunoelectrophoresis proved to be a valuable technique for use in systematic serology. To obtain an image of protein similarity (serological correspondence) the sum of total rocket heights proved useful. Rocket patterns made it possible to measure individual immunoprecipitin systems which were identical or partially identical to those systems detected in reference reactions, and quantification of all available rocket heights proved valuable for evaluating similarities.

The obtained data proved valuable also when compared with data obtained from other disciplines by earlier taxonomists. The inclusion of diverse types of data aid in developing taxonomic profiles for each taxon, which could be used in comparative studies to establish similarities which could be interpreted in terms of relationships.

References

Axelsen NH, Kröll J, Weeke B (eds) (1975) A manual of quantitative immunoelectrophoresis. Oslo, Norway

Brunner F, Fairbrothers DE (1979) Bull Torrey Bot Club 106:97–103
Burger WC (1975) Taxon 24:45–50
Chung T (1957) Korean flora. Shin-ji-sa, Seoul, Korea
Fairbrothers DE (1980) Taxon 79:412–416
Lee TB (1961) Res Bull Korean Agric Soc 7:87–108
Lee YS (1977) System Bot 2:169–179
Lee YS (1981a) System Bot 6:113–125
Lee YS (1981b) J Korean Plant Taxon 11:77–84
Lee YS, Dickinson DB (1979) Am J Bot 66:245–252
Makino T (1961) Makino's new illustrated flora of Japan. Hokuryukan, Tokyo
Melchior H (ed) (1964) Engler's Syllabus der Pflanzenfamilien II. Bornträger, Berlin-Nikolassee
Nakai T (1952) Bull Nat Sci Mus, Japan
Oersted AS (1871) K Dan Vidensk Selsk Skr 9:334–370
Petersen FP (1983) In: Jensen U, Fairbrothers DE (eds) Proteins and nucleic acids in plant systematics. Springer, Berlin Heidelberg New York
Petersen FP, Fairbrothers DE (1979) Taxon 4:230–241
Schwarz O (1936–1937) Notizbl Bot Gart Mus Berlin-Dahlem 13:1–22, 495–496
Smith PM (1972) Ann Bot (London) 36:1–30

Serological Data and Current Plant Classifications

The Importance of Modern Serological Research for Angiosperm Classification

R. DAHLGREN[1]

Abstract. The importance of the results from serological research during the last 30 years, for drawing conclusions concerning phylogenetic relationships among major plant groups are discussed in connection with their role in taxonomy on higher levels. In drawing these conclusions the serological evidence is evaluated in conjunction with that obtained from comparative studies in the fields of macro-morphology, embryology, chemistry (secondary metabolites), and other fields used in the construction of angiosperm classifications. Explanations are presented as to why it has been important to understand the methods and materials used in serological investigations when evaluating their results. When serological investigations include an adequate number of suitable taxa, relevant methods are used, and when the results are considered in conjunction with evidence from the other disciplines, such data can be extremely useful, as is demonstrated by several of the examples which are presented. If *only* serological data are used, the taxa studied are not representative or too few, and the results are not placed in the context of previous findings, the conclusions can be too far-reaching and either incorrect or inadequate. A review is given of a variety of serological studies distributed over the major groups of the flowering plants, with the emphasis on relationships between families and orders. The review indicates that serological contributions are being incorporated to an increasing degree in the basic evidence for conclusions on relationships between families and family groups, and in the material utilized in constructing modern systems of classification. These data have sometimes supported some views, and contradicted others, in matters of controversy between classification systems. They may also reveal new alternative interpretations for consideration of evolution.

1 Introduction

During the last 30 years protein chemistry, and in particular serology, has contributed extensive information of great potential use to taxonomy at various levels. In a survey of serological contributions Fairbrothers (1983) and Fairbrothers and Petersen (1983) calculated that approximately 600 taxa have been included in about 200 sero-taxonomic publications during this period. By studying the references in the literature they also estimated the influence of serology on the angiosperm classifications of Takhtajan, Cronquist, Thorne and Dahlgren, in that order, claiming that the influence has been greatest on Takhtajan's and least on my own classification. Their estimation is somewhat misleading as actually it mostly reflects the proliferation of references presented by the four authors. I have, indeed, been considerably influenced by the results of both the Fairbrothers and Jensen schools, though I have evidently failed to

[1] Botanical Museum, University of Copenhagen, Gothersgade 130, 1123-Copenhagen, Denmark

Proteins and Nucleic Acids in Plant Systematics
ed. by U. Jensen and D.E. Fairbrothers

cite them to a sufficient extent, mainly because their results often did not greatly deviate from evaluations based on evidence from other sources. This is regrettable, because it is precisely the correspondence between serological (and other protein chemical) evidence and evidence from various other fields that needs to be analyzed and stressed, and contributions should not only be discussed when results are opposed to the views expressed, but also when they substantiate a point of view.

However, when serological data are used for taxonomic and phylogenetic conclusions, the credibility of such data has to be examined carefully. Of special interest are the bases of the serological experiments. They are discussed in some of the contributions of this volume (Fazekas de St. Groth, Fairbrothers and Petersen, and Jensen and Grumpe). Additionally the calculations of the serological data are evaluated by Lester et al. However, questions remain, especially concerning the evidence of the particular serological method and technique, which are difficult to understand for those taxonomists who do not use such procedures.

Thus different data resulting from source, composition, and condition of injected material should be considered when evaluating the data, as well as the number and period of injections. For those taxonomists not using serological methods it may be difficult to judge from the published material, to which extent deviating conditions could have influenced the results. Is it always taken for granted that the seeds (or other plant material) are mature and comparable, and the association of the antigenic proteins to lipids or phenolic substances is having no influence? Is it possible to evaluate serological reactivity of antigenic material if it originates from different tissues, e.g. from perisperm, endosperm or embryo tissue? This point becomes important for the question of the relationship between *Nymphaea* (copious perisperm, some endosperm, small embryo) and *Nelumbo* (storage of seed proteins in the embryo only), which has been discussed by Simon (1970).

Further, the implications of a positive versus a negative test are not always clear. One rule says that a strong positive test implies shared determinants and consequently at least partial identity of proteins, while a negative test implies only a lack of reaction. Nevertheless, the absence of any reaction or a weak reaction is generally considered a criterion of distant relationship and the decision must rest on this. I shall return to this problem later.

The choice of taxa and the number of taxa to be included in an investigation are of vital importance for its significance. Seed or pollen material from the most suitable taxa to be compared serologically may be difficult to acquire. The choice is often conventional, and the potentially most relevant taxa (as judged on other evidence such as from embryology or non-protein chemistry) may have been omitted because seed material has not been available, and thus results will be incomplete. A few taxa chosen from each family in an interordinal study may prove inadequate and the taxonomist may be tempted to draw premature conclusions. With insufficient data a general taxonomic evaluation and judgement remains difficult.

Since botanists with limited serological background will probably turn to the serological conclusions for information, instead of studying the data, it is of the utmost importance for serologists to pay greater attention to the wording of the conclusions, including the necessary reservations and sources of error, also making sure not to omit potentially important details. Where serological results prove to be of significance it is relevant to look for characters, morphological for instance, that accord with the serological findings. With this combination links between plant groups have often been indicated, whether or not they have been proposed by previous researchers.

In the following text I shall survey a number of the most outstanding serological results that relate to more than one family. Actually, serological studies are perhaps of greatest use for establishing relationships at intergeneric and intrafamilial levels (i.e. in large to moderately-sized families), such as in Ranunculaceae (Jensen 1967, 1968), Berberidaceae (Jensen 1974), Poaceae (many papers), Brassicaceae (Kolbe 1982), and Fabaceae (Cristofolini 1980, 1983).

In order to enable the reader to survey the many orders mentioned in the following text, a diagram of them (in accordance with Dahlgren (1980)) is presented here in Fig. 1.

2 Monocotyledons

2.1 General Problems

The co-occurrence of a single cotyledon and sieve tube plastids with cuneate (triangular) protein crystalline bodies (Behnke 1969, 1981, Behnke and Dahlgren 1976) indicates that the monocotyledons are a monophyletic group. Both of these properties are extremely rare in dicotyledons. The larger constellations of orders in the monocotyledons and their interrelationships are currently being discussed (Huber 1969, 1977, Dahlgren and Clifford 1981, 1982, Dahlgren 1983, Thorne 1981, 1983) on the basis of a great number of characteristics. Some of these have been surveyed in recent years, for instance anatomical characters (Tomlinson 1969, Cutler 1969 etc.), embryology (Wunderlich 1959, Hamann 1962, 1974), seed structure (Huber 1969), UV-fluorescence in cell walls (Harris and Hartley 1980), epicuticular wax structures (Barthlott and Frölich 1983, Behnke and Barthlott 1983) and stoma ontogeny (Rasmussen unpublished). The distribution of a great number of characteristics in the monocotyledons is presented and discussed in Dahlgren and Clifford (1982).

Thus the classification of monocotyledons based on their main features is developing rapidly, and the sparse serological contributions to the monocotyledons (apart from the grasses) can be viewed against a large body of evidence from other sources. Here, I shall discuss two papers by Chupov and Kutiavina (1978, 1981) on the Liliiflorae and one by Lee and Fairbrothers (1972) on the relationships of Typhales.

2.2 Liliiflorae

The family Liliaceae was previously broadly circumscribed, as is still the practice in many floras, and in some classifications such as Thorne's (1976, 1981). It is now more often split, to the greatest extent perhaps by Huber (1969) who acknowledges more than 30 separate families that can be aggregated to some major complexes of ordinal rank, e.g. the "dioscoreoid", "haemodoreoid", "asparagoid", and "colchicoid" Liliiflorae. These have been treated more formally as orders e.g. in Dahlgren

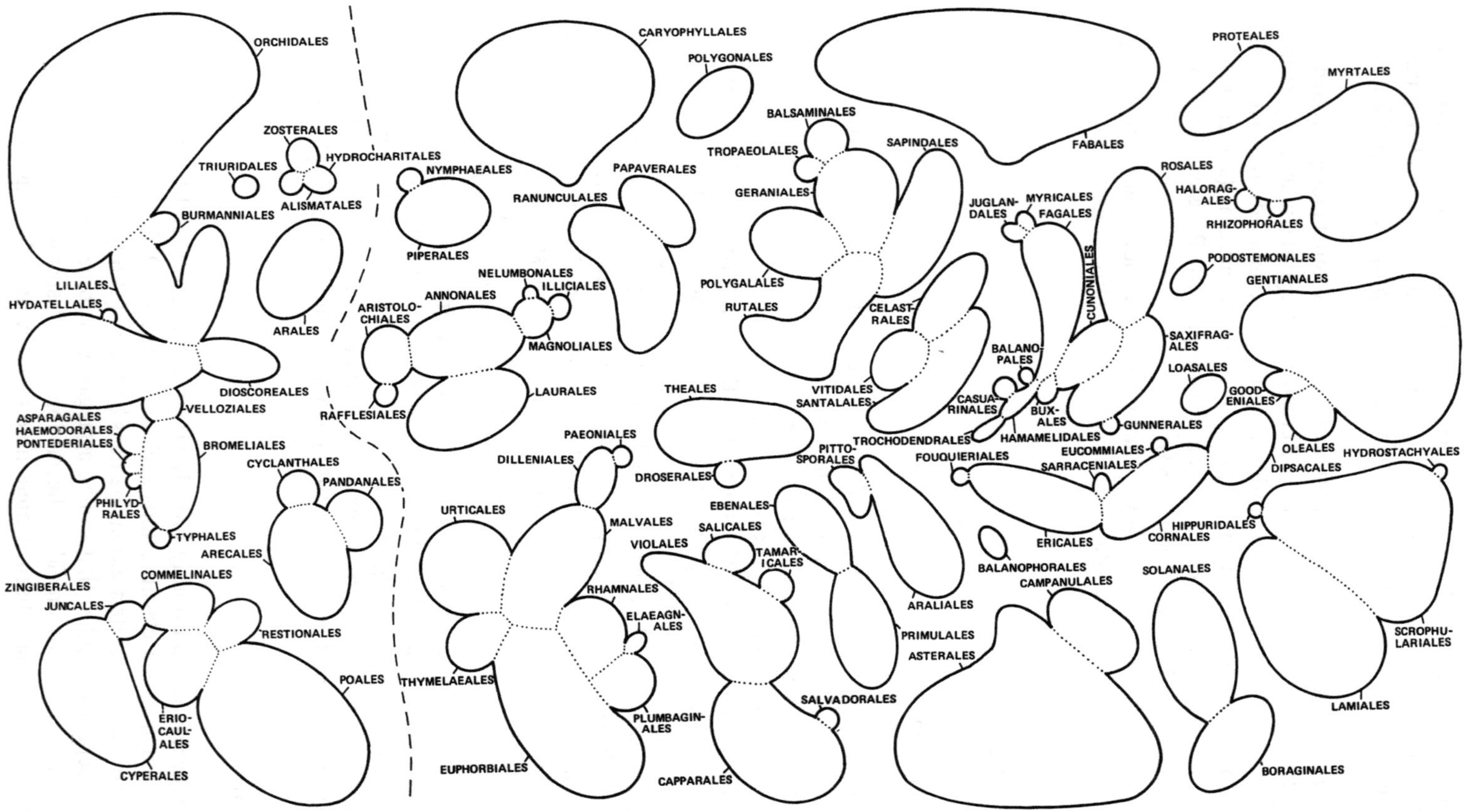

Fig. 1. Diagram illustrating the angiosperm orders according to Dahlgren (1980). This is the classification of orders and superorders as conceived in the present article

(1975a, 1980) and Dahlgren and Clifford (1981, 1982), but a more inclusive ordinal concept with a single order Liliales is more common (see also Dahlgren and Clifford 1982, p. 327). In separating the orders Dioscoreales, Asparagales, and Liliales s.str. from each other (corresponding to Huber's "dioscoreoid", "asparagoid", and "colchicoid" Liliiflorae, respectively) a variety of criteria are used (Huber 1969), the most important being those related to the seed. Thus, one important criterion of the Asparagales is that they have either a baccate fruit (with unpigmented seeds) or a capsular fruit with black phytomelan-coated seeds, or (rarely, as in Asparagaceae and Phormiaceae) a baccate fruit with phytomelan-coated seeds. In the asparagalean seeds the inner integument (tegmen sensu Corner, 1976) has collapsed to form a thin membrane while the outer integument (testa sensu Corner) either often disintegrates (when the fruits are berries) or is provided with a phytomelan coat (when the fruit is capsular and the seeds liberated). This type of seed is unique in angiosperms and is therefore believed to have arisen only once during the course of evolution. On the other hand the phytomelan coat can probably be easily lost, as in a few Hyacinthaceae, some Australian sclerophyll taxa and various secondarily or tertiarily baccate-fruited groups of genera, e.g. in certain Amaryllidaceae. Phytomelan seeds (p) or berries (b) characterize the asparagalean families as follows (nomenclature according to Dahlgren 1983): Smilacaceae, b; Philesiaceae, b; Geitonoplesiaceae, p (b); Convallariaceae, b; Asparagaceae (p) b; Herreriaceae, p; Dracaenaceae, b; Nolinaceae, b; Asteliaceae, p (b); Dasypogonaceae –; Calectasiaceae –p; Xanthorrhoeaceae –; Agavaceae, p (b); Hypoxidaceae, p; Tecophilaeaceae, p; Cyanastraceae –; Phormiaceae, p (b); Eriospermaceae – (hairy seeds); Asphodelaceae, p; Anthericaceae, p; Hemerocallidaceae, p; Funkiaceae, p; Hyacinthaceae, p; Alliaceae, p; Amaryllidaceae, p (rarely b).

In Liliales s.str. the seed coat is better developed with the tegmen retained as a distinct layer and the phytomelan coat never present (the fruits are only very rarely baccate in this order). The seeds are often pigmented, and are then either reddish-brown or brown (the pigments are phlobaphenes). With this circumscription the order Liliales includes Colchicaceae, Tricyrtidaceae, Calochortaceae, Liliaceae s.str., Alstroemeriaceae, Melanthiaceae, Iridaceae, and Geosiridaceae. These families also differ as a rule from Asparagales in having spotted tepals, nectaries at the tepal base (instead of nectaries in the septa of the ovary as in most Asparagales), extrors anthers and elongated stem internodes and dispersed leaves (see Huber 1969, p. 511). There are, however, numerous exceptions as regards each of these characteristics. Yet it seems that Huber's division is largely a natural one, though as far as I know few taxonomists have accepted his ideas.

What does the serological evidence presented by Chupov and Kutiavina (1978, 1981) contribute to the classification of the Liliiflorae? Though these authors cite Huber's treatment of the group from 1969 they do not seem to have been in any way influenced by this treatment. In each of these investigations they used antisera made from 8 species, and as the degree of precipitation varied considerably it may be advisable to stress the differences in precipitation for one antiserum at a time. An immunoelectrophoretic technique was used for separating the storage proteins of seeds, and the intensity of the reactions was evaluated visually by recording the number and intensity of the bands formed, calculated as a percentage of the reference-reaction.

In the 1978 study, antisera were used from species of the asparagalean genera *Asparagus* (Asparagaceae), *Dracaena* (Dracaenaceae), *Bulbine* (Asphodelaceae), *Hemerocallis* (Hemerocallidaceae), *Nothoscordon* (Alliaceae), and *Galanthus* (Amaryllidaceae) and of the lilialean genera *Lilium* (Liliaceae s.str.) and *Veratrum* (Melanthiaceae). These were tested against 60 liliifloran species dispersed over the orders (sensu Dahlgren) Dioscoreales (4 species), Asparagales (44 species), and Liliales (12 species).

I shall here relate these results to the classification of Huber (1969) and Dahlgren (1980). Chupov and Kutiavina (1978) found that there was generally little or no precipitation between the antisera of *Asparagus, Dracaena, Hemerocallis* or *Galanthus* and antigenic material of the taxa of Liliales (Liliaceae, Colchicaceae, Melanthiaceae, Iridaceae) and Dioscoreales (Trilliaceae, Dioscoreaceae). However, they noted reactions with a variable number of asparagalean taxa, reasonable support thus being obtained for upholding the order Asparagales in Huber's (1969) concept. *Bulbine* and *Nothoscordon* partly agreed with the above genera, but also showed weak serological affinity with some lilialean taxa. *Lilium* (Liliales) serum reacted with antigenic material of *Fritillaria, Cardiocrinum*, and *Tulipa* only, all members of the Liliaceae s.str. which are characterized by possessing bulbs, *Fritillaria* type embryo sac, nucellus without parietal cell and seeds lacking phytomelan. *Veratrum* antiserum, however, reacted with antigenic material of various species in all three others, although the reaction was often moderate. This is somewhat difficult to interpret. There could be methodical reasons (see below, and Chupov and Kutiavina 1981) or the protein may have determinants in common with more genera of Liliiflorae than the other genera used for antiserum. The fact that the reactions between antisera of other liliiflorous genera and antigenic material of *Veratrum* were weak or lacking counterbalances this, however.

Some noteworthy results from this study can be mentioned. Thus *Dracaena* (Dracaenaceae) reacted strongly with taxa of *Nolina* (Nolinaceae), *Cordyline* (Asteliaceae), *Maianthemum* and *Polygonatum* (Convallariaceae), but less strongly with *Yucca* or *Agave* (Agavaceae), which supports Huber's evidence for a position among the other berry-fruited Asparagales. Only a weak reaction or none at all was obtained between the six antisera and antigenic material of the Dioscoreaceae. *Bulbine* (Asphodelaceae) antiserum reacted most strongly with *Asphodelus, Asphodeline, Kniphofia, Eremurus* (Asphodelaceae), *Hosta* (Funkiaceae), and *Cordyline* (Asteliaceae); *Nothoscordon* (Alliaceae) reacted most strongly with *Agapanthus, Allium*, and *Milla* (Alliaceae), *Eremurus* (Asphodelaceae), and *Asparagus* (Asparagaceae). The reactions obtained with *Hemerocallis* and *Galanthus* antisera were weak and difficult to evaluate.

Proceeding to Chupov and Kutiavina's 1981 investigation which gives an account of immuno-precipitating systems for species of the asparagalean genera *Ruscus* (Asparagaceae), *Yucca* (Agavaceae), *Phormium* (Phormiaceae), *Hosta* (Funkiaceae), *Scilla* (Hyacinthaceae), and *Ornithogalum* (Hyacinthaceae), the dioscorealean genus *Dioscorea* (Dioscoreaceae) and the palm genus *Rhopalostylis* in reactions with antigenic material of 64 species of Liliiflorae. Results comparable to those of the previous investigations (1978) were obtained, with far stronger reactions within Asparagales than between asparagalean taxa and taxa of Liliales or Dioscoreales (Fig. 2). Moreover, antigenic material of *Veratrum* did not react to any great extent with antisera of the asparagalean genera. Antiserum of *Rhopalostylis* reacted (weakly) with anti-

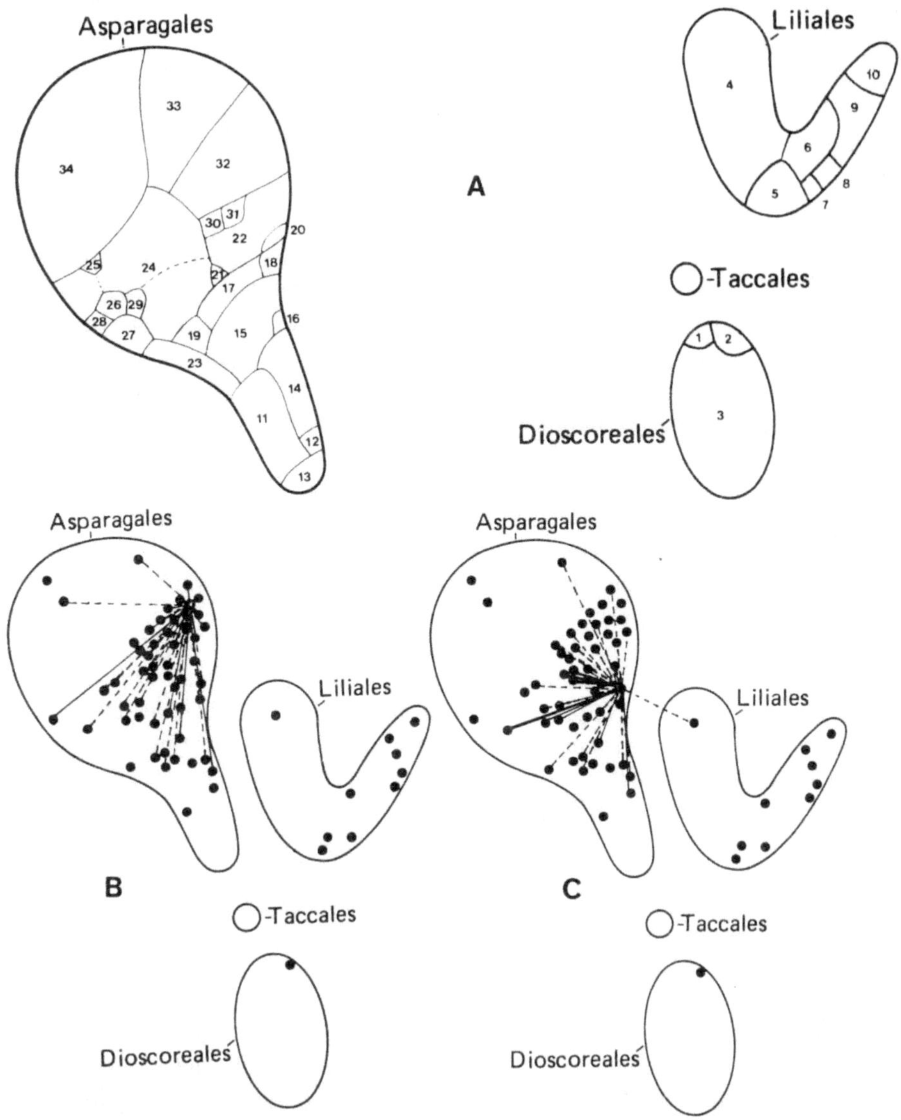

Fig. 2A–C. Taxonomic interpretation of the serological results of Chupov and Kutiavina (1981).
A Some orders of the Liliiflorae as circumscribed and depicted in Dahlgren and Clifford (1982). *1* Stemonaceae; *2* Trilliaceae; *3* Dioscoreaceae; *4* Iridaceae; *5* Colchicaceae; *6* Alstroemeriaceae; *7* Tricyrtidaceae; *8* Calochortaceae; *9* Liliaceae s.str.; *10* Melanthiaceae; *11* Smilacaceae; *12* Petermanniaceae; *13* Philesiaceae; *14* Convallariaceae; *15* Asparagaceae; *16* Herreriaceae; *17* Dracaenaceae; *18* Doryanthaceae; *19* Dasypogonaceae; *20* Phormiaceae; *21* Xanthorrhoeaceae; *22* Agavaceae; *23* Hypoxidaceae; *24* Asphodelaceae; *25* Aphyllanthaceae; *26* Dianellaceae; *27* Tecophilaeaceae; *28* Cyanastraceae; *29* Eriospermaceae; *30* Hemerocallidaceae; *31* Funkiaceae; *32* Hyacinthaceae; *33* Alliaceae; *34* Amaryllidaceae (a splitting approach has been chosen here).
B Serological reactions between antisera of *Ornithogalum montanum* (Hyacinthaceae) and antigenic material of ca. 60 species (*dots*) from different families.
C Serological reactions between antisera of *Phormium colensoi* (Phormiaceae) and antigenic material of ca. 60 species (*dots*) from different families.
——— strong serological reaction (= value > 30 according to Chupov and Kutiavina)
_____ distinct serological reaction (15–30)
– – – weak serological reaction (5–10)
no line no serological reaction

genic material of the liliiflorous genera *Doryanthes* (Doryanthaceae) and *Hypoxis* (Hypoxidaceae) only.

A number of hyacinthaceous genera (*Scilla, Chionodoxa, Hyacinthus, Bellevalia,* etc.) displayed strong mutual affinity, but reacted only moderately with *Bowiea* and hardly at all with *Camassia*, though both are members of the same family, Hyacinthaceae.

Ruscus (Asparagaceae subfam. Ruscoideae) showed strong serological affinity with other genera of the same subfamily and with *Convallaria* (Convallariaceae) but less affinity with *Asparagus*, which may lend support to Huber's view that the Ruscoideae should form a separate family. *Hosta* (Funkiaceae), which has been shown to have a karyotype almost identical with that of genera of Agavaceae, but otherwise differs conspicuously from them, proved to have strong serological affinity with *Agave* and *Yucca* as well as with *Camassia* (though much less with other Hyacinthaceae). This is interesting since some taxonomists consider that *Hosta* is related to Agavaceae, whereas others place it next to Hyacinthaceae.

Phormium, which I refer to Phormiaceae (incl. Dianellaceae) on the basis of the similar embryology, e.g. the simultaneous microsporogenesis (also found in *Doryanthes* and *Hemerocallis*) and also the similar leaves, shows strong serological affinity with *Dianella*. *Phormium* also displays serological affinity with *Hemerocallis* though not to *Hosta*. Cave (1955) has also demonstrated embryological difference between these last two genera.

Dioscorea (Dioscoreaceae) showed no clear serological affinity with any of the asparagalean or lilialean taxa tested, not even with *Trillium* (Trilliaceae, Dioscoreales) or *Smilax* (Smilacaceae, Asparagales) with both of which it is sometimes associated.

It is obvious that there is no random serological affinity between the genera of Liliiflorae. I find that our ordinal rank for Asparagales is supported by these results, while the homogeneity of Liliales is less obvious and the Dioscoreales are perhaps heterogeneous.

It could be concluded from Chupov and Kutiavina's investigation of 1978 that Liliaceae s.str., Dioscoreaceae, and Amaryllidaceae are serologically isolated, although this is shown less clearly for Amaryllidaceae in Chupov and Kutiavina (1981). Technical reservations must be made. Whereas antisera of genera such as *Phormium* (Phormiaceae) and *Dracaena* (Dracaenaceae) showed considerable reaction with taxa of a few families only, which is ideal in a serological investigation, the non-specific reactions of the *Veratrum* (Melanthiaceae) antiserum leave the interpretor with a feeling of uncertainty as to its significance.

2.3 Position of the Typhales

The order Typhales has long been open to discussion as regards its position. In our multi-character analysis (Dahlgren and Clifford 1982) we found that Typhaceae, Sparganiaceae, Philydraceae, Haemodoraceae, and Pontederiaceae share most or all of the following attributes: (1) stem with vessels having scalariform perforation plates, (2) paracytic stomata, (3) presence of oxalate raphides, (4) amoeboid tapetum, (5) starchy endosperm, and (6) helobial endosperm formation (with a very small chalazal chamber). This is corroborated by the fact that the cell walls of these taxa (unlike Asparagales, Liliales or Dioscoreales) show UV-fluorescence (Harris and Hartley 1980) and may have a common structure of the epicuticular wax (Barthlott and Frölich 1983, Behnke and Barthlott 1983).

Lee and Fairbrothers (1972) tested the serological affinities between *Sparganium* and *Typha* (Typhales) and taxa of the families Agavaceae, Araceae, Arecaceae, Asparagaceae, Commelinaceae, Cyperaceae, Juncaceae, Nolinaceae, Pandanaceae, and Poaceae. Apart from the strong mutual reaction between *Sparganium* and *Typha* measurable serological reactions were found only with taxa of Agavaceae (*Yucca* and *Agave*), Nolinaceae (*Nolina*), and Asparagaceae (*Asparagus*), all members of the order Asparagales. The reactions, however, were weak. From this one can conclude, as did Lee and Fairbrothers, that Typhales arose from the liliiflorous stock. However, on the basis of the characters enumerated above, the Asparagales are probably more closely related to the Haemodorales, Pontederiales, and Philydrales. Unfortunately no member of any of these orders was included in the serological investigation. About ten years ago few phylogenists would propose that Typhales could be related to any of them. Thus here as elsewhere the omission of strategic groups may lead to conclusions that are irrelevant or at least need to be supplemented. Now I may be seriously wrong in believing that a comparison between *Sparganium* or *Typha* and, for example, any of the genera *Pontederia, Philydrum,* and *Wachendorfia* would give stronger reactions, but from relevant comparative data I should predict that this would be so.

A complete lack of serological cross-reactions is of little value for phylogenetic conclusions. For example, Lee et al. (1975) found no serological reactions between any of the genera *Eleusine* (Poaceae), *Flagellaria* (Flagellariaceae), *Joinvillea* (Joinvilleaceae), *Hanguana* (Hanguanaceae), *Juncus* (Juncaceae), and *Scirpus* (Cyperaceae). This may not necessarily mean that they are all very distantly related although this would be the natural conclusion to draw. Some of the genera they studied (*Flagellaria, Joinvillea, Eleusine*) share certain distinct morphological characteristics, which supports a common ancestry not too long ago.

3 Dicotyledons

3.1 Magnoliiflorae and Ranunculiflorae

As is the case with monocotyledons, extensive serological investigations have been made on only a small proportion of the dicotyledonous families. Among the best-known investigations are those by Jensen and his collaborators dealing with the Ranunculaceae and allied taxa (Jensen 1967, 1968, Jensen and Penner 1980). In Ranunculaceae the results can be correlated with the distribution of attributes including (1) presence of honey-leaves, (2) floral symmetry, (3) fruit types, (4) number of integuments in the ovaries, (5) chromosome number, (6) chromosome size, (7) occurrence of protoanemonin, (8) accumulation of tertiary benzylisoquinoline bases, (9) occurrence of diterpene alkaloids, and (10) occurrence of cyanogenic compounds. These and other attributes in combination with the serological results (as shown in Jensen (1968) and Frohne and Jensen (1973, 1979)) provide an excellent basis for a sound and natural subdivision of the Ranunculaceae.

Frohne (1962) and Jensen (1967, 1968) have in addition presented serological evidence in favour of grouping Ranunculales next to Papaverales (or in the same order as in Thorne 1981). This is also indicated by the presence of special types of benzylisoquinoline alkaloids, as well as by a great number of morphological features including the trimerous basic floral symmetry, the numerous stamens, the atactostely and the embryological make-up. It can be noted here that serological data also lend support to the division of the former "Rhoeadales" into Papaverales s.str. and Capparales, Frohne (1962) having demonstrated that there is almost no serological reaction between Papaveraceae and Brassicaceae. This division has also independently become obvious from the different contents of secondary metabolites, Papaverales containing benzylisoquinoline alkaloids and Capparales containing glucosinolates. The superficial similarity in floral organization is probably the result of convergent evolution (Merxmüller and Leins 1967).

As regards the Magnoliiflorae, serological results from seed proteins may support a subdivision of Magnoliales s.lat. into minor orders (Fairbrothers and Petersen this volume). The serological results in this taxonomic group should have excellent opportunities to be considered in relation to evidence from the occurrence of alkaloids, of endosperm and perisperm in the ripe seed, sieve-tube plastid inclusions and a body of other evidence.

Jensen and Büttner (1981) and Jensen (unpublished) have also found that storage protein antisera of *Magnolia* cross-react strongly with homologous proteins of other Magnoliiflorae, but also with those in the Corniflorae, which are generally regarded as being quite advanced. As such results are obtained repeatedly a taxonomic re-evaluation can be expected. It should be noted, for example, that primitive vessel types with oblique end walls and scalariform perforation plates are frequent in these orders and the presumably primitively cellular endosperm formation (see Wunderlich 1959) are almost universal. Anyway, this examplifies a case where serological results also might induce an overall investigation of the possible relationships between particular taxonomic groups. The Corniflorae (see below) in various ways form a central group in the dicotyledons, although in my opinion they are grossly misinterpreted by most taxonomists. The superorder is fairly uniform, which becomes successively obvious from the fields of morphology (Huber 1963), embryology (Philipson 1974, Dahlgren 1975a,b), iridoid chemistry (Jensen et al. 1975), and now serology. Ericales, which are probably related to Theales (Thorne 1981), are even more diffusely delimited from the Cornales, which merge into Dipsacales. In most classifications Ericales, Cornales, and Dipsacales are placed widely apart, for example in different subclasses(!), Dilleniidae, Rosidae, and Asteridae, respectively, by Cronquist (1981) and Takhtajan (1980). Chemically the Corniflorae are characterized by having carbocyclic iridoids and (rarely) secoiridoids. Ellagitannins occur in some Ericales and other kinds of tannins occur in this order and in Cornales, while tannins are largely substituted by caffeic acid in Dipsacales. It is unlikely, in my opinion, that Corniflorae are actually derived from Theales or other groups with multistaminal androecia placed in Cronquist's and Takhtajan's Dilleniidae (see Dahlgren 1983). Nor are there any embryological or chemical links with extant Rosidae except very possibly Hamamelidales. The links are with part of the present sympetalous groups (Gentianiflorae, Lamiiflorae) and with some minor groups such as Loasaceae (= Loasiflorae), all rather specialized complexes.

Thus the origin of the Corniflorae must be sought among primitive dicotyledons, presumably with long vessels having scalariform perforation plates, with petaliferous diplo- or obdiplostemonous flowers, with seeds having copious endosperm and small embryos and with cellular endosperm formation. Such a group would not be dissimilar to Hamamelidales or Cunoniales. Important innovations in Corniflorae were the reduc-
and with cellular endosperm formation. Such a group would not be dissimilar to Hamamelidales or Cunoniales. Important innovations in Corniflorae were the reduc-
tion of the ovules to become unitegmic and tenuinucellar and the specialization in chemical defence by producing iridoids. If thus Corniflorae has evolved from such a primitive group of dicotyledons, perhaps parallel to complexes like Theiflorae, Araliiflorae, and Rosiflorae, then its serological links with the Magnoliiflorae would seem rather challenging.

Simon (1970, 1971) contributed serological evidence towards establishing the relationships of Nymphaeales and Nelumbonales (=*Nelumbo*). *Nelumbo* antiserum gave little or no reaction with the antigenic material of the genera of Nymphaeaceae tested, but gave a clear reaction with antigenic material of *Magnolia* and also a (mostly weak) reaction with antigenic material of other taxa of Magnoliiflorae and taxa among the monocotyledons. The reactions of Nymphaeaceae antisera resembled those of *Nelumbo* antiserum, although the reaction with *Magnolia* seed protein was weaker. The study suggests that *Nelumbo* should be associated with Magnoliales rather than with Nymphaeales or Ranunculales. It has also been pointed out that its position in Nymphaeales is false in the light of most kinds of evidence. Apparently *Nelumbo* is fairly isolated, but belongs to the Magnoliiflorae. (In my classification of 1980 I also transferred it to this position partly on Simon's evidence.)

Within the Nymphaeaceae, *Nymphaea* and *Nuphar* are serologically closer to each other than each genus is to *Victoria* or *Euryale* (Simon 1971). In Simon's study unfortunately no member of the order Piperales (Saururaceae and Piperaceae) was included, although they share with Nymphaeales the general seed structure, i.e. a copious perisperm and an apical endosperm enclosing a small embryo, a unique type of seed indicating a common origin and little likelihood of having evolved by convergence.

3.2 Violiflorae – Malviflorae

Among other selected contributions to serology I shall comment on a study on Violales by Kolbe and John (1979). Although in some of the families used for this study only single or a few species have been included in the serological tests, the results are of significance and indicate that Violales should be divided into several groups of families. The two major groups are (1) the Flacourtiaceae, Violaceae, Passifloraceae, and Turneraceae, a group with chiefly hypogynous flowers, and (2) the Cucurbitaceae, Begoniaceae, and Datiscaceae, with epigynous flowers. These appear to form natural orders, and it will be a future task to reveal whether a comprehensive comparative study of many characters can sustain this division.

The families showing weak, or none at all, serological affinity to these two family groups are also of great interest. The Cistaceae, Bixaceae, and Cochlospermaceae, which proved to be serologically isolated from the other Violales, are undoubtedly best placed in the order Malvales on the basis of stellate trichomes, mucilage cells and androecium structure, as well as their flavonoid chemistry (Gornall et al. 1979, Young 1981). Serological tests between the above three families and the Tiliaceae, Sterculiaceae, Dipterocarpaceae, Malvaceae, and Bombacaceae are needed.

Further, antiserum of a member of Loasaceae gave no serological reaction with antigenic material of any of the other families of Violales even in its broadest circumscription. However, they reacted, though weakly, with taxa of the Polemoniaceae, Hydrophyllaceae, Campanulaceae, Primulaceae, Theaceae, and Ericaceae. Previously (Dahlgren 1975a,b, 1980) I have stressed the occurrence in Loasaceae of unitegmic tenuinucellar ovules with cellular endosperm formation and extensive endosperm haustoria (Garcia 1962). This is a syndrome of embryological characters typical for many Corniflorae and Lamiiflorae and for a part of the Gentianiflorae, groups which are known to quite often contain iridoid compounds. Now iridoids are not only present but also abundant and variable in the Loasaceae and include secoiridoids (Jensen, Nielsen, and Damtoft, personal communication), which are otherwise found mainly in the Gentianiflorae and Corniflorae (Cornales, Dipsacales). On the basis of all this evidence it would seem probable that the Loasaceae were related to Dipsacales or Cornales. The solaniflorean and campanulalean taxa also show certain embryological similarities to the Loasaceae, but they lack iridoids, though this could of course be a secondary loss. It would be highly interesting to test the Loasaceae against various of the above mentioned taxa.

Kolbe (1978) tested the serological affinity between a number of glucosinolate-containing families and found mutual reaction between taxa of Brassicaceae, Capparaceae, Tovariaceae, and Resedaceae, but not between any of these and the likewise glucosinolate-bearing Moringaceae or Tropaeolaceae. It may seem that these last two families should be excluded from Capparales if the order is strictly circumscribed. Floral structure, pollen morphology and other details also diverge, although seed characters as well as the presence of myrosin cells and glucosinolates support their inclusion in the order.

3.3 The "Amentifers"

The "Amentiferae" sensu Engler form a more or less artificial group of orders which is partly retained in some modern systems of classification (as the subclass "Hamamelidae"), for example by Cronquist (1981) and Takhtajan (1980). The crucial point is to what degree the Amentiferae are a homogeneous assemblage of families or consists of groups of families that have converged as the result of adaptation to wind-pollination. For example it is not entirely clear whether Juglandales or Myricales are at all closely related to Fagales. In his classifications Thorne (1968, 1974, 1976, 1981) has consistently emphasized the evidence for a link between Juglandales-Myricales and the wind-pollinated Anacardiaceae. This is seen particularly in growth habit, foliar morphology, stem anatomy, fruits, pollen grains, and basic chromosome number. Among the Anacardiaceae, genera such as *Pistacia* and *Amphipterygium* would be those that most closely approach Juglandaceae and Myricaceae. However, Cronquist and Takhtajan point to a number of conspicuous morphological, embryological, pollen morphological, and chemical similarities between Myricales, Juglandales, and Fagales, such as the unusual state of chalazogamy, the *Normapolles* pollen type and the abundance of tannins. Thus the serological approach to this problem made by Petersen and Fairbrothers (1978, 1979, 1983) was most welcome. Their investigation involved species of *Carya* and *Juglans* of the Juglandaceae, *Myrica* and *Comptonia* of the Myricaceae, *Quercus* and *Fagus* of the Fagaceae, and *Rhus* and *Toxicodendron* of the Anacardiaceae. A great serological similarity was detected between the taxa of Juglandaceae and those of Myricaceae, and the genera of these two families on the average showed better correspondence with *Quercus* than with *Rhus* (though none to *Fagus* and *Toxicodendron*). The conclusion was that Juglandaceae and Myricaceae would be better placed in conjunction with Fagales than with Anacardiaceae (Rutales or Sapindales), which supports the Cronquist and Takhtajan classifications in this respect. These studies now have been further expanded (Petersen, this volume). They strongly support that Myricales and Juglandales are to be placed close together with the fagalean families (Fagaceae, Corylaceae, and Betulaceae), while their serological reaction with families such as Anacardiaceae, is weak, indicating remote relationship. More genera, such as *Rhus* (Anacardiaceae) have been included since the preliminary reports. Admittedly, already the first published results from this investigation influenced my decision to place Juglandales and Myricales closer to Fagales (Dahlgren 1980) after having previously (1975) followed Thorne in placing the first two orders next to Sapindales.

Brunner and Fairbrothers (1979) found strong serological correspondence between *Alnus-Betula* (Betulaceae) and *Carpinus-Corylus-Ostrya* (Corylaceae), indicating this grouping. They concluded that Corylaceae could best be included in Betulaceae. The number of tepals, stamens, and carpels in the flowers, for example, may make it practical to maintain the Corylaceae as a family, even though the close relationships find additional support.

Further studies by Petersen and Fairbrothers (1983) show that *Leitneria* (Leitneriaceae), another "amentifer" with variable taxonomic history behind it, has close serological affinity with *Ailanthus* of the rutalean family Simaroubaceae, whereas *Amphipterygium* (Anacardiaceae) is close to *Rhus* in the same family. In my classification (Dahlgren 1980) Anacardiaceae and Leitneriaceae are placed in Sapindales next to Rutales. The borderline between these orders is diffuse, and there may be good reasons for treating them as a single order.

Chupov (1978) used pollen albumin antisera from species of *Populus* (Salicaceae) and *Corylus* (Corylaceae) for testing the affinities of these families with certain genera considered to be related to them. He also concluded that Fagaceae, Betulaceae, and Juglandaceae form a closely related group of families. Juglandaceae were not considered to be at all closely related to Rutales. From this order Chupov tested *Dictamnus* of the Rutaceae only, which is insufficient. In addition, Chupov found that Salicales *(Populus)* antiserum cross-reacted with antigenic material of juglandaceous and fagaceous taxa, but as it also reacted with antigenic material of *Paeonia, Rubus*, and *Philadelphus* it seems that nonspecific determinants might have been involved, in which case no decisive conclusions should be drawn.

3.4 Saxifragales and Corniflorae

Grund and Jensen (1981) have presented a taxonomically important investigation on the Saxifragales, which also tackles some controversial differences between the current systems of classification.

The concepts of the family Saxifragaceae and the order Saxifragales (often included in Rosales) vary to an embarrassing degree today. In particular the relationships of some of the generic groups that have previously often been placed within the family Saxifragaceae are open to discussion.

Takhtajan (1980) includes 25 families in his Saxifragales, for instance Saxifragaceae, Grossulariaceae, Crassulaceae, and some families that are placed in Cunoniales, Pittosporales, and Droserales in other classifications. Cronquist (1981) and Thorne (1981) both include Saxifragales in their broadly circumscribed Rosales. In all these three classifications the family Saxifragaceae is placed next to, or includes, the Hydrangeaceae, Columelliaceae, Montiniaceae, and Escalloniaceae: Takhtajan, for instance, treats the Escalloniaceae as a separate saxifragalean family, Cronquist includes them in Grossulariaceae, and Thorne places them in Saxifragaceae.

However, as early as 1963 Huber defined the order Cornales to include Cornaceae, Philadelphaceae (= Hydrangeaceae), Styracaceae, Symplocaceae, Escalloniaceae, Diapensiaceae, and Aquifoliaceae. The order Cornales with this circumscription was based on wood, leaf, flower, and ovule characters; I find the embryological characters most decisive: the ovules are unitegmic, the nucellus lacks a parietal tissue, the endosperm formation is cellular and the seeds contain copious endosperm. The order Cornales has subsequently been enlarged in my own classifications (1975, 1980) to include many more families which are obviously allied to those named above with the exception of Styracaceae. New characteristics, such as the frequent occurrence of iridoid compounds, have been added to the definition of Cornales, which is unfortunately somewhat diffusely delimited since our knowledge of a number of provisional members is incomplete. Core families are the Hydrangeaceae, Escalloniaceae, Icacinaceae, Aquifoliaceae, Symplocaceae, Cornaceae, and Stylidiaceae, all excepting Aquifoliaceae being known to contain one or more species with iridoids.

The evidence for separating the Hydrangeaceae, Escalloniaceae, and some other generic groups from Saxifragales has been virtually ignored by Takhtajan, Cronquist, and Thorne. There is, however, a varying amount of support from floral ontogeny (Gelius 1967), seed structure (Krach 1976), pollen morphology (Hideaux and Ferguson 1976), and chemistry (Jensen et al. 1975, Gershenzon and Mabry 1983). When Grund and Jensen (1981) investigated the serological reactions of seed proteins from taxa often regarded as members or relatives of Saxifragales the serological evidence was thus awaited with interest. Their studies show that the genera *Heuchera, Astilbe, Rodgersia, Mitella, Tiarella, Tellima,* and *Tolmia* (though not *Chrysosplenium*) showed a similar, close serological affinity with *Saxifraga*. All these genera belong to the family Saxifragaceae s.str., whose members generally have bitegmic ovules and a parietal tissue and have never been found to contain iridoids. *Ribes* (Grossulariaceae) and *Hamamelis* (Hamamelidaceae) also showed clear serological affinity with this group of genera, whereas the genera of Crassulaceae tested, and in particular Hydrangeaceae and Parnassiaceae, showed less affinity with it.

The fact that Crassulaceae, placed in the same order as Saxifragaceae s.str. in all current systems of classification, shows only moderate serological affinity with Saxifragaceae is not unexpected with regard to its distinctness in various respects, viz. pronounced succulence, the isomerous gynoecium, the presence of characteristic scales at the carpel bases and the lack of endosperm in ripe seeds. The possibility of placing Crassulaceae in a separate order should be considered.

The serological affinity between Saxifragaceae and *Hamamelis* (Hamamelidaceae, Hamamelidales) makes the idea of further serological studies, involving more genera, most tempting. It suggests placing Hamamelidales in the vicinity of Saxifragales, and at least in the same superorder. This is not so in Takhtajan's, Cronquist's, and Thorne's classifications, although the subclasses Rosidae (with Saxifragales-Rosales) and Hamamelidae (with Hamamelidales) are generally considered to have a common ancestor with a *Hamamelis*-like floral structure.

The Hydrangeaceae *(Deutzia, Hydrangea, Kirengeshoma, Philadelphus)*, Escalloniaceae *(Escallonia)*, and Cornaceae *(Cornus)* in Grund and Jensen's (1981) investigation show close mutual serological affinities, but their affinity with the Saxifragaceae s.str. is weak, which lends support to the evidence from embryology, morphology, and iridoid chemistry mentioned above, and Grund and Jensen consider this warrants placing Hydrangeaceae and Escalloniaceae in Cornales. In their study Roridulaceae showed greater serological affinity with the families Hydrangeaceae, Escalloniaceae, and Cornaceae than did the Ericaceae *(Erica tetralix)*, which suggests a place in Cornales. Other evidence argues for placing Roridulaceae *(Roridula)* in Ericales: they are mycorrhizal plants (Marloth 1925) with scalariform perforation plates, a racemose inflorescence and poricidal anthers. On the other hand *Roridula* lacks the distinct endosperm haustoria found in Ericales (Vani-Hardev 1972) and has no pollen tetrads, which argues for cornalean affinity. A position in either of the closely allied orders Ericales or Cornales is currently the most appropriate choice, particularly as *Roridula* has also been shown to contain iridoid compounds (Jensen et al. 1975).

Parnassia (Parnassiaceae) is problematic as regards its closest relationships. Grund and Jensen (1981) found no serological reaction with the Saxifragaceae, from which *Parnassia* also differs in the nuclear endosperm formation, the parietal placentation and the variable, often digitate staminodia. Serological tests of *Parnassia* against various Theales (e.g. *Hypericum*) or against the Droseraceae might be more rewarding.

Grund and Jensen's investigation (1981) gives some evidence as regards the distinctness and the relationships of Cornales. A close serological affinity between Cornaceae and Caprifoliaceae as well as with *Sambucus* and *Viburnum* (both generally included in Caprifoliaceae) was, for example, evident. Major serological investigations have previously been made on Cornales and their relatives, viz. by the Fairbrothers group. In a serological investigation on the affinities between *Davidia, Corokia, Griselinia, Camptotheca,* and *Nyssa* Fairbrothers (1977) and Brunner and Fairbrothers (1978) found that antigen material of all five genera reacted with *Cornus* antiserum, *Nyssa* showing the strongest, and *Griselinia* perhaps the weakest reaction. All these cornalean genera except *Nyssa* are known to contain iridoids (Jensen et al. 1975), and on various grounds are considered to be related. The main problem is how to circumscribe the families in this group, and though serology gives a certain amount of help it is only by means of a multidisciplinary study that the limits of these families can be satisfactorily determined. In such a study, carried out by several specialists, the genus *Aralidium* was unequivocally found to be cornalean though sufficiently distinct to deserve separate family status. Within this study Fairbrothers (1980) found that *Aralidium* showed serological affinity with *Nyssa*, though not to *Aralia* or *Angelica* (Araliales). In the same study Jensen and Nielsen (1980) discovered iridoids in *Aralidium*.

Investigations by Hillebrand and Fairbrothers (1965, 1970a,b) indicate that *Viburnum* in particular, but also *Sambucus* are serologically distinct from the genera of Caprifoliaceae s.str. studied, and that the Caprifoliaceae approach some species of *Cornus* (Cornaceae) and also show affinity with *Nyssa* (Nyssaceae). This is in agreement with the similarities in chemistry found between the Cornales and Dipsacales (see Dahlgren 1975a, 1977b, 1980, 1983, Jensen et al. 1975) with the Nyssaceae and Cornaceae often producing seco-iridoids as do families within the Dipsacales and Gentianales.

In a further investigation Lee and Fairbrothers (1978) found that the Rubiaceae show close serological affinity with the Cornaceae and Nyssaceae as well as with Gentianaceae, Caprifoliaceae, and Asclepiadaceae, less so with the Apocynaceae and Dipsacaceae. The relatively strong serological correspondence between representatives of Rubiaceae, Cornaceae, and Nyssaceae contrasted with the negative correspondence with genera of the Apiaceae.

The papers by Grund and Jensen (1981) and by Lee and Fairbrothers (1978) support several features of angiosperm classification that are also based on other evidence (Fig. 3):

(1) The Saxifragales should primarily be restricted to tanniniferous taxa that lack the ability to produce iridoids and generally are characterized by bitegmic ovules. They show affinity with Rosales and Hamamelidales, and are distant from Cornales. Saxifragales should include Saxifragaceae, Grossulariaceae and, with some hesitation, Crassulaceae but not the Hydrangeaceae or Escalloniaceae.

(2) The two last-mentioned families belong to Cornales, an order where most taxa have unitegmic tenuinucellar ovules, cellular endosperm formation and seeds with copious endosperm. Iridoids are common in these families, but ellagitannins are rare.

(3) Cornales in this sense borders on Dipsacales, and taxa such as Caprifoliaceae, Adoxaceae, Sambucaceae or Viburnaceae (if the last two are treated as families) should be placed in one or the other of the two orders, or the orders could be fused into one.

(4) Serological evidence, as with the distribution of seco-iridoids, indicates that the Gentianiflorae and Cornales are closely related. However, Gentianales (incl. Rubiaceae) can be well defined by its nuclear endosperm formation; presence of intraxylar phloem and epigynous flowers provide further help in the actual placing of its families.

3.5 Araliales

Evidence from morphology, embryology, and chemistry indicates that the Araliales (Araliaceae and Apiaceae) are most closely related to the Asteraceae and Campanula-

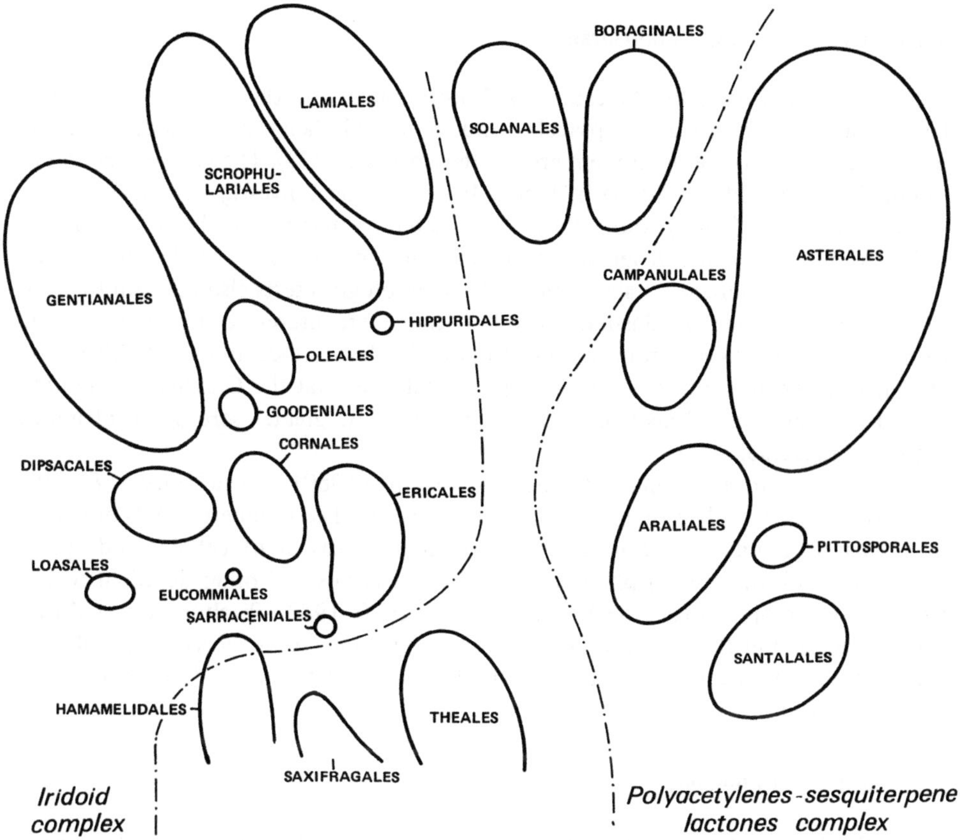

Fig. 3. Diagram showing the orders of, primarily, sympetalous dicotyledons arranged according to relationships as judged from morphological, embryological, phytochemical, and other evidence. As will be commented in the text the evidence from serological investigations tend to support various features in this classification

ceae. Previously (Dahlgren 1977a,b) I have emphasized the indications that the "Sympetalae" have differentiated along two or three main lines, one primarily iridoid-bearing line (Corniflorae-Loasiflorae-Lamiiflorae-Gentianiflorae), one polyacetylene and sesquiterpene lactone-bearing line (Araliiflorae-Asteriflorae) and one line (which may be a derivative of either of the other two) with neither of these compounds but (fairly often) with either tropane or necine alkaloids (Solaniflorae). Though this is probably a gross oversimplification of the truth some support is derived from morphology and embryology. The first to report chemical evidence for such a division was Hegnauer (1964). The faint serological reactions between Rubiaceae or its relatives and the Araliales are in agreement with this philosophy.

3.6 Gentianiflorae and Lamiiflorae

The Oleales have basically the same floral morphology and the same embryological features as the Cornales and Dipsacales, and are iridoid plants (with a special type of iridoid belonging to the oleuropein group (Jensen et al. 1975). They are placed in my Gentianiflorae (Dahlgren 1975a,b, 1980). In a serological investigation dealing with Oleaceae Piechura (1980) found they were related to members of both Gentianales and Scrophulariales, which lends support to a line of evolution parallel to those of these orders (Piechura and Fairbrothers 1979). Serological tests also confirm that the subfamilies Oleoideae and Jasminoideae in Oleaceae are discrete groups, and that the former probably evolved from the ancestors of the latter (Piechura and Fairbrothers 1983). Further he concludes, on serological evidence, that the Oleaceae are remote from Rhamnales and Celastrales, both of which show greater serological affinities with Saxifragaceae.

Another controversial matter tackled by Piechura (1980) is the affinity of *Buddleja*. This genus is usually either included in Loganiaceae (Leeuwenberg 1980) or placed in a separate family near Loganiaceae, but on the basis of its cellular endosperm formation, lack of internal phloem, lack of seco-iridoids and other details it would fit better in Scrophulariales (Dahlgren 1975a,b, 1980, 1983). Piechura reported that "Bignoniaceae and Buddlejaceae, although sharing some (serological) relationship to the Oleaceae, have greater affinities with the Scrophulariaceae and are included in Scrophulariales".

3.7 Theales and Ebenales

The affinities and homogeneity of the orders Theales and Ebenales have long been matters of great uncertainty. Although I shall not be able to summarize and evaluate the problems in grouping the families referred to these orders, it is interesting to consider the serological evidence presented in two papers by Kolbe and John (Kolbe and John 1980, John and Kolbe 1980).

As regards Ebenales (Kolbe and John 1980), species of *Diospyros* (Ebenaceae) did not cross-react with any of the genera of Ebenaceae studied (viz. *Styrax, Halesia, Pterostyrax*) nor with those of Sapotaceae (viz. *Sideroxylon, Mimusops*), while some cross-reaction between certain taxa of the last two families was obtained. The Styracaceae showed stronger reactions with *Actinidia* (Actinidiaceae) than with the genera of Sapotaceae, but also reacted with Clethraceae, Ericaceae, Primulaceae, and (weakly) Myrisinaceae. Similarly, the Sapotaceae showed affinity with extra-ebenalean taxa, such as *Actinidia*, Theaceae, and Primulaceae, but not with material of other families of Ebenales. The conclusion drawn from serological data alone, is that Ebenales is not a natural group in the commonly circumscribed sense. Also in floral morphology and embryological characters Ebenales is heterogenous. A future drastic re-classification of the families currently placed in Ebenales is almost inevitable (note the above-mentioned approach by Huber 1963). An acceptable position for the Symplocaceae, which are generally placed in Ebenales, seems to be in Cornales if one looks to morphology and embryology and considers the production of iridoids in *Symplocos glauca* (Jensen and Nielsen personal communication).

John and Kolbe (1980) have also tested serologically the links between Theales and certain other groups. They have chosen a few species, sometimes only single species, from each family (and order). I thus find some difficulty in evaluating fully the significance of their results, which suggest a close affinity between Theales *(Camellia)* and taxa of Ebenales, Ericales, and Primulales.

In assessing these data I am surprised to note the strong serological correspondence between Ericales and the embryologically divergent Theales and Primulales. All these complexes have in common tenuinucellar ovules, but the Theales and Primulales have bitegmic ovules and generally nuclear endosperm formation; they also lack iridoids, though ellagi-tannins occur in all three orders. It will be necessary to assess the merits of these results in the future. The long-suspected affinity between the sympetalous Primulales and the choripetalous Theales based on embryology (see above), chemistry, leaf teeth (Hickey and Wolfe 1975) and other data, is corroborated by serological results.

4 Concluding Comments

In spite of the limitations and reservations referred to above the serological data are important and fascinating. It is tempting for a taxonomist (as it has been for me here) to focus attention on new projects devised to solve urgent problems or to clarify points of controversy.

Other methods, including numerical analyses of character states and cladistics, also make use of a limited set of data leading to superficially objective though incomplete inferences. I am convinced that classifications based on these methods will not always coincide with one based on serological reactions, but by wise integration one may come closer to the factual phylogenetic pattern than we are today. Circumscriptions of genera and families will probably be less influenced by results from protein chemistry. With the classification at ordinal and superordinal levels, and with the grouping of genera in families, however, it may be necessary to circumscribe and arrange the groups according to all kinds of evidence considered to reflect evolutionary relationships.

A severe warning is delivered by Cronquist (1983) against constructing phylogenetic classifications based on concealed evidence such as serological data, amino acid sequence data, biosynthetical pathways of secondary metabolites and small details in embryology and other microstructural evidence. In doing so we may deviate to such an extent from a phenetically useful classification that the practical taxonomist would be happy to completely ignore our classification in favour of an artificial one that will prove useful in the field and in a normally equipped laboratory. The next stage might be to have one type of classification for the phylogenist and another for the pheneticist, which is what has in fact already taken place. In floras and practical handbooks there is a tendency for even present-day authors to follow the Englerian classification, while at academic institutions the approach is more phylogenetical.

5 Suggestions for Future Serological Studies for Solving Macrosystematic Problems

Certain acute problems met with in angiosperm classification remain unsolved. Supposed relationships between certain families and orders may be based upon little evidence, and data can be controversial or inadequate for admitting a clear assessment of affinities. In such a situation serological investigations may contribute evidence that can lead to a better understanding and that may reveal expected or unexpected links between families. Here I shall mention a number of challenging problems that await investigation. Genera of the families in the left-hand column should be tested, with their antisera, against genera of the families to the right, those which are either presumed to be closely allied to them or which show conspicuous similarities that have previously been considered to be of phylogenetic importance.

Rafflesiaceae	against Aristolochiaceae and Annonaceae
Saururaceae and Piperaceae	against Cabombaceae, Ceratophyllaceae, Nymphaeaceae, Chloranthaceae, and Annonaceae
Lactoridaceae	against Aristolochiaceae, Annonaceae, Saururaceae, Piperaceae, Chloranthaceae, Lardizabalaceae, and Winteraceae
Glaucidiaceae *(Glaucidium)*	against Paeoniaceae, Ranunculaceae *(Hydrastis)*, Berberidaceae *(Podophyllum)*
Polygonaceae	against Lardizabalaceae, Menispermaceae, Berberidaceae, Plumbaginaceae, Limoniaceae, and Caryophyllaceae
Plumbaginaceae (sensu stricto)	against Limoniaceae *(Armeria, Limonium, Acantholimon)*
Dilleniaceae	against Paeoniaceae, Glaucidiaceae, Cochlospermaceae, Bixaceae, Cistaceae, Crossosomataceae, Cunoniaceae, and Rosaceae
Euphorbiaceae	against Thymelaeaceae, Malvaceae, Sterculiaceae, Ulmaceae, Buxaceae, Didymelaceae, and Rhamnaceae
Thymelaeaceae	against Euphorbiaceae, Elaeagnaceae, Rhamnaceae, Proteaceae, and Combretaceae
Loasaceae	against Dipsacaceae, Cornaceae, Hydrangeaceae, Rubiaceae, Boraginaceae, and Hydrophyllaceae
Gyrostemonaceae	against Resedaceae, Tovariaceae, Capparaceae, Brassicaceae, Batidaceae, and Sapindaceae
Salvadoraceae	against Gyrostemonaceae, Capparaceae, Euphorbiaceae *(Drypetes)*, and Celastraceae
Salicaceae	against Tamaricaceae, Frankeniaceae, Flacourtiaceae, Cistaceae, and Betulaceae
Droseraceae	against Parnassiaceae, Clusiaceae *(Hypericum)*, Elatinaceae, and Saxifragaceae
Coridaceae *(Coris)*	against Primulaceae and Lythraceae
Trochodendraceae	against Tetracentraceae, Cercidiphyllaceae, Hamamelidaceae, Betulaceae, Winteraceae, Cunoniaceae, Rosaceae, and Buxaceae
Geissolomataceae	against Hamamelidaceae, Myrothamnaceae, Oleaceae, Buxaceae, and Penaeaceae

Buxaceae	against Didymelaceae, Daphniphyllaceae, Hamamelidaceae, Balanopaceae, and Euphorbiaceae
Podostemonaceae	against various groups, incl. Saxifragaceae, Hydrostachyaceae, Haloragaceae, and Balanophoraceae
Fabaceae and other Fabales	against Sapindaceae, Staphyleaceae, Connaraceae, Chrysobalanaceae, Amygdalaceae, and Rosaceae
Chrysobalanaceae	against Ochnaceae Surianaceae, Connaraceae, and Rosaceae
Haloragaceae	against Araliaceae, Rhizophoraceae (different subfamilies), Lythraceae, and Cornaceae
Tropaeolaceae	against Limnanthaceae, Balsaminaceae, Geraniaceae, Brassicaceae (e.g. *Stanleya*), and Resedaceae
Polygalaceae	against Vochysiaceae, Malpighiaceae, Krameriaceae, Sapindaceae, and Fabaceae
Olacaceae or Loranthaceae	against Santalaceae, Viscaceae, Araliaceae, Apiaceae, Pittosporaceae, and Icacinaceae
Vitidaceae	against Rhamnaceae, Olacaceae, Santalaceae, Rutaceae, Myrsinaceae, and Ebenaceae
Pittosporaceae	against Tremandraceae, Byblidaceae, Araliaceae, Buxaceae, Daphniphyllaceae, and Cunoniaceae
Campanulaceae	against Lobeliaceae (incl. *Cyphea*), Asteraceae: tribuș Lactuceae, Goodeniaceae, Stylidiaceae, Solanaceae, and Boraginaceae
Eucommiaceae	against Cornaceae, Escalloniaceae, Hamamelidaceae (incl. *Liquidambar*), Cercidiphyllaceae, and Araliaceae
Symplocaceae	against Aquifoliaceae, Icacinaceae, Escalloniaceae, Ebenaceae, and Sapotaceae
Sarraceniaceae	against Clethraceae, Cyrillaceae, Pyrolaceae, Theaceae, and Ranunculaceae
Menyanthaceae	against Cornaceae, Escalloniaceae, Caprifoliaceae, Oleaceae, Loganiaceae, Gentianaceae, and Solanaceae
Adoxaceae	against Sambucaceae, Viburnaceae, Caprifoliaceae, Cornaceae, and Hydrangeaceae
Hippuridaceae	against Scrophulariaceae, Plantaginaceae, Cornaceae, Haloragaceae, and Gunneraceae
Aponogetonaceae	against Butomaceae, Hydrocharitaceae, Juncaginaceae, Alismataceae, Araceae, and Nymphaeaceae
Najadaceae	against Potamogetonaceae, Juncaginaceae, and Zannichelliaceae
Araceae (Pothoideae)	against Dioscoreaceae, Trichopodaceae, Aponogetonaceae, and Saururaceae
Lemnaceae	against Araceae (different groups, incl. *Pistia*)
Burmanniaceae	against Philydraceae, Orchidaceae, Iridaceae, and Melanthiaceae
Velloziaceae	against Bromeliaceae, Pontederiaceae, Haemodoraceae, Hypoxidaceae, and Agavaceae
Typhaceae and Sparganiaceae	against Pontederiaceae, Philydraceae, Musaceae, and Agavaceae
Commelinaceae	against Eriocaulaceae, Bromeliaceae, Pontederiaceae, Restionaceae, and Juncaceae

Cyclanthaceae (*Cyclanthus*)	against other Cyclanthaceae, Arecaceae, Pandanaceae, and Araceae

Acknowledgements. Much of the information presented in this article has been generously contributed by U. Jensen and D. Fairbrothers, who have also read the article critically. M. Greenwood Petersson helped with the English.

References

Barthlott W, Frölich D (1983) Plant Syst Evol 142:171–185
Behnke H-D (1969) Planta 126:31–54
Behnke H-D (1981) Ber Dtsch Bot Ges 94:647–662
Behnke H-D, Barthlott W (1983) Nord J Bot 3:43–66
Behnke H-D, Dahlgren R (1976) Bot Not 129:289–295
Brunner F, Fairbrothers DE (1978) Serol Mus Bull 53:2–5
Brunner F, Fairbrothers DE (1979) Bull Torrey Bot Club 106:97–103
Cave MS (1955) Phytomorphology 5:247–253
Chupov VS (1978) Bot Zh (Leningrad) 63:1579–1585
Chupov VS, Kutiavina NG (1978) Bot Zh (Leningrad) 63:473–493
Chupov VS, Kutiavina NG (1981) Bot Zh (Leningrad) 66:75–81
Corner EJH (1976) The seeds of dicotyledons. Cambridge Univ Press, London New York Melbourne
Cristofolini G (1980) In: Bisby FA, Vaughan JG, Wright CA (eds). Chemosystematics: Principles and practice. Academic Press, London New York, pp 269–288
Cristofolini G, Peri P (1983) In: Jensen U, Fairbrothers DE (eds) Proteins and nucleic acids in plant systematics. Springer, Berlin Heidelberg New York
Cronquist A (1981) An integrated system of classification of flowering plants. Columbia Univ Press, New York
Cronquist A (1983) Nord J Bot 3:75–83
Cutler D (1969) In: Metcalfe CR (ed) Anatomy of the monocotyledons, vol IV. Juncales. Clarendon Press, Oxford
Dahlgren R (1975a) Bot Not 128:148–180
Dahlgren R (1975b) Bot Not 128:181–197
Dahlgren R (1977a) Plant Syst Evol Suppl 1:253–283
Dahlgren R (1977b) Publ Cairo Univ Herb 7–8:83–102
Dahlgren R (1980) Bot J Linn Soc 80:91–124
Dahlgren R (1983) Nord J Bot 3:119–149
Dahlgren R, Clifford HT (1981) Ber Dtsch Bot Ges 94:203–227
Dahlgren R, Clifford HT (1982) The monocotyledons. A comparative study. Academic Press, London New York
Dahlgren R, Fairbrothers DE, Nielsen BJ, Jensen SR (1979) Invitation to an international "Cornalean Project". Offset circular, Copenhagen
Fairbrothers DE (1977) Ann Mo Bot Gard 64:147–160
Fairbrothers DE (1980) Taxon 29:412–416
Fairbrothers DE (1983) Nord J Bot 3:35–41
Fairbrothers DE, Petersen FP (1983) In: Jensen U, Fairbrothers DE (eds) Proteins and nucleic acids in plant systematics. Springer, Berlin Heidelberg New York
Fazekas de St. Groth S (1983) In: Jensen U, Fairbrothers DE (eds) Proteins and nucleic acids in plant systematics. Springer, Berlin Heidelberg New York
Frohne D (1962) Planta Med 10:283–297
Frohne D, Jensen U (1973) Systematik des Pflanzenreichs unter besonderer Berücksichtigung chemischer Merkmale und pflanzlicher Drogen. G Fischer, Stuttgart

Frohne D, Jensen U (1979) Ibid, 2nd edn
Garcia V (1962) In: Plant embryology, a symposium, CSIR, New Delhi, pp 157–161
Gelius L (1967) Bot Jahrb Syst 87:253–303
Gershenzon J, Mabry TJ (1983) Nord J Bot 3:5–34
Gornall RJ, Bohm BA, Dahlgren R (1979) Bot Not 132:1–30
Grund C, Jensen U (1981) Plant Syst Evol 137:1–22
Hamann U (1962) Willdenowia 2:639–768
Hamann U (1974) Ber Dtsch Bot Ges 77:45–54
Harris PJ, Hartley RD (1980) Biochem Syst Ecol:153–160
Hegnauer R (1964) Chemotaxonomie der Pflanzen, vol III. Birkhäuser, Basel
Hickey LJ, Wolfe JA (1975) Ann Mo Bot Gard 62:538–589
Hideaux MJ, Ferguson IK (1976) In: Ferguson IK, Muller J (eds) The evolutionary significance of the exine. Academic Press, London New York, pp 327–377
Hillebrand GR, Fairbrothers DE (1965) Am J Bot 52:648
Hillebrand GR, Fairbrothers DE (1970a) Am J Bot 57:810–815
Hillebrand GR, Fairbrothers DE (1970b) Brittonia 22:125–133
Huber H (1963) Die Verwandschaftsverhältnisse der Rosifloren. Mitt Bot Staatssamml Muenchen 5:1–48
Huber H (1969) Die Samenmerkmale und Verwandtschaftsverhältnisse der Liliiflorae. Mitt Bot Staatssamml Muenchen 8:219–538
Huber H (1977) Plant Syst Evol Suppl 1:285–298
Jensen SR, Nielsen BJ (1980) Taxon 29:409–411
Jensen SR, Nielsen BJ, Dahlgren R (1975) Bot Not 128:148–180
Jensen U (1967) Ber Dtsch Bot Ges 80:621–624
Jensen U (1968) Bot Jahrb Syst 88:204–310
Jensen U (1974) In: Bendz G, Santesson J (eds) Chemistry in botanical classification. Academic Press, London New York, pp 217–226
Jensen U, Büttner C (1981) Taxon 30:404–419
Jensen U, Grumpe B (1983) In: Jensen U, Fairbrothers DE (eds) Proteins and nucleic acids in plant systematics. Springer, Berlin Heidelberg New York
Jensen U, Penner R (1980) Biochem Syst Ecol 8:161–170
John J, Kolbe K-P (1980) Biochem Syst Ecol 8:241–248
Kolbe P (1978) Bot Jahrb Syst 99:468–489
Kolbe P (1982) Plant Syst Evol 140:39–55
Kolbe P, John J (1979) Bot Jahrb Syst 101:3–15
Kolbe P, John J (1980) Biochem Syst Ecol 8:241–248
Krach JE (1976) Bot Jahrb Syst 97:1–60
Lee DW, Fairbrothers DE (1972) Taxon 21:39–44
Lee DW, Yap KP, Liew FY (1975) Bot J Linn Soc London 70:77–81
Lee YS, Fairbrothers DE (1978) Taxon 27:159–185
Leeuwenberg AJM (1980) Die natürlichen Pflanzenfamilien, 28 b I. Angiospermae: Ordnung Gentianales. Fam Loganiaceae. Duncker and Humbolt, Berlin
Lester RN, Roberts PA, Lester C (1983) In: Jensen U, Fairbrothers DE (eds) Proteins and nucleic acids in plant systematics. Springer, Berlin Heidelberg New York
Marloth R (1925) The flora of South Africa, 2. Cambridge
Merxmüller H, Leins P (1967) Bot Jahrb Syst 86:113–129
Petersen FP, Fairbrothers DE (1978) Serol Mus Bull 53:10
Petersen FP, Fairbrothers DE (1979) Syst Bot 4:230–241
Petersen FP, Fairbrothers DE (1983) Syst Bot 8:134–148
Philipson WR (1974) Bot J Linn Soc London 68:89–108
Piechura JE (1980) PhD thesis, Univ Wisconsin
Piechura JE, Fairbrothers DE (1979) Bot Soc Am Misc Publ 157:65
Piechura JE, Fairbrothers DE (1983) Am J Bot 70:780–789
Simon J-P (1970) Aliso 7:243–261
Simon J-P (1971) Aliso 7:325–350

Takhtajan A (1980) Bot Rev 46:225–359
Thorne RF (1968) Aliso 6:57–66
Thorne RF (1974) Brittonia 25:395–405
Thorne RF (1976) Evol Biol 9:35–106
Thorne RF (1981) In: Young DA, Seigler DS (eds) Phytochemistry and angiosperm phylogeny. Praeger, New York
Thorne RF (1983) Nord J Bot 3:85–117
Tomlinson PB (1969) In: Metcalfe CR (ed) Anatomy of the monocotyledons, II: Palmae. Clarendon Press, Oxford
Vani-Hardev (1972) Beitr Biol Pflanz 48:339–351
Wunderlich R (1959) Oesterr Bot Z 106:203–293
Young DA (1981) In: Young DA, Seigler DS (eds) Phytochemistry and angiosperm phylogeny. Praeger, New York, pp 205–232

Symposium Statements and Conclusions

U. JENSEN, D.E. FAIRBROTHERS, and D. BOULTER

1 The Bases for Phenetic and Phylogenetic Schemes

Discussions including phylogenetic and taxonomic botany require a basic explanation of the data, rationale, and evaluation employed in the construction of the diagrams (trees). Therefore, it is especially important to know whether phenetic or phylogenetic schemes are being presented. In this respect a precise and appropriate definition of the terms used is important.

In Systematic Botany the term "species" is used for those individual plants which are phenetically similar and provide actual or potential genetic interchange in a population. This term is confusing when it is also used to indicate different *kinds* of particular macromolecules, for example, DNA-species, RNA-species, globulin-species. This usage, although common, may be misunderstood to mean taxonomic species defined by properties of DNA, RNA or globulins.

A similar caution is required in the use of the term "relationship". It is incorrect to speak of related morphological or physiological processes when referring to similar morphological or physiological processes from related taxa. In the sense of genealogies only taxa, e.g., species, populations or individuals are closely or less closely related, according to the greater or more remote cladistic connections in their evolutionary history. In phenetic comparisons, similarities, not relationships, are investigated and we should not speak of relationships between pollen proteins, ferredoxins, etc., when we mean phenetic similarities. Dendrograms represent phenetic schemes, i.e., the character similarity of the taxa investigated whether these be molecules or morphological characters. Such dendrograms can, of course, be used to discuss hypothetical phylogenies.

However, in some cases the intention is to obtain a representation of the phylogenetic distance of the taxa compared and cladistic methods of analysis are used; for example the amino acid sequence method with sequence data from homologous proteins in different taxa (Boulter this volume), or the ancestral cladistic analysis of serological data as exemplified by Lester et al. in this volume.

In all cases it is essential that originators of schemes, whether these purport to be phenetic, based on phyletic concepts or mixed phenetic and phyletic schemes, state clearly the criteria on which the scheme is based and the objectives sought.

Proteins and Nucleic Acids in Plant Systematics
ed. by U. Jensen and D.E. Fairbrothers

2 Usefulness of Proteins and Nucleic Acids in Systematics

It is generally agreed that evolution is a result of random gene changes acted on by natural selection and genetic drift, and that phenotypic morphological characters are as important for systematics as are gene structures. However, as far as phyletic studies, the comparison of semantides (information carrying molecules), i.e., DNA, RNA, proteins, have often been considered of special significance.

DNA Level

A relatively small proportion of the DNA of eukaryotes codes for proteins, and much of the DNA has other functions. Thus, some DNA may play a structural role while other DNA structures may be involved with storage, replication, and evolution. Much of the repetitious DNA has no known function, and it has been suggested that "selfish" DNA exists with no function other than that of its own replication and increase in preponderance in the genome.

Gene structure is more complicated in eukaryotes than in prokaryotes and in addition to coding sequences there are normally introns, flanking control sequences and intergenic spacer regions which may or may not be transcribed; pseudo-genes and processed genes also occur; clearly nucleotide sequences with different functions may be of varying usefulness in taxonomic investigations. Thus whilst there is no reason to suppose that DNA data will not be more valuable even in studies of phylogenetic relationship than protein data, the proof of this statement will require additional data than are presently available (Beyreuther this volume).

Up to 99% of the nuclear DNA in eukaryotic plants may not be transcribed, is redundant, and highly repetitive. Differences in DNA base compositions are important systematic indicators in lower plants only. Repeated DNA "families" are simple or interspersed often in a complex way, and they are either species-specific or common to larger or smaller groups of common descent. Hybridizing cRNA from particular satellite or cloned repetitive DNA sequences gives precise information about karyotype location and homologies within and between species. These all indicate cycles of amplifications, interspersions, and replacements of DNA sequences, and are mechanisms of speciation and evolutionary divergence (Ehrendorfer this volume).

RNA Level

The RNA molecules of the cell can be conveniently classified into functional nucleic acids, such as ribosomal RNA, transfer RNA, etc. and messenger RNA molecules which specify proteins. Techniques such as hybridization and restriction mapping of RNA molecules has been conducted (Kössel et al. this volume); but, insufficient data prevent the assessment of the significance of these data for systematic purposes. Kössel et al. discuss various technical aspects of the advantages of using RNA as well as discussing the selection of the type of RNA molecule to be sequenced. For prokaryotes, on the other hand, valuable RNA trees have been published already (Sprinzl this volume).

Protein Level

Proteins are the end product of the transcription and translation of the genetic material. They fulfil many important biological functions, but of the various protein classes enzymes are the most numerous. Since proteins connect two important evolutionary levels, i.e., genetic expression and natural selection (proteins are part of the phenotype) they contain very valuable information for helping to understand phylogeny and taxonomy. Their usefulness however, is restricted by the fact that the functional evolution of many metabolic important proteins occurred in prokaryotes before the evolution of eukaryotes (Sprinzl this volume). Thus in eukaryotes, and especially in higher plants, relatively unimportant changes have been fixed in evolution, and parallel and back mutations of functionally less important residues have occurred frequently. It has also become apparent that many proteins undergo considerable post-translational modifications, and this must be taken into account when deciding the importance of some residue positions.

Throughout this symposium the question has been raised as to which protein type and which protein property should be selected to answer specific systematic questions. Although only preliminary answers are available, it is possible to make some useful statements by comparing, for example, enzymes of primary metabolism with the plant storage proteins. Thus, enzymes like cytochrome c, plastocyanin, and ferredoxin underwent their functional evolution primarily in prokaryotes, whereas the storage proteins probably arose later in phylogeny. Storage proteins are also much less subject to functional constraints compared with the primary metabolic enzymes, and are likely therefore to show greater variability. Therefore, the rubisco small subunit (SS) which is coded by nuclear genes, and the large subunit (LS) coded by chloroplast genes are used to differentiate between species or infraspecific taxa, and are also useful for indicating parentage involved in hybridization. Thus the SS and LS compositions together, or separately, are useful as taxonomic markers or as indicators of the relative age of taxa in plant phylogeny. Whether this is related to the fact that the protein consists of subunits, thereby possessing greater variability, has not been ascertained (Wildman this volume). Studies have shown that the number of isozymes reflects the number of subcellular compartments in which a particular catalytic reaction is required. For glycolysis and the oxidative pentose phosphate pathways the number is two. The changes in the number of isozymes in diploid plants brought about by duplications of coding gene loci facilitate the application of electrophoretic evidence above the species level. Isozyme number is also useful in determining ploidy level independent of chromosome number (Gottlieb this volume). Both Gottlieb and Sengbusch (this volume) indicate that the cytosolic and plastid variants of isozymes are exposed to different selective forces.

In considering the protein property to be investigated it is important to differentiate between those characteristics most useful in phylogeny, and those most useful in taxonomic studies (both Dahlgren and Jensen and Grumpe this volume).

In *phylogenetic* research the protein characters to be compared must be easily analyzed by methods suited to the construction of cladistic schemes. Such methods have been proposed for serological data (Lester et al. this volume), but the main application is with amino acid sequence data as discussed by Boulter (this volume).

For *taxonomic* research the protein(s) of an adequate number of taxa must be compared, and therefore the techniques must not be too time-consuming. In practice, immunological, electrophoretic, and immunoelectrophoretic methods have proven to be the most feasible. Although it is not possible to generalize, electrophoretic data from pollen and seed proteins have proved useful to differentiate infraspecific taxa, while immunological data have proved to be most valuable when comparing conspecific, congeneric or confamiliar taxa; see the chapter by Hurka (this volume) for electrophoretic data, and the chapters by Cristofolini, Fairbrothers and Petersen, Lee, Lester et al., and Smith (this volume), for serological data. This statement is also valid for pollen proteins (Petersen this volume).

Dahlgren (this volume) has demonstrated the importance of serological data for helping to formulate conclusions concerning phylogenetic relationships in connection with their role in the taxonomy of higher levels. He also emphasizes the importance of understanding the methods and materials used in serological investigations before coming to valid conclusions. He clearly demonstrates that serological data support some views and contradict others, as well as open new alternative taxonomic interpretations.

Thus, the structures of proteins and nucleic acids supply a virtually inexhaustible data set applicable to systematics, and it is inconceivable not to use these semantides in much of the modern systematic research. However, it is not possible to predict with certainty, that the use of this information will establish the pathways of plant evolution and solve the "Abominable Mystery" posed several times by C. Darwin and others.

3 Logistic Aspects of the Use of Nucleic Acids in Plant Systematics

The possibility now exists to make comparisons of the nucleotide sequences of homologous genes from different taxa. We should be aware however, of the *size of this task* and ask if the "biological return" justifies the effort.

Several plant genes from particular species have now been cloned, and in principle one or more of these could be used to "pull out" the equivalent gene from the isolated DNA of another plant. This would require the DNA to be isolated, restricted and cloned into a lambda or cosmid bank, operations which might occupy an experienced worker for 6 or more months. The next step, the selection of the gene from the bank, would not be a simple routine matter, and the stringency of hybridization conditions needed to achieve specific isolation might take some time to establish. Although gene sequencing itself is relatively fast, as pointed out by several speakers at this symposium, the establishment of a physical restriction map, a prior requisite to sequencing, is also required. Compared with amino acid sequencing of proteins, where only coding sequences exist, gene sequencing involves sequencing through spacers, introns, flanking sequences, pseudo-genes, as well as the coding sequences available for direct comparisons. Initially then we consider a time scale of approximately 2 years, per gene, per taxon, per researcher, although with experience this period of time may be reduced. The gene chosen to study preferably should be a "single" copy gene, otherwise the problem of identifying orthologous, as opposed to paralogous homologous, genes will greatly complicate interpretations. However, we can hopefully anticipate that future advances in techniques could reduce these time scales for conducting such research.

Subject Index

MIX
Papier aus verantwortungsvollen Quellen
Paper from responsible sources
FSC® C105338

If you have any concerns about our products,
you can contact us on
ProductSafety@springernature.com

In case Publisher is established outside the EU,
the EU authorized representative is:
Springer Nature Customer Service Center GmbH
Europaplatz 3, 69115 Heidelberg, Germany

Printed by Libri Plureos GmbH
in Hamburg, Germany